T0318153

CAMBRIDGE MONOGRAPHS ON
MATHEMATICAL PHYSICS

General editors: P. V. Landshoff, D. R. Nelson, D. W. Sciama, S. Weinberg

TWISTOR GEOMETRY AND FIELD THEORY

Cambridge Monographs on Mathematical Physics

*Issued as a paperback

TWISTOR GEOMETRY
AND
FIELD THEORY

R. S. WARD

Department of Mathematics, University of Durham

RAYMOND O. WELLS, JR.

Department of Mathematics, Rice University, Houston, Texas

CAMBRIDGE
UNIVERSITY PRESS

Published by the Press Syndicate of the University of Cambridge
The Pitt Building, Trumpington Street, Cambridge CB2 1RP
40 West 20th Street, New York, NY 10011-4211, USA
10 Stamford Road, Oakleigh, Melbourne 3166, Australia

First published 1990
First paperback edition 1991
Reprinted 1995

British Library cataloguing in publication data

Ward, R. S. (Richard Samuel)
Twistor geometry and field theory.
1. Twistor theory
I. Title   II. Wells, R. O.
530.1′42   QC173.75.T85

Library of Congress cataloguing in publication data

Ward, R. S. (Richard Samuel), 1951–
Twistor geometry and field theory / R. S. Ward and R. O. Wells, Jr.
p. cm.
Bibliography: p.
1. Twistor theory. 2. Geometry, Integral. 3. Field theory (Physics)
4. Integral transforms. I. Wells, R. O. (Raymond O'Neil), 1940– . II. Title.
QC173.75.T85W37 1988   512′.33–dc19
87-31179
CIP

ISBN 0 521 26890 7 hardback
ISBN 0 521 42268 X paperback

Transferred to digital printing 2004

RPS

To

ROGER PENROSE

Contents

Part III The Penrose transform

Preface

The authors are happy and relieved that this book has finally seen the light of day after a six-year gestation. It evolved from lectures given by the first author at Stony Brook and by the second author at Boulder some six years ago. At the time, we decided to combine our two somewhat different points of view about twistor theory into a full-scale book on the subject. We feel that it represents the major developments over the past two decades. A few topics are not covered: For example, the twistor approaches to quantum field theory, the quasilocal mass formula, and the work by Donaldson and others on invariants of four-manifolds (the last two developed while this book was being written).

The manuscript went through several typographical incarnations, via typescripts and an older computer code to the implementation of the complete book in TEX, which was carried out by the staff of Rosenlaui Publishing Services, Inc., under a contract with Cambridge University Press. The authors would like to thank the staff of Rosenlaui and CUP for all of their hard work. In particular, we thank Adrianne Hurndell and Emily Nghiêm of Rosenlaui for their painstaking efforts. The art work and copyediting was done in England by CUP staff, and their fine work is greatly appreciated.

We would like to thank Gerald Harnett, who attended the lectures of the second author in Boulder, read much of the manuscript, and made very helpful suggestions on improving the earlier chapters; and Harry Braden, who commented on the algebraic topology and spectral sequence sections and whose criticism was very useful.

We want to express our appreciation to David Tranah, the mathematics editor at Cambridge University Press, who first suggested to us a book on this topic, and who has been very supportive throughout the project. We apologize to him that it took somewhat longer to finish than any of us had anticipated, and thank him for his patience.

The authors would like to thank the Institute of Theoretical Physics at the State University at Stony Brook, the University of Colorado at Boulder and the Stanislaw Ulam Visiting Professorship, the Research Foundation of the University of Durham, and the National Science

Foundation for their support throughout this lengthy project. In addition, we would like to express our appreciation to the staffs in the Departments of Mathematics at our respective universities for all of their assistance.

Finally, we would like to acknowledge our considerable debt to Roger Penrose. His inspiration to us, as to numerous other researchers, has been a primary motivation. His encouragement and advice has been helpful to both of us throughout the course of our work in twistor theory, and in particular in the conception and writing of this book.

Richard Ward
Durham, England

Raymond O. Wells, Jr.
Houston, Texas

July 1989

The second author would like to dedicate this book to his wife, Rena Schwarze Wells. Throughout the entire process of producing this book, she has been a tremendous support in so many ways: from helping type the first few chapters in Boulder and bringing drafts to Colorado, to China, and to California, among other places, to continuous editorial updating and overseeing the typesetting and final production at Rosenlaui Publishing Services. Her dedication and skill have been an essential ingredient in the book. Her support, encouragement, and unflappable nature in spite of great strains, created by this author's propensity to try to do too many things at once, have been a backbone for our family, our marriage, and our friendship over many years.

Raymond O. Wells, Jr.
Houston, Texas

July 1989

Introduction

In 1967 Roger Penrose introduced the notion of twistors and twistor geometry in a seminal paper entitled 'Twistor Algebra'. Since that time, there has developed a significant literature and research movement in many parts of the world, devoted to understanding the ramifications of his original proposals. The fundamental thesis of the 'twistor program' is to replace the usual background space-time, in which most of the phenomena of modern physics is presented, by a new background space of twistors. In particular, the physical phenomena and the equations describing them are to be reinterpreted in this new background space, with the intent of gaining new insight into them. There is some analogy to introducing 'momentum space' and using the Fourier transform to transform back and forth from space-time to momentum space. A principal aim of this book is to present a systematic study of the main developments in the applications of twistor geometry to problems arising from theoretical physics.

There are several aspects to the twistor program. First, it has been quite successful in giving new insight into various nonlinear classical field equations which are of interest to the physics community, as well as a new point of view for the much better understood classical linear field theories. Namely, new families of solutions to the nonlinear equations have been found, often with a complete classification of solutions in specific cases, which had not been amenable to more traditional approaches. The equations which have been amenable to the new techniques include those of Maxwell, Dirac, Weyl, Einstein, Yang and Mills, and many of their numerous coupled versions. This has been of interest to theoretical physicists attempting to use these solutions in quantum field theory, classical relativity theory, and in recent attempts to quantize gravity.

The second aspect of the twistor program is the fact that the mathematics which it uses is at the same time both very classical and very modern. It includes the classical nineteenth century geometric studies of Grassmann, Plücker, and Klein, in which they studied projective algebraic geometry and developed a geometry of lines, planes, etc., in projective space. It includes parametrizations of the set of all lines

1

in a given space, or of all planes in a given space, etc. This subject
is well-known to algebraic geometers, but is not well-known to the
general mathematical public, and it plays a crucial role in the geometry of twistors. In 1949 J. Leray invented the notions of sheaf
theory, sheaf cohomology, and spectral sequences in a study of hyperbolic partial differential equations. These became standard tools of
modern algebraic topology, several complex variables, and algebraic
geometry, after J. P. Serre and H. Cartan showed how to reinterpret
classical phenomena in terms of the new (at that time) language. At
about the same time, the notions of fiber bundles and vector bundles
took their place in the repertoire of modern geometers, allowing the
ideas of E. Cartan and others to be developed in a more systematic
fashion. All of these notions, which have become a part of modern
geometry, play an important role in twistor geometry.

Thus has arisen a problem for researchers who are interested in
understanding the twistor program. On the one hand, there are theoretical and mathematical physicists who are familiar with the classical
field theory, but not so familiar with the methodology and techniques
of the classical and modern geometry alluded to above. On the other
hand, there are geometers working on problems in topology, algebraic geometry, and several complex variables, who are interested in
understanding the application of the geometric ideas and techniques
which they have at their disposal, but have little familiarity with the
principal ideas and fundamental results in the very traditional subject of classical field theory and its modern relativistic and gauge field
variations.

In response to this problem, one additional purpose of this book
is to attempt to develop, in one setting, sufficient language and techniques from both of these traditionally distinct intellectual disciplines, so that future researchers will be able to pursue the next
generation of problems with less difficulty in translating from one
language to another. The book is divided into three parts. Part I is
a study of classical and modern geometry, Part II is an introduction
to the language of classical field theory, and Part III is an up-to-date
description of the Penrose transforms from twistor space to space-
time, which yield solutions of the field theories discussed in Part II
by means of the geometric ideas introduced in Part I.

In Part I we present an overview of classical and modern geometry, where the choice of topics is dictated by their eventual use
in the context of twistor geometry. We present them so that the

mathematically trained reader can learn the fundamentals from the book. We give an indication of references to standard texts for certain details which would carry us too far afield from the main thread of development. The fundamental topics covered include the theory of projective spaces, Grassmannian manifolds, and their generalizations, as well as the correspondences between them, one of the fundamental ideas to be utilized. We also develop the theory of bundles and characteristic classes, emphasizing the differential-geometric point of view pioneered by Chern. We develop some of the fundamental notions of integral geometry, which were first formulated in the work of Radon (and his predecessors), used for transforming functions (and their generalizations such as bundles and sheaves) from one space to another by an integration-over-the-fiber technique (this is what is meant by 'integral geometry').

In Part II we discuss the linear field theories along with their group invariance properties. These are the prototypes of the nonlinear field theories of gauge theory and relativity theory. Here the basic ideas are developed so that the reader who has not been exposed to these topics can master the fundamental points and obtain some facility in reading the contemporary physics literature as well as some of the standard references on the subject. The physical background and significance of the classical field theories (for instance the relation to experimental physics) is not discussed in any detail, and the reader is referred to the physics literature review articles. Thus, the languages of both the mathematician and the physicist are employed, in an attempt to develop a consistent and uniform frame of reference.

In Part III a case-by-case analysis of the Penrose transform is developed in the various field theories where it has been successful. By a *Penrose transform* we mean an integral-geometric transform of data on some twistor-geometric space to corresponding data on space-time (where space-time refers generically to four-dimensional Euclidean or Minkowski space or its complexified and/or curved version). Here we give a significant amount of detail, as these Penrose transforms are prototypes of Penrose transforms which have not yet been computed and which might be the basis for future study. In Part III we freely use the language of both Parts I and II.

Each chapter begins with an overview of its contents. At the end of each section, we have sets of relevant exercises. Throughout the chapters, numerous examples are discussed, which will help to illuminate the new notions as they are introduced, and which will be used in conjunction with the exercises. At the end of the book, one finds a comprehensive bibliography of both the historical and contemporary literature.

Part I

Geometry

1

The Klein correspondence

Manifolds were first introduced into mathematics by Riemann in the 1850s in the form of Riemann surfaces. Since that time they have come to play an increasingly important role in mathematics and physics, as well as in other branches of science. In this chapter we shall introduce a class of manifolds which are important for our development of twistor geometry. These manifolds include complex projective space and its various generalizations, and were studied extensively in the nineteenth century by Grassmann, Plücker, and Klein. Each manifold in this class is homogeneous with natural groups of motions similar to the Euclidean motions of Euclidean space or a standard sphere.

1.1 Projective space and Grassmannian manifolds

We shall let $\mathbf{R}$, $\mathbf{C}$, and $\mathbf{H}$ denote the real numbers, complex numbers, and quaternions, respectively. The cartesian products $\mathbf{R}^n$, $\mathbf{C}^n$, and $\mathbf{H}^n$ are the fundamental local models for the manifolds we want to study. We shall consider principally complex-valued functions unless specified otherwise. A differentiable function f defined on an open subset of $\mathbf{R}^n$ is *smooth* or *differentiable* if partial derivatives of all orders exist and are continuous. A differentiable function f defined on an open set of $\mathbf{C}^n$ is said to be *holomorphic* if it satisfies the Cauchy–Riemann equations

$$\frac{\partial f}{\partial \bar{z}^j} = 0, \quad j = 1, \ldots, n,$$

where $z^j = x^j + iy^j$, $j = 1, \ldots, n$ are coordinates and where

$$\frac{\partial}{\partial z^j} = \frac{1}{2} \left(\frac{\partial}{\partial x^j} - i \frac{\partial}{\partial y^j} \right), \qquad \frac{\partial}{\partial \bar{z}^j} = \frac{1}{2} \left(\frac{\partial}{\partial x^j} + i \frac{\partial}{\partial y^j} \right).$$

We define a *manifold M* of real dimension n to be a topological space which is locally homeomorphic to an open subset of $\mathbf{R}^n$, i.e.,

7

for each $p \in M$ there is a neighborhood U of p, an open set $U' \subset \mathbf{R}^n$, and a homeomorphism $\varphi_U : U \to U'$. We refer to U as a *coordinate chart*, and φ_U as a *coordinate mapping*. In practice we shall usually not distinguish between the sets U and U', and functions on U will be expressed in terms of the coordinates of U'. A *differentiable manifold* is a manifold, as above, with the additional property that if U and V are any two coordinate charts which have nonempty intersection, then the induced mapping $\varphi_U \circ \varphi_V^{-1}$ should be a differentiable mapping with a differentiable inverse.

A function $f : U \to \mathbf{R}$ is *differentiable* if $f \circ \varphi_U^{-1}$ is a differentiable function on the open set $U' \subset \mathbf{R}^n$. A function f on any open subset of M is differentiable if its restriction to any coordinate chart is differentiable. Since the transition functions from overlapping coordinate charts are differentiable, this definition makes sense. We denote the set of differentiable functions on an open set U of a differentiable manifold by $\mathcal{E}(U)$. We see that $\mathcal{E}(U)$ is, in a natural way, an *algebra* over the complex numbers, as the sum and product of differentiable functions are again differentiable functions, and we can multiply such functions by complex numbers.

A *complex manifold* has local coordinate charts which are open subsets of $\mathbf{C}^n$, and the induced transition functions from one coordinate chart to another should be a holomorphic vector-valued function with holomorphic inverse. A function defined on an open set of a complex manifold is *holomorphic* if its restriction to any coordinate chart composed with the coordinate mapping is holomorphic. We denote the algebra of holomorphic functions on an open set U of a complex manifold by $\mathcal{O}(U)$ (noting once again that the product of two holomorphic functions is holomorphic, etc.). More generally, if M and N are two complex manifolds, then a *mapping* $f : M \to N$ is *holomorphic* if restriction and composition with coordinate charts induce a holomorphic vector-valued function.

A *complex submanifold* N of an n-dimensional complex manifold M is a closed subset N with the property that near each point $p \in N$, there is a neighborhood U of p and r holomorphic functions $f^1, \ldots, f^r$ defined in U so that:

(a) $N \cap U = \{q \in U : f^j(q) = 0, \quad j = 1, \ldots, r\}$;

(b) In local coordinates $(z^1, \ldots, z^n)$ near $q \in N$ the matrix $\frac{\partial f^j}{\partial z^k}$ has maximal rank at q.

One sees that N is a complex manifold of dimension $d = n - r$ (Exercise 1.1.1), and we say that N has *codimension* r in M.

We shall now give some fundamental examples of both differentiable and complex manifolds which will play an important role in the remainder of this book.

Example 1.1.1 (Euclidean space and submanifolds): Let

$$S^n = \{(x^1, \ldots, x^{n+1}) \in \mathbf{R}^{n+1} : (x^1)^2 + \cdots + (x^{n+1})^2 = 1\}.$$

Then S^n is the *standard n-sphere*. By the implicit function theorem, it is a differentiable manifold of dimension n. If we consider *product manifolds* we have a *torus* $T^n = S^1 \times \cdots \times S^1$, or manifolds of mixed type $S^1 \times S^3$, etc. One can also consider complex submanifolds of $\mathbf{C}^n$ given by, for instance,

$$M = \{(z^1, \ldots, z^n) \in \mathbf{C}^n : (z^1)^2 + \cdots + (z^n)^2 = 1\}.$$

We note that this is a complex manifold of dimension $n - 1$ (by the complex-analytic version of the implicit function theorem), but M is *not* compact in $\mathbf{C}^n$. The only compact complex submanifolds of $\mathbf{C}^n$ are points (zero-dimensional manifolds), as one can see easily (cf. Wells 1980). On the other hand one can consider real submanifolds of $\mathbf{C}^n$ given by

$$M = \{|z^1|^2 + \cdots + |z^n|^2 = 1\},$$

which is a sphere of type S^{2n-1}, and there are noncompact hyperboloids of the type (for $k < n$)

$$M = \{|z^1|^2 + \cdots + |z^k|^2 - |z^{k+1}|^2 - \cdots - |z^n|^2 = 1\}.$$

Now we want to discuss the most important examples of manifolds which are not defined as submanifolds of Euclidean space, namely *projective spaces* and their generalizations. We shall concentrate on complex projective spaces as they are most important for our applications and also their topological properties are simpler. We shall start with the simplest example which will be a prototype for the somewhat more complicated Grassmannian manifolds and flag manifolds discussed further below.

Example 1.1.2 (Projective space): Let V be a complex finite-dimensional vector space, and let[1]

$$\mathbf{P}(V) := \{\text{one-dimensional subspaces of } V\}.$$

---

[1] The notation ":=" means that the object on the left is by definition equal to the object on the right.

We call $\mathbf{P}(V)$ the *projective space of the vector space V*. If $V = \mathbf{C}^{n+1}$, then we set $\mathbf{P}_n(\mathbf{C}) = \mathbf{P}(\mathbf{C}^{n+1})$ and refer to $\mathbf{P}_n(\mathbf{C})$ as *n-dimensional complex projective space*. We shall see that $\mathbf{P}_n(\mathbf{C})$ can be equipped in a natural manner with the structure of a connected compact complex manifold of dimension n, thus justifying the dimensional aspect of the above definition. We shall usually write simply $\mathbf{P}_n$ unless there is some danger of confusion with real projective space $\mathbf{P}_n(\mathbf{R})$, which will be considered hardly at all in this book. We give $\mathbf{P}_n$ a topology by noting that there is a projection

$$\mathrm{pr} : \mathbf{C}^{n+1} - \{0\} \to \mathbf{P}_n, \tag{1.1.1}$$

given by $\mathrm{pr}(Z) = \mathbf{C}Z$, for $Z \neq 0$, where $\mathbf{C}Z$ denotes all complex multiples of Z. We shall also denote $\mathrm{pr}(Z)$ by $[Z]$, i.e., $[tZ] = [Z]$, for all $t \in \mathbf{C}, t \neq 0$. Then $\mathbf{P}_n$ can be viewed as equivalence classes of points in $\mathbf{C}^{n+1} - \{0\}$, where two points in $\mathbf{C}^{n+1} - \{0\}$ are equivalent if they lie on the same complex line. We then give $\mathbf{P}_n$ the *quotient topology* under such an equivalence, namely, a set $U \subset \mathbf{P}_n$ is *open* if and only if $\mathrm{pr}^{-1}(U)$ is open. Then pr becomes an open and a continuous mapping. We note that S^{2n+1} is a sphere in $\mathbf{C}^{n+1}$ which is mapped by pr onto $\mathbf{P}_n$, since every line passes through a point of S^{2n+1}. Thus the image of S^{2n+1} must be compact. We want to find coordinate charts for $\mathbf{P}_n$. Let

$$U_i = \{[z^0, \dots, z^n] : z^i \neq 0\}.$$

It is clear that $\{U_i\}$ is a cover of $\mathbf{P}_n$ by $(n+1)$ open sets. Consider the mapping

$$\varphi_i : (w^1, \dots, w^n) \in \mathbf{C}^n \mapsto [w^1, \dots, w^{i-1}, 1, w^{i+1}, \dots, w^n]$$

Then we see that $\varphi_i : \mathbf{C}^n \to U_i$ is a homeomorphism and thus provides a coordinate chart for $\mathbf{P}_n$.

One needs to check that the transition functions $\varphi_j^{-1} \circ \varphi_i$ are holomorphic mappings (see Exercise 1.1.2), and that $\mathbf{P}_n$ then becomes a well-defined complex manifold with a covering of $(n+1)$ coordinate charts. We shall have much more to say about this example in the course of the book, but for the time being it is the fundamental and simplest example of a compact complex manifold. We note that for $n = 1$, $\mathbf{P}_1$ is *diffeomorphic* to the two-sphere S^2 (the Riemann sphere) and this can be identified with the one-point compactification of the complex plane $\mathbf{C} \cup \{\mathrm{pt}\}$ (Exercise 1.1.3).

Example 1.1.3 (Grassmannian manifold):   Let V be a fixed n-dimensional complex vector space. We define the *Grassmannian manifold of k-dimensional subspaces of V* to be:

$$\mathbf{G}_k(V) := \{k\text{-dimensional subspaces of } V\}.$$

Just as for projective space we can equip $\mathbf{G}_k(V)$ with the structure of a compact complex manifold of dimension $k(n-k)$. For $k = 1$, we see that $\mathbf{G}_1(V) = \mathbf{P}(V) \cong \mathbf{P}_{n-1}(\mathbf{C})$, which we studied in the previous example. We need to find the neighborhood structure and coordinate charts, just as for projective space. For this we need generalized homogeneous coordinates. Let $\mathbf{C}^{n \times k}$ be the $n \times k$ complex-valued matrices, let $\mathbf{C}_*^{n \times k}$ be the $n \times k$ matrices of maximal rank, and let $\mathrm{GL}(r, \mathbf{C})$ denote the invertible $r \times r$ matrices, as usual. Choose a basis for V so that $V \cong \mathbf{C}^n$, where we think of $\mathbf{C}^n$ as being column vectors. We shall then consider

$$\mathbf{G}_{k,n}(\mathbf{C}) := \mathbf{G}_k(\mathbf{C}^n),$$

which for simplicity we shall write $\mathbf{G}_{k,n}$ (one can define $\mathbf{G}_{k,n}(\mathbf{R})$, etc., in the same manner).

Now consider the mapping

$$\mathrm{pr} : \mathbf{C}_*^{n \times k} \to \mathbf{G}_{k,n},$$

given by

$$\mathrm{pr}(m) := \mathrm{span}\{\text{columns of } m\},$$

and we shall also denote, as in the case of projective space, $\mathrm{pr}(m) = [m]$. Thus we see that

$$\mathrm{pr}(m) = [m] = [m \cdot \mathrm{GL}(k, \mathbf{C})],$$

as the right-action on m by $\mathrm{GL}(k, \mathbf{C})$ corresponds to changes in the basis of the subspace $[m] \in \mathbf{G}_{k,n}$. On the other hand we see that there is a transitive (left-) action of $\mathrm{GL}(n, \mathbf{C})$ on $\mathbf{G}_{k,n}$ given by

$$(g, [m]) \in \mathrm{GL}(n, \mathbf{C}) \times \mathbf{G}_{k,n} \mapsto [gm].$$

This is easily seen to be well-defined and transitive (Exercise 1.1.4). Note that this includes the special case of the transitive action of $\mathrm{GL}(n+1, \mathbf{C})$ on $\mathbf{P}_n$. We say that a set $U \subset \mathbf{G}_{k,n}$ is *open* if and only if $\text{pr}^{-1}(U)$ is open in $\mathbf{C}_*^{n \times k}$. Just as in the case of projective space, one sees easily that $\mathbf{G}_{k,n}$ is a compact space (Exercise 1.1.5).

Let

$$Z = (Z^{ij}) \in \mathbf{C}^{(n-k) \times k} \cong \mathbf{C}^{k \times (n-k)},$$

and consider the mapping

$$\varphi_1 : \mathbf{C}^{(n-k) \times k} \to \mathbf{G}_{k,n},$$

given by

$$Z \mapsto \begin{bmatrix} Z \\ I_k \end{bmatrix}, \tag{1.1.2}$$

where I_k is the $k \times k$ identity matrix. We recall that the right-hand side of (1.1.2) is the *span* of the $n \times k$ matrix

$$\begin{pmatrix} Z \\ I_k \end{pmatrix},$$

and this is consistent with the usual notation $[z^0, \ldots, z^n]$ (represented as a row vector) for homogeneous coordinates in n-dimensional projective space. One can then check that φ_1 is a homeomorphism and thus is a coordinate chart for $\mathbf{G}_{k,n}$. There is a finite subgroup of $GL(n, \mathbf{C})$ of permutation matrices which map all $k \times k$ submatrices of an $n \times k$ matrix to the position of the $k \times k$ submatrix occupying the last k rows. Denoting such a permutation matrix by T, one sees that

$$\varphi_T := T \circ \varphi_1 : \mathbf{C}^{k \times (n-k)} \to \mathbf{G}_{k,n}$$

are also homeomorphisms and thus coordinate charts. This system of coordinate charts covers $\mathbf{G}_{k,n}$, and since the transition functions are again biholomorphic, $\mathbf{G}_{k,n}$ is equipped with the structure of a compact complex manifold.

We now want to discuss infinitesimal structures of various types associated with a manifold. The first and most important is the *tangent bundle* to a manifold M. The tangent bundle to a manifold M is the disjoint union of the tangent spaces to the manifold at different points of M. The tangent space to M at a point $x \in M$ is the vector space of directional derivatives at that point, which we shall define precisely below.

Suppose M is a differentiable manifold, then for any open set $U \subset M$ a *vector field* X on U is a derivation of the algebra $\mathcal{E}(U)$ of differentiable functions on U, i.e., X is a C-linear mapping

$$X : \mathcal{E}(U) \to \mathcal{E}(U)$$

satisfying the product rule,

$$X(fg) = fX(g) + X(f)g.$$

If $U \subset \mathbf{R}^n$, and x^i are local coordinates in U, then it is easy to see that

$$\frac{\partial}{\partial x^1}, \cdots, \frac{\partial}{\partial x^n}$$

are all vector fields, and any vector field X on U is of the form

$$X = \sum_{i=1}^n a^i \frac{\partial}{\partial x^i}, \quad a^i \in \mathcal{E}(U).$$

Then for $p \in U$,

$$X_p := \sum_{i=1}^n a^i(p) \frac{\partial}{\partial x^i},$$

which is a classical *directional derivative* at p.

We define the tangent space to a differentiable manifold M to be the set of all tangent vectors at p, and denote this by $T_p(M)$. It is clear that $T_p(M)$ is a vector space (over $\mathbf{C}$). If f is a differentiable mapping

$$f : M \to N,$$

where M and N are differentiable manifolds, then there is an induced mapping on tangent spaces denoted by

$$df_p : T_p(M) \to T_{f(p)}(N).$$

This is defined simply by composition of locally defined smooth functions near p and $f(p)$. Namely, if $X \in T_p(M)$, then

$$Y(g) := df_p(X)(g) := X(g \circ f),$$

where g is a smooth function defined near $f(p)$. It is easy to see that Y defined in this manner is a derivation and, therefore, that the mapping df_p is well-defined. Moreover, one can check that df_p is a *linear mapping*. We call df_p the *derivative of f at the point p*. The derivative mapping df_p is the *linear approximation at p* to the (generally nonlinear) mapping f. If f is a diffeomorphism, then df_p is an isomorphism of vector spaces. So, in particular, if

$$\varphi : U \subset \mathbf{R}^n \to M$$

is a coordinate chart, then $T_p(U) = T_p(\mathbf{R}^n)$ and

$$d\varphi_p : T_p(\mathbf{R}^n) \to T_{\varphi(p)}(M)$$

is an isomorphism. One can check easily (Exercise 1.1.6) that

$$T_p(\mathbf{R}^n) \cong \operatorname{span}_{\mathbf{C}} \left\{ \frac{\partial}{\partial x^1}, \ldots, \frac{\partial}{\partial x^n} \right\},$$

and hence $T_p(M)$ is an n-dimensional vector space (we could also define the real vector space $T_p(M)^{\mathbf{R}}$ of derivations of real-valued smooth functions defined near p, and then one has

$$T_p(M) = T_p(M)^{\mathbf{R}} \otimes_{\mathbf{R}} \mathbf{C};$$

cf. Exercise 1.1.7).

If $f : M \to N$ is a differentiable mapping, and $(x^1, \ldots, x^n)$ are local coordinates near $p \in M$, where $p = (0, \ldots, 0)$, and $(y^1, \ldots, y^k)$ are local coordinates near $p' = f(p) \in N$, then f can be represented near p by

$$y^j = f^j(x^1, \ldots, x^n).$$

Similarly, tangent vectors near p and p' are of the form

$$\sum_{i=1}^n a^i(x) \frac{\partial}{\partial x^i}, \quad a^i \text{ smooth},$$

$$\sum_{i=1}^k b^i(y) \frac{\partial}{\partial y^i}, \quad b^i \text{ smooth}.$$

The derivative mapping is then represented in terms of the bases

$$\left\{ \frac{\partial}{\partial x^1}, \ldots, \frac{\partial}{\partial x^n} \right\}, \quad \left\{ \frac{\partial}{\partial y^1}, \ldots, \frac{\partial}{\partial y^k} \right\}$$

by the $k \times n$ matrix

$$\frac{\partial f^i}{\partial x^j}(x),$$

and the mapping df_q is given by

$$\begin{pmatrix} a^1(x) \\ \vdots \\ a^n(x) \end{pmatrix} \mapsto \begin{pmatrix} \frac{\partial f^1}{\partial x^1}(x) & \cdots & \frac{\partial f^1}{\partial x^n}(x) \\ \vdots & \ddots & \vdots \\ \frac{\partial f^k}{\partial x^1}(x) & \cdots & \frac{\partial f^k}{\partial x^n}(x) \end{pmatrix} \begin{pmatrix} a^1(x) \\ \vdots \\ a^n(x) \end{pmatrix} = \begin{pmatrix} b^1(y) \\ \vdots \\ b^k(y) \end{pmatrix},$$

where $y = f(x)$, for x near 0. This is the classical *Jacobian matrix* of a differentiable mapping of a vector-valued function of several variables.

The vector fields

$$\frac{\partial}{\partial x^1}, \ldots, \frac{\partial}{\partial x^n}$$

span the tangent space to $\mathbf{R}^n$ at any point of $\mathbf{R}^n$. In general, on a given manifold M, it is not possible to find linearly independent vector fields $\{X_1, \ldots, X_n\}$ which span the tangent space at each point of M. Locally, near any point of M, this is possible. Namely, if M is a manifold, then near a point p there is a coordinate chart U' containing the point p, and a diffeomorphism $\varphi : U \to U'$, where $U \subset \mathbf{R}^n$. Then let

$$X_j = d\varphi\left(\frac{\partial}{\partial x^j}\right),$$

and the X_j will be, at each point $q \in U'$, n linearly independent tangent vectors which span $T_q(M)$, $q \in U'$. Later in Chapter 2 we shall discuss fiber bundles in general, and shall see that these local vector fields spanning the tangent space at each point are the basic ingredients in having a vector bundle structure, but we shall postpone discussion of that until later. The *tangent bundle* $T(M)$ is the disjoint union of the tangent spaces $T_p(M)$, for $p \in M$, and we see that these vector spaces can be described by locally defined linearly independent (at each point) vector fields.

So far we have considered manifolds as abstract spaces with local coordinate systems. An important class of complex manifolds are the submanifolds of complex projective space. If we consider a complex submanifold $N \subset \mathbf{P}_n$, then N is given *a priori* as the local zeros of holomorphic functions. It is a beautiful theorem of Chow (see, e.g., Gunning and Rossi 1965), that N can in fact be described as the zeros of homogeneous polynomials in $\mathbf{C}^{n+1}$. Namely, there are homogeneous polynomials $p_1, \ldots, p_r$ in $\mathbf{C}^{n+1}$ such that

$$N = \{[z^0, \ldots, z^n] : p_j(z^0, \ldots, z^n) = 0, \quad j = 1, \ldots, r\}.$$

We call such a submanifold a *projective algebraic manifold*, and Chow's theorem asserts that any complex submanifold of $\mathbf{P}_n$ is projective algebraic. We shall not need this result, but we remark that there are many complex manifolds which cannot be embedded as a submanifold of $\mathbf{P}_n$ (in 1954 Kodaira gave a complete differential-geometric characterization of projective algebraic manifolds; cf. Wells 1980). This contrasts with the classical theorem of Whitney that any differentiable manifold of dimension n can be embedded in $\mathbf{R}^{2n+1}$ (cf. Sternberg 1965).

Our last example in this section will be of a very specific kind of projective algebraic manifold, namely a hypersurface defined by a single homogeneous quadratic equation.

Example 1.1.4 (Quadric in $\mathbf{P}_n$): Let q_{ij} be a complex symmetric matrix and define the bilinear form

$$q(Z, W) = \sum_{ij} q_{ij} Z^i W^j$$

for $Z, W \in \mathbf{C}^{n+1}$. Then q is *nondegenerate* if $q(Z, W) = 0$, for all $W \in \mathbf{C}^{n+1}$, implies that $Z = 0$. It is well known that this is equivalent to $\det(q_{ij}) \neq 0$. The only invariant of q, up to a complex linear change of variables, is the *rank*, i.e., the number of nonzero diagonal entries in a diagonalization. If we define

$$Q = \{[Z] \in \mathbf{P}_n : q(Z, Z) = 0\}, \qquad (1.1.3)$$

then Q is a projective algebraic variety in $\mathbf{P}_n$. Two such quadrics are *equivalent* (by a linear change of homogeneous coordinates) if and only if the defining quadratic forms have the same rank.

We can calculate

$$\frac{\partial q}{\partial Z^i} = 2 \sum_j q_{ij} Z^j.$$

So the *singular set* of Q is precisely given by

$$\ker q : \mathbf{C}^{n+1} \to \mathbf{C}^{n+1},$$

considered as a linear mapping. Thus Q is *nonsingular* (at all points) if and only if q is nondegenerate, i.e., if q has maximal rank $= n + 1$.

If we let $\tilde{Q} \subset \mathbf{C}^{n+1}$ be the cone over $Q \subset \mathbf{P}_n$ defined by (1.1.3) (i.e., $\tilde{Q} = \{Z \in \mathbf{C}^{n+1} : q(Z, Z) = 0\}$), then we can define the tangent space to $\tilde{Q}$ at $\tilde{p} \in \mathbf{C}_*^{n+1}$ by differentiating the equation (1.1.3) obtaining

$$T_{\tilde{p}}(\tilde{Q}_{\tilde{p}}) = \{Z \in \mathbf{C}^{n+1} : \sum q_{ij} a^j Z^i = 0\}, \qquad (1.1.4)$$

where $\tilde{p} = (a^0, \ldots, a^n)$. The equation (1.1.4) is a single homogeneous linear equation and thus defines a projective hyperplane in $\mathbf{P}_n$ which we shall call the *tangent plane* to Q at $p = [a^0, \ldots, a^n]$, and we denote this by

$$\overline{T}_p(Q) = \{[Z] \in \mathbf{P}_n : \sum q_{ij} a^j Z^i = 0\}. \qquad (1.1.5)$$

We see that $\overline{T}_p(Q)$ is a compact submanifold of $\mathbf{P}_n$ which is tangent to Q at p by construction. One can show that $\overline{T}_p(Q)$ is the *completion* of $T_p(Q)$ in $\mathbf{P}_n$, where $T_p(Q)$ is considered as an affine subspace of a standard coordinate chart containing p. We shall not need this fact, but our notation $\overline{T}_p$ is indicative of this.

More abstractly, it follows from (1.1.5) that if

$$Q = \{[Z] : q(Z, Z) = 0\},$$

then

$$\overline{T}_p(Q) = \{[Z] : q(Z, p) = 0\}. \qquad (1.1.6)$$

This characterization of the tangent plane to a quadric Q will be important for us later.

One interesting question about quadrics which has been studied for some time by algebraic geometers is: which projective linear spaces are contained in a given quadric? We mention the following classical theorem in this direction (see, e.g., Griffiths and Harris 1978, p. 735).

Theorem 1.1.5. *A smooth quadric Q of dimension n contains no projective linear subspaces of dimension strictly greater than $n/2$; on the other hand, if*

(a) $n = 2m + 1$ *is odd, then Q contains an irreducible k-dimensional family of m-planes where $k = (m + 1)(m + 2)/2$;*

(b) $n = 2m$ is even, then Q contains two irreducible, k-dimensional families of m-planes, where $k = m(m + 1)/2$, and moreover, for any two m-planes $P, P' \subset Q$, $\dim(P \cap P') \cong m \mod(2)$ if and only if P and P' belong to the same family.

In particular, if Q has dimension four as a submanifold of $\mathbf{P}_5$, then Q contains two distinct families of two-planes, each parametrized by $\mathbf{P}_3$. We shall study this case and these families of planes (called α-planes and β-planes by Felix Klein) in more detail in §1.4. In particular we shall see a proof of this theorem in this particular case.

Exercises for Section 1.1

1.1.1. Show that any complex submanifold N of a complex manifold M is a complex manifold.

1.1.2. Compute the transition functions $\varphi_j^{-1} \circ \varphi_i$ for projective space in Example 1.1.2, and show that they are holomorphic mappings from $\mathbf{C}^n - \{w_i = 0\}$ to $\mathbf{C}^n - \{w_j = 0\}$.

1.1.3. Show that $\mathbf{P}_1(\mathbf{C})$ is diffeomorphic to S^2 and, by stereographic projection, to the extended complex plane $\mathbf{C} \cup \{\infty\}$.

1.1.4. Show that the left-action of $\mathrm{GL}(n, \mathbf{C})$ on $\mathbf{G}_{k,n}$ given by $[m] \mapsto [gm]$, for $g \in \mathrm{GL}(n, \mathbf{C})$ is:

(a) well-defined, i.e., if $p = [m] \in \mathbf{G}_{k,n}$, and $p \mapsto g \cdot p := [gm]$, then $g \cdot p$ is well-defined independent of the choice of matrix representative m for p;

(b) transitive, i.e., if $p_1 = [m_1]$, and $p_2 = [m_2]$ are two points of $\mathbf{G}_{k,n}$, then there is a $g \in \mathrm{GL}(n, \mathbf{C})$ such that $p_2 = g \cdot p_1$.

1.1.5. Show that $\mathbf{G}_{k,n}$ is a compact complex manifold. (Hint: find a compact family of matrices in $\mathbf{C}_*^{n \times k}$ which map surjectively onto $\mathbf{G}_{k,n}$.)

1.1.6. Show that any $\{\frac{\partial}{\partial x^1}, \ldots, \frac{\partial}{\partial x^n}\}$ is a basis for the set of directional derivatives X_p at a point, and from this that any vector field X defined on $U \subset \mathbf{R}^n$ has the form

$$X = \sum_{i=1}^{n} a^i \frac{\partial}{\partial x^i}, \quad a^i \in \mathcal{E}(U).$$

1.1.7. The *complexification* of a vector space V defined over the real numbers $\mathbf{R}$, is defined to be the tensor product over the real numbers $\mathbf{R}$ of V with $\mathbf{C}$, denoted by $V \otimes_{\mathbf{R}} \mathbf{C}$ (see, e.g., Lang 1971 for a discussion of vector space, fields, rings, modules, and their tensor products). Show that $\mathbf{R}^n \otimes_{\mathbf{R}} \mathbf{C} = \mathbf{C}^n$. Use this to show that $T_p(M) = T_p(M)^{\mathbf{R}} \otimes_{\mathbf{R}} \mathbf{C}$.

1.2 Twistor manifolds and correspondences

In the previous section we defined projective space and a Grassmannian manifold as the set of one-dimensional and k-dimensional subspaces of a fixed vector space V. We now generalize this notion slightly by considering sets of nested sequences of subspaces of V of fixed dimensions. More precisely, let V be a fixed complex vector space of dimension n, and let $\{d_1, \ldots, d_m\}$ be a sequence of positive integers satisfying $1 \leq d_1 < d_2 < \cdots < d_m < n$. We define

$$\mathbf{F}_{d_1 \ldots d_m}(V) := \{(S_1, \ldots, S_m) : S_j \text{ are subspaces of}$$
$$V \text{ of dimension } d_j, S_1 \subset S_2 \subset \cdots \subset S_m\} \quad (1.2.1)$$

and this is called a *flag manifold* of type $(d_1, \ldots, d_m)$. These flag manifolds are all compact complex manifolds with a covering by coordinate charts given by matrices which generalize the matrix coordinates of a Grassmannian manifold in a natural manner. We shall see explicit examples of this later, and leave the general case to the reader (Exercise 1.2.1).

We now want to use the notion of flag manifold to define the fundamental complex manifolds of *twistor geometry*. We let T denote a fixed four-complex-dimensional vector space. This vector space T will be called the space of twistors or *twistor space*. In due course we shall also equip T with various real structures Φ, where Φ will denote a specific antilinear involution, a specific Hermitian form, or some such similar object on T. Twistor spaces arise in various contexts, as we shall see, but for a *fixed* twistor space there is a canonical geometric theory derivable from it. This is analogous to developing quantum mechanics by starting with a *fixed* Hilbert space.

Now let T be a fixed twistor space. We can associate with T a family of flag manifolds of the form (1.2.1), where we let V be the fixed four-complex-dimensional vector space T, and we shall write

$\mathbf{F}_{12} = \mathbf{F}_{12}(\mathbf{T})$, etc., omitting the explicit dependence on the vector space $\mathbf{T}$. Consider now three special twistor manifolds $\mathbf{F}_{12}$, $\mathbf{F}_1$, and $\mathbf{F}_2$. There is a natural diagram

$$
\begin{array}{ccc}
 & \mathbf{F}_{12} & \\
{}_{\mu}\swarrow & & \searrow{}^{\nu} \\
\mathbf{F}_1 & & \mathbf{F}_2
\end{array}
\qquad (1.2.2)
$$

where $\mu(S_1, S_2) = S_1$ and $\nu(S_1, S_2) = S_2$ are natural 'projection' mappings. As will be clear from the local representations for the mappings μ and ν, one finds that both of these mappings are surjective holomorphic mappings of maximal rank. We call (1.2.2) a *double fibration*, and we shall see further examples of such double fibrations throughout the book. This will be our fundamental example of such a double fibration, and we shall now develop some of its basic properties which will be prototypes of similar properties for other double fibrations.

In general, if one has two spaces A and B, then the notion of a *mapping* from A to B is an assignment of a point of B to each point of A. A more general notion than mapping is that of a *correspondence* c from A to B. A correspondence c is an assignment for each point $p \in A$ of a *subset* $c(p) \subset B$. A double fibration such as (1.2.2) always yields a natural correspondence (denoted by $\Rightarrow$), namely

$$
\mathbf{F}_1 \overset{c}{\Rightarrow} \mathbf{F}_2
\qquad (1.2.3)
$$

where $c(p) := \nu \circ \mu^{-1}$. There is also a natural 'inverse correspondence' given by $c^{-1}(q) := \mu \circ \nu^{-1}$. The notation c and c^{-1} are simply a convention to keep track of the direction of the correspondence. Since we shall have occasion to use the correspondence in (1.2.3) quite a lot, we need a more compact notation to keep track of the elements involved. At the same time we shall replace the point p above by an arbitrary subset. Let us denote

$$
\begin{aligned}
\mathbf{P} &:= \mathbf{F}_1 = \mathbf{F}_1(\mathbf{T}) \cong \mathbf{P}_3(\mathbf{C}) \\
\mathbf{M} &:= \mathbf{F}_2 = \mathbf{F}_2(\mathbf{T}) \cong \mathbf{G}_{2,4}(\mathbf{C}) \\
\mathbf{F} &:= \mathbf{F}_{12} = \mathbf{F}_{12}(\mathbf{T})
\end{aligned}
$$

and these will be called:

> $\mathbf{P}$ — *projective twistor space;*
>
> $\mathbf{M}$ — *compactified complexified Minkowski space;*
>
> $\mathbf{F}$ — *the correspondence space between* $\mathbf{P}$ *and* $\mathbf{M}$.

We shall see later why **M** is designated in this way. We mention that real Minkowski space can be naturally realized as a four-real-dimensional subset of the four-complex-dimensional twistor manifold **M**. The fundamental diagram (1.2.2), becomes, with this new notation,

$$\begin{array}{ccc} & \mathbf{F} & \\ {}^{\mu}\swarrow & & \searrow^{\nu} \\ \mathbf{P} & & \mathbf{M} \end{array} \qquad (1.2.4)$$

The double fibration (1.2.4) will allow us to transform information from the space **P** (or subsets of **P**) to the space **M** (or subsets of **M**). The essence of twistor geometry is understanding the transfer of information from the 'twistor space' **P** (and its natural generalizations) to the 'Minkowski space' **M**. In principle (and quite specifically later in this book), one can represent solutions of a variety of differential equations of classical field theory which are naturally defined on **M** as an integral formula in terms of data defined on **P**. For the present we want to confine our attention to the *geometry* of (1.2.4) and leave the issues of differential equations and integral formulas to later chapters.

If A is a subset of **P**, then we denote the *correspondence of* A *under* (1.2.4) by $\tilde{A}$ or $A\tilde{}$ defined by

$$\tilde{A} := \nu \circ \mu^{-1}(A) \subset \mathbf{M}. \qquad (1.2.5)$$

Similarly, if A is a subset of **M** we denote the *inverse correspondence of* A *under* (1.2.4) by $\hat{A}$ or $A\hat{}$ where

$$\hat{A} := \mu \circ \nu^{-1}(A) \subset \mathbf{P}. \qquad (1.2.6)$$

Thus we have schematically the associations

$$\begin{array}{ccc} \mathbf{P} \overset{c}{\Rightarrow} \mathbf{M} & & \mathbf{P} \overset{c^{-1}}{\Leftarrow} \mathbf{M} \\ \cup & & \cup \\ A \mapsto \tilde{A} & & \hat{B} \leftarrowtail B \end{array} \qquad (1.2.7)$$

One should think of ' $\tilde{}$ ' and ' $\hat{}$ ' as *geometric transform symbols* similar to $f \mapsto \hat{f}$, which often denotes the Fourier transform of a suitably defined function. When we consider functions and their generalizations on these spaces, we shall have an analogue of the Fourier transform to which we shall extend this notation in a natural manner. Let us look now at special cases of this geometric transform.

Proposition 1.2.1. *Let $p \in$ P, $q \in$ M; then:*

(a) $\tilde{p} \cong$ P$_2$,

(b) $\hat{q} \cong$ P$_1$.

Proof: (a) By definition we have that $p = S_1^0$, where S_1^0 is a fixed one-dimensional subspace of T. Thus

$$\tilde{p} = \{S_2 \subset \mathsf{T} : \dim S_2 = 2 \text{ and } S_1^0 \subset S_2\}.$$

Let $Z_0 \in S_1^0$ be a nonzero vector and choose a basis for T of the form $\{Z_0, Z_1, Z_2, Z_3\}$. Letting $[w^1, w^2, w^3]$ be homogeneous coordinates for P$_2$, define the mapping

$$w = [w_1, w_2, w_3] \mapsto S_2^w = \mathrm{span}\{Z_0, w^1 Z_1 + w^2 Z_2 + w^3 Z_3\}.$$

It is easy to check that each such S_2^w contains S_1^0 and that all such subspaces arise in this manner. Thus we have a complex-analytic isomorphism $\tilde{p} \cong$ P$_2$.

(b) This is somewhat simpler. We note that, for $q = S_2^0$, where S_2^0 is a fixed two-dimensional subspace of T, one has

$$\hat{q} = \{S_1 \subset \mathsf{T} : S_1 \subset S_2^0\}.$$

But $S_2^0 \cong$ C^2, and hence $\hat{q}$ is isomorphic to the set of one-dimensional subspaces of C^2, which is the very definition of P$_1$. ■

In order to do analysis on our twistor manifolds, we shall need local coordinate representations of these mappings and correspondences. First we choose coordinates for M. Letting

$$z = (z^{jk}) \cong \mathbf{C}^{2 \times 2} \overset{\varphi}{\mapsto} \begin{bmatrix} iz \\ I_2 \end{bmatrix} \tag{1.2.8}$$

be a coordinate mapping for M as in (1.1.2) (where $i = \sqrt{-1}$ is introduced for convenience), we define the coordinate chart on M

$$\mathsf{M}^I := \varphi(\mathbf{C}^{2 \times 2}) \cong \mathbf{C}^4.$$

We call M^I *affine complexified Minkowski space*, noting that it is simply one of six choices of standard coordinate charts for M^I. The symbol I stands for 'infinity', and denotes the fact that the 'points at infinity—I' have been removed from **M**. We shall say more about this later. Let

$$\mathbf{P}^I := (\mathbf{M}^I)^{\hat{}} = \mu \circ \nu^{-1}(\mathbf{M}^I),$$
$$\mathbf{F}^I := \nu^{-1}(\mathbf{M}^I).$$

These are called the *affine parts* of projective twistor space and the associated correspondence space, respectively (although they are certainly not affine spaces in the usual sense of algebraic geometry). One sees readily that

$$\mathbf{F}^I \cong \mathbf{M}^I \times \mathbf{P}_1. \tag{1.2.9}$$

Namely, if $v = [v^0, v^1]$ are homogeneous coordinates for $\mathbf{P}_1$ and $z = (z^{jk})$ are homogeneous coordinates for $\mathbf{M}^I$ as given by (1.2.8), then consider the mapping

$$(z, [v]) \mapsto \left(\begin{bmatrix} iz \\ I_2 \end{bmatrix} v, \begin{bmatrix} iz \\ I_2 \end{bmatrix} \right)$$
$$= \left(\begin{bmatrix} izv \\ v \end{bmatrix}, \begin{bmatrix} iz \\ I_2 \end{bmatrix} \right) \tag{1.2.10}$$
$$= (S_1^{z,[v]}, S_2^{z,[v]}) \in \mathbf{F},$$

where zv is matrix multiplication and we consider v as a column vector. One checks readily that (1.2.10) is an isomorphism, thus verifying (1.2.9).

The projection $\mu : \mathbf{F}^I \to \mathbf{P}^I$ is given readily by means of (1.2.9) and (1.2.10) and has the form (using again row vectors for homogeneous coordinates for projective space)

$$(z, [v]) \overset{\mu}{\mapsto} [izv, v] \in \mathbf{P}^I.$$

Therefore our double fibration (1.2.4), in terms of these coordinates, has the form

$$(z, [v]) \in \mathbf{F}$$

$$[izv, v] \in \mathbf{P} \qquad\qquad z \in \mathbf{M} \tag{1.2.11}$$

In particular, we note that $\mathbf{F}$ is five-dimensional, and by choosing a coordinate covering for $\mathbf{P}_1$, we obtain local coordinates for both $\mathbf{F}^I$ and $\mathbf{P}^I$. We shall have numerous occasions to use these coordinates for the double fibration (1.2.4) later in this book. In the next chapter we shall see how these coordinates can be chosen to be in a natural sense *spinors* satisfying certain group invariance properties.

We want to conclude this section by indicating briefly how the correspondence (1.2.4) and its coordinate representation (1.2.11) can be generalized to other settings which will be useful in later applications. Namely, consider the naturally defined correspondences

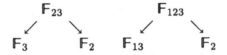

both of which relate F_3 and F_{13} to F_2 ($= M$) in the same way that F_1 (projective twistor space) does. Now, F_3 is naturally *dual* to F_1 in the sense that

$$F_3 = F_3(T) \cong F_1(T^*),$$

using the natural duality between lines and hyperplanes in a vector space. We call F_3 *dual projective twistor space* (it is canonically isomorphic to $P_3(T^*)$), and we denote it by P^*. The twistor manifold $F^* := F_{23}$ is an associated correspondence space, and we write

$$
\begin{array}{ccc}
 & F^* & \\
\swarrow & & \searrow \\
P^* & & M
\end{array}
\qquad (1.2.12)
$$

Similarly, we set $A := F_{13}$, which we call *ambitwistor space* (it has both dual aspects of F_1 and F_3 embedded naturally in it), and let $G = F_{123}$ be the corresponding correspondence space. Thus we have also

$$
\begin{array}{ccc}
 & G & \\
\swarrow & & \searrow \\
A & & M
\end{array}
\qquad (1.2.13)
$$

We shall refer to any of these spaces by their generic labeling F_{13}, etc., whenever this is convenient, but it is helpful to get rid of the subscripts for computational purposes in a given context.

We have for (1.2.12) and (1.2.13) analogues of Proposition 1.2.1. Namely for $p \in M$, we denote the corresponding sets in P, P^*, and A by $(p)\hat{}_P$, $(p)\hat{}_{P^*}$, and $(p)\hat{}_A$ and we have the following proposition whose proof is left to the reader.

Proposition 1.2.2. (a) *If* $p \in \mathsf{M}$ *then*

$$(p)\hat{}_{\mathsf{P}} \cong \mathsf{P}_1, \qquad (p)\hat{}_{\mathsf{P}^*} \cong \mathsf{P}_1, \qquad (p)\hat{}_{\mathsf{A}} \cong \mathsf{P}_1 \times \mathsf{P}_1.$$

(b) *If* $q \in \mathsf{P}$, $r \in \mathsf{P}^*$, $s \in \mathsf{A}$, *then*

$$\tilde{q} \cong \mathsf{P}_2, \qquad \tilde{r} \cong \mathsf{P}_2, \qquad \tilde{s} \cong \mathsf{P}_1.$$

We shall see that the family of two-planes of the form $\tilde{q}$ for $q \in \mathsf{P}$ (to be called α-planes) is disjoint from the family of two-planes of the form $\tilde{r}$, for $r \in \mathsf{P}^*$ (which will be called β-planes). These properties are considered in more detail in §1.4. For all of these situations we have coordinate representations similar to (1.2.8), and we shall develop suitable notation for these coordinates as they are needed.

Exercises for Section 1.2

1.2.1. Show that the flag manifolds $\mathsf{F}_{d_1 \ldots d_m}(V)$ are compact complex manifolds of dimension

$$d = d_1(n - d_1) + (d_2 - d_1)(n - d_2)$$
$$+ (d_3 - d_2)(n - d_3) + \cdots + (d_m - d_{m-1})(n - d_m),$$

where $n = \dim_{\mathsf{C}}(V)$.

1.2.2. Show that there is an embedding

$$f : \mathsf{A} \to \mathsf{P} \times \mathsf{P}^*,$$

given by $(L_1 \subset L_3) \mapsto (L_1, L_3)$, and that $f(\mathsf{A})$ is defined by a quadratic equation

$$V = \{[Z^\alpha], [W_\beta] \in \mathsf{P} \times \mathsf{P}^* : \sum_\alpha Z^\alpha W_\alpha = 0\},$$

where (Z^α) and (W_β) are dual coordinates for T and T^*.

1.2.3. Prove Proposition 1.2.2.

1.3   The Plücker embedding of M into P_5

If V is a complex vector space, then associated with V is the *exterior algebra* of V denoted by $\bigwedge^* V$, where

$$\bigwedge^* V = \sum \bigwedge^k V,$$

and $\bigwedge^k$ is the kth exterior power of the vector space V, that is, the skew-symmetric k-fold tensor product of V with itself. We recall that if V has a basis $\{e_1, \ldots, e_n\}$, then $\bigwedge^k V$ has a basis given by

$$\{e_{i_1} \wedge \cdots \wedge e_{i_k}\}, \quad i_1 < \cdots < i_k.$$

Therefore, if $\omega \in \bigwedge^k V$, then ω can be written in the form

$$\omega = \sum_{i_1 < \cdots < i_k} z^{i_1 \cdots i_k} (e_{i_1} \wedge \cdots \wedge e_{i_k}),$$

and $z^{i_1 \cdots i_k}$ is skew-symmetric in its indices.

Consider the special case of $\dim V = 4$, and define a mapping

$$\mathrm{pl} : \mathbf{G}_2(V) \to \mathbf{P}(\textstyle\bigwedge^2 V)$$

by

$$\mathrm{pl}([Z, W]) = [Z \wedge W] \in \mathbf{P}(\textstyle\bigwedge^2 V), \qquad (1.3.1)$$

where Z and W are linearly independent vectors in V, and $[Z, W]$ and $[Z \wedge W]$ denote $\mathrm{span}(Z, W)$ and $\mathrm{span}(Z \wedge W)$ in $\mathbf{G}_2(V)$ and $\mathbf{P}(\bigwedge^2 V)$, respectively. Letting $\{e_1, \ldots, e_4\}$ be a basis for V, and letting (z^{ij}) be coordinates for $(\bigwedge^2 V)$ as above, then we have the following theorem.

Theorem 1.3.1. *The mapping* pl *in* (1.3.1) *is an embedding and the image* $Q_4 = \mathrm{pl}(\mathbf{G}_{2,4})$ *is a projective algebraic hypersurface of degree two in* $\mathbf{P}_5$ *given by*

$$Q_4 = \{z^{ij} : z^{12}z^{34} - z^{13}z^{24} + z^{14}z^{23} = 0\}, \qquad (1.3.2)$$

where $[z^{ij}]$ *are homogeneous coordinates for* $\mathbf{P}(\bigwedge^2 V)$.

The hypersurface Q_4 is called the *Klein quadric* in $\mathbf{P}_5$ and was the object of a fundamental study by Klein in the late nineteenth century. We shall develop numerous properties of Q_4 and use them in our study of the complex Grassmannian manifold $G_{2,4}$, which is complex-analytically equivalent to Q_4.

Remark: In general, the mapping

$$[Z^1, \ldots, Z^k] \rightarrow [Z^1 \wedge \cdots \wedge Z^k]$$

gives an embedding (the *Plücker embedding*) of $G_k(V)$ into $\mathbf{P}(\bigwedge^k V)$. Now, if $\dim V = n$, then

$$\dim G_k = k(n - k),$$

$$\dim \mathbf{P}(\textstyle\bigwedge^k V) = \binom{k}{n} - 1.$$

For instance, if $k = 2$, we have

$$\dim G_2(V) = 2(n - 2) \quad \text{(linear in } n\text{)},$$
$$\dim \mathbf{P}(\textstyle\bigwedge^2 V) = \tfrac{1}{2}n(n - 1) - 1 \quad \text{(quadratic in } n\text{)},$$

so $\mathrm{pl}(G_{2,n})$ is, in general, of high codimension and defined by homogeneous polynomials also of degree two. The case studied by Klein is the simplest of these.

Proof: Given $\omega = \sum z^{ij} e_i \wedge e_j$, we have to characterize those ω which are *decomposable*, i.e., where $\omega = Z \wedge W$, for $Z, W \in V$. The following lemma will then complete the proof of the theorem.

Lemma 1.3.2. *Let V be a finite-dimensional vector space. Let $\omega \in \bigwedge^2 V$; then ω is decomposable if and only if $\omega \wedge \omega = 0$.*

Proof: If ω is decomposable , then it is immediate that $\omega \wedge \omega = 0$. To show the sufficiency of this condition, it is convenient to use the interior product operator. If $\theta \in V^*$, the *interior product* by θ is the map $\theta \lrcorner : \bigwedge^p V \rightarrow \bigwedge^{p-1} V$ defined by

$$\langle \theta \lrcorner W, \psi \rangle_{p-1} = \langle W, \theta \wedge \psi \rangle_p,$$

where $\langle \, , \, \rangle_q : \bigwedge^q V \times \bigwedge^q V^* \rightarrow \mathbf{C}$ is the duality pairing (using $\mathbf{R}$ for real vector spaces), and where $W \in \bigwedge^p V$, and $\psi \in \bigwedge^{p-1} V^*$. That is, $\theta \lrcorner$ is the transpose of left exterior multiplication by θ. It is an antiderivation of the exterior algebra, i.e.,

$$(\theta \lrcorner)(U \wedge W) = (\theta \lrcorner U) \wedge W + (-)^p U \wedge (\theta \lrcorner W)$$

when $U \in \bigwedge^p V$.

If $\omega \in \bigwedge^2 V$ is nonzero, then there is a $\theta \in V^*$ such that $\theta \lrcorner \omega \neq 0$ (since $\langle \theta \lrcorner \omega, \psi \rangle = \langle \omega, \theta \wedge \psi \rangle = 0$ for all θ, ψ implies that $\omega = 0$). Now, $\omega \wedge \omega = 0$ implies that

$$0 = (\theta \lrcorner)(\omega \wedge \omega) = 2(\theta \lrcorner \omega) \wedge \omega,$$

which implies by Cartan's lemma (Exercise 1.3.1) that

$$\omega = 2(\theta \lrcorner \omega) \wedge U,$$

for some $U \in V$.   ∎

Exercises for Section 1.3

1.3.1. Let V be a finite-dimensional vector space, and let $\eta \in \bigwedge^p V$ be a p-vector, and let $\varphi \in V$ be a nonzero one-vector such that $\varphi \wedge \eta = 0$; then show that there exists a $(p-1)$-vector ζ satisfying $\eta = \varphi \wedge \zeta$ (Cartan's lemma).

1.3.2. Show that if Q is any nonsingular quadric surface defined in $\mathbf{P}_5$, then there is an embedding $\mathrm{pl'} : \mathbf{G}_{2,4} \to \mathbf{P}_5$ such that $\mathrm{pl'}(\mathbf{G}_{2,4}) = Q$.

1.4 The linear geometry of the Klein quadric

In the last section we saw how the complex Grassmannian $\mathbf{G}_{2,4}$ was embedded in $\mathbf{P}_5$ and how its image could be identified with a quadric hypersurface $Q_4 \subset \mathbf{P}_5$, which we called the Klein quadric. In this section we want to study the relationship of Q_4 to linear submanifolds of $\mathbf{P}_5$ (lines, planes, etc.). We recall that a *hyperplane* in $\mathbf{P}_5$ is the zero set of a nontrivial homogeneous function of degree one and is hence a four-dimensional complex submanifold of $\mathbf{P}_5$. A *plane* in $\mathbf{P}_5$ is a two-dimensional complex submanifold defined by the zero set of three linearly independent linear functions of the homogeneous coordinates, and a *line* in $\mathbf{P}_5$ is a one-dimensional complex manifold which is the zero set of four linearly independent linear functions of the homogeneous coordinates. Our intuitive notions of points, lines, planes, etc., from elementary geometry carry over nicely to the projective setting, and this was developed quite precisely by the eighteenth- and nineteenth-century geometers, who developed projective geometry in considerable detail, both from the synthetic (i.e., deriving from axioms) and the analytic or algebraic point of view (as we do here).

If $X \in T_p(Q_4)$ is a tangent vector, then we shall say that X is *tangent* to a submanifold M of Q_4 passing through p if X is in the image of the natural derivative mapping

$$T_p(M) \to T_p(Q_4)$$

given by the embedding $M \to Q_4$, and specifically this defines what we mean by a tangent vector to Q_4 being tangent to a given line or plane in $\mathbf{P}_5$.

Let us recall the twistor correspondences from §1.2:

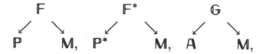

and recalling that $\mathbf{M} = \mathbf{F}_2 \cong \mathbf{G}_{2,4}$, we see that $\mathbf{M}$ is embedded in $\mathbf{P}_5$ by the Plücker embedding and is complex-analytically equivalent to the Klein quadric Q_4, i.e.,

$$\mathbf{M} \xrightarrow[\cong]{\mathrm{pl}} Q_4 = \{[z] : z \wedge z = 0\} \subset \mathbf{P}_5. \tag{1.4.1}$$

Let $\mathbf{T}$ be a fixed four-dimensional twistor space as in §1.2, and consider the bilinear mapping

$$\textstyle\bigwedge^2 \mathbf{T} \times \bigwedge^2 \mathbf{T} \to \bigwedge^4 \mathbf{T} \cong \mathbf{C}, \quad (\omega, \eta) \mapsto \omega \wedge \eta = \langle \omega, \eta \rangle \, \mathrm{vol},$$

where $\mathrm{vol} = Z_1 \wedge Z_2 \wedge Z_3 \wedge Z_4$ is some fixed nonzero four-form in $\bigwedge^4 \mathbf{T}$ and thus gives a basis for this one-dimensional vector space (here $Z_1, \ldots, Z_4$ is some basis for $\mathbf{T}$). The bilinear pairing $(\omega, \eta) \mapsto \langle \omega, \eta \rangle$ is nondegenerate (see Exercise 1.4.1) and hence gives an explicit isomorphism between $\bigwedge^2 \mathbf{T}$ and $(\bigwedge^2 \mathbf{T})^*$. It follows that any hyperplane in $\mathbf{P}(\bigwedge^2 \mathbf{T}) \cong \mathbf{P}_5$ can be described (in homogeneous coordinates) by $H = H(\omega_0)$, where

$$H(\omega_0) = \{\omega \in \textstyle\bigwedge^2 \mathbf{T} : \omega \wedge \omega_0 = 0\} \tag{1.4.2}$$

for some fixed $\omega_0 \in \bigwedge^2 \mathsf{T}$. Thus $\omega \wedge \omega_0 = 0$ is a single linear equation for the hypersurface $H(\omega_0)$, expressed in terms of the exterior algebra (which will turn out to be quite convenient). Note that we are expressing points of $\mathbf{P}(\bigwedge^2 \mathsf{T})$ homogeneously in an equation like (1.4.2), i.e., $H(z_0) = \{$points in $\mathbf{P}(\bigwedge^2 \mathsf{T})$ whose homogeneous coordinates ω satisfy the equation in (1.4.2)$\}$. We shall use this convention repeatedly.

We now want to consider intersections of the quadric Q_4 with hypersurfaces, planes, and lines. First, a hyperplane section s of Q_4 is the intersection of a hyperplane H with Q_4, i.e., for some $\omega_0 \in \bigwedge^2 \mathsf{T}$,

$$s = s(\omega_0) = Q_4 \cap H(\omega_0) = \{\omega \in \bigwedge^2 \mathsf{T} : \omega = Z_1 \wedge Z_2, \omega \wedge \omega_0 = 0\}.$$

Of particular interest for us later will be the hyperplane sections $s(\omega_0)$, where $[\omega_0] \in Q_4$, i.e., $\omega_0 = Z_1 \wedge Z_2$, for some $Z_1, Z_2 \in \mathsf{T}$.

Consider the set of all lines in $\mathbf{P}_5$. Generically, the one-dimensional lines will intersect the four-dimensional manifold Q_4 in isolated points. However, there is a distinguished subset of the lines in $\mathbf{P}_5$, each of which is properly contained in the quadric Q_4. We shall say that a line L in $\mathbf{P}_5$ is *null* if L is contained in Q_4. We shall see that there is precisely a five-parameter family of such null lines which is a subfamily of the eight-parameter family of all lines in $\mathbf{P}_5$. We shall say that a tangent vector $X \in T_p(\mathsf{M}) \cong T_p(Q_4)$ is *null* if X is tangent to a null line in Q_4. This will then give us a natural *conformal structure* on M, defined by the embedding of M into $\mathbf{P}_5$ via the Klein quadric. A *conformal structure* on a manifold in general is the prescription of a *null cone* (cone of null lines) defined by a quadratic function in the tangent space at each point of the manifold. We shall discuss the notion of conformal structure and conformal invariance in more detail in §§1.6 and 4.4, and in Chapters 5 and 6. We shall first develop the fundamental geometric properties of null lines in Q_4, which will then be useful for the understanding of this conformal structure.

Let $\overline{T}_p(Q_4)$ be the tangent plane to the quadric Q_4 as defined in Example 1.1.4. Suppose $y \in \overline{T}_p(Q_4) \cap Q_4$, then

$$q(y,y) = 0,$$
$$q(y,p) = 0, \qquad\qquad (1.4.3)$$
$$q(p,p) = 0,$$

where q is the quadratic form defining Q_4. It follows that

$$q(tp + (1-t)y, tp + (1-t)y) = 0,$$

and hence that the line $\overline{py}$ joining p to y is contained in Q_4, and thus clearly

$$\overline{py} \subset \overline{T}_p(Q_4) \cap Q_4.$$

Therefore, we find that

$$\overline{T}_p(Q_4) \cap Q_4 = \{\text{null lines passing through } p\}.$$

We observe that this shows that we could have defined the null lines passing through p to be the intersection of the quadric with its tangent plane at that point.

If S is a subspace of T of dimension k, then we shall denote by $[S]_\mathsf{P}$ and $[S]_\mathsf{M}$ the corresponding point set in P and M and delete the subscripts if the context is clear. This is a type of 'homogeneous coordinates'. For instance, if S_2 is a two-plane in T, then $[S_2]_\mathsf{P}$ will be a line in P, and $[S_2]_\mathsf{M}$ will be a point in M.

Now consider a point $p \in Q_4$, then $p = [Z_1 \wedge Z_2]$ for some $Z_1, Z_2 \in \mathsf{T}$. Now let

$$s(p) = \{q \in \mathsf{M} : \hat{q} \cap \hat{p} \neq \emptyset\}.$$

Thus $s(p)$ is a subset of M containing the point p. We claim that

$$s(p) = \{q \in Q_4 : q \wedge Z_1 \wedge Z_2 = 0\}. \tag{1.4.4}$$

Namely, if $q \in Q_4$, then $q = [Z \wedge W]$, and $Z \wedge W \wedge Z_1 \wedge Z_2 = 0$ means that the planes

$$S_2 = [Z, W], \qquad S_2' = [Z_1, Z_2]$$

must intersect in a one-dimensional subspace $S_1 \subset \mathsf{T}$, and hence, if $q = [Z, W]$, $p = [Z_1, Z_2]$, then

$$\hat{q} = [S_2], \quad \hat{p} = [S_2'], \quad \text{and} \quad \hat{q} \cap \hat{p} = [S_1] \neq \emptyset.$$

The converse is similar, and hence (1.4.4) is shown. Thus $s(p)$ is a hyperplane section. But then from (1.4.4) we note that

$$\overline{T}_p(Q_4) = \{\omega : \omega \wedge p = 0\}. \tag{1.4.5}$$

This follows from (1.1.6) and (1.4.1) where we are using $q(\omega, \eta) = \omega \wedge \eta$. Using (1.4.3) with (1.4.4) we find that

$$s(p) = \overline{T}_p(Q_4) \cap Q_4 = \{\text{null lines in } Q_4 \text{ passing through } p\}.$$

Thus we conclude with the important assertion that for $p, q \in \mathbf{M}$,

$$\hat{p} \cap \hat{q} \neq \emptyset \text{ in } \mathbf{P} \Leftrightarrow p, q \text{ lie on a null line in } \mathbf{M}. \qquad (1.4.6)$$

We shall say that a plane $N \subset \mathbf{P}_5$ is a *null plane* if any line L in N is null. It is shown in the exercises that null lines and null planes can be characterized infinitesimally by requiring simply that their tangent vectors be null at each point, but it is simpler to work directly with the geometric objects.

We see that if $p \in \mathbf{P}$, $q \in \mathbf{P}^*$, then $\tilde{p}$ and $\tilde{q}$ are both planes ($\cong \mathbf{P}_2$) embedded in $\mathbf{M}$, and similarly, if $f \in \mathbf{A}$, then $\tilde{f}$ is a complex line embedded in $\mathbf{M}$. We define an *α-plane* in $\mathbf{M}$ to be an embedded plane of the form $\tilde{p}$ for $p \in \mathbf{P}$ and a *β-plane* in $\mathbf{M}$ to be of the form $\tilde{q}$ for $q \in \mathbf{P}^*$. We now have the following important theorem which relates the null lines and null planes in Q_4 to the α- and β-planes defined by the correspondences.

Theorem 1.4.1.
   (a) *All α- and β-planes in $\mathbf{M}$ are null planes.*
   (b) *Any null plane is either an α-plane or a β-plane.*
   (c) *The null lines in $\mathbf{M}$ are precisely the five-dimensional family of lines parametrized by $\mathbf{A}$ of the form $\tilde{f}$ for $f \in \mathbf{A}$.*
   (d) *Any null line in $\mathbf{M}$ is the intersection of an α-plane and a β-plane.*

Proof: We shall first show that any plane $N \subset Q_4 \subset \mathbf{P}_5$ is a null plane. Now any line $L \subset N$ is contained in Q_4 and hence is a null line, thus N is null. We have seen earlier that an α-plane or a β-plane is given by an embedding of $\mathbf{P}_2$ into $\mathbf{M}$. We shall show that the composition of such an embedding of $\mathbf{P}_2$ into $\mathbf{M}$ with the Plücker embedding of $\mathbf{M}$ into $\mathbf{P}_5$ gives a linear embedding of $\mathbf{P}_2$ into $\mathbf{P}_5$, and thus the α-planes and β-planes are planes in $\mathbf{P}_5$ which are contained in Q_4 and hence are then clearly null.

Now consider, for $p_0 = [S_1^0] \in \mathbf{P}$, for a fixed one-dimensional subspace $S_1^0 \subset \mathbf{T}$,

$$\tilde{p}_0 := (p_0)^\sim = \{x \in \mathbf{M} : x = [S_2], S_1^0 \subset S_2\},$$

where S_2 is a two-dimensional subspace of $\mathbf{T}$. Rewriting this homogeneously in terms of exterior algebra, we see that

$$\tilde{p}_0 = \{\omega = Z_1 \wedge Z_2 : Z_1 \wedge Z_2 \wedge Z_3 = 0, S_1^0 = [Z_3]\},$$

where $Z_1 \wedge Z_2 \wedge Z_3 = 0$ expresses the fact that Z_3 is linearly dependent on the vectors Z_1 and Z_2, i.e.,

$$S_1^0 \subset [Z_1, Z_2] = \text{span}\{Z_1, Z_2\}.$$

We now choose a basis $\{Z_3, W_1, W_2, W_3\}$ for T and define a mapping

$$\tilde{p}_0 \cong \mathbf{P}_2 \xrightarrow{m} \mathbf{P}(\textstyle\bigwedge^2 \mathsf{T}) \cong \mathbf{P}_5,$$

given by

$$[w^1, w^2, w^3] \mapsto [Z_3 \wedge (w^1 W_1 + w^2 W_2 + w^3 W_3)].$$

It is clear then that m is a linear mapping from $\mathbf{P}_2$ to $\mathbf{P}_5$ of the desired type.

For $h_0 \in \mathbf{P}^* = \mathsf{F}_3$ we proceed in a similar manner. We have that $h_0 = [S_3^0]$, for a fixed $S_3^0 \subset \mathsf{T}$. Thus,

$$\tilde{h}_0 := (h_0)^{\sim} = \{x = [S_2] \in \mathbf{M} : S_2 \subset S_3^0\}.$$

We let $h_0 = [Z_1, Z_2, Z_3] = \text{span}\{Z_1, Z_2, Z_3\}$, for suitable vectors Z_1, Z_2, Z_3 in T. Now

$$\textstyle\bigwedge^1 \mathsf{T} \times \bigwedge^3 \mathsf{T} \to \bigwedge^4 \mathsf{T} \cong \mathbf{C}$$

gives again a nondegenerate pairing so that we can identify

$$\textstyle\bigwedge^3 \mathsf{T} \cong (\bigwedge^1 \mathsf{T})^* = \mathsf{T}^*.$$

Therefore we can identify h_0 with the point

$$[Z_1 \wedge Z_2 \wedge Z_3] \in \mathbf{P}(\textstyle\bigwedge^3 \mathsf{T}) \cong \mathbf{P}(\mathsf{T}^*).$$

The pairing

$$\textstyle\bigwedge^2 \mathsf{T} \times \bigwedge^2 \mathsf{T} \to \bigwedge^4 \mathsf{T} \cong \mathbf{C}$$

allows us to identify

$$\textstyle\bigwedge^2 \mathsf{T} \cong (\bigwedge^2 \mathsf{T})^* \cong \bigwedge^2 \mathsf{T}^*,$$

and hence the family of two-dimensional subspaces

$$S_2 \subset S_3^0 \subset \mathsf{T}$$

corresponds to a family of two-planes

$$(S_3^0)^\perp \subset S_2^\perp \subset T^*.$$

The previous linear embedding of the two-dimensional family of two-planes in T containing a given one-plane applies to this situation in the dual space, yielding a commutative embedding diagram

$$
\begin{array}{ccc}
P_2 \cong \tilde{h}_0 & \xrightarrow{\ m^*\ } & P(\wedge^2 T^*) \\
\ i \downarrow & & \shortparallel \\
M & \xrightarrow{\ \mathrm{pl}\ } & P(\wedge^2 T)
\end{array}
$$

where i is the inclusion mapping and m^* is a (projective-) linear mapping.

Similarly, if $\tilde{f}_0$ is the projective line in M corresponding to $f_0 \in A$, then we have (letting $f_0 = [S_1^0, S_3^0]$)

$$\tilde{f}_0 = \{x = [S_2] : S_1^0 \subset S_2 \subset S_3^0\}.$$

If we let $S_1^0 = [Z_1], S_3^0 = [Z_1, Z_2, Z_3]$, then define the mapping m^{**},

$$\tilde{f}_0 \cong P_1 \xrightarrow{\ m^{**}\ } P(\wedge^2 T) \cong P_5$$

by

$$m^{**}([w^1, w^2]) = [Z_1 \wedge (w^1 Z_2 + w^2 Z_3)].$$

Thus we see that A parametrizes a five-dimensional family of null lines of the form $\tilde{f}_0 \subset M$ (they are null because their images under m^{**} above lie in Q_4!).

Now we shall show that any null line is of the form $\tilde{f}_0$. Let $L \subset M$ be a null line and let a and b be distinct points on L. Then by (1.4.6), $\hat{a} \cap \hat{b} \neq \emptyset$ in P. Let $p = \hat{a} \cap \hat{b}$ and let h be the span of the two lines $\hat{a}$ and $\hat{b}$ in P. Then the pair (p, h) determines a point $f_0 \in A$, as above, and we see that $\tilde{f}_0$ contains the points a and b and hence must coincide with the line L originally given.

Now we want to show part (b). Let N be a null plane in M. We note that for any $p \in N$, $N \subset \overline{T}_p(Q_4) \cap Q_4$. Thus, if p_1, p_2, and p_3 are three noncollinear points in N, we see that

$$N \subset Q_4 \cap \overline{T}_{p_1}(Q_4) \cap \overline{T}_{p_2}(Q_4) \cap \overline{T}_{p_3}(Q_4)$$
$$= \{p \in M : \hat{p} \cap \hat{p}_i \neq \emptyset, \quad i = 1, 2, 3\}.$$

Now the line $\overline{p_i p_j} \subset N \subset Q_4$, and thus $\hat{p}_i \cap \hat{p}_j$ is not empty and contains a point p_{ij}. There are only two cases possible, due to the noncollinearity of p_1, p_2, and p_3:

(i) p_{12}, p_{23}, and p_{13} are *distinct*, and thus $\hat{p}$ intersects $\hat{p}_i$, $i = 1, 2, 3$, if and only if $\hat{p}$ is contained in the hyperplane h spanned by the three points p_{12}, p_{23}, and p_{13}.

(ii) $p_{12} = p_{23} = p_{13}$, and hence $\hat{p}_1$, $\hat{p}_2$, and $\hat{p}_3$ cannot be coplanar. Thus $\hat{p}$ will intersect $\hat{p}_1$, $\hat{p}_2$, $\hat{p}_3$ if and only if $\hat{p}$ passes through $p = p_{12}$.

We see that case (i) says that

$$N = \{z : \hat{z} \subset h\},$$

which is a β-plane, while case (ii) says that

$$N = \{z : p \in \hat{z}\},$$

which is an α-plane.

Part (d) now follows easily from the fact that if a null line L is of the form $\tilde{f}$ for $f = [S_1, S_3]$, then clearly

$$\tilde{f} = [S_1]^\sim \cap [S_3]^\sim,$$

which is an α-plane intersecting a β-plane as desired. ∎

Exercises for Section 1.4

1.4.1. Show that the bilinear pairing

$$(\omega, \eta) \mapsto \langle \omega, \eta \rangle$$

given by

$$(\omega, \eta) \mapsto \omega \wedge \eta = \langle \omega, \eta \rangle \, \text{vol}$$

is nondegenerate, where $\text{vol} = Z_1 \wedge Z_2 \wedge Z_3 \wedge Z_4$ is some fixed nonzero four-form in $\bigwedge^4 \mathsf{T}$.

1.4.2. Let P be a one- or two-dimensional submanifold of $Q_4 \subset \mathbf{P}_5$, and suppose that for each point $p \in P$, $T_p(P)$ consists only of null vectors. Then show that P is a linear submanifold of $\mathbf{P}_5$, and is, in fact, a null line or a null plane depending on the dimension of P. (Hint: use ordinary differential equations and Frobenius' theorem (Sternberg 1965) to translate the infinitesimal data to the global behavior.)

1.5   Group actions and homogeneous structures

The manifolds $\mathbf{P}, \mathbf{M}, \mathbf{F}$, etc., which have been studied in the previous sections, are all examples of homogeneous spaces. A *homogeneous space* is a manifold M equipped with a specified type of transformation group G which acts transitively on the manifold M. For instance, if $M = E^2$, the classical Euclidean plane, then E^2 is a homogeneous space with respect to the group G_E of translations and rotations. From a geometric point of view, all points in E^2 appear the same, and hence one calls such a space homogeneous. The elements of the group G_E depend continuously (and, in fact, real-analytically) on three parameters, the two translation parameters and the rotation angle. In general, an abstract group G, which is also a real-analytic manifold such that the group multiplication properties are continuous with respect to the topology of the manifold structure, is called a *Lie group*. Since their introduction by Sophus Lie in the late nineteenth century, Lie groups, their associated homogeneous spaces, and their representation theory have played an increasingly important and central role in both mathematics and physics. In this book we shall use very little of the abstract theory of Lie groups and homogeneous spaces, but shall be interested in some specific examples. For a thorough treatment of Lie groups, their associated Lie algebras, and the general theory and classification of symmetric spaces, see the important treatise by Helgason (1978).

The basic examples of Lie groups we shall need are the *classical linear groups*, which are groups of matrices defined by some specific property. For instance, if $M_n(\mathbf{R})$ and $M_n(\mathbf{C})$ denote $n \times n$ real and complex matrices, respectively, and ^tA denotes the transpose of a matrix, as usual, then we have the following well-known examples of Lie groups:

$$\mathrm{GL}(n, \mathbf{R}) = \{g \in M_n(\mathbf{R}) : \det g \neq 0\} \quad \text{— } \textit{general linear group;}$$

$$\mathrm{GL}(n, \mathbf{C}) = \{g \in M_n(\mathbf{C}) : \det g \neq 0\} \quad \text{— } \textit{complex general linear group;}$$

$$\mathrm{SL}(n, \mathbf{C}) = \{g \in M_n(\mathbf{C}) : \det g = 1\} \quad \text{— } \textit{special linear group;}$$

$$\mathrm{O}(n) = \{g \in M_n(\mathbf{R}) : {}^tAA = I_n\} \quad \text{— } \textit{orthogonal group;}$$

$$\mathrm{O}(p, q) = \{g \in M_n(\mathbf{R}) : g \text{ preserves}$$
$$|x^1|^2 + \cdots + |x^p|^2 - |x^{p+1}|^2 - \cdots - |x^n|^2$$

as a linear transformation of $\mathbf{R}^n$} — *pseudoorthogonal*
group of type (p, q);
$$\mathrm{SU}(p, q) = \{g \in M_n(\mathbf{C}) : \det g = 1, \ g \text{ preserves}$$
$$|Z^1|^2 + \cdots + |Z^p|^2 - |Z^{p+1}|^2 - \cdots - |Z^n|^2$$
as a linear transformation of $\mathbf{C}^n$} — *special unitary group*
of type (p, q).

There are many others, and they are all a part of the general classification of semisimple real or complex groups, initiated by Killing and E. Cartan at the turn of the century. We shall define precisely any specific group we have occasion to use.

For any Lie group G there is an associated *Lie algebra* $\mathfrak{g}$. Abstractly, $\mathfrak{g}$ is the vector space of tangent vectors to G at e, $T_e(G)$, where e is the identity element of G, and where $T_e(G)$ is equipped (infinitesimally) with the structure of G. The algebra multiplication is denoted by $X, Y \mapsto [X, Y]$, and this product is anticommutative and satisfies

$$[X, [Y, Z]] + [Y, [Z, X]] + [Z, [X, Y]] = 0 \quad \text{(the Jacobi identity)}.$$

Associated with any Lie algebra there is a Lie group, given by an *exponential mapping* $\exp : \mathfrak{g} \to G$. Just as the tangent space to a manifold gives a linearized picture of a curved space, the Lie algebra of a Lie group gives a linearized picture of the Lie group, but, moreover, one can recapture the Lie group in the neighborhood of the identity by means of the exponential mapping (see Helgason 1978).

For our purposes, the exponential mapping is given by the classical power series for the exponential function when applied to the classical linear groups. For instance, if $G = \mathrm{SL}(n, \mathbf{C})$ is a specific Lie group, then $\mathfrak{g} = \mathfrak{sl}(n, \mathbf{C})$, the Lie algebra of $n \times n$ matrices with $\mathrm{tr}(X) = 0$, equipped with the Lie algebra product $[X, Y] = XY - YX$ (the commutator). If $X \in \mathfrak{sl}(n, \mathbf{C})$, defined as above, then

$$g = e^{tX} = I_n + tX + \frac{t^2 X^2}{2} + \cdots + \frac{t^n X^n}{n!} + \cdots$$

is, for any $t \in \mathbf{C}$, a convergent power series which defines an element $g \in \mathrm{SL}(n, \mathbf{C})$. One can see that in such a formula $\det g = 1$, if and only if $\mathrm{tr}(X) = 0$ (Exercise 1.5.1). Thus the Lie algebra elements are represented by the linear terms of the Taylor expansion of the Lie group element near the identity. The theory of geodesics and ordinary differential equations gives an abstract exponential mapping from a Lie algebra of a Lie group back to the original Lie group. We shall speak of Lie algebras and Lie groups quite generally, as it is convenient to do so. However, we shall always have in mind specific examples such as these matrix groups, even if our assertions are true at a more general and abstract level (not *all* Lie groups are standard matrix groups, therein lies the difficulty!).

If G and H are two Lie groups and H is a closed subgroup of G, then

$$G/H = \{gH : g \in G\}$$

is the coset space of equivalence classes (left cosets) of the group G modulo the group H. If H is a normal subgroup, then G/H will again be a (Lie) group. However, if H is not necessarily a normal subgroup, then G/H can be given the structure of a *manifold M* (without the structure of a group) and in such a manner that G acts *transitively* on M (for any points $x, y \in M$ there is a $g \in G$ such that $g(x) = y$). In other words, G/H can be made into a homogeneous space (the action of G on G/H is by left multiplication of the cosets). For instance, if $O(n)$ is the orthogonal group acting on $\mathbf{R}^n$, then consider the quotient space

$$M = O(n)/O(n-1),$$

where $O(n-1)$ is a subgroup of $O(n)$, given by the embedding

$$M \mapsto \begin{pmatrix} 1 & 0 & \cdots & 0 \\ 0 & & & \\ \vdots & & M & \\ 0 & & & \end{pmatrix}$$

where M is an $(n-1) \times (n-1)$ matrix. We can choose a unique representative of $O(n)/O(n-1)$ by letting $g \mapsto g_1$, where g_1 is the first column of g. Then $g_1 \in \mathbf{R}^n$ and g_1 is a unit vector. Thus we have a mapping

$$O(n) \xrightarrow{\tilde{\varphi}} S^{n-1} \subset \mathbf{R}^n,$$

which factors nicely into

$$O(n)/O(n-1) \xrightarrow{\varphi} S^{n-1}.$$

One can check easily that, in fact, φ is one-to-one and onto. Thus, we can see that $O(n)/O(n-1)$ can be given the structure of a smooth manifold and clearly the group $O(n)$ acts transitively on the $(n-1)$-sphere $S^{n-1} \subset \mathbf{R}^n$. Abstractly, any quotient of a Lie group by a closed Lie subgroup can be given a nice manifold structure (Helgason 1978). It is often useful to view a given homogeneous space (such as S^{n-1}) as the specific quotient of two Lie groups. Such a representation is, in fact, not unique, as we shall see below.

In this example we started with the pair of Lie groups G and H and produced the space. In the other direction, can we be given a homogeneous space M and a given transitive group action G acting on M and find such a pair of Lie groups? The answer is yes; one simply takes any random point $p \in M$ and defines

$$H = H_p = \{g \in G : g(p) = p\},$$

the *isotropy group* at p. Then it will follow that H is indeed a closed Lie subgroup of G and that $M \cong G/H$ are manifolds. We shall now see specific examples of such a homogeneous space structure for our twistor-geometric spaces.

Letting T be twistor space, we define $\mathrm{GL}(\mathsf{T})$ to be the set of invertible linear transformations of T. For a choice of basis of T we see that $\mathrm{GL}(\mathsf{T}) \cong \mathrm{GL}(4, \mathbf{C})$. If T is equipped with a nonzero 'volume element' $\Omega \in \bigwedge^4 \mathsf{T}^*$, then $\mathrm{SL}(\mathsf{T})$ is the subset of $\mathrm{GL}(\mathsf{T})$ which preserves Ω, and, again with a choice of bases, $\mathrm{SL}(\mathsf{T}) \cong \mathrm{SL}(4, \mathbf{C})$. Suppose Φ is a nondegenerate Hermitian form on T of type (p, q) (i.e., with respect to a suitable basis, Φ is diagonal with p positive diagonal entries and q negative diagonal entries). The Hermitian form Φ determines a volume form $\Omega = (\mathrm{Im}\,\Phi) \wedge (\mathrm{Im}\,\Phi)$, where $\mathrm{Im}\,\Phi \in \bigwedge^2 \mathsf{T}^*$ is an alternating two-form. Thus, given Φ, we can define $\mathrm{SU}(\mathsf{T}, \Phi)$ to be the subset of $\mathrm{GL}(\mathsf{T})$ which preserves Φ and Ω. It follows that if Φ has type $(2, 2)$, for instance, then

$$\mathrm{SU}(\mathsf{T}, \Phi) \cong \mathrm{SU}(2, 2) = \{g \in M_4(\mathbf{C}) : \det g = 1,\ g \text{ preserves}$$
$$|Z^0|^2 + |Z^1|^2 - |Z^2|^2 - |Z^3|^2\}.$$

Not every twistor space which arises has a canonical basis, or a canonical form Φ, and it is necessary to have the abstract notions $\mathrm{GL}(\mathsf{T})$, $\mathrm{SU}(\mathsf{T}, \Phi)$, etc. In practice, we shall assume there is a basis given and work with the specific matrix groups, unless there is a specific invariance question which arises.

We shall consider a twistor space T equipped with a basis and consider the matrix group $\mathrm{SL}(4, \mathbb{C})$ acting on T. Then this group is transitive on $\mathsf{T} - \{0\}$ and, moreover, given any two flags

$$[S_1, \ldots, S_m], [S_1', \ldots, S_m'] \in \mathsf{F}_{d_1 \ldots d_m},$$

one can find $g \in \mathrm{SL}(4, \mathbb{C})$ so that $g(S_1) = S_1', \ldots, g(S_m) = S_m'$. This is a simple case of extension of bases and renormalizing the resulting linear mapping by dividing by the resulting determinant (see Exercise 1.5.2). Thus $\mathrm{SL}(4, \mathbb{C})$ acts transitively on this twistor manifold. Moreover, if

$$
\begin{array}{ccc}
 & \mathsf{F}_{12} & \\
\mu \swarrow & & \searrow \nu \\
\mathsf{F}_1 & & \mathsf{F}_2
\end{array}
$$

is one of our standard correspondence diagrams, the group action is *equivariant*, i.e., is compatible with the above mappings in the natural sense that

$$\mu(g \cdot f) = g \cdot (\mu f), \quad \text{for } f \in \mathsf{F}_{12}, \ g \in \mathrm{SL}(4, \mathbb{C})$$

and $g \cdot x$ denotes the action of $\mathrm{SL}(4, \mathbb{C})$ on the element $x \in \mathsf{F}_{12}$ or F_1 (Exercise 1.5.3) (and similarly for the other mapping). The other twistor correspondences are all equivariant also.

Consider the space F_{12} and let

$$p = \begin{bmatrix} 1 \\ 0 \\ 0 \\ 0 \end{bmatrix}$$

be a specific point of F_1, and let

$$H_p = \{g \in \mathrm{SL}(4, \mathbb{C}) : g(p) = p\}.$$

Then it follows readily that

$$H_p = \left\{ \begin{pmatrix} * & * & * & * \\ 0 & * & * & * \\ 0 & * & * & * \\ 0 & * & * & * \end{pmatrix} \subset \mathrm{SL}(4, \mathbb{C}) \right\}, \tag{1.5.1}$$

where $*$ denotes a generic complex number. Thus

$$\mathbf{F}_1 \cong \mathbf{P}_3 \cong \mathrm{SL}(4, \mathbf{C})/H_p,$$

where H_p is given by (1.5.1). One can find homogeneous space representations for all of the twistor manifolds; the structure of the corresponding H_p simply becomes more complicated. If we consider $\mathrm{U}(4)$ acting on $\mathbf{T}$, we find out that $\mathrm{U}(4)$ acts transitively on subspaces of $\mathbf{T}$ also, and one can deduce that

$$\mathbf{F}_1 \cong \mathbf{P}_3 \cong \mathrm{U}(4)/[H_p \cap \mathrm{U}(4)]$$

where

$$H_p \cap \mathrm{U}(4) = \left\{ \begin{pmatrix} \mathrm{U}(1) & 0 & 0 & 0 \\ 0 & & & \\ 0 & & \mathrm{U}(3) & \\ 0 & & & \end{pmatrix} \right\},$$

i.e., $H_p \cap \mathrm{U}(4) \cong \mathrm{U}(1) \times \mathrm{U}(3)$, and hence

$$\mathbf{F}_1 \cong \mathbf{P}_3 \cong \mathrm{U}(4)/[\mathrm{U}(1) \times \mathrm{U}(3)],$$

which is very similar to the representation of $S^4 = \mathrm{O}(4)/\mathrm{O}(3)$ discussed earlier. We note that $\mathrm{SL}(4, \mathbf{C})$ and H_p are noncompact *complex Lie groups*, i.e., Lie groups with a complex manifold structure, whereas $\mathrm{U}(4)$ and $\mathrm{U}(3)$ are compact Lie groups (which do not happen to have a complex structure). In general, if a space is the quotient of two complex Lie groups, it will have the structure of a complex manifold, and if a space is the quotient of two compact Lie groups, it will be a compact space. The space $\mathbf{F}_1 \cong \mathbf{P}_3$ is indeed a compact complex manifold as we well know, and we could have defined $\mathbf{F}_1$ in terms of the above Lie groups had we desired to do so. Sometimes this is convenient when the groups are given, and an explicit description of the space is not too simple (which, of course, is not the case with $\mathbf{P}_3$).

Suppose $\mathbf{T}$ is equipped with an Hermitian form Φ of type $(2, 2)$; then we consider the subgroup $G = \mathrm{SU}(\mathbf{T}, \Phi) \cong \mathrm{SU}(2, 2)$, and its action on the diagram

$$\begin{array}{ccc} & \mathbf{F}_{12} & \\ {}^{\mu}\!\swarrow & & \searrow^{\nu} \\ \mathbf{F}_1 & & \mathbf{F}_2 \end{array} \qquad (1.5.2)$$

It turns out that G does *not* act transitively on these manifolds, as SU(4) does, for instance. In fact, if $p \in \mathsf{F}_1$, say, and we consider

$$O_p = \{g \cdot p : g \in \mathrm{SU}(2,2)\},$$

then we call O_p the *orbit* of p with respect to the action of the group, and, as it turns out, O_p can be either an open set or a closed set of lower dimension, depending on the choice of p. For a fixed orbit O_p, the restriction of the action of G to O_p is transitive at p (which is quite evident from the definition). We shall describe explicitly the orbits of G, as they play an important role in our later work. We shall simply define certain subsets of F_1, etc., and then show that they are indeed orbits of G. We shall work exclusively with (1.5.2) above, but the same discussion will hold for the other correspondences.

Since Φ is an Hermitian form on T, it follows that for each vector $Z \in \mathsf{T}, \Phi(Z)$ is real and hence is positive, negative, or zero. Here $\Phi(Z)$ denotes the real number $\Phi(Z, Z)$. Correspondingly, we define for any subspace $S \subset \mathsf{T}$ of dimension ≤ 2,

$$\Phi(S) > 0 \iff \Phi(Z) > 0, \quad \text{for all} \quad Z \in S, Z \neq 0,$$
$$\Phi(S) = 0 \iff \Phi(Z) = 0, \quad \text{for all} \quad Z \in S,$$
$$\Phi(S) < 0 \iff \Phi(Z) < 0, \quad \text{for all} \quad Z \in S, Z \neq 0.$$

If $\dim S = 3$, then we note that the form Φ induces an Hermitian form Φ^* on T^*, and if we let $S^\perp \subset \mathsf{T}^*$ be the annihilator of S, $(S^\perp := \{W \in \mathsf{T}^* : \langle W, Z \rangle = 0, Z \in S\})$, then we say that $\Phi(S) > 0$ if $\Phi^*(S^\perp) > 0$, and similarly for the cases $\Phi(S) < 0$ and $\Phi(S) = 0$. This is necessary since there are no three-dimensional subspaces S of T on which $\Phi|_S$ is definite or zero. We then define

$$\mathsf{F}^+_{d_1 \ldots d_m} = \{[S_1, \ldots, S_m] : \Phi(S_j) > 0, \quad j = 1, \ldots, m\},$$
$$\mathsf{F}^0_{d_1 \ldots d_m} = \{[S_1, \ldots, S_m] : \Phi(S_j) = 0, \quad j = 1, \ldots, m\}, \quad (1.5.3)$$
$$\mathsf{F}^-_{d_1 \ldots d_m} = \{[S_1, \ldots, S_m] : \Phi(S_j) < 0, \quad j = 1, \ldots, m\}.$$

We shall speak of *positive, isotropic*, and *negative vectors, subspaces*, and *flags*, respectively.

Lemma 1.5.1. *The group $G = \mathrm{SU}(\mathsf{T}, \Phi)$ maps each of the flag manifolds in (1.5.3) into itself, and is moreover transitive on each of them.*

Proof: Suppose $Z \in \mathsf{T}$ and $\Phi(Z) > 0$, then $\Phi(gZ) = \Phi(Z)$, since g preserves Φ, and hence $\Phi(gZ) > 0$. Thus it is clear that a positive flag is carried to a positive flag by the action of g, and similarly for negative and isotropic flags. To show the transitivity, we shall choose a special case for convenience, as it will illustrate the general case. Namely, we shall show that G acts transitively on F_{12}^+. Suppose $[S_1, S_2]$ and $[S_1', S_2']$ are two flags in F_{12}^+. We shall say that an ordered basis $\{Z_i : i = 0, 1, 2, 3\}$ for T is *orthonormal with respect to* Φ if

$$
\begin{aligned}
\Phi(Z_i, Z_j) &= 0, \quad i \neq j, \\
\Phi(Z_0, Z_0) &= \Phi(Z_1, Z_1) = 1, \\
\Phi(Z_2, Z_2) &= \Phi(Z_3, Z_3) = -1.
\end{aligned}
$$

Since Φ restricted to S_2 is positive-definite, we can choose a basis $\{Z_0, Z_1\}$ for S_2 satisfying

$$
\begin{aligned}
Z_0 &\in S_1, \\
\Phi(Z_0, Z_1) &= 0, \\
\Phi(Z_0, Z_0) &= \Phi(Z_1, Z_1) = 1.
\end{aligned}
$$

Similarly, we can choose a basis $\{Z_0', Z_1'\}$ for S_2' satisfying

$$
\begin{aligned}
Z_0' &\in S_1', \\
\Phi(Z_0', Z_1') &= 0, \\
\Phi(Z_0', Z_0') &= \Phi(Z_1', Z_1') = 1.
\end{aligned}
$$

Extending $\{Z_0, Z_1\}$ and $\{Z_0', Z_1'\}$ to orthonormal bases for T with respect to Φ, we see that the mapping defined by taking the first basis to the second is an element of G and maps $[S_1, S_2]$ to $[S_1', S_2']$, as desired. The general case is a simple extension of this argument. ∎

We now want to look at an important special case of the above geometric situation. We recall that we have set

$$
\begin{aligned}
\mathsf{P} &= \mathsf{F}_1, \\
\mathsf{M} &= \mathsf{F}_2, \\
\mathsf{F} &= \mathsf{F}_{12}.
\end{aligned}
$$

We now assume we have an Hermitian form Φ of signature $(2, 2)$ on T and define

$$
\begin{aligned}
P &= \mathsf{F}_1^0, \\
M &= \mathsf{F}_2^0, \\
F &= \mathsf{F}_{12}^0.
\end{aligned}
$$

We thus have a double fibration

$$P \qquad\qquad M \qquad\qquad (1.5.4)$$

each 'term' of which is a submanifold of the corresponding complex manifold in

$$\mathsf{F}$$
$$\mu \swarrow \qquad \searrow \nu$$
$$\mathsf{P} \qquad\qquad \mathsf{M}$$

and the mappings μ and ν are the restrictions of the corresponding holomorphic mappings (using the same notation). The equations defining P, F, and M are not holomorphic ones, and hence we do not expect these to be complex submanifolds of P, F, and M, respectively. They turn out to be smooth real submanifolds with various properties relating to the ambient complex manifolds. In the following proposition are listed some of the most important properties.

Proposition 1.5.2. *P, M, and F are closed orbits in P, M, and F of $G = \mathrm{SU}(\mathsf{T}, \Phi)$ of real dimension five, four, and six, respectively; and the fibers of μ and ν are diffeomorphic to S^1 and S^2, respectively.*

Proof: The fact that P, M, and F are closed is immediate from their definition as being the zero sets of continuous functions. To see that they are orbits of G, we must observe that, first of all, G maps each of these spaces to themselves, since the isotropy condition is preserved by the group G. To know that G is transitive on, say, M, we proceed as follows. Let $p = S_2$ and $p' = S_2'$ be two points of M. We have to choose two bases B and B' for T which are adapted to the subspaces S_2 and S_2' and such that the mapping defined by taking basis elements of B to basis elements of B' provide an element of G. We shall indicate briefly how to choose a basis B. Namely, if S_2 is null, we can find two null vectors N_1 and N_2 spanning S_2 and satisfying $\Phi(N_1, N_2) = 1$. We can find additional null vectors N_3 and N_4 in T satisfying $\Phi(N_3, N_4) = 1$, and such that $\Phi(N_3, S_2) = \Phi(N_4, S_2) = 0$. By letting

$$Z_0 = \sqrt{2}(N_1 + N_2),$$
$$Z_1 = \sqrt{2}(N_1 - N_2),$$
$$Z_3 = \sqrt{2}(N_3 + N_4),$$
$$Z_4 = \sqrt{2}(N_3 - N_4),$$

we find that

$$\Phi(Z_0) = -\Phi(Z_1) = \Phi(Z_2) = -\Phi(Z_3) = 1,$$

and

$$\Phi(Z_i, Z_j) = 0, \quad i \neq j.$$

Thus we have an adapted basis, where Z_0, Z_1 span S_2. Choosing a similar basis for S_2' will then clearly complete the proof. The proofs of the transitivity of G on P and F are quite similar and are omitted.

∎

Exercises for Section 1.5

1.5.1. Show that if

$$g = e^{tX} = I_n + tX + \frac{t^2 X^2}{2} + \cdots + \frac{t^n X^n}{n!} + \cdots,$$

where $X \in M_n(\mathbf{C})$, then det $g = 1$ if and only if $\mathrm{tr}(X) = 0$.

1.5.2. Show that $\mathrm{SL}(4, \mathbf{C})$ acts transitively on $\mathbf{F}_{d_1 \ldots d_m}$.

1.5.3. Show that the action of $\mathrm{SL}(4, \mathbf{C})$ on the double fibration

$$
\begin{array}{ccc}
 & \mathbf{F}_{12} & \\
{\scriptstyle\mu}\swarrow & & \searrow{\scriptstyle\nu} \\
\mathbf{F}_1 & & \mathbf{F}_2
\end{array}
$$

is equivariant.

1.6   Minkowski and Euclidean space

The theory of special relativity was discovered almost simultaneously at the beginning of this century by Einstein, Lorentz, and Poincaré. Somewhat later, Minkowski showed how the phenomenon of special relativity could be interpreted as a four-dimensional space with an indefinite (Lorentz) metric. This is now referred to as *Minkowski space M^4*, and is discussed more precisely in the next several paragraphs. It is a homogeneous space with a natural group of motions (the Poincaré group) and is an important part of the fabric of modern physics.

More recently, due to developments in quantum field theory over the past 20 years, one has seen an increasing use in physics of the much older concept of *Euclidean four-space* E^4, which is also homogeneous with a natural group of motions. In this section we shall study both of these spaces, their associated conformal groups of transformations, and their natural (conformal) compactifications. We shall see that the natural compactification of M^4 is, in fact, M, the isotropic submanifold of the four-complex-dimensional twistor manifold M discussed in the previous section. In addition, we shall see how the natural compactification of E^4 (which is simply a four-sphere) is also embedded in M in a homogeneous (or equivariant) manner.

Minkowski space M^4 is defined to be $\mathbf{R}^4$ equipped with a real symmetric nondegenerate bilinear form B_M which in diagonal form is represented by a matrix of the form

$$\begin{pmatrix} 1 & 0 & 0 & 0 \\ 0 & -1 & 0 & 0 \\ 0 & 0 & -1 & 0 \\ 0 & 0 & 0 & -1 \end{pmatrix}$$

We say that a bilinear form has type $(+---)$ if this is the case (with a similar definition for other types). We can use B_M to define the Minkowski inner product of vectors in M^4, namely

$$(x,y)_M := B_M(x,y), \quad x,y \in M^4.$$

Similarly, we define *Euclidean four-space* to be $\mathbf{R}^4$ equipped with a positive-definite bilinear form B_E (which is thus of type $(++++)$), and we define the Euclidean inner product (in the usual manner) by

$$(x,y)_E = B_E(x,y), \quad x,y \in E^4.$$

As we have defined them, M^4 and E^4 are both vector spaces. In fact, we have defined Minkowski space and Euclidean four-space *with a choice of origin*. Both M^4 and E^4 should be considered as *affine spaces*, a generalization of vector spaces with inner products in which there is no distinguished origin. Rather than formally define this notion here (see for instance, Shirokov and Shirokov 1962), we shall assume a choice of origin, as we have done, and later take into account the changes of origin as we need to. In general relativity, as we shall see later, one studies a general four-manifold X whose tangent space (which is $\cong \mathbf{R}^4$) at each point is equipped with a Minkowski inner product as we are describing here.

As the reader probably knows, one defines $O(1, 3)$ to be the group of linear transformations $g : M^4 \to M^4$ such that $(gx, gy)_M = (x, y)_M$, for all $x, y \in M^4$. In the context of Minkowski space, we call $O(1, 3)$ the *Lorentz group* L_M. The Lorentz group is not connected (see Exercise 1.6.1), just as the orthogonal group is not, and we let $L_M^0 = O(1, 3)_0$ be the connected component of the identity element; this is the *restricted Lorentz group*. Let T_M be the group of *translations* in M^4, and let P_M be the semidirect product of L_M and T_M as transformation groups, i.e., P_M is generated by composition of elements of L_M and T_M. We call P_M the *Poincaré group*. One of the fundamental hypotheses of special relativity is that laws describing any physical system on M^4 (considered as space-time with appropriate scale-factors) should be *invariant* under the Poincaré group P_M.

One of the theories describable on M^4 is Maxwell's theory of electromagnetism described by Maxwell's equations (see Chapter 4). It was noticed by Bateman that Maxwell's equations are invariant under a larger group than P_M, and this led to the discovery of the *conformal group* C_M of Minkowski space. This is defined by adjoining to P_M additional transformations of the form

$$x \mapsto rx, \quad r \in \mathbf{R} \quad \text{(dilations)}, \tag{1.6.1a}$$

$$x \mapsto \frac{-x}{\|x\|_M^2}, \quad \text{(inversions)}, \tag{1.6.1b}$$

where $\|x\|_M^2 = (x, x)_M$ is the Minkowski norm on the vector $x \in M^4$. One can show that L_M, P_M, and C_M are all Lie groups which depend continuously on 6, 10, and 15 parameters, respectively, or in more modern language, have dimensions as manifolds of 6, 10, and 15 (see Exercise 1.6.2).

We note that, if we choose coordinates (x^0, x^1, x^2, x^3) in M^4, then the inversions defined by (1.6.1b) are rational functions of the variables, and with singularities on the set

$$C_M = \{x \in M^4 : \|x\|_M = 0\}.$$

We call C_M the *light cone* or *null cone* at the origin. In special relativity (with suitable scales), massless particles such as photons move at the speed of light in the direction of vectors x which satisfy $\|x\|_M = 0$. These are called *light rays*, and the cone C is a cone of light rays, hence the name *light cone* (see Figure 1.6.1).

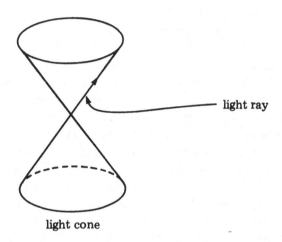

light ray

light cone

Fig. 1.6.1. Light cone.

We now want to represent Minkowski space with a specific useful type of coordinate system. Consider $H(2)$, the space of 2×2 Hermitian matrices, and define the mapping (for a particular choice of origin in M^4)

$$M^4 \overset{\cong}{\to} H(2)$$

by

$$M^4 \ni x = (x^0, x^1, x^2, x^3) \mapsto$$
$$\tilde{x} = \frac{1}{\sqrt{2}} \begin{pmatrix} x^0 + x^3 & x^1 - ix^2 \\ x^1 + ix^2 & x^0 - x^3 \end{pmatrix} \tag{1.6.2}$$
$$= \begin{pmatrix} x^{00} & x^{01} \\ x^{10} & x^{11} \end{pmatrix} = (x^{jk}) \in H(2),$$

and we see that

$$\det \tilde{x} = \tfrac{1}{2} \|x\|_M^2.$$

If we complexify M^4 and extend $(\, , \,)_M$ by complex-linearity, then we obtain a four-dimensional complex vector space equipped with a nondegenerate complex-bilinear form

$$(z, w) = z^0 w^0 - z^1 w^1 - z^2 w^2 - z^3 w^3.$$

We call this extension of Minkowski space to the complex domain *complexified Minkowski space,* and denote it by M_C^4. We see that (1.6.2) extends to a complex isomorphism

$$M_C^4 \stackrel{\cong}{\to} \mathbf{C}^{2\times2} \cong \mathbf{C}^4,$$

where

$$M_C^4 \ni z = (z^0, z^1, z^2, z^3) \mapsto$$
$$\tilde{z} = \frac{1}{\sqrt{2}} \begin{pmatrix} z^0 + z^3 & z^1 - iz^2 \\ z^1 + iz^2 & z^0 - z^3 \end{pmatrix} \tag{1.6.3}$$
$$= \begin{pmatrix} z^{00} & z^{01} \\ z^{10} & z^{11} \end{pmatrix} = (z^{jk}) \in \mathbf{C}^{2\times2}.$$

We shall let x denote either $x = (x^a)$ or $x = (x^{jk})$ as a point of M^4, and similarly, $z = (z^a)$ or (z^{jk}), for points of M_C^4, depending on the context, dropping the superfluous notation $\tilde{x}$ and $\tilde{z}$ in (1.6.2) and (1.6.3).

Now M_C^4 is a vector space, and as such there is a natural identification of the tangent space to M_C^4 at any point with the vector space itself, i.e., if $(a^0, a^1, a^2, a^3) \in M_C^4$, then

$$X = \sum_{j=0}^{3} a^j \frac{\partial}{\partial z^j}$$

is an element of the complex tangent space (holomorphic directional derivative) at any point $p \in M_C^4$.

Recalling our earlier discussion about the light cone C, we shall say that a real vector $x \in M^4$ is *null* if

$$(x, x)_M = \|x\|_M^2 = 0,$$

i.e., if $(x^0)^2 - (x^1)^2 - (x^2)^2 - (x^3)^2 = 0$. Similarly, a complex vector $z \in M_C^4$ is said to be *null* if

$$(z, z)_{M_C^4} = \|z\|_{M_C^4}^2 = (z^0)^2 - (z^1)^2 - (z^2)^2 - (z^3)^2 = 0.$$

The set of null vectors is called the *real light cone* or *complex light cone* (at the origin) in these two respective cases. More abstractly, there is a light cone defined at each point as a proper subset of the tangent space at each point given by the above identification between the vector spaces and their real and complex tangent spaces, respectively. This is indicated in Figure 1.6.2.

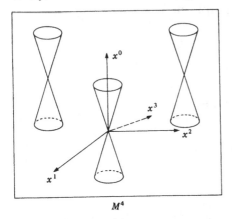

Three-real-dimensional light cone C_p at each point $p \in M^4$, translates of light cone at origin defined by $(x^0)^2 - (x^1)^2 - (x^2)^2 - (x^3)^2 = 0$.

M^4

(a)

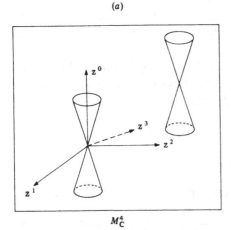

Three-real-dimensional light cone C_p at each point $p \in M^4$, translates of light cone at origin defined by $(z^0)^2 - (z^1)^2 - (z^2)^2 - (z^3)^2 = 0$.

$M_{\mathbb{C}}^4$

(b)

Fig. 1.6.2. Translates of light cones.

Letting $z^a = x^a + iy^a$, we see that $M^4 \subset M_{\mathbb{C}}^4$ is given by $y^a = 0$. If $p \in M^4$, then the real light cone $C_p \subset M_{\mathbb{C}}^4$ is the restriction of the light cone $\widetilde{C}_p$ in $M_{\mathbb{C}}^4$ to M^4, i.e.,

$$C_p = \widetilde{C}_p \cap \{z^a : y^a = 0, \quad a = 0, 1, 2, 3\},$$

and we can regard $\widetilde{C}_p$ as being a *complexification* of the real light cone.[2]

---

[2] In general, one can talk about complexifications of real manifolds or manifolds with singularities, but they need not be unique, except infinitesimally transversal to real manifolds.

A real manifold is said to have a *conformal structure* if there is a smoothly varying family of light cones (null cones) in the tangent spaces at each point, defined pointwise by the vanishing of a non-degenerate real quadratic form. Thus in M^4, the family of light cones C_p above is a conformal structure on M^4. Similarly, a *holomorphic conformal structure* on a complex manifold is a holomorphically varying family of null cones in each holomorphic tangent space which is defined by nondegenerate holomorphic quadratic forms at each point. Clearly, the family of light cones $\widetilde{C}_p$ defined above gives such a structure on $M_{\mathbb{C}}^4$. In relativity theory, a real conformal structure (or we could simply say 'light cone structure') determines the causality of space-time, i.e., which events are in the future or past of other events, and which events are *causally related* to one another. This is important from the point of view of predicting outcomes of experiments when relativistic considerations are taken into account. All reasonable physical theories must contain such a causal structure, and the dynamics of the physical theory must preserve the structure. As such, it is an important foundational structure of any physical theory, a particular case being special relativity, which has the causal structure or conformal structure given by the real light cones in M^4 above.

We now want to show that M^4 and $M_{\mathbb{C}}^4$ can be identified in a canonical fashion with specific twistor manifolds derived from a twistor space T and in such a way that the conformal structure on M^4 and $M_{\mathbb{C}}^4$ is equivalent to conformal structures on the twistor manifolds induced from the null line structure on the Klein quadric Q_4. This will also yield natural conformal compactifications of M^4 and $M_{\mathbb{C}}^4$, i.e., larger compact manifolds which have a conformal structure which restricts to the given one. The action of the conformal group will also extend naturally to the compactification, as we shall see. Let us look at an example of a conformal compactification in classical function theory to see what is involved in the general situation. Consider the complex plane $\mathbb{C}$ and two groups of transformations on $\mathbb{C}$, the *affine transformations*

$$z \mapsto az + b, \quad a, b \in \mathbb{C}, \quad a \neq 0$$

and the *Möbius transformations*

$$z \mapsto \frac{az + b}{cz + d}, \quad a, b, c, d \in \mathbb{C}, \quad ad - bc \neq 0.$$

The affine transformations act transitively on C, but the Möbius transformations have poles if $c \neq 0$, and are singular at such points, and hence do not map the complex plane to itself. If we add a single point $\{\infty\}$ (called the point at infinity) to the complex plane in the usual manner, we obtain the compact complex manifold $\overline{C} = C \cup \{\infty\} \cong S^2 \cong P_1(C)$. We see that the Möbius transformations are transitive on $\overline{C}$ taking the pole to $\{\infty\}$ and the point $\{\infty\}$ to some finite point. The group of Möbius transformations are conformal mappings in the classical sense of preserving angles infinitesimally. We say that the space C has been extended to $\overline{C}$ in such a way that the conformal transformations acting (with singularities) on C are well-defined and act transitively on $\overline{C}$. We refer to such an extension as a *conformal compactification* (that is to say, there is a group of conformal transformations[3] of the original space which extend as conformal transformations to the compactified space). Looking at the Minkowski conformal groups, we see that the inversions are singular precisely on a light cone; so, in order to mimic the one-point compactification of C, we should add a 'light cone at ∞', in such a way that the conformal transformations extend and become well-defined. We shall now see how this can be done.

Consider a twistor space T equipped with a real Hermitian form Φ of signature $(+ + - -)$. Choose coordinates (Z^0, Z^1, Z^2, Z^3) so that the Hermitian form Φ has the matrix form

$$\begin{pmatrix} 0 & I_2 \\ I_2 & 0 \end{pmatrix}, \tag{1.6.4}$$

that is to say,

$$\Phi(Z, Z) = Z^0 \overline{Z^2} + Z^1 \overline{Z^3} + Z^2 \overline{Z^0} + Z^3 \overline{Z^1}.$$

We shall denote the twistor space equipped with this Hermitian form and this choice of coordinates by T^α, and its dual space by T_α. The dual coordinates are given by $W_\alpha = (W_0, W_1, W_2, W_3)$, where

$$Z_0 = \overline{Z^2}, \quad Z_1 = \overline{Z^3}, \quad Z_2 = \overline{Z^0}, \quad Z_3 = \overline{Z^1}.$$

Thus we can write in terms of these variables the duality pairing

---

[3] We have not defined conformal transformation yet; we are only working with examples so far. The general concept will be discussed in the later chapters.

between T^α and its dual T_α by writing[4] for $Z = Z^\alpha \in \mathsf{T}^\alpha$ and $W = W_\alpha \in \mathsf{T}_\alpha$,

$$\langle Z, W \rangle = Z \cdot W = Z^\alpha W_\alpha.$$

The quadratic form $\Phi(Z)$ can then be expressed as $Z^\alpha Z_\alpha$ or simply the shorthand $Z \cdot Z$, both of which we shall have the occasion to use later on in the book. The space T^α is the classical twistor space (along with its dual space T_α) as originally introduced by Penrose in 1967 (see Penrose and MacCallum 1973).

Using the coordinates on T^α, we can now let $G = \mathrm{SU}(\mathsf{T}^\alpha) \cong \mathrm{SU}(2,2)$. Now G acts on M by

$$g \cdot \begin{pmatrix} z \\ w \end{pmatrix} = \begin{pmatrix} A & B \\ C & D \end{pmatrix} \begin{pmatrix} z \\ w \end{pmatrix} = \begin{pmatrix} Az & + & Bw \\ Cz & + & Dw, \end{pmatrix}$$

where $\begin{pmatrix} z \\ w \end{pmatrix}$ are homogeneous coordinates for M, i.e.,

$$\begin{pmatrix} z \\ w \end{pmatrix} \in \mathbf{C}_*^{4 \times 2}, \quad \text{where } z, w \in \mathbf{C}^{2 \times 2}.$$

Letting M^I be the coordinate chart

$$z \longmapsto \begin{pmatrix} iz \\ I_2 \end{pmatrix},$$

as before, we let $\mathsf{M}^+ = \mathsf{F}_2^+$, $\mathsf{M}^- = \mathsf{F}_2^-$, and $M = \mathsf{F}_2^0$. We then have the following two propositions which show how the open sets $\mathsf{M}^\pm$ (which depend on the form Φ) relate to the coordinate chart M^I (which depend on the choice of coordinates on T^α which give Φ the special form (1.6.4)).

---

[4] We use the Einstein summation convention: for any two appropriate mathematical objects Q^α and P_β indexed over some range with upper and lower subscripts, we set

$$Q^\alpha P_\alpha := \sum_\alpha Q^\alpha P_\alpha.$$

Proposition 1.6.1. *The open sets* $\mathbf{M}^+$ *and* $\mathbf{M}^-$ *are contained in* $\mathbf{M}^I$.

Proof: If $\mathbf{M}^+ \not\subset \mathbf{M}^I$, then there is some subspace $S_2 \subset T^\alpha$ such that $[S_2] \in \mathbf{M}^+$, but $[S_2] \notin \mathbf{M}^I$. By the definition of $\mathbf{M}^I$, it follows then that S_2 must contain a vector of the form

$$Z = \begin{pmatrix} a \\ b \\ 0 \\ 0 \end{pmatrix}.$$

But then

$$\Phi(Z, Z) = \begin{pmatrix} \bar{a} & \bar{b} & 0 & 0 \end{pmatrix} \begin{pmatrix} 0 & I_2 \\ I_2 & 0 \end{pmatrix} \begin{pmatrix} a \\ b \\ 0 \\ 0 \end{pmatrix} = 0,$$

and hence $Z \notin \mathbf{M}^+$. It follows that $\mathbf{M}^+ \subset \mathbf{M}^I$. Similarly, one shows that $\mathbf{M}^- \subset \mathbf{M}^I$.   ∎

We already knew that $\mathbf{M}^+$ and $\mathbf{M}^-$ were open connected orbits of G, but now we know they are both contained in the same coordinate chart. What about $\mathbf{M}^0 = M$; is it contained in the same coordinate chart? The answer is no, as we shall see below, but we can see what $\mathbf{M}^+$, $\mathbf{M}^-$, and $M \cap \mathbf{M}^I$ look like in terms of the coordinates on $\mathbf{M}^I$. Let $>>$ denote positive-definiteness of a matrix, let $z = z^{ij}$ be coordinates on $\mathbf{M}^I$ as before, and let $z = x - iy$, where x and y are Hermitian 2×2 matrices.

Proposition 1.6.2.

(a) $\mathbf{M}^+ = \{z = x - iy : y >> 0\}$,

(b) $\mathbf{M}^- = \{z = x - iy : y << 0\}$,

(c) $M \cap \mathbf{M}^I = \{z = x - iy : y = 0\}$.

Remark: The domains $\mathbf{M}^\pm$ in Proposition 1.6.2 are *tube domains* with *distinguished boundary* $M \cap \mathbf{M}^I$ (see, e.g., Bochner and Martin 1948). They are generalizations of the classical upper and lower half-plane (see Figures 1.6.3 and 1.6.4).

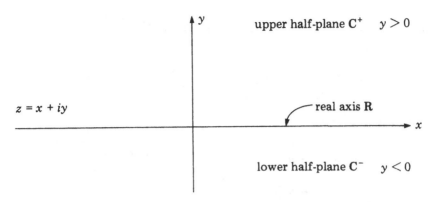

Fig. 1.6.3. Upper and lower half-planes in **C**.

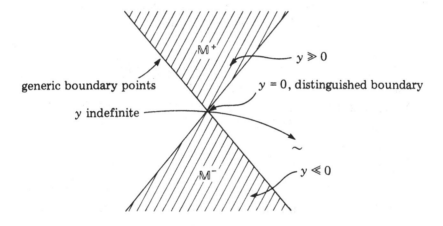

Fig. 1.6.4. $\mathbf{C}^{2\times2}$ with generalized half-planes and distinguished boundary.

We see that $\mathbf{M}^{\pm}$ have topological boundaries which are of real dimension seven generically; these are the points where y has a zero eigenvalue. On the other hand, the distinguished boundary is important in that one has a maximum principle and a Cauchy integral formula for boundary values on $M \cap M^I$ and integration over same, by analogy with the classical results in one-complex-variable theory (see Bochner and Martin 1948).

Proof: In terms of the coordinates of $\mathbf{M}^I$, we see that if

$$v \in \begin{bmatrix} iz \\ I_2 \end{bmatrix},$$

then v is a linear combination of the columns of $\begin{pmatrix} iz \\ I_2 \end{pmatrix}$, i.e.,

$$v = \begin{pmatrix} iz \\ I_2 \end{pmatrix} w, \quad \text{where } w = \begin{pmatrix} w^1 \\ w^2 \end{pmatrix}, \quad w^1, w^2 \in \mathbf{C}.$$

Thus

$$\begin{aligned} \Phi(v,v) &= w^*(-iz^*, I_2) \begin{pmatrix} 0 & I_2 \\ I_2 & 0 \end{pmatrix} \begin{pmatrix} iz \\ I_2 \end{pmatrix} w, \\ &= w^*(i(z - z^*))w, \\ &= 2w^* y w. \end{aligned} \tag{1.6.5}$$

Thus $z \in \mathbf{M}^+$ if and only if $y \gg 0$. The assertions for $\mathbf{M}^-$ and $M \cap \mathbf{M}^I$ follow immediately from (1.6.5) also. ∎

What is the null cone structure induced on $M \cap \mathbf{M}^I$ from the Klein quadric? Let $p \in M \cap \mathbf{M}^I$, and define C_p to be the set of null lines in $\mathbf{M}$ passing through p restricted to $M \cap \mathbf{M}^I$. We now look at the Plücker embedding of $\mathbf{M}$ into $\mathbf{P}_5$ given by

$$Z \mapsto Z \wedge Z.$$

Suppose $[Z] \in \mathbf{M}^I$, then we compute the six independent wedge products in this mapping in terms of 2×2 determinants, obtaining

$$\begin{bmatrix} iz \\ I_2 \end{bmatrix} \wedge \begin{bmatrix} iz \\ I_2 \end{bmatrix} = \begin{bmatrix} iz^{11} & iz^{12} \\ iz^{21} & iz^{22} \\ 1 & 0 \\ 0 & 1 \end{bmatrix} \wedge \begin{bmatrix} iz^{11} & iz^{12} \\ iz^{21} & iz^{22} \\ 1 & 0 \\ 0 & 1 \end{bmatrix}$$

$$= \begin{bmatrix} -z^{11}z^{22} + z^{21}z^{12} \\ -iz^{12} \\ iz^{11} \\ -iz^{22} \\ iz^{21} \\ 1 \end{bmatrix} \in \mathbf{P}(\wedge^2 \mathbf{T}) \cong \mathbf{P}_5. \tag{1.6.6}$$

The null lines in Q_4 are the lines contained in Q_4 as we have defined previously in §1.4. We see from (1.6.6) that $Q_4 \cap \mathbf{M}^I$ is parametrized by (letting $[y^0, \ldots, y^5]$ be homogeneous coordinates for $\mathbf{P}_5$)

$$y^0 = -(z^{11}z^{22} - z^{21}z^{12}),$$
$$y^1 = -iz^{12},$$
$$y^2 = iz^{11},$$
$$y^3 = -iz^{22},$$
$$y^4 = iz^{21},$$
$$y^5 = 1.$$

We can also consider $(y^0, y^1, y^2, y^3, y^4)$ as affine coordinates in $\mathbf{P}_5$. Thus, $Q_4 \cap \mathbf{M}^I$ is the graph in $\mathbf{C}^5$ of

$$y^0 = -(z^{11}z^{22} - z^{21}z^{12}),$$

graphed over the hyperplane with coordinates $(z^{11}, z^{12}, z^{21}, z^{22})$. See Figure 1.6.5.

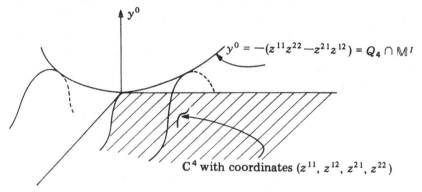

Fig. 1.6.5. $Q_4 \cap \mathbf{M}^I$ as a graph over $\mathbf{C}^4$.

Thus, we see that $Q_4 \cap \mathbf{M}^I \subset \{y_5 \neq 0\} \cong \mathbf{C}^5$, an affine coordinate system for $\mathbf{P}_5$. So, null lines on $\mathbf{M}$ restricted to $Q_4 \cap \mathbf{M}^I$ will be simply straight lines in $\mathbf{C}^5$ (coordinates $(y^0, \ldots, y^4)$) which lie in $Q_4 \cap \mathbf{M}^I$.

Let us compute the null cone at the origin of $M \cap \mathbf{M}^I$ in these coordinates. At the origin we see that

$$\overline{T}_0(Q_4) \cap \mathbf{M}^I = \{(y^0, y^1, \ldots, y^4) : y^0 = 0\}.$$

Thus the null cone at $(0, \ldots, 0)$ is given by

$$N_0 = \{y^0 = 0\} \cap \{y^0 = -(z^{11}z^{22} - z^{21}z^{12})\} = \{\det(z^{ij}) = 0\}.$$

But if we use (1.6.3) to realize $M_{\mathbf{C}}^4$ as $\mathbf{C}^{2\times2}$, the 2×2 complex matrices, we see that the light cone of complexified Minkowski space at the origin coincides with the null cone

$$\overline{T}_0(Q_4) \cap Q_4 \cap \mathbf{M}^I,$$

since

$$\begin{aligned}
\|z\|_{M_{\mathbf{C}}^4}^2 &= (z^0)^2 - (z^1)^2 - (z^2)^2 - (z^3)^2 \\
&= 2(z^{11}z^{22} - z^{12}z^{21}) \\
&= 2\det z
\end{aligned}$$

by (1.6.3).

We shall now see that the transformation group G acting on $M \cap \mathbf{M}^I$ corresponds to the conformal group C_M acting on M^4. Since both group actions preserve the conformal structure of their respective spaces, and their null cone structures agree at a given point, it will follow that they have the same null cone structure everywhere. Thus we shall be able to identify M^4 with $M \cap \mathbf{M}^I$ equipped with its null cone structure induced from Q_4. The equivalence of the transformation groups is given in the following discussion.

Recall that $G = \mathrm{SL}(\mathsf{T}^\alpha) \cong \mathrm{SU}(2,2)$ acts on $\mathbf{M}$ by

$$\mathbf{M} \ni \begin{bmatrix} z \\ w \end{bmatrix} \mapsto g \cdot \begin{bmatrix} z \\ w \end{bmatrix} = \begin{pmatrix} A & B \\ C & D \end{pmatrix} \begin{bmatrix} z \\ w \end{bmatrix} = \begin{bmatrix} Az + Bw \\ Cz + Dw \end{bmatrix},$$
$$\text{for } g = \begin{pmatrix} A & B \\ C & D \end{pmatrix} \in G, \quad (1.6.7)$$

where $\begin{bmatrix} z \\ w \end{bmatrix}$ are homogeneous coordinates for $\mathbf{M}$. In $\mathbf{M}^I$ we have the action of g on a point $z \in \mathbf{M}^I$ given by

$$g \cdot \begin{bmatrix} iz \\ I_2 \end{bmatrix} = \begin{bmatrix} iAz + B \\ iCz + D \end{bmatrix} = \begin{bmatrix} (iAz + B)(iCz + D)^{-1} \\ I_2 \end{bmatrix}.$$

Thus we find, explicitly in terms of the coordinates on $\mathbf{M}^I$,

$$z \in \mathbf{C}^{2\times2} \mapsto g \cdot z = -i(iAz + B)(iCz + D)^{-1},$$

which is a rational function of z, with singularities at the points where $iCz + D$ is not invertible. This is completely analogous to the derivation of Möbius transformations from the action of $SL(2, \mathbf{C})$ on $\mathbf{P}_1$, restricted to an affine coordinate chart of $\mathbf{P}_1$.

Now G preserves Φ and hence we must have that

$$\begin{pmatrix} A^* & C^* \\ B^* & D^* \end{pmatrix} \begin{pmatrix} 0 & I_2 \\ I_2 & 0 \end{pmatrix} \begin{pmatrix} A & B \\ C & D \end{pmatrix} = \begin{pmatrix} 0 & I_2 \\ I_2 & 0 \end{pmatrix},$$

which implies that

$$A^*C = -C^*A, \quad A^*D + C^*B = I_2, \quad B^*D = -D^*B,$$

and we also have $\det g = 1$.

Define a subgroup $\tilde{P} \subset G$ by

$$\tilde{P} = \left\{ g \in G : g = \begin{pmatrix} A & B \\ 0 & D \end{pmatrix} \right\}.$$

Then we have that

$$A^*D = I_2, \quad B^*D = -D^*B, \quad \det AD = 1,$$

or

$$D = (A^*)^{-1}, \quad (D^*)^{-1}B^* = -BD^{-1}, \quad \det A(A^*)^{-1} = 1.$$

Eliminating the variable D from the expression for $\tilde{P}$, we find that the action of $\tilde{P}$ on $\mathbf{M}^I$ is given by

$$z \mapsto (Az - iB)((A^*)^{-1})^{-1} = AzA^* - iBA^*. \tag{1.6.8}$$

We now look at a subgroup of $\tilde{P}$ defined by

$$P = \left\{ \begin{pmatrix} A & B \\ 0 & D \end{pmatrix} \in \tilde{P} : A \in SL(2, \mathbf{C}) \right\},$$

then the action (1.6.8) for $g \in P$ consists of an action of $SL(2, \mathbf{C})$ on $\mathbf{M}^I$ plus translations by the Hermitian matrices iBA^* (that iBA^* is Hermitian follows from the above identities, Exercise 1.6.3). Now restrict the action of P to $M \cap \mathbf{M}^I = H(2)$, which by (1.6.2) we can identify with M^4, and we have that

$$g = \begin{pmatrix} A & B \\ 0 & (A^*)^{-1} \end{pmatrix}, \quad z \in H(2) \mapsto AzA^* - iBA^*.$$

The action of $SL(2, \mathbf{C})$ on $H(2)$ given by

$$z \mapsto AzA^*, \quad A \in SL(2, \mathbf{C})$$

induces a mapping

$$L : M^4 \rightarrow M^4$$

by (1.6.2). As before, we note that

$$2 \det z = 2 \det z^{ij} = (z^0)^2 - (z^1)^2 - (z^2)^2 - (z^3)^2$$

from (1.6.3), using both notations for points in $M_{\mathbf{C}}^4$. Since $\det A = \det A^* = 1$ (by the definition of $SL(2, \mathbf{C})$), it follows that

$$\det z = \det(AzA^*),$$

and hence that L preserves the Minkowski quadratic form

$$(x^0)^2 - (x^1)^2 - (x^2)^2 - (x^3)^2,$$

and thus must be an element of the Lorentz group. In fact, it is an element of the restricted Lorentz group (which is the connected component of the identity which preserves the time and space orientations). It is not too difficult to see that any element of the restricted Lorentz group arises in this manner, and is unique up to multiplication of A by -1 (see Exercise 1.6.4). Thus we have identified P as a two-to-one covering group of the restricted Poincaré group, and the action of P on M^4 coincides with the action of the restricted Poincaré group on M^4.

Now *dilations* correspond to mappings of the form

$$g_\rho = \begin{pmatrix} \rho I_2 & 0 \\ 0 & \rho^{-1} I_2 \end{pmatrix} \in \tilde{P}, \quad \rho \in \mathbf{R},$$

where

$$g \cdot z = \rho^2 z.$$

The *inversions* are given by matrices in G of the form (this does not form a subgroup!)

$$\left\{ \begin{pmatrix} 0 & B \\ C & 0 \end{pmatrix} : BC^* = -I_2 \right\} \quad \text{or} \quad \left\{ \begin{pmatrix} 0 & B \\ (-B^*)^{-1} & 0 \end{pmatrix} \right\}.$$

In other words, an inversion is a mapping of the form

$$z \mapsto -iB(i(-B^*)^{-1}z)^{-1} = Bz^{-1}B^*.$$

Let

$$ds^2 = dz^{11}dz^{22} - dz^{12}dz^{21}$$

be the metric form on $\mathbf{M}^I$ given by the determinant (equivalent to the Minkowski metric on $M_\mathbf{C}^4$), then the dilations and the inversions preserve this form up to a multiple. Namely, if

$$z \mapsto \rho^2 z,$$

then

$$ds^2 \mapsto \rho^4 ds^2,$$

or, if

$$z \mapsto Bz^{-1}B^* = g(z),$$

then

$$
\begin{aligned}
g^*(ds^2) &= \det(dg(z)) \\
&= \det(Bd(z^{-1})B^*) \\
&= \det(-Bz^{-1}dz\,z^{-1}B^*) \\
&= \det(BB^*)(\det z)^{-2}\det dz \\
&= \det(BB^*)(\det z)^{-2}ds^2.
\end{aligned}
$$

In both cases ds^2 transforms by a multiple which depends (in the second case) on the point z at which one is making the transformation. This is known as a conformal change of scale of the metric form, and metric forms and conformal rescalings will be discussed more generally in Chapters 4 and 6. This is simply given as an illustrative example at this point.

We now want to identify the conformal compactification of M^4 using the identification of M^4 with $M \cap \mathbf{M}^I \subset \mathbf{M}$. We see that M is a four-dimensional real submanifold of $\mathbf{M}$ which contains $M \cap \mathbf{M}^I \cong M^4$ as an open subset. Moreover, the group G acts transitively on M, and the restriction of this action to $M \cap \mathbf{M}^I$ is a family of rational mappings which can be identified with the action of the conformal group on M^4. Thus we call M the *conformal compactification of* M^4. Let us now see what M looks like as a differentiable manifold.

Recall that if we have the classical upper half-plane in the complex plane, then there is a Möbius transformation which maps the upper half-plane to the unit disc, and which is usually referred to as the *Cayley transform*. The Cayley transform is given by

$$z \overset{c}{\mapsto} \frac{z - i}{z + i} = w,$$

and it essentially moves the point at ∞ on the Riemann sphere into the lower half-plane, so that the resulting unit disc (holomorphically equivalent to the original upper half-plane) is exhibited with all of its boundary (a circle) in one coordinate chart (the w-plane) (see Figure 1.6.6). This picture shows us that the compactification of the boundary of $\mathbf{C}^+$ in the first coordinate chart is given in the second coordinate chart by the circle S^1 which is the full boundary of $\mathbf{C}^+$.

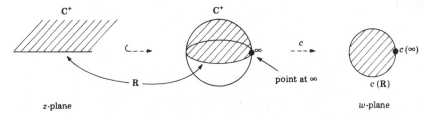

Fig. 1.6.6. The Cayley transform c of the upper half-plane $\mathbf{C}^+$.

We can do the same thing by using the realization given in Proposition 1.6.2 of $\mathbf{M}^+$ as a generalized upper half-plane. Namely, we know that

$$\mathbf{M}^+ = \{z = x - iy \in \mathbf{C}^{2 \times 2} : x \text{ and } y \text{ are Hermitian and } y \gg 0\}$$

Now $\mathbf{M}^+$ is an unbounded domain in $\mathbf{M}^I$ and $M \cap \mathbf{M}^I$ is part of the boundary of $\mathbf{M}^+$. We want to map $\mathbf{M}^+$ into a bounded domain in a different coordinate chart to see what the full boundary looks like. The manifold M will then be a subset of the full boundary of $\mathbf{M}^+$ which we shall be able to identify.

Let

$$c : \mathbf{M}^+ \to \widetilde{\mathbf{M}}^+ \tag{1.6.9}$$

be defined by

$$w = c(z) = i(z + iI_2)(z - iI_2)^{-1}.$$

Then

$$\mathsf{M}^+ = \{z : y \gg 0\} \xrightarrow[\cong]{c} \widetilde{\mathsf{M}}^+ = \{w : I_2 - w^* w \gg 0\}.$$

The topological boundary $\partial \widetilde{\mathsf{M}}^+$ is given by the set of those w which have $I_2 - w^* w$ positive-semidefinite, with at least one zero eigenvalue. The subset of $\partial \widetilde{\mathsf{M}}^+$ given by those w such that $I_2 - w^* w$ has *all* zero eigenvalues, i.e., such that $I_2 - w^* w = 0$, is a distinguished part of the boundary of $\widetilde{\mathsf{M}}^+$ which can be identified with M. Namely, we see that $c(M \cap \mathsf{M}^I)$ is an open subset of the connected manifold

$$\{w : I_2 = w^* w\}, \tag{1.6.10}$$

which we recognize as being the same as the Lie group $U(2)$. Thus we see that M can be identified with $U(2)$ since both M and $U(2)$ are connected closed orbits of G (the action on $U(2)$ is given by conjugation by c), and we can use c to identify the open set $M \cap \mathsf{M}^I$ with an open subset of $U(2)$. We formulate this as a proposition.

Proposition 1.6.3. (a) $M = U(2) \subset c(\mathsf{M}^I)$ *and the action of G on M is given by*

$$w \in U(2) \mapsto cgc^{-1} w,$$

where $g = \begin{pmatrix} A & B \\ C & D \end{pmatrix}$ *acts on* M *as in* (1.6.7).

(b) $M \cong S^1 \times S^3 / \mathbf{Z}_2$, *where $\mathbf{Z}_2$ acts naturally on the circle S^1 by*

$$r \mapsto r + m, \quad m \in \mathbf{Z}_2,$$

and where $r \mapsto e^{i\pi r}$ parametrizes the circle S^1.

Proof: Part (a) was discussed above. To see part (b) we define a homomorphism

$$\eta : S^1 \times SU(2) \to U(2)$$

by

$$(e^{i\theta}, T) \mapsto e^{i\theta} T.$$

We see from the usual orthogonality relations for the coefficients of the matrices in $SU(2)$ that $SU(2)$ is diffeomorphic to S^3 (see Exercise 1.6.5). So this gives

$$S^1 \times S^3 \xrightarrow{\eta} U(2).$$

The map η is surjective and has a kernel given by

$$\left\{ \begin{pmatrix} e^{i\theta}a & e^{i\theta}b \\ e^{i\theta}c & e^{i\theta}d \end{pmatrix} = \begin{pmatrix} 1 & 0 \\ 0 & 1 \end{pmatrix} \right\}.$$

Since $\begin{pmatrix} a & b \\ c & d \end{pmatrix} \in \mathrm{SU}(2)$, we have $ad - bc = 1$, and since $b = c = 0$, we obtain $ad = 1$. Thus, $a = d = \pm 1$, and $\theta = 0$ or π. Therefore,

$$\mathrm{U}(2) \cong S^1 \times \mathrm{SU}(2)/\ker \eta \cong S^1 \times S^3/\mathbf{Z}_2,$$

with the $\mathbf{Z}_2$ action on the factor S^1. ∎

Remark: The natural parametrization of $\mathrm{U}(2)$ is as we have given it above. But, in fact, $S^1 \times S^3/\mathbf{Z}_2$ is diffeomorphic to $S^1 \times S^3$ (see Exercise 1.6.6), but this diffeomorphism does not preserve the group action of G on M as it does on $S^1 \times S^3/\mathbf{Z}_2$, as parametrized above.

We now turn to a realization of the compactification of E^4 in compactified complexified Minkowski space **M**. For this, we start with a twistor space (for instance, the space T^α we used above) and introduce a different real structure (we shall ignore the Hermitian form Φ for the time being). Let **i**, **j**, **k** be the generators of the ring of quaternions **H**, i.e.,

$$\mathbf{H} = \{x^0 + \mathbf{i}x^1 + \mathbf{j}x^2 + \mathbf{k}x^3\},$$

where $x^j \in \mathbf{R}$, $\mathbf{i}^2 = \mathbf{j}^2 = \mathbf{k}^2 = -1$, $\mathbf{ij} = \mathbf{k} = -\mathbf{ji}$, and all cyclic permutations of these identities. Define, for Z^α being coordinates on T^α as before, a mapping

$$\mathsf{T}^\alpha \overset{\cong}{\to} \mathbf{H}^2 \tag{1.6.11}$$

by

$$Z^\alpha \mapsto (w^0, w^1) := (Z^0 + Z^1\mathbf{j}, Z^2 + Z^3\mathbf{j}).$$

Let $\sigma : \mathbf{H}^2 \to \mathbf{H}^2$ be left multiplication by $-\mathbf{j}$. This induces a conjugate-linear mapping $\sigma : \mathbf{C}^4 \to \mathbf{C}^4$, of the form

$$\sigma(Z^0, Z^1, Z^2, Z^3) = (\overline{Z^1}, -\overline{Z^0}, \overline{Z^3}, -\overline{Z^2}).$$

Note the analogy to the more standard conjugation on $\mathbf{C}^4$ given by

$$(Z^0, \ldots, Z^3) \mapsto (\overline{Z^0}, \overline{Z^1}, \overline{Z^2}, \overline{Z^3}).$$

The mapping σ induces a conjugate-linear mapping $\sigma : \mathbf{P} \to \mathbf{P}$. The standard conjugation has fixed points which are the 'real points' of $\mathbf{P}$. On the other hand, σ has no fixed points in $\mathbf{P}$ at all, but it does have complex projective lines which are invariant under σ. These are called *real lines* with respect to the involution σ induced from the quaternionic structure on T^α.

The mapping $\mathsf{T}^\alpha \to \mathbf{H}^2$ induces a mapping of projective spaces

$$\pi : \mathbf{P}(\mathsf{T}^\alpha) \to \mathbf{P}(\mathbf{H}^2),$$

where we define $\mathbf{P}(\mathbf{H}^2)$ to be the left-(quaternionic) scalar multiples of a given two-vector in $\mathbf{H}^2$, i.e.,

$$\mathbf{P}(\mathbf{H}^2) \cong (\mathbf{H}^2 - \{0\})/ \sim, \qquad (1.6.12)$$

where $(a, b) \sim (c, d) \Leftrightarrow (a, b) = \lambda(c, d), 0 \neq \lambda \in \mathbf{H}$. It follows from (1.6.12) that $\mathbf{P}(\mathbf{H}^2)$ is covered by two coordinate charts, each diffeomorphic to $\mathbf{H} \cong \mathbf{R}^4$, and hence $\mathbf{P}(\mathbf{H}^2) \cong S^4$ (see Exercise 1.6.7; this is completely analogous to $\mathbf{P}(\mathbf{C}^2) \cong S^2$).

One can check that the fibers of the mapping π are complex projective lines embedded in $\mathbf{P}(= \mathbf{P}(\mathsf{T}^\alpha))$. Moreover, these are precisely the real lines mentioned above (see Exercise 1.6.8). Now we know that M parametrizes *all* of the projective lines in $\mathbf{P}$, and by the above consideration we have a projection $\pi : \mathbf{P} \to S^4$ whose fibers are a distinguished four-real-parameter family of projective lines in $\mathbf{P}$. Thus we have a natural embedding of the four-sphere in the parameter space M of all lines, as indicated in the following diagram:

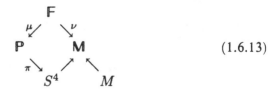

$$(1.6.13)$$

Thus, both $M(\cong S^1 \times S^3)$ and S^4 are embedded naturally in complexified compactified Minkowski space M. In fact, they are embedded as *totally real submanifolds* of M (see Exercise 2.2.8). This means, in this case, that locally their tangent spaces look like $\mathbf{R}^4$ embedded in its complexification $\mathbf{C}^4$, i.e., $\mathbf{M}$ is a complexification of both M and S^4 in a different manner.

We already know that M is the closed orbit of $G \cong \mathrm{SU}(2,2)$. One can also see that S^4 is the closed orbit of the Lie group $G_E \subset \mathrm{SL}(\mathsf{T}^\alpha)$, where $G_E \cong \mathrm{Spin}(5,1)$.[5] Both of these real Lie groups have $\mathrm{SL}(\mathsf{T}) \cong \mathrm{SL}(4,\mathbf{C})$ as its unique Lie group complexification. The orbits S^4 and M are related at the Lie algebra level and in local coordinates by a transformation of the form $t \to it$, where t represents the time coordinate. This essentially changes the Minkowski inner product to a Euclidean inner product. Explicitly, we shall look at this for the coordinate systems used earlier.

We know that affine Minkowski space and affine Euclidean space both look like $\mathbf{R}^4$, so let us look at S^4 and M intersecting the same coordinate chart $\mathbf{M}^I$, where we use the coordinates (z^0, z^1, z^2, z^3) on $\mathbf{M}^I$ related to the matrix coordinates z^{ij} by (1.6.3). We let $z^a = x^a + iy^a$, where x^a denotes coordinates of M^4. Now real affine Minkowski space is given by

$$M^4 = M \cap \mathbf{M}^I = \{z \in \mathbf{M}^I : y^0 = y^1 = y^2 = y^3 = 0\},$$

so that M^4 has coordinates (x^0, x^1, x^2, x^3), as usual. Quite analogously, one finds that (Exercise 1.6.9) $S^4 \cap \mathbf{M}^I$ is given by

$$S^4 \cap \mathbf{M}^I = \{z \in \mathbf{M}^I : y^0 = x^1 = x^2 = x^3 = 0\},$$

so it has coordinates x^0, y^1, y^2, y^3. Thus the mapping

$$\begin{pmatrix} z^0 \\ z^1 \\ z^2 \\ z^3 \end{pmatrix} \mapsto -i \begin{pmatrix} iz^0 \\ z^1 \\ z^2 \\ z^3 \end{pmatrix} \tag{1.6.14}$$

maps $S^4 \cap \mathbf{M}^I$ linearly onto $M \cap \mathbf{M}^I$, that is,

$$(x^0, iy^1, iy^2, iy^3) \mapsto (x^0, y^1, y^2, y^3).$$

The mapping (1.6.14) is a multiple of the physicists' mapping

$$(t, x, y, z) \to (it, x, y, z),$$

---

[5] See §3.5 for a discussion of the group $\mathrm{Spin}(p,q)$.

in passing from Minkowski geometry to Euclidean geometry. It is clear that this mapping, which is a diffeomorphism of $\mathbf{R}^4$ to $\mathbf{R}^4$, does not extend to a mapping of

$$S^4 \to S^1 \times S^3,$$

since they are not homeomorphic.

Exercises for Section 1.6

1.6.1. Show that the Lorentz group L_M has four components as a topological space.

1.6.2. Show that P_M and C_M are Lie groups with dimensions 10 and 15, respectively. How many connected components do these groups have?

1.6.3. Show that the matrices iBA^* in (1.6.8) are Hermitian matrices.

1.6.4. Show that any element of the restricted Lorentz group is represented by

$$z \mapsto AzA^*, \quad A \in \mathrm{SL}(2, \mathbf{C}),$$

where $z \in H(2)$, is the vector space of Hermitian 2×2 matrices, and that, moreover, there is a two-to-one covering mapping

$$\mathrm{SL}(2, \mathbf{C}) \to L_M.$$

1.6.5. Show that $\mathrm{SU}(2)$ is diffeomorphic to S^3.

1.6.6. Show that $(S^1 \times S^3)/\mathbf{Z}_2$ is diffeomorphic to $S^1 \times S^3$.

1.6.7. Show that $\mathbf{P}(\mathbf{H}^2)$ is diffeomorphic to S^4.

1.6.8. Show that a projective line l in $\mathbf{P}$ is real, i.e., invariant under the involution σ given by the quaternionic structure on $\mathbb{T}^\alpha$, if and only if $l = \pi^{-1}(x)$, where $\pi : \mathbf{P} \to S^4$ is induced by the mapping (1.6.11).

1.6.9. Show that

$$S^4 \cap \mathbf{M}^I = \{z \in \mathbf{M}^I : y^0 = x^1 = x^2 = x^3 = 0\}.$$

2

Fiber bundles

Vector fields have been an important part of mathematics and physics for several centuries. Indeed, the phrase 'field theory' stems from the notion of a vector field in its various physical interpretations. One knows that the global properties of vector fields are related to the geometry of the manifold on which they are defined. For instance, the familiar observation that one cannot 'comb the hair on a billiard ball' is a restatement of the fact that any smooth global vector field on the two-sphere S^2 must vanish somewhere. The notion of fiber bundle, in particular vector bundle and principal bundle, has evolved in the past 40 years as a geometric notion which contains an important link between the global geometry of a manifold and the analysis of functions, namely, vector fields and their generalization on the manifolds. In this chapter we shall introduce the reader to these notions and some of their fundamental properties.

2.1 Vector bundles and principal bundles

Let M be a real differentiable manifold. Locally, M is diffeomorphic to an open subset of $\mathbf{R}^n$, but globally this certainly need not be the case. Consider a differentiable mapping

$$f : M \to \mathbf{R}^k;$$

then $f = (f^1, \ldots, f^k)$ is a vector-valued function on M. We can consider the *graph of* f,

$$\gamma_f : M \to M \times \mathbf{R}^k,$$

where $\gamma_f(x) = (x, f(x))$, which we pictorially represent in Figure 2.1.1 as in the familiar case of one real variable. Note that if we let

$$\pi : M \times \mathbf{R}^k \to M$$

be the natural projection, then

$$\pi \circ \gamma_f(x) = x, \quad \text{for all } x \in M. \tag{2.1.1}$$

Thus, it is quite clear that understanding vector-valued functions is equivalent to understanding mappings of the form

$$\gamma : M \to M \times \mathbf{R}^k, \quad \text{where } \pi \circ \gamma = \text{id}. \tag{2.1.2}$$

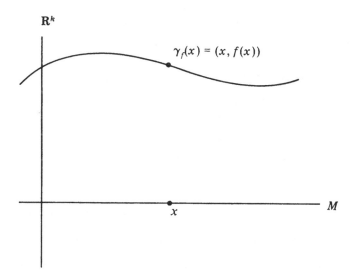

Fig. 2.1.1. Graph of a vector-valued function f.

Let us look at the mapping (2.1.1). For each $x \in M$, $\pi^{-1}(x) = \{x\} \times \mathbf{R}^k \cong \mathbf{R}^k$ is a vector space. We think of $M \times \mathbf{R}^k$ as being *fibered* over the manifold M with *fibers* isomorphic to $\mathbf{R}^k$ as illustrated in Figure 2.1.2, and the mapping γ of the form (2.1.2) maps M to a *cross section* of this fibering, i.e., to each point $x \in M$ there is precisely one point $\gamma(x)$ in the fiber 'over the point' x (see Figure 2.1.2). This is a somewhat radical geometric reinterpretation of the elementary notion of vector-valued functions, but it allows us to immediately define the notion of vector bundle, which is a natural generalization of this interpretation of vector-valued functions.

We shall define a *real vector bundle of rank k* over a manifold M to be a manifold E and a mapping π, where

$$\pi : E \to M$$

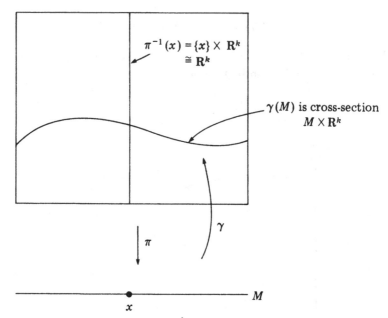

Fig. 2.1.2. $M \times \mathbf{R}^k$ fibered over M.

and satisfying:

(2.1.3)(a) π is onto,

(2.1.3)(b) for each $x \in M$, $E_x := \pi^{-1}(x)$ is equipped with the structure of a real vector space of dimension k, i.e., there is a mapping $h_x : E_x \overset{\cong}{\to} \mathbf{R}^k$, for each $x \in M$,

(2.1.3)(c) for each $x \in M$, there is a neighborhood U of x and a mapping
$$h_U : \pi^{-1}(U) \to U \times \mathbf{R}^k,$$
such that
$$h_U|_{E_x} : E_x \to \{x\} \times \mathbf{R}^k$$
is $\mathbf{R}$-linear.

We shall refer to E as the *total space* of the bundle, M as the *base space* of the bundle, π as the *bundle mapping*, and E_x as the *fiber over the point* x. In normal usage, we shall refer to E itself as a vector bundle over (or on) M, the mapping π to the base space (manifold) M being understood as part of the data of E (these are illustrated pictorially in Figure 2.1.3).

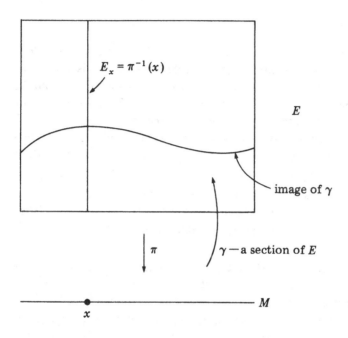

Fig. 2.1.3. A vector bundle E fibered over its base space M.

We shall say that the vector bundle E is *differentiable* if the mapping π and the local trivializations h_U are differentiable, of maximal rank and diffeomorphisms, respectively. We define a *complex vector bundle of rank k* in the same manner by replacing $\mathbf{R}^k$ by $\mathbf{C}^k$, and R-linear by C-linear in the definitions. In this case we can also speak of *holomorphic vector bundles*, where the mapping π and the local trivializations h_U are holomorphic of maximal rank and biholomorphic, respectively. One could also talk about real-analytic vector bundles, etc., but our primary concern will be continuous, differentiable, and holomorphic vector bundles.

Thus we see that a vector bundle is a family of vector spaces parametrized by a manifold M and which locally, near any point of M, is a product of the parameter space with a vector space. Before we give examples and properties, let us define the generalization of vector-valued functions (as described in (2.1.2)), which will be important in the examples also. Let $E \to M$ be a vector bundle; then a *section* γ of E is a continuous mapping

$$\gamma : M \to E,$$

satisfying $\pi \circ \gamma = $ id (see Figure 2.1.3). If E is a differentiable vector bundle, we can talk about *differentiable sections* if γ is a differentiable mapping, and similarly, if E is a holomorphic vector bundle, we have the notion of holomorphic section of E. Before we go on to examples, we have one last piece of terminology. A *line bundle* is a very commonly used expression for a vector bundle of rank one which we shall adopt (complex line bundle, holomorphic line bundle, etc.).

Example 2.1.1: The products $M \times \mathbf{R}^k$ or $M \times \mathbf{C}^k$ are real and complex vector bundles of rank k, as we saw in the above discussion. These are called *trivial vector bundles* of the appropriate class. One of the basic problems in bundle theory is to decide how far from being trivial a given bundle is.

Example 2.1.2: Let M be a differentiable manifold, and let $x \in M$. Consider $\widetilde{\mathcal{E}}_x$ to be the set of all real-valued differentiable functions defined near the point x. On this set we introduce the equivalence relation: for $f_1, f_2 \in \widetilde{\mathcal{E}}_x$, we say $f_1 \sim f_2$ if there is some neighborhood U of x such that $f_1|_U = f_2|_U$. Let $\mathcal{E}_x = \widetilde{\mathcal{E}}_x / \sim$ be the set of equivalence classes. Then it is easy to see that $\mathcal{E}_x$ forms an *algebra* over $\mathbf{R}$. In other words, one can add or multiply two equivalence classes by adding or multiplying representatives on common domains. The equivalence classes are called *germs*, and they each denote a common localization at x of many distinct functions with different domains, all containing the point x. The algebra $\mathcal{E}_x$ is called the *algebra of germs of differentiable functions* at x.

If we consider the subalgebra $\mathcal{A}_x$ of $\mathcal{E}_x$ consisting of those germs with a real-analytic representative at x, then one readily sees that each germ f in $\mathcal{A}_x$ is determined uniquely by its power series expansion, and the algebra $\mathcal{A}_x$ is algebraically isomorphic to the algebra of convergent power series at the point x. In general, we cannot use power series to represent germs of functions, but we can for holomorphic and real-analytic germs.

We define a *tangent vector* at x to be a *derivation* of the algebra $\mathcal{E}_x$, i.e., an $\mathbf{R}$-linear mapping

$$D : \mathcal{E}_x \to \mathbf{R},$$

satisfying

$$D(fg) = (Df)g(x) + f(x)Dg,$$

where we note that the *value* $f(x)$ of a germ f at x is well-defined.

We let $T_x(M)$ be the set of all tangent vectors to M at x. It is easily checked that $T_x(M)$ is a vector space for each $x \in M$. We note that if M' is a second manifold such that M and M' coincide near x, then $T_x(M) = T_x(M')$, since the algebra $\mathcal{E}_x$ depends only on M (or M') near x. For instance, if $U \subset \mathbf{R}^n$, then $T_x(U) = T_x(\mathbf{R}^n)$. Using this remark, we shall see now that $T_x(M) \cong T_x(\mathbf{R}^n)$, if M is n-dimensional. Namely, if M is an n-dimensional manifold, then for $x \in M$, there is a coordinate mapping

$$h : U \xrightarrow{\cong} U' \subset \mathbf{R}^n.$$

The mapping h will pull back functions defined on U' by $h^* f = f \circ h$, and hence induces an isomorphism of algebras

$$h^* : \mathcal{E}_{h(x)} \xrightarrow{\cong} \mathcal{E}_x,$$

and consequently an isomorphism of tangent spaces

$$h_* : T_x(M) \xrightarrow{\cong} T_{h(x)}(\mathbf{R}^n).$$

Now

$$\left\{ \frac{\partial}{\partial x^1}, \ldots, \frac{\partial}{\partial x^n} \right\}$$

are easily seen to be derivations of $\mathcal{E}_y$. Moreover, for any $y \in \mathbf{R}^n$, as is well known from advanced calculus, any directional derivative

$$D : \mathcal{E}_y \to \mathbf{R}$$

is of the form

$$D = \sum_{j=1}^{n} c^j \frac{\partial}{\partial x^j},$$

that is,

$$\left\{ \frac{\partial}{\partial x^1}, \ldots, \frac{\partial}{\partial x^n} \right\}$$

forms a basis for $T_y(\mathbf{R}^n)$ (Exercise 1.1.6). Hence,

$$D_j = (h_*)^{-1} \left(\frac{\partial}{\partial x^j} \right), \quad j = 1, \ldots, n,$$

is a basis for $T_x(M)$, and thus $T_x(M)$ is an n-dimensional vector space for each $x \in M$.

Letting

$$T(M) = \bigcup_{x \in M} T_x(M)$$

be the disjoint union of the tangent spaces, we have a family of vector spaces of rank n parametrized by M, and there is a natural mapping $\pi : T(M) \to M$. To make $T(M) \overset{\pi}{\to} M$ into a vector bundle of rank n, we need to give $T(M)$ a differentiable structure so that π is a smooth mapping, and so that we have $\pi^{-1}(U) \cong U \times \mathbf{R}^n$ for a system of open sets $\{U\}$ which covers M. Using a coordinate chart U, we see that $\pi^{-1}(U) \cong T(U')$, $U' \subset \mathbf{R}^n$, and one sees that there is a bijection, linear on fibers, of the form

$$T(U') \cong U \times \mathbf{R}^n$$

by mapping

$$(x, D_x) \to (x; c^1, \ldots, c^n)$$

where

$$D_x = \sum_{j=1}^{n} c^j \frac{\partial}{\partial x^j}.$$

One can check that this induces a differentiable manifold structure on $T(M)$, so that $T(M)$ becomes a vector bundle of rank n over M. The *sections* of M will be *vector fields* on M (see Chapter 1), i.e., a section

$$\gamma : M \to T(M)$$

will be: for each x, γ_x is a directional derivative acting on functions locally defined near x, and this is the definition of a vector field.

Below we shall see further examples of vector bundles and line bundles, but first we want to discuss a few concepts concerning descriptions of vector bundles and the relationships between vector bundles, which will clarify the discussion of examples. The language of vector spaces essentially transfers intact to become the language of vector bundles. Consider a fixed base manifold M and vector bundles E and F over M. A *homomorphism* L from E to F (continuous, differentiable, or holomorphic), is a mapping (continuous, differentiable, or holomorphic) $L : E \to F$, such that L restricted to the fiber

E_x is a linear mapping to the fiber F_x. The vector bundle E is a *sub-bundle* of F if E is a submanifold of F and E_x is a linear subspace of F_x for each $x \in M$. Just as for vector spaces, we have the notion of a *vector bundle monomorphism* (one-to-one linear mapping on the fibers), *vector bundle epimorphism* (surjective linear mapping on the fibers), and a *vector bundle isomorphism* (one-to-one and surjective on the fibers). Given two vector bundles E and F, we can form new vector bundles. For instance, we can form the dual bundle, the direct sum of two bundles, the tensor product of two bundles, and the symmetric tensor product of two bundles, which would be denoted with the usual notation

$$E^*, \quad E \oplus F, \quad E \otimes F, \quad E \odot F,$$

or, if E is a subbundle of F, one can form the quotient bundle F/E. We also can form the exterior product of a vector bundle of rank n,

$$\bigwedge^k E, \quad k = 0, \ldots, n.$$

In each case, one defines the derived bundles fiberwise by the linear algebra operation indicated, e.g.,

$$(E \oplus F)_x := E_x \oplus F_x,$$

obtaining an appropriate family of vector spaces parametrized by M. The problem in each case is to obtain the topology, manifold structure, and local trivializations, which must be derived from the corresponding information from the input bundles.

A useful description of vector bundles is in terms of *transition functions*, which allows one to carry out the program outlined above, and it is very important in its own right. Consider a vector bundle $E \to M$, and let $\{U_\alpha\}$ be an open covering of M such that $E|_{U_\alpha}$ is locally trivial with (local) trivializations

$$h_\alpha : \pi^{-1}(U_\alpha) \to U_\alpha \times \mathbf{R}^k.$$

Suppose $U_\alpha \cap U_\beta \neq \emptyset$; then we can consider the induced mappings

$$
\begin{array}{c}
U_\alpha \cap U_\beta \times \mathbf{R}^k \\
{\scriptstyle h_\alpha} \nearrow \\
\pi^{-1}(U_\alpha \cap U_\beta) \qquad \downarrow g_{\alpha\beta} := h_\beta \circ h_\alpha^{-1} \\
{\scriptstyle h_\beta} \searrow \\
U_\alpha \cap U_\beta \times \mathbf{R}^k
\end{array}
$$

where

$$g_{\alpha\beta} : U_\alpha \cap U_\beta \times \mathbf{R}^k \to U_\alpha \cap U_\beta \times \mathbf{R}^k \qquad (2.1.4)$$

are called the *transition functions* for the vector bundle E with respect to the covering $\{U_\alpha\}$. Note that $g_{\alpha\beta}$ maps from the U_α-trivialization to the U_β-trivialization (the order is important). By the definition of $g_{\alpha\beta}$ we see that $g_{\alpha\beta}$ is an isomorphism of the trivial bundle $U_\alpha \cap U_\beta \times \mathbf{R}^k$ onto itself, and we can express $g_{\alpha\beta}(x, v)$ as

$$g_{\alpha\beta}(x, v) = (x, \tilde{g}_{\alpha\beta}(x)v),$$

where

$$\tilde{g}_{\alpha\beta} : U_\alpha \cap U_\beta \to \mathrm{GL}(k, \mathbf{R}) \qquad (2.1.5)$$

is a nonsingular matrix-valued function defined on $U_\alpha \cap U_\beta$. Any $g_{\alpha\beta}$ will define a corresponding $\tilde{g}_{\alpha\beta}$ and conversely, and we shall drop the '~' and let $g_{\alpha\beta}$ denote either the bundle isomorphism (2.1.4) or the matrix-valued mapping (2.1.5) depending on the context. Most often we shall refer to the matrix-valued mapping (2.1.5). Thus, given a bundle E, one determines a set of transition functions $g_{\alpha\beta}$. Since local trivializations are by no means uniquely specified for a given bundle, a bundle may have different sets of transition functions, just as a manifold can have distinct sets of coordinate systems and transition functions. One thinks of a local trivialization $U_\alpha \times \mathbf{R}^k$ for a bundle as being a 'local coordinate system' for the bundle even if the open set $U_\alpha \subset M$ is not necessarily a coordinate chart for the base manifold M. We do have 'honest' coordinates for the linear structure in the fibers, though. The transition functions $g_{\alpha\beta}$ for a bundle E satisfy the relation

$$g_{\alpha\beta} \circ g_{\beta\gamma} \circ g_{\gamma\alpha} = \mathrm{id} \quad \text{on} \quad U_\alpha \cap U_\beta \cap U_\gamma, \qquad (2.1.6)$$

which follows directly from their definition in terms of local trivializations. We can use $g_{\alpha\beta}$ satisfying (2.1.6) to construct vector bundles, as the following proposition shows.

Proposition 2.1.3. *Let M be a manifold, let $\{U_\alpha\}$ be an open covering of M, and let $g_{\alpha\beta}$ be a collection of matrix-valued functions $g_{\alpha\beta}$: $U_\alpha \cap U_\beta \to \mathrm{GL}(k, \mathbf{R})$ satisfying:*

$$g_{\alpha\beta} \circ g_{\beta\gamma} \circ g_{\gamma\alpha} = \mathrm{id} \quad \text{on} \quad U_\alpha \cap U_\beta \cap U_\gamma.$$

Then there is a vector bundle E of rank k with local trivializations

$$h_\alpha : E|_{U_\alpha} \to U_\alpha \times \mathbf{R}^k,$$

such that $g_{\alpha\beta} = h_\beta \circ h_\alpha^{-1}$.

Proof: Define $\widetilde{E}$ to be the disjoint union

$$\widetilde{E} = \bigcup_\alpha U_\alpha \times \mathbf{R}^k.$$

As such, $\widetilde{E}$ is a manifold (and in particular a trivial vector bundle over the disjoint union of the U_α). Define an equivalence relation on $\widetilde{E}$ by saying that

$$(x, v) \in U_\alpha \times \mathbf{R}^k$$

and

$$(y, w) \in U_\beta \times \mathbf{R}^k$$

are *equivalent*, written

$$(x, v) \sim (y, w),$$

if and only if

$$x = y,$$
$$w = g_{\alpha\beta}(x)v.$$

To prove that $\sim$ is an equivalence relation, one needs (2.1.6). Let

$$E = \{\text{equivalence classes of } \widetilde{E}\} = \widetilde{E}/\sim .$$

If $(x, v) \in U_\alpha \times \mathbf{R}^k$ for some α, let $[(x, v)] \in E$ be the associated equivalence class. Then there is a natural mapping $E \to M$ given by $[(x, v)] \mapsto x$. Moreover, if $x \in M$, then $x \in U_\alpha$ for some α, and

$$\pi^{-1}(x) = \{[(x, v)] : v \in \mathbf{R}^k\} \cong \mathbf{R}^k.$$

Thus, the fibers of the mapping $E \overset{\alpha}{\to} M$ are vector spaces. Moreover, if $x \in U_\alpha$, then $E|_{U_\alpha} := \pi^{-1}(U_\alpha)$ is mapped to $U_\alpha \times \mathbf{R}^k$ by

$$[(x, v)] \mapsto (x, v),$$

and this is a homeomorphism which is an $\mathbf{R}$-linear isomorphism on the fibers (which we can call h_α). And it follows that $g_{\alpha\beta} = h_\beta \circ h_\alpha^{-1}$.

The $\{h_\alpha\}$ provide the local coordinates and local trivializations which makes E a vector bundle. ∎

This proposition is true in the differentiable and holomorphic categories also. We call any set of functions $g_{\alpha\beta}$ satisfying (2.1.6) *transition functions*, and the bundle constructed in Proposition 2.1.3 is a canonical example of a vector bundle which has the given transition functions. Thus we can use transition functions to define an isomorphism class of vector bundles (namely, all vector bundles isomorphic to the vector bundle constructed in the proposition). We shall often ignore the distinction between isomorphic vector bundles and simply speak of 'the vector bundle defined by $\{g_{\alpha\beta}\}$' or 'the transition functions of a given vector bundle'. In our examples we shall often have explicit (or implicitly defined) transition functions to describe a bundle.

For instance, to describe the direct sum of two bundles $E \oplus F$, we let $\{g_{\alpha\beta}\}$ and $\{h_{\alpha\beta}\}$ be transition functions for vector bundles E and F of rank m and n, respectively. Then the matrices

$$k_{\alpha\beta} = \begin{pmatrix} g_{\alpha\beta} & 0 \\ 0 & h_{\alpha\beta} \end{pmatrix} \tag{2.1.7}$$

will be of rank $m + n$ and will satisfy the compatibility conditions and hence define a vector bundle. The crucial issue here is that if V and W are vector spaces of rank m and n, and

$$g : V \to V,$$
$$h : W \to W,$$

are isomorphisms, then there is an *induced* isomorphism

$$g \oplus h : V \oplus W \to V \oplus W.$$

In our example above,

$$V = \{x\} \times \mathbf{R}^k, \quad W = \{x\} \times \mathbf{R}^n,$$
$$g = g(x), \quad h = h(x),$$
$$g \oplus h = k(x).$$

In general, if one considers

$$V \otimes W \quad \text{(tensor product)},$$
$$\wedge^* V \quad \text{(exterior algebra)},$$
$$\odot^n V \quad \text{(symmetric tensor product)},$$
$$V^* \quad \text{(dual space)},$$

then g and h induce naturally

$$g \otimes h : V \otimes W \to V \otimes W,$$
$$\lambda^* : \wedge^* V \to \wedge^* V,$$
$$\odot^n g : \odot^n V \to \odot^n V,$$
$$g^* : V^* \to V^*.$$

The *formulas* for some of these induced mappings in terms of g and h (and a choice of basis) are simply more complicated than the elementary block-diagonal matrix k in (2.1.7), but *conceptually* this is sufficient to ensure that the vector bundles $E \otimes F$, $\wedge^* E$, $\odot^n E$, etc., are well-defined vector bundles in terms of given vector bundles E and F, say.

Thus, if we have a specific example of a vector bundle (for instance, the tangent bundle $T(M) \to M$, for a differentiable manifold M), then this process generates a whole sequence of associated bundles. We can consider $T^*(M)$, the *cotangent bundle* (the dual tangent bundle), or $\wedge^k T^*(M)$, the kth exterior power of $T^*(M)$, as well as many others.

We denote for any vector bundle $E \to M$ the *sections* over an open set $U \subset M$ by $\Gamma(U, E)$. These are considered to be continuous, smooth, or holomorphic sections, depending on the discussion and on the nature of E. If we need to be more precise, we can write

$$\Gamma^k(U, E), \quad \Gamma^\infty(U, E), \quad \Gamma^\omega(U, E),$$

by analogy with the classical notation for functions

$$C^k(U), \quad C^\infty(U), \quad C^\omega(U) \quad (C^\omega \text{ meaning real-analytic}).$$

Holomorphic functions on an open set are denoted by $\mathcal{O}(U)$, so we can write simply $\mathcal{O}(U, E)$ for the holomorphic sections of a holomorphic vector bundle when we need to be this precise.

Now we can identify the global sections of $\bigwedge^k T^*(M)$, denoted by

$$\Gamma(E, \bigwedge^k T^*(M))$$

with the classical notion of (smooth) *differential forms on M of degree* k, usually denoted by $\mathcal{E}^k(M)$. To see that this is the case, we recall that one definition of differential forms is an alternating mapping defined on k-fold products of vector fields. Namely, if $\omega \in \mathcal{E}^k(M)$, then for vector fields $X_1, \ldots, X_k$ on M, $\omega(X_1, \ldots, X_k) \in C^\infty(M)$, and ω is alternating in the X_j. Thus, for

$$x \in M, \quad X_1(x), X_2(x), \ldots, X_k(x) \in T_x(M),$$

the map

$$\omega(x) : T_x(M) \times \cdots \times T_x(M) \to \mathbf{R}$$

is alternating. But this means that $\omega(x) \in \bigwedge^k T_x^*(M)$, i.e., $\omega(x)$ is an element of the fiber of $\bigwedge^k T^*(M)$ at the point x, and this is precisely what a section of $\bigwedge^k T^*(M)$ should be. Since sections and differential forms vary smoothly from point to point, one can make a complete identification of $\mathcal{E}^k(M)$ with $\Gamma(M, \bigwedge^k T^*(M))$.

More generally, we can consider the *algebra of vector bundles* generated by $T(M)$ and $T^*(M)$, and the tensor product. This is called the *tensor algebra* of M, and contains the tangent bundle and cotangent bundle, and the exterior product bundle as special bundles in this collection of bundles. The *sections* of any bundle in the tensor algebra are called *tensor fields* (generalizations of functions, vector fields, and differential forms) and are classically described in terms of indices, $\tau_{mnp\ldots}^{ijk\ldots}$, where the number of raised indices corresponds to the number of tensor products of $T(M)$ with itself, and the number of lower indices corresponds to the number of tensor products of $T^*(M)$ with itself. The classical notation is very convenient for codifying exactly which tensor product bundles we are interested in. This tensor algebra of bundles and an abstraction of the classical index notation will be discussed in more detail in §2.3.

Let us now give some more examples of vector bundles.

Example 2.1.4 (Universal bundle): Consider the Grassmannian manifold $\mathbf{G}_{k,n}(\mathbf{C}) = \mathbf{G}_{k,n}$. Form the trivial vector bundle

$$\mathbf{G}_{k,n} \times \mathbf{C}^n$$

and recall that each point $x \in \mathbf{G}_{k,n}$ is of the form $x = [S]$, where S is a k-dimensional subspace of $\mathbf{C}^n$. Let us denote the subspace corresponding to the point $x \in \mathbf{G}_{k,n}$ by S_x. Thus, for $x \in \mathbf{G}_{k,n}$ we have $S_x \subset \mathbf{C}^n$. Let $U_{k,n}$ be the disjoint union of the subspaces S_x for $x \in \mathbf{G}_{k,n}$. Then we see that there is a natural inclusion in the trivial bundle

$$U_{k,n} \to \mathbf{G}^{k,n} \times \mathbf{C}^n,$$

given by

$$S_x \mapsto \{x\} \times \mathbf{C}^n,$$

since S_x is a subspace of $\mathbf{C}^n$. This gives a natural topology to $U_{k,n}$, and there is a continuous projection mapping

$$U_{k,n} \xrightarrow{\pi} \mathbf{G}_{k,n},$$

given by the mapping

$$S_x \mapsto x, \quad \text{for } x \in \mathbf{G}_{k,n}.$$

Thus, we have the ingredients of a vector bundle if we can find a local trivialization. To do this, we simply use a local coordinate chart $U_0 \subset \mathbf{G}_{k,n}$ of the form

$$z \in \mathbf{C}^{(n-k)\times k} \mapsto \begin{bmatrix} z \\ I_k \end{bmatrix} \in \mathbf{G}_{k,n}.$$

If $x = \begin{bmatrix} z \\ I_k \end{bmatrix} \in \mathbf{G}_{k,n}$, then any vector $v \in S_x$ uniquely determines a vector $\zeta \in \mathbf{C}^k$ by

$$v = \begin{pmatrix} z \\ I_k \end{pmatrix} \zeta$$

(for S_x). Thus the mapping

$$(z, \zeta) \in \mathbf{C}^{(n-k)\times k} \times \mathbf{C}^k \mapsto \begin{pmatrix} z \\ I_k \end{pmatrix} \zeta \in S_x$$

gives a trivialization of $U_{k,n}$ over U_0. The mappings are all holomorphic, and thus we see that $U_{k,n}$ is a holomorphic vector bundle over the complex manifold $\mathbf{G}_{k,n}$. We shall return to this example again in later discussions.

Recall that a section of a vector bundle $E \xrightarrow{\pi} M$ is a continuous mapping $\gamma : M \to E$ satisfying $\pi \circ \gamma = \text{id}$, i.e., γ maps a point $x \in M$ into the fiber E_x over the point x in a continuous fashion. This is an abstract notion. How can one work with sections in a more concrete fashion for explicit bundles? As noted above, a differential form ω of degree k on a differentiable manifold M is a section of the bundle $\bigwedge^k T^*(M)$. If $(x^1, \ldots, x^n)$ are local coordinates near a point p on M, then a local basis for $T(M)$ is

$$\left\{ \frac{\partial}{\partial x^1}, \ldots, \frac{\partial}{\partial x^n} \right\}$$

(that is, for each point q near p, these vector fields form a basis for $T_q(M)$). We define

$$\{dx^1, \ldots, dx^n\}$$

to be the *dual basis* for $T^*(M)$ with respect to these coordinates. Thus,

$$\{dx^{i_1} \wedge \cdots \wedge dx^{i_k}, \quad \text{for } 1 \leq i_1 < \cdots < i_k \leq n\}$$

is a local basis for $\bigwedge^k T^*(M)$ near p, and any differential form ω is expressible as a linear combination of this basis in the form

$$\omega = \sum_{i_1 < \cdots < i_k} f_{i_1 \ldots i_k} dx^{i_1} \wedge \cdots \wedge dx^{i_k},$$

where $f_{i_1 \ldots i_k}$ are smooth functions (locally defined) which are skew-symmetric in the indices. Thus, we can think of a differential form locally as the vector-valued function $\{f_{i_1 \ldots i_k}\}$. In some other coordinates system there will be a similar expression, say

$$\omega = \sum g_{i_1 \ldots i_k} dy^{i_1} \wedge \cdots \wedge dy^{i_k},$$

and if the coordinate systems are related (i.e., the charts intersect), then one finds that

$$g_{i_1 \ldots i_k} = \sum_{j_1 < \cdots < j_k} \left(\frac{\partial y^{j_1}}{\partial x^{i_1}} \right) \cdots \left(\frac{\partial y^{j_k}}{\partial x^{i_k}} \right) f_{j_1 \ldots j_k}.$$

That is, the coefficients of the differential form are related in the usual fashion by the chain rule. The coefficient matrix

$$\left(\frac{\partial y^{j_1}}{\partial x^{i_1}} \right) \cdots \left(\frac{\partial y^{j_k}}{\partial x^{i_k}} \right)$$

is simply the transition function matrix for these overlapping charts for the bundle $\bigwedge^k T^*(M)$. This is a general phenomenon, as is indicated by the following elementary proposition.

Proposition 2.1.5. *Let $E \to M$ be a vector bundle of rank k with transition functions $\{g_{\alpha\beta}\}$ with respect to a covering U_α of M. Consider a collection of vector-valued functions*

$$f_\alpha : U_\alpha \to \mathbf{R}^k,$$

satisfying on $U_\alpha \cap U_\beta$,

$$f_\alpha(x) = g_{\alpha\beta}(x) f_\beta(x).$$

Then $\{f_\alpha\}$ determines canonically a section f of E, and all sections of E arise this way.

Proof: Let

$$E|_{U_\alpha} \xrightarrow{h_\alpha} U_\alpha \times \mathbf{R}^k$$

be the trivializations which give rise to the transition functions $\{g_{\alpha\beta}\}$.
   The functions f_α induce sections

$$\gamma_\alpha : U_\alpha \to U_\alpha \times \mathbf{R}^k$$

by

$$\gamma_\alpha(x) = (x, f_\alpha(x)),$$

as before. Define

$$\tilde{f}_\alpha = h_\alpha^{-1} \circ \gamma_\alpha,$$

and then one checks that

$$\tilde{f}_\alpha|_{U_\alpha \cap U_\beta} = \tilde{f}_\beta|_{U_\alpha \cap U_\beta},$$

and hence, letting

$$f = \tilde{f}_\alpha|_{U_\alpha} \quad \text{on} \quad U_\alpha, \quad \text{for each } \alpha,$$

we have a well-defined section of E. It is clear that any section gives rise in this manner to a collection of functions f_α, and the proposition is proved. ∎

   Thus we can think of sections of vector bundles as locally defined vector-valued functions which satisfy suitable compatibility conditions on overlaps. The classical examples of vector fields and differential forms clearly are of this form.

We have seen how vector bundles arise from the tangent bundle structure of a manifold, and now we shall look at two other ways in which vector bundles arise naturally: from the representation theory of Lie groups, and from algebraic geometry.

Example 2.1.6 (Homogeneous vector bundles): Let $M = G/H$ be a homogeneous manifold as described in §1.5, where G and H are Lie groups (with H being a closed subgroup of G). Let $\rho : H \to GL(V)$ be a fixed finite-dimensional representation of H on some vector space V. Then we can define a vector bundle on M with rank = dimension (V) in a canonical fashion, as follows. Consider the product $G \times V$ and define an equivalence relation '$\sim$' on $G \times V$ by setting

$$(g_1, v_1) \sim (g_2, v_2) \Longleftrightarrow g_2 = g_1 h \text{ for } h \in H, \text{ and } v_2 = \rho(h^{-1}) \cdot v_1.$$

Let E_ρ be defined as the set of equivalence classes in $G \times V$ with respect to '$\sim$'. There are natural mappings

$$G \times V \to E_\rho \xrightarrow{\pi} G/H = M,$$

given by

$$(g, v) \mapsto [(g, v)] \mapsto gH.$$

Consider a local coordinate system for G/H, near the identity $e \in G$. This consists of a submanifold N of G defined near e and such that $T_e(N) \cap T_e(H) = \{0\}$ and $T_e(G) = T_e(N) \oplus T_e(H)$, that is, the submanifold N is *transversal at* e to the submanifold $H \subset G$ (as in Figure 2.1.4).

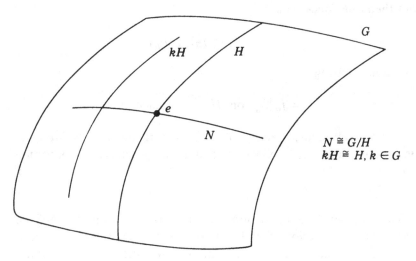

Fig. 2.1.4. A homogeneous manifold G/H.

The points of N parametrize the cosets nH for n close to e and give a local coordinate system for G/H near eH. The mapping

$$N \times V \to E_\rho|_N,$$
$$(g,v) \mapsto [(g,v)]$$

is a homeomorphism and maps fibers to fibers, thus giving a trivialization of E_ρ over N. Such a parametrization of G/H exists near any point by translation of the submanifold N, and hence E_ρ is a vector bundle.

If the subgroup H is compact, one can classify the finite-dimensional representations of H in terms of the Lie algebra structure and characters of H by a classical theorem of Hermann Weyl (see, e.g., Wallach 1973). This classification yields a classification of the corresponding homogeneous vector bundles, but we shall not get any further into this now.

Example 2.1.7 (Divisors): Let M be a complex manifold, and let $\{f_\alpha\}$ be a collection of holomorphic functions defined on an open covering $\{U_\alpha\}$ such that f_α/f_β is nonsingular and nonvanishing on $U_\alpha \cap U_\beta$. We say that f_α and f_β have the same zeros and poles on the overlap region, and we call $\{f_\alpha\}$ a *divisor*. Classically, on a Riemann surface (or simply on the complex plane), a divisor is simply a set of points p_i with multiplicities m_i, where $m_i \in \mathbf{Z}$. One classical problem in function theory was to find a global meromorphic function which had the prescribed zeros with multiplicities (these are poles for negative multiplicity). The well-known theorem of Weierstrass tells us that this is indeed possible. In several complex variables one can ask similar questions: do there exist global meromorphic functions f on M such that f/f_α is nonvanishing and holomorphic on U_α for all α? There are different answers depending on the geometry of M (see, e.g., Weil 1958, Hirzebruch 1966), but one of the fundamental observations in this regard (due to Weil) is that a divisor determines a holomorphic line bundle. Namely, if f_α is given, then define $g_{\alpha\beta} = f_\alpha/f_\beta$ on $U_\alpha \cap U_\beta$, and it follows that

$$g_{\alpha\beta} : U_\alpha \cap U_\beta \to \mathbf{C}_* = \mathbf{C} - \{0\} = \mathrm{GL}(1, \mathbf{C}).$$

Moreover, it is clear that

$$g_{\alpha\beta} \circ g_{\beta\gamma} \circ g_{\gamma\alpha} = \mathrm{id},$$

and hence $g_{\alpha\beta}$ defines a holomorphic line bundle by Proposition 2.1.3. These holomorphic line bundles play an important role in the Riemann–Roch theorem of Riemann surface theory and its generalization by Hirzebruch in the 1950s and Atiyah and Singer in the 1960s (see Hirzebruch 1966). We shall see a special case of this in the next example.

Example 2.1.8 (Hyperplane section bundle in projective space): Let H be a hyperplane in $\mathbf{P}_n$, that is,

$$H = \{(z^0, \ldots, z^n) : \sum a_j z^j = 0, \ 0 \neq (a^0, \ldots, a^n) \in \mathbf{C}^{n+1}\}.$$

Consider the divisor $\{f_\alpha\}$ defined by

$$f_\alpha = a_0 \frac{z^0}{z^\alpha} + \cdots + a_\alpha + \cdots + a_n \frac{z^n}{z^\alpha} \quad \text{in} \quad U_\alpha.$$

The zero set of this divisor is precisely the original hyperplane H. Then we see that

$$g_{\alpha\beta} = \frac{f_\alpha}{f_\beta} = \frac{z^\beta}{z^\alpha} \quad \text{on} \quad U_\alpha \cap U_\beta,$$

and this is a well-defined holomorphic line bundle on $\mathbf{P}_n$ called the *hyperplane section bundle*, since it is the line bundle associated with any hyperplane in $\mathbf{P}_n$. We denote this line bundle simply by $H \to \mathbf{P}_n$.

If $L \to M$ is any line bundle over a manifold M, we define

$$L^m := \underbrace{L \otimes \cdots \otimes L}_{m \text{ factors}},$$
$$L^{-1} := L^*,$$

so that any power L^m, $m \in \mathbf{Z}$ is well-defined. These are all line bundles since the tensor products of one-dimensional spaces are still one-dimensional. If $\{g_{\alpha\beta}\}$ are the transition functions for L, then $\{g_{\alpha\beta}^m\}$ are the transition functions for L^m, as is easy to verify (Exercise 2.1.1). Thus in the case of the hyperplane section bundle, we find that H^m has transition functions

$$g_{\alpha\beta}^m = \left(\frac{z^\beta}{z^\alpha}\right)^m.$$

Consider the universal line bundle

$$U_{1,n+1} \to \mathbf{P}_n,$$

a special case of the universal bundle over a Grassmannian manifold. Then one finds that (Exercise 2.1.2) the transition functions for $U_{1,n+1}$ are

$$\frac{z^\alpha}{z^\beta} = \left(\frac{z^\beta}{z^\alpha}\right)^{-1},$$

and hence we can identify $U_{1,n+1} \cong H^* = H^{-1}$.

What are the holomorphic sections of this class of bundles? We let $\mathbf{C}[z^0, \dots, z^n]$ denote the ring of polynomials in $\mathbf{C}^{n+1}$ and $\mathbf{C}[z^0, \dots, z^n]_m$ be the subset of homogeneous polynomials of degree m. We have the following description of the sections of H^m:

$$\Gamma(\mathbf{P}_n, H^m) \cong \mathbf{C}[z^0, \dots, z^n]_m, \quad m \geq 0,$$
$$\Gamma(\mathbf{P}_n, H^m) = 0, \quad m < 0.$$

The proof of this is outlined in Exercise 2.1.3 and depends on Laurent expansions of the local representation of a section in the standard open covering of $\mathbf{P}_n$.

In the above paragraphs we have described vector bundles and some of their properties. One of the crucial aspects of a vector bundle of rank k was the transition functions:

$$g_{\alpha\beta} : U_\alpha \cap U_\beta \to \mathrm{GL}(k, \mathbf{R}).$$

Recall that they were used for identifying two different trivializations,

$$
\begin{array}{ccc}
& U_\alpha \times \mathbf{R}^k \ni & (x, v) \\
\nearrow & & \\
E|_{U_\alpha \cap U_\beta} & \downarrow g_{\alpha\beta} & \downarrow \\
\searrow & & \\
& U_\beta \times \mathbf{R}^k \ni & (x, g_{\alpha\beta}(x)v)
\end{array}
\qquad (2.1.8)
$$

We can use the transition functions from a given vector bundle to form general bundles with a similar local product structure. In (2.1.8) we have $g_{\alpha\beta}(x) \cdot v$, and the observation is that we can replace the vector v with any mathematical object w for which a multiplication $g_{\alpha\beta}(x) \cdot w$ makes sense.

Consider

$$
\begin{array}{ccc}
U_\alpha \times \mathrm{GL}(k,\mathbf{R}) & \ni & (x,w) \\
\downarrow g_{\alpha\beta} & & \downarrow \\
U_\beta \times \mathrm{GL}(k,\mathbf{R}) & \ni & (x, g_{\alpha\beta}(x)w)
\end{array}
\qquad (2.1.9)
$$

where $g_{\alpha\beta}(x)w$ is matrix multiplication. We can form

$$
\widetilde{P} = \bigcup_\alpha U_\alpha \times \mathrm{GL}(k,\mathbf{R})
$$

and let

$$
P = \widetilde{P}/\sim,
$$

where $(x,w) \in U_\alpha \times \mathrm{GL}(k,\mathbf{R})$ is equivalent ('$\sim$') to $(y,z) \in U_\beta \times \mathrm{GL}(k,\mathbf{R})$ if and only if

$$
\begin{aligned}
x &= y, \\
z &= g_{\alpha\beta}(x)w.
\end{aligned}
$$

Thus we see by analogy to vector bundles that P is a manifold with a projection mapping

$$
P \xrightarrow{\pi} M,
$$

and

$$
P_x := \pi^{-1}(x) \cong \mathrm{GL}(k,\mathbf{R}),
$$

and P is locally trivial, i.e., for each $x \in M$ there is a neighborhood U of x, such that

$$
P|_U \cong U \times \mathrm{GL}(k,\mathbf{R}).
$$

We call such a manifold P with these properties a *principal bundle with structure group* $\mathrm{GL}(k,\mathbf{R})$. Thus, for each vector bundle there is an associated principal bundle with structure group $\mathrm{GL}(k,\mathbf{R})$ or $\mathrm{GL}(k,\mathbf{C})$, depending on whether the vector bundle was a real or a complex vector bundle. Conversely, given such a principal bundle, one can recover the original vector bundle, since one has the same transition functions; schematically there is a one-to-one correspondence (of equivalence classes):

$$
\begin{array}{ccc}
\text{Vector Bundle} & & \text{Principal Bundle} \\
\text{fiber} \cong \mathbf{R}^k & \longleftrightarrow & \text{fiber} \cong \mathrm{GL}(k,\mathbf{R}) \\
& \searrow \quad \nearrow & \\
& \{g_{\alpha\beta}\} &
\end{array}
$$

A *principal bundle over a manifold* M with structure group G, where G is a Lie group, is a space P with a mapping $P \to M$ with fibers $P_x \cong G$, with local trivializations

$$P|_{U_\alpha} \cong U_\alpha \times G,$$

and with transition functions $g_{\alpha\beta} \in U_\alpha \cap U_\beta \to G$

$$P|_{U_\alpha \cap U_\beta} \begin{array}{c} \nearrow \\ \\ \searrow \end{array} \begin{array}{l} U_\alpha \times G \ni \quad (x, g) \\ \\ U_\beta \times G \ni \quad (x, g_{\alpha\beta} \cdot g) \end{array}$$

Here the primary difference between a general principal bundle and one associated to a vector bundle is that the transition functions are required to have values in the given Lie group.

We can go one step further and suppose that P is a principal G-bundle for any Lie group G, and let $\rho : G \to \mathrm{GL}(V)$ be a representation of G on some vector space V. Then we can construct a vector bundle associated with P and ρ, denoted by E_ρ, with the transition functions being simply

$$\rho(g_{\alpha\beta}) : U_\alpha \cap U_\beta \to \mathrm{GL}(V).$$

Thus a vector bundle generates a single principal bundle, while a principal bundle and a representation can generate a vector bundle. Given a $\mathrm{GL}(k, \mathbf{R})$-principal bundle, then the associated vector bundle described above is simply the vector bundle associated with the regular representation, i.e., left-matrix multiplication of vectors.

In Yang–Mills theory one studies vector bundles and principal bundles associated with specific classical groups, e.g., $\mathrm{SU}(2)$, $\mathrm{SU}(3)$, etc. An important vector bundle associated with such a principal bundle is that given by the *adjoint representation* of G on its Lie algebra $\mathfrak{g}$, i.e.,

$$\mathrm{ad} : G \to \mathrm{Hom}(\mathfrak{g})$$

given by

$$\mathrm{ad}(g)X = gXg^{-1}.$$

For instance, if P is an $\mathrm{SU}(2)$-bundle over a manifold M, then $E = E_{\mathrm{ad}}$ will have fibers isomorphic to $\mathfrak{su}(2) \cong \mathbf{R}^3$, which is generated by the three Pauli matrices

$$\sigma_1 = \begin{pmatrix} 0 & 1 \\ 1 & 0 \end{pmatrix}, \quad \sigma_2 = \begin{pmatrix} 0 & -i \\ i & 0 \end{pmatrix}, \quad \sigma_3 = \begin{pmatrix} 1 & 0 \\ 0 & -1 \end{pmatrix}.$$

We shall see more examples of principal bundles and their uses in later sections and shall content ourselves here with their definition and note that they have many formal properties in common with vector bundles. In general, a *fiber bundle* is like a principal bundle and a vector bundle, only the fibers need not necessarily have algebraic structure. It is only necessary that a Lie group be able to *act* on the fibers. For instance, the two-sphere S^2 is neither a vector space nor a Lie group, but if one has an SO(3)-principal bundle on a manifold M, for instance, then one has transition functions $g_{\alpha\beta} : U_\alpha \cap U_\beta \to$ SO(3), and we can form a space

$$\widetilde{B} = \bigcup_\alpha U_\alpha \times S^2,$$

with an equivalence relation as before:

$$(x, p) \in U_\alpha \times S^2 \quad \text{and} \quad (y, q) \in U_\beta \times S^2$$

are equivalent if $x = y$, $q = g_{\alpha\beta}(x)p$, where $S^2 \subset \mathbf{R}^3$ and SO(3) acts on S^2 by the usual rotation in three-space. Thus we can form $B = \widetilde{B}/\sim$, and we have a *fiber bundle* and transition functions belonging to SO(3). A general fiber bundle will have a Lie group for the transition functions, and a typical fiber F on which the transition functions can act. The classic book by Steenrod (1951) is still one of the best introductions to the subject of fiber bundles in general.

In later sections we shall develop other properties of bundles (either vector, principal, or more general fiber bundles) as the need arises.

Exercises for Section 2.1

2.1.1. Let L be a line bundle with transition functions $\{g_{\alpha\beta}\}$ with respect to a covering $\{U_\alpha\}$. Show that $L^m := L \otimes \cdots \otimes L$ (m factors) has transition functions for the same covering given by $\{g_{\alpha\beta}^m\}$.

2.1.2. Show that the transition functions for the universal bundle $U_{1,n+1} \to \mathbf{P}_n(\mathbf{C})$ have the form

$$g_{\alpha\beta} = \frac{z^\alpha}{z^\beta},$$

with respect to the covering $\{U_\alpha\}$, where

$$U_\alpha = \{[z^0, \ldots, z^n] : z^\alpha \neq 0\}.$$

2.1.3. Show that

$$\Gamma(\mathbf{P}_n, H^m) \cong \mathbf{C}[z^0, \ldots, z^n]_m, \quad m \geq 0,$$

where $\mathbf{C}[z^0, \ldots, z^n]_m$ denotes the homogeneous polynomials in $\mathbf{C}^{n+1}$ of degree m. Moreover, show that

$$\Gamma(\mathbf{P}_n, H^m) = 0, \quad m < 0.$$

(Hint: Consider a section σ of $\Gamma(\mathbf{P}_n, H^m)$ and restrict σ to two coordinate charts U_α and U_β. Expand each restriction in a Laurent series, and use the transition functions for H^m to express one local representation in terms of the other. This will give constraints on the Laurent coefficients which will yield the desired results.)

2.1.4. Let E be a vector bundle over a manifold X, then we define $\det E := \bigwedge^r E$, where $r = \operatorname{rank} E$. Show that if $\{g_{\alpha\beta}\}$ are transition functions for the bundle E, then $\{\det g_{\alpha\beta}\}$ are transition functions for $\det E$ (hence the notation).

2.1.5. Show that if

$$0 \to U \to V \to W \to 0$$

is a short exact sequence of vector bundles, then

$$\det V \cong \det U \otimes \det W.$$

2.1.6. If V is a vector bundle of rank r and V^* is its dual bundle, then show that
(a) $V \otimes V^*$ is a trivial bundle of rank r^2,
(b) $\det(V \otimes V^*)$ is a trivial line bundle.

2.1.7. Let $U \to \mathbf{G}_{k,n}$ be the universal bundle of $\mathbf{G}_{k,n}$. Considering U to be a subbundle of the trivial bundle $\mathbf{G}_{k,n} \times \mathbf{C}^n$, and Q to be the quotient bundle $Q := (\mathbf{G}_{k,n} \times \mathbf{C}^n)/U$, we have the exact sequence of vector bundles

$$0 \to U \to \mathbf{G}_{k,n} \times \mathbf{C}^n \to Q \to 0. \qquad (*)$$

Show that:
(a) $T(\mathbf{G}_{k,n}) \cong \operatorname{Hom}(U, Q)$,
(b) $\det(T^*(\mathbf{G}_{k,n})) \cong (\det U)^n$,

where det of a bundle is defined in Exercise 2.1.4. (Hint: (a) Consider a one-parameter family of subspaces $S_t \in \mathbf{G}_{k,n}$, and represent S_t as the graph, for t near 0, of a linear function f_t. Then note that $\frac{dS_t}{dt}|_{t=0}$ can be used to represent an element of $\mathrm{Hom}(S_0, Q)$; cf. Milnor and Stasheff 1974. (b) Use Exercises 2.1.5, 2.1.6 and the exact sequence in (∗); cf. Wells 1982b.)

2.2 Differential forms

Let M be a differentiable manifold. Associated with M is the algebra of differential forms which has been mentioned briefly in §2.1. We now want to discuss in more detail the structure of this algebra and its relation to the global geometry of the manifold M. If $T(M)$ is the tangent bundle to M as before, then let, for $p = 0, \ldots, n = \dim M$,

$$\textstyle\bigwedge^p(M) := \bigwedge^p T^*(M) \otimes_{\mathbf{R}} \mathbf{C}$$

be the complexification of the exterior product of the cotangent bundle $T^*(M)$. The transition functions for $\bigwedge^p(M)$ are induced from the transition functions for $T(M)$ as indicated in §2.1. We define the *complex vector space $\mathcal{E}^p(M)$ of differential forms of degree p on M* to be the smooth sections of $\bigwedge^p(M)$, i.e.,

$$\mathcal{E}^p(M) := \Gamma^\infty(M, \textstyle\bigwedge^p(M)).$$

We call these the *p-forms* on M. If $\{dx^1, \ldots, dx^n\}$ is a local frame for $T^*(M)$ near some point of M, then any $\omega \in \mathcal{E}^p(M)$ can be expressed in the form

$$\omega = {\sum_{|I|=p}}' f_I dx^I, \qquad (2.2.1)$$

where $I = (i_1, \ldots, i_p)$ is a multiindex, $|I|$ is the number of indices in I,

$$f_I := f_{i_1 \ldots i_p},$$
$$dx^I := dx^{i_1} \wedge \cdots \wedge dx^{i_p},$$

and $\sum'$ denotes the sum over multiindices satisfying $1 \le i_1 < \cdots < i_p \le n$. Thus (2.2.1) is shorthand for

$$\sum_{1 \le i_1 < \cdots < i_p \le n} f_{i_1 \ldots i_p} dx^{i_1} \wedge \cdots \wedge dx^{i_p},$$

which was discussed in §2.1.

The *exterior derivative mapping*

$$d : \mathcal{E}^p(M) \to \mathcal{E}^{p+1}(M)$$

is a linear mapping defined for $p = 0, \ldots, n$ which satisfies the following properties:

(a) $d(\alpha \wedge \beta) = d\alpha \wedge \beta + (-1)^{\deg \alpha} \alpha \wedge d\beta$, where α is a homogeneous differential form of degree deg α,

(b) if $f \in \mathcal{E}(M)$, and if X is a vector field on M, then for $x \in M$, $df_x \in T_x^*(M)$, and $df_x(X_x) = (Xf)(x)$, where $X_x \in T_x(M)$ is the tangent vector at x induced by X,

(c) $d^2\alpha = d \circ d(\alpha) = 0$, for any $\alpha \in \mathcal{E}^p(M)$. \hfill (2.2.2)

One can show that there is a unique linear mapping d which satisfies (2.2.2) (see, e.g., Helgason 1978, p. 22). One can derive from these operational properties of d two formulas for the calculation of d, a local formula and a global formula, each of which is important in its own right. The local formula has the form

$$d\omega = \sum_{|I|=p}' \sum_{j=1}^{n} \frac{\partial f_I}{\partial x^j} dx^j \wedge dx^I, \hspace{2em} (2.2.3)$$

when ω is of the form (2.2.1). Note that if ω^i, $i = 1, \ldots, n$ is a local frame for $T^*(M)$, then $d\omega^i$ need not be zero. The special frame $\{dx^1, \ldots, dx^n\}$ constructed from a coordinate system $\{x^1, \ldots, x^n\}$ as dual to the local vector fields $\{\frac{\partial}{\partial x^1}, \ldots, \frac{\partial}{\partial x^n}\}$ does have the property, however, that $d(dx^i) = 0$. Using this fact and the properties (2.2.2), we see that (2.2.3) is valid (Exercise 2.2.1).

On the other hand, let $\mathcal{V}(M) := \Gamma(M, T(M))$ be the vector space of smooth vector fields on M. Now $\mathcal{V}(M)$ is also a module over the ring of smooth functions $\mathcal{E}(M)$, and as such we can consider $\mathcal{E}(M)$-multilinear mappings of the p-fold cartesian product

$$\omega : \mathcal{V}(M) \times \cdots \times \mathcal{V}(M) \to \mathcal{E}(M).$$

In particular we can consider those ω of this form which are alternating, i.e.,

$$\omega(X_1, \ldots, X_i, \ldots, X_j, \ldots, X_p) = -\omega(X_1, \ldots, X_j, \ldots, X_i, \ldots, X_p).$$

The $\mathcal{E}(M)$-module of alternating p-linear forms can be identified with $\mathcal{E}^p(M)$ in a natural manner, and using this identification we can use the properties of (2.2.1) to derive the formula, for $\omega \in \mathcal{E}^p(M)$,

$$d\omega(X_1, \ldots, X_{p+1})$$
$$= \frac{1}{p+1} \sum_j (-1)^{j-1} X_j \omega(X_1, \ldots, \widehat{X}_j, \ldots, X_{p+1})$$
$$+ \frac{1}{p+1} \sum_{i<j} (-1)^{i+j} \omega([X_i, X_j], X_1, \ldots, \widehat{X}_i, \ldots \widehat{X}_j, \ldots, X_{p+1}),$$

$$(2.2.4)$$

where the '$\,\widehat{\,}\,$' denotes deletion of the corresponding vector field, and $[X_i, X_j] = X_i X_j - X_j X_i$ is the commutator (Lie bracket) of the two vector fields X_i and X_j.

The basic idea in the existence and uniqueness of d is to assume existence, deriving the local formula (2.2.3), thus obtaining uniqueness. The global formula (2.2.4) can then be shown to satisfy (2.2.2) and thus provides the existence proof.

We shall see many examples of differential forms and their applications in the course of the book. We shall content ourselves here with one simple example since it indicates the relation between differential forms and the topology of a space. Consider $M = \mathbf{R}^2 - \{0\}$, and let $\omega = d\theta$, where $\theta = \tan^{-1}(y/x)$ is the polar angle in polar coordinates. We know that θ is a multivalued function on M, but its exterior derivative $\omega = d\theta$ is well-defined since any two branches of θ differ by a constant. In fact one sees that

$$\omega = \frac{-y\,dx + x\,dy}{x^2 + y^2}, \qquad (2.2.5)$$

which is smooth on $\mathbf{R}^2 - \{0\}$, but is singular at the origin. We observe that ω satisfies the equation $d\omega = 0$, since locally (choosing a branch of θ) $\omega = d\theta$ and $d\omega = d^2\theta = 0$ (one could just calculate $d\omega$ from (2.2.5) above, of course). Globally, there is no smooth function f defined on M satisfying $df = \omega$. If there were, then for some such f we would have $df - d\theta = 0$, and hence $\theta - f$ is a constant, which would force θ to be single-valued, which is not the case.

This example leads one to consider *closed forms on M* (ω satisfying $d\omega = 0$ on M), and *exact forms on M* (ω satisfying $\omega = d\tau$, for some τ smooth on M). Since any exact form is closed ($d^2 = 0$), we can form the quotient vector space:

$$H_{\mathrm{DR}}^r(M) := \text{closed } r\text{-forms/exact } r\text{-forms}, \qquad (2.2.6)$$

which is called the rth *de Rham cohomology group of the manifold M*. For $M = \mathbf{R}^2 - \{0\}$, our example above shows that $H^1_{\mathrm{DR}}(\mathbf{R}^2 - \{0\})$ is not trivial. In fact, we shall see later that $H^1_{\mathrm{DR}}(\mathbf{R}^2 - \{0\})$ is one-dimensional, and the equivalence class defined by ω in (2.2.6) gives a basis for this vector space. The one-dimensionality of $H^1_{\mathrm{DR}}(\mathbf{R}^2 - \{0\})$ is a measure of the fact that $\mathbf{R}^2 - \{0\}$ has one 'hole' in its topological make-up. The de Rham groups, in general, turn out to be a reflection of the global nature of the topology. Their dimensions (when finite) are, in fact, topological invariants, a subject we shall discuss in more detail in Chapter 3.

We can now formalize the notion of a de Rham group in the following manner, which will turn out to be quite useful to us. A *complex* S^* is a sequence of linear spaces and mappings

$$S^0 \xrightarrow{L_0} S^1 \xrightarrow{L_1} \cdots \xrightarrow{L_m} S^m \qquad (2.2.7)$$

which satisfy $L_j \circ L_{j-1} = 0$, $j = 1, \ldots, m$. The *cohomology groups of a complex S^** are defined to be the quotient spaces[1]

$$H^q(S^*) := \frac{\ker L_q : S^q \to S^{q+1}}{\operatorname{im} L_{q-1} : S^{q-1} \to S^q}, \qquad (2.2.8)$$

$q = 1, 2, \ldots$.

As an example we have the *de Rham complex* of differential forms on a differentiable manifold M:

$$O \to \mathcal{E}^0(M) \xrightarrow{d} \mathcal{E}^1(M) \xrightarrow{d} \cdots \xrightarrow{d} \mathcal{E}^p(M) \xrightarrow{d} \cdots \xrightarrow{d} \mathcal{E}^n(M) \to 0,$$

and the associated cohomology groups

$$H^q_{\mathrm{DR}}(M) = H^q(\mathcal{E}^*(M)), \qquad q = 0, \ldots, n,$$

which coincide with the de Rham cohomology groups of M, which were defined above in (2.2.6).

We define

$$b^r_{\mathrm{DR}} = \dim_{\mathbf{C}} H^r_{\mathrm{DR}}(M), \qquad r = 0, 1, 2, \ldots, \qquad (2.2.9)$$

---

[1] These cohomology groups are vector spaces, inheriting the vector space structure from S^*, and are called 'groups' by tradition. In general, if S^* has some suitable algebraic structure, e.g., abelian groups, modules over a ring, etc., then the cohomology groups have this same structure.

and we shall see later that if M is compact, then b^r_{DR} is always finite and vanishes for $r > n = \dim M$. For instance, we shall see that for a sphere S^n,

$$b^0_{\mathrm{DR}}(S^n) = b^n_{\mathrm{DR}}(S^n) = 1,$$
$$b^r_{\mathrm{DR}}(S^n) = 0, \quad r \neq 0, n, \tag{2.2.10}$$

or, for a torus $T^n = S^1 \times \cdots \times S^1$ (n factors), one will find out that

$$b^0_{\mathrm{DR}}(T^n) = b^n_{\mathrm{DR}}(T^n) = 1,$$
$$b^1_{DR}(T^n) = b^{n-1}_{DR}(T^n) = n, \tag{2.2.11}$$
$$b^r_{DR}(T^n) = 0, \quad r \neq 0, 1, n-1, n.$$

These are most easily seen when we learn how to calculate topological invariants in Chapter 3.

As mentioned before, one of the important features of the de Rham complex is that it measures how many 'holes' or 'global twistings' are in the manifold M. One reason it is specifically global in nature is that locally these groups are always zero. Namely, one has, in contrast to (2.2.10) and (2.2.11) above,

$$b^0_{DR}(\mathbf{R}^n) = 1,$$
$$b^r_{DR}(\mathbf{R}^n) = 0, \quad r = 1, 2, \ldots.$$

This is a consequence of the famous Poincaré lemma, which we shall formulate and prove later in this section. Before we do this, we want to look at some general properties of differential forms and consider differential forms on complex manifolds.

Differential forms were introduced into geometry as mathematical objects on a manifold which one could integrate, just as one can integrate functions on open sets of $\mathbf{R}^n$. To understand integration of differential forms we need to understand restriction and pullbacks of differential forms.

Suppose $f : M \to N$ is a smooth mapping of differentiable manifolds. Then, as we saw in §2.1, there is a tangent mapping

$$f_* : T(M) \to T(N),$$

and thus an induced dual mapping on the cotangent bundles

$$f^* : T^*(N) \to T^*(M),$$

which in turn induces a linear mapping

$$f^* : \bigwedge^p(N) \to \bigwedge^p(M),$$

and this induces in turn

$$f^* : \mathcal{E}^p(N) \to \mathcal{E}^p(M),$$

the *pullback mapping* of differential forms. This abstract point of view quickly gives an abstract picture of the pullback mapping, but we recall that if locally

$$f = (f^1, \ldots, f^m) : \mathbf{R}^n \to \mathbf{R}^m,$$
$$\omega = \sum_{|I|=p} a_I(y) dy^I,$$

for local coordinates $(y^1, \ldots, y^m)$ of $\mathbf{R}^m$, then

$$f^*\omega(x) = {\sum_{|I|=p}}' \alpha_i(f(x)) df^I(x).$$

This allows us to define the *restriction* of differential forms to submanifolds as the pullback under the inclusion mapping of the submanifold into the ambient space.

If ω is a p-form defined on an open subset U of $\mathbf{R}^p$, then

$$\omega = f(y) dy^1 \wedge \cdots \wedge dy^p,$$

and we define

$$\int_U \omega = \int_U f(y) dy^1 \ldots dy^p, \qquad (2.2.12)$$

where $dy^1 \ldots dy^p$ represents the Lebesgue measure on $\mathbf{R}^p$. Therefore, if we have a submanifold M of an open set $U \subset \mathbf{R}^n$, and

$$\varphi : B \subset \mathbf{R}^p \to \mathbf{R}^n$$

is a coordinate chart for M, then we can define

$$\int_{\varphi(B)} \omega = \int_B \varphi^*\omega.$$

This integral might not converge, but it certainly will if ω has compact support when restricted to $\varphi(B)$. If we cover M by $V_i = \varphi_i(B_i)$, a locally finite covering of such charts V_i, and let χ_i be a partition of unity on M with respect to the covering V_i, then we define

$$\int_M \omega = \sum_i \int_{B_i} \varphi_i^* \chi_i \omega.$$

This is a well-defined notion of integration of ω on M (Exercise 2.2.2) which generalizes (2.2.12) to a curved submanifold of Euclidean space. More generally, this same definition applies even if M is a manifold which is not embedded in $\mathbf{R}^n$. Thus we can integrate differential forms of degree p over manifolds of dimension p.

We now show in what sense the de Rham groups are locally trivial. Let $B \subset \mathbf{R}^n$ be a rectangular box of the form $B = I_1 \times \cdots \times I_n$ where each $I_i = (a_i, b_i)$ is an open interval. We have the following lemma due to Poincaré.

Lemma 2.2.1 (Poincaré). *For any $r = 1, \ldots, n$, let ω be an r-form defined on B satisfying $d\omega = 0$, then there exists an $(r-1)$-form η such that $\omega = d\eta$ on B, i.e.,*

$$H_{\mathrm{DR}}^r(B) = 0, \quad r = 1, \ldots, n.$$

Proof: The proof is by induction. Let $F^m = F^m(\mathcal{E}^r(B))$ be the subspace of differential forms of degree r which are of the form

$$\omega = \sideset{}{'}\sum_{1 \leq i_1 < \cdots < i_r \leq m} f_{i_1 \ldots i_r} dx^{i_1} \wedge \cdots \wedge dx^{i_r}.$$

In other words, $\omega \in F^m$ only involves $dx^1, \ldots, dx^m$. We shall prove the lemma by induction on m, for $m = r, \ldots, n$.

For $m = r$, we see that ω is of the form

$$\omega = f(x) dx^1 \wedge \cdots \wedge dx^r,$$

where $f \in C^\infty(B)$. Define

$$\eta = \left\{ \int_0^{x^1} f(t, x^2, \ldots, x^r) dt \right\} dx^2 \wedge \cdots \wedge dx^r,$$

and we see that, by the fundamental theorem of calculus, $d\eta = \omega$, as desired.

Suppose that for $p \geq r$,

$$\omega \in F^p \text{ and } d\omega = 0 \Rightarrow \text{there is an } \eta \in F^p \text{ such that } d\eta = \omega \tag{2.2.13$_p$}$$

is valid. Then we want to show that $(2.2.13_{p+1})$ is true. Then for $p = n$, the lemma will be proved. Let us assume $(2.2.13_p)$ and let $\omega \in F^{p+1}$ and suppose that $d\omega = 0$. Decompose ω into

$$\omega = dx^{p+1} \wedge \alpha + \beta,$$

where $\alpha, \beta \in F^p$. Now

$$0 = d\omega = dx^{p+1} \wedge d\alpha + d\beta.$$

Since $\alpha, \beta \in F^p$, it follows that the coefficients of α satisfy

$$\frac{\partial}{\partial x^j}(\text{coefficient}) = 0, \quad j = p+2, \ldots, n, \tag{2.2.14}$$

i.e., the coefficients are constant in the $(x^{p+2}, \ldots, x^n)$-directions. Now we want to solve the equation

$$d\gamma = dx^{p+1} \wedge \alpha + \delta,$$

for some $\delta \in F^p$. If

$$\alpha = \sideset{}{'}\sum_{1 \leq i_1 < \ldots < i_r \leq p} f_{i_1 \ldots i_r} dx^{i_1} \wedge \cdots \wedge dx^{i_r},$$

then let

$$\gamma = \sum \left[\int_0^{x^{p+1}} f_{i_1 \ldots i_p}(x^1, \ldots, x^p, t) dt \right] dx^{i_1} \wedge \cdots \wedge dx^{i_p}.$$

By calculating we see that

$$d\gamma - dx^{p+1} \wedge \alpha \in F^p$$

as desired. Now let

$$\varphi = \omega - d\gamma,$$

and we see that φ is closed and that $\varphi \in F^p$. By the induction hypothesis, $\varphi = d\psi$, where $\psi \in F^p$. Therefore we find that

$$\omega = d(\gamma + \psi), \quad \gamma + \psi \in F^p \subset F^{p+1},$$

and hence the lemma is proved.   ∎

If M is a complex manifold, then we have the holomorphic tangent bundle $T(M)$ whose local sections are holomorphic vector fields of the form

$$X = \sum_{i=1}^{r} f^i \frac{\partial}{\partial z^i},$$

where $(z^1, \ldots, z^n)$ are local coordinates. On the other hand, if we consider real coordinates on the same manifold $(x^1, y^1, \ldots, x^n, y^n)$, where $z^j = x^j + iy^j$, $j = 1, \ldots, n$, then any real vector field on M is of the form

$$X_{\mathbf{R}} = \sum_{j=1}^{n} u^j \frac{\partial}{\partial x^j} + v^j \frac{\partial}{\partial y^j},$$

where u^j and v^j are real-valued smooth functions. We define

$$\frac{\partial}{\partial z^j} := \frac{1}{2}\left(\frac{\partial}{\partial x^j} - i\frac{\partial}{\partial y^j}\right)$$

$$\frac{\partial}{\partial \bar{z}^j} := \frac{1}{2}\left(\frac{\partial}{\partial x^j} + i\frac{\partial}{\partial y^j}\right),$$

(2.2.15)

as before, and rewrite $X_{\mathbf{R}}$ above as

$$X_{\mathbf{R}} = \sum_{j=1}^{n} f^j \frac{\partial}{\partial z^j} + \bar{f}^j \frac{\partial}{\partial \bar{z}^j},$$

where $f^j = u^j + iv^j$, $j = 1, \ldots, n$.

Consider the holomorphic tangent bundle $T(M)$ of a complex manifold M; then we have trivializations given by coordinate charts

$$T(M)|_{U_\alpha} \cong U_\alpha \times \mathbf{C}^n,$$

and holomorphic transition matrices

$$U_\alpha \times \mathbf{C}^n \xrightarrow{g_{\alpha\beta}} U_\beta \times \mathbf{C}^n,$$

where

$$g_{\alpha\beta}(z) = \left(\frac{\partial z_\alpha^i}{\partial z_\beta^j}\right),$$

and $(z_\alpha^1, \ldots, z_\alpha^n)$ are the holomorphic coordinates in U_α, and $(z_\beta^1, \ldots, z_\beta^n)$ are the holomorphic coordinates in U_β. If we consider the real and imaginary parts of $\{z_\alpha^j\}$ and $\{z_\beta^j\}$ as real coordinate systems for M of the form

$$z_\alpha^j = u^j + iv^j, \quad z_\beta^j = x^j + iy^j, \qquad j = 1, \ldots, n,$$

then we see that $T(M)$ has the structure of a real vector bundle of (real) rank $2n$,

$$\begin{array}{ccc}
U_\alpha \times \mathbf{C}^n & \cong & U_\alpha \times \mathbf{R}^{2n} \\
\downarrow g_{\alpha\beta} & & \downarrow \tilde{g}_{\alpha\beta} \\
U_\beta \times \mathbf{C}^n & \cong & U_\beta \times \mathbf{R}^{2n}
\end{array}, \qquad (2.2.16)$$

where $\tilde{g}$ is a $2n \times 2n$ block matrix

$$\begin{pmatrix}
\frac{\partial u}{\partial x} & \frac{\partial u}{\partial y} \\
\frac{\partial v}{\partial x} & \frac{\partial v}{\partial y}
\end{pmatrix}$$

where $\frac{\partial u}{\partial x} = \frac{\partial u^i}{\partial x^j}$, etc., are $n \times n$ block matrices. The fact that $T(M)$ is a holomorphic vector bundle is equivalent to the fact that the Cauchy–Riemann equations (in $\mathbf{R}^{2n}$)

$$\frac{\partial u}{\partial x} = \frac{\partial v}{\partial y},$$

$$\frac{\partial v}{\partial x} = -\frac{\partial u}{\partial y},$$

are satisfied. Thus we can identify in this manner the holomorphic and real ($\mathbf{C}^\infty$) tangent bundles

$$T(M)_{\text{hol}} \cong_{\mathbf{R}} T(M)_{\mathbf{C}^\infty}.$$

Now $T(M)_x$ is a complex vector space for each $x \in M$, and as such the mapping $v \mapsto iv$ is well-defined for each $v \in T(M)_x$. If we consider $T(M)_x$ as a real vector space by (2.2.16) then the mapping $t \mapsto it$ induces a mapping

$$J_x : T(M)_x^{\mathbf{R}} \mapsto T(M)_x^{\mathbf{R}}.$$

We call this the *almost-complex tensor* of M induced from the complex structure of M: If we have a local basis for $T(M)$,

$$\left(\frac{\partial}{\partial z^1}, \ldots, \frac{\partial}{\partial z^n}\right),$$

then $t = \begin{pmatrix} f^1 \\ \vdots \\ f^n \end{pmatrix}$ corresponds to a vector field at x,

$$t = \sum_{i=1}^n f^j \frac{\partial}{\partial z^j} \mapsto it = \sum_{j=1}^n (if^j)\frac{\partial}{\partial z^j}.$$

Hence in terms of real coordinates, we see that, setting $f^j = u^j + iv^j$,

$$t = \begin{pmatrix} u \\ v \end{pmatrix} = \begin{pmatrix} u^1 \\ \vdots \\ u^n \\ v^1 \\ \vdots \\ v^n \end{pmatrix} \mapsto \begin{pmatrix} -v^1 \\ \vdots \\ -v^n \\ u^1 \\ \vdots \\ u^n \end{pmatrix} = \begin{pmatrix} 0 & -I_n \\ I_n & 0 \end{pmatrix}\begin{pmatrix} u \\ v \end{pmatrix}.$$

Expressed equivalently, but in terms of trivializations of the tangent bundle, the local basis $\frac{\partial}{\partial z^1}, \ldots, \frac{\partial}{\partial z^n}$ induces, letting $\times i$ denote multiplication by i,

$$\begin{array}{ccc} U \times \mathbf{C}^n & \ni & (x, \zeta^1, \ldots, \zeta^n) \\ \downarrow \times i & & \times i \downarrow \\ U \times \mathbf{C}^n & \ni & (x, i\zeta^1, \ldots, i\zeta^n) \end{array}$$

$$T(M)|_U \cong$$

which, letting the fiber coordinates $\zeta^j = \xi^j + i\eta^j$, where $(\xi^1, \ldots, \xi^n, \eta^1, \ldots, \eta^n)$ are fiber coordinates for the associated real tangent bundle,

$$\begin{array}{ccc} U \times \mathbf{R}^{2n} & \ni & (x, \xi^1, \ldots, \xi^n, \eta^1, \ldots, \eta^n) \\ J_x \downarrow & & \downarrow \\ U \times \mathbf{R}^{2n} & \ni & (x, -\eta^1, \ldots, -\eta^n, \xi^1, \ldots, \xi^n) \end{array}$$

So in block form, $J = \begin{pmatrix} 0 & -I_n \\ I_n & 0 \end{pmatrix}$, as before, and we note that $J^2 = -I_{2n}$, corresponding to $i^2 = -1$.

Let $T(M)^C := T(M) \otimes_{\mathbf{R}} \mathbf{C}$, the *complexification* of the bundle $T(M)$. Any real linear mapping on a real vector space extends uniquely as a complex-linear mapping to the complexification of the vector space. So J extends from $T(M)$ to $T(M)^C$ as a complex-linear mapping. Moreover, the extended mapping, still denoted by J, satisfies $J^2 = -I$ on $T(M)^C$. Since $T(M)_x^C$ is a complex vector space, we see that J_x has eigenvalues $\pm i$. Let, at $x \in M$,

$$T^{1,0}(M)_x = (+i)\text{-eigenspace of } J_x$$
$$T^{0,1}(M)_x = (-i)\text{-eigenspace of } J_x,$$

and let $T^{1,0}(M)$ and $T^{0,1}(M)$ be the vector bundles formed from these families of eigenspaces. It follows that

$$T(M)^C = T^{1,0}(M) \oplus T^{0,1}(M)$$

is a bundle decomposition of the complexified tangent bundle into these two 'eigenbundles' of the linear operator J. Both $T^{1,0}(M)$ and $T^{0,1}$ are complex bundles of complex rank n. This is easy to see since the conjugation mapping (well-defined on the complexification)

$$T(M)^C \xrightarrow{Q} T(M)^C$$

given by

$$v \mapsto Q(v) := \bar{v},$$

satisfies $\overline{T^{1,0}} := Q(T^{1,0}) = T^{0,1}$, and conversely.

The complexification of the dual holomorphic tangent bundle to M has an analogous decomposition. On the one hand J induces on $T^*(M)^C$ by duality J^* which will also satisfy $(J^*)^2 = -I$, and one can proceed as before. On the other hand, one can define

$$T_{1,0}^*(M)_x = \{\omega \in T^*(M)_x^C : \omega(it) = i\omega(t), \quad \text{for } t \in T(M)_x\}$$
$$T_{0,1}^*(M)_x = \{\omega \in T^*(M)_x^C : \omega(it) = -i\omega(t), \quad \text{for } t \in T(M)_x\}.$$

Then one finds that these are the eigenspaces of J^* and that one has the vector bundle decomposition

$$T^*(M)^C = T_{1,0}^*(M) \oplus T_{0,1}^*(M).$$

Consider now the algebra $\mathcal{E}^*(M)$ of differential forms on the complex manifold M, and recall that these forms are complex-valued, i.e., they are sections of

$$\bigwedge^p(M) = \bigwedge^p T^*(M)^{\mathbf{R}} \otimes \mathbf{C}, \quad p = 0, \ldots, 2n = \dim_{\mathbf{R}} M.$$

Now define

$$\bigwedge^{p,q}(M) = \bigwedge^p T^*_{1,0}(M) \otimes \bigwedge^q T^*_{0,1}(M),$$

and we see that

$$\bigwedge^r(M) = \sum_{p+q=r} \bigwedge^{p,q}(M). \tag{2.2.17}$$

Taking sections of these bundles, we obtain differential forms on the complex manifold of type (p,q). Namely, we set

$$\mathcal{E}^{p,q}(M) := \Gamma^\infty(M, \bigwedge^{p,q}(M)).$$

Locally, we see that any $\omega \in \mathcal{E}^{p,q}(M)$ is of the form

$$\omega = \sum_{\substack{1 \le i_1 < \cdots < i_p \le n \\ 1 \le j_1 < \cdots < j_q \le n}} f_{i_1 \ldots i_p, j_1 \ldots j_q} dz^{i_1} \wedge \cdots \wedge dz^{i_p} \wedge d\bar{z}^{j_1} \wedge \cdots \wedge d\bar{z}^{j_q},$$

This follows from the fact that local frames for $T^*_{1,0}$ and $T^*_{0,1}$ are given by $\{dz^1, \ldots, dz^n\}$ and $\{d\bar{z}^1, \ldots, d\bar{z}^n\}$, respectively (with respect to some coordinate system $(z^1, \ldots, z^n)$). We use the shorthand notation

$$\omega = {\sum_{\substack{|I|=p \\ |J|=q}}}' f_{IJ} dz^I \wedge d\bar{z}^J,$$

where the "$'$" indicates increasing multiindices as before.

The decomposition of bundles given in (2.2.17) translates to a decomposition of differential forms of degree r into the direct sum of differential forms of type (p,q), i.e.,

$$\mathcal{E}^r(M) = \sum_{p+q=r} \mathcal{E}^{p,q}(M). \tag{2.2.18}$$

Let us denote by $\pi_{p,q}$ the natural projection operator

$$\mathcal{E}^r(M) \xrightarrow{\pi_{p,q}} \mathcal{E}^{p,q}(M)$$

given by the direct sum (2.2.18). We can now define the complex analogues of the exterior derivative operator. Define the operators

$$\mathcal{E}^{p,q}(M) \xrightarrow{\partial} \mathcal{E}^{p+1,q}(M),$$

$$\mathcal{E}^{p,q}(M) \xrightarrow{\bar{\partial}} \mathcal{E}^{p,q+1}(M)$$

by

$$\partial := \pi_{p+1,q} \circ d,$$
$$\bar{\partial} := \pi_{p,q+1} \circ d.$$

In general, if we compute $d\omega^{p,q}$, where ω is of type (p,q), then $d\omega^{p,q}$ has total degree $p + q + 1$, and would be of the form (using (2.2.18)),

$$d\omega^{p,q} = \varphi^{p+q+1,0} + \varphi^{p+q,1} + \cdots$$
$$+ \varphi^{p+1,q} + \varphi^{p,q+1} + \cdots + \varphi^{0,p+q+1}, \qquad (2.2.19)$$

where $\varphi^{r,s}$ are suitable forms of type (r,s). We have defined $\partial\omega^{p,q}$ and $\bar{\partial}\omega^{p,q}$ to be the terms $\varphi^{p+1,q}$ and $\varphi^{p,q+1}$ in the expansion (2.2.19). It is a remarkable fact that all of the other terms are zero. Namely, we have the following proposition.

Proposition 2.2.2. *On a complex manifold M*

$$d = \partial + \bar{\partial},$$

and moreover,

$$\partial^2 = \bar{\partial}^2 = \partial\bar{\partial} + \bar{\partial}\partial = 0,$$

as operators acting on (p,q)-forms on M.

Proof: We shall use the local representation of (p,q)-forms and the local form of d acting on forms. First, if $\omega = f$ is a scalar function, then clearly

$$d = \partial + \bar{\partial},$$

since $\mathcal{E}^1(M) = \mathcal{E}^{1,0}(M) \oplus \mathcal{E}^{0,1}(M)$. This gives us a local formula for ∂ and $\bar{\partial}$ acting on functions. Namely, we have

$$dz^j = dx^j + idy^j, \quad \text{and} \quad d\bar{z}^j = dx^j - idy^j, \qquad (2.2.20)$$

in terms of local holomorphic coordinates $(z^1, \ldots, z^n)$. Now let (x^j, y^j) be real coordinates with $z^j = x^j + iy^j$, and write

$$df = \sum_{i=1}^{n} \frac{\partial f}{\partial x^j} dx^j + \frac{\partial f}{\partial y^j} dy^j$$

$$= \sum_{i=1}^{n} \frac{\partial f}{\partial z^j} dz^j + \frac{\partial f}{\partial \bar{z}^j} d\bar{z}^j,$$

using (2.2.15) and (2.2.20) to rearrange the terms into the latter form. It follows that, acting on functions,

$$\partial f = \sum_{j=1}^{n} \frac{\partial f}{\partial z^j} dz^j,$$

$$\bar{\partial} f = \sum_{j=1}^{n} \frac{\partial f}{\partial \bar{z}^j} d\bar{z}^j.$$

Now let

$$\omega = \sum_{\substack{|I|=p \\ |J|=q}}' f_{IJ} dz^I \wedge d\bar{z}^J$$

be a general (p, q)-form near some point of M. Then

$$d\omega = \sum_{\substack{|I|=p \\ |J|=q}}' (df_{IJ}) \wedge dz^I \wedge d\bar{z}^J$$

$$= \sum_{\substack{|I|=p \\ |J|=q}}' (\partial f_{IJ}) \wedge dz^I \wedge d\bar{z}^J + \sum_{\substack{|I|=p \\ |J|=q}}' (\bar{\partial} f_{IJ}) \wedge dz^I \wedge d\bar{z}^J,$$

and these two terms belong to $\mathcal{E}^{p+1,q}$ and $\mathcal{E}^{p,q+1}$, respectively.

The second part of the proposition follows from $d = \partial + \bar{\partial}$ and the fact that $d^2 = 0$ as well as that (2.2.18) is a direct sum. Namely, for $\omega \in \mathcal{E}^{p,q}$,

$$d^2\omega = (\partial + \bar{\partial})^2\omega = \partial^2\omega + (\partial\bar{\partial} + \bar{\partial}\partial)\omega + \bar{\partial}^2\omega = 0. \qquad (2.2.21)$$

But $\partial^2\omega$ has type $(p+2, q)$, $(\partial\bar{\partial}+\bar{\partial}\partial)\omega$ has type $(p+1, q+1)$, and $\bar{\partial}^2\omega$ has type $(p, q + 2)$. So, since they belong to separate summands in the direct sum (2.2.18), the individual terms in (2.2.21) also vanish. ∎

The operator $\bar{\partial}$ is a very important one on a complex manifold. We note that a smooth function f on a complex manifold M satisfies

$\bar{\partial} f = 0$ if and only if f is holomorphic. This is a local statement, and we see that locally

$$\bar{\partial} f = \sum_{j=1}^{n} \frac{\partial f}{\partial \bar{z}^j} d\bar{z}^j = 0$$

if and only if

$$\frac{\partial f}{\partial \bar{z}^j} = 0, \quad j = 1, \ldots, n,$$

which are the Cauchy–Riemann equations. Hence we call $\bar{\partial}$ the Cauchy–Riemann operator. The inhomogeneous Cauchy–Riemann equations

$$\frac{\partial f}{\partial \bar{z}^j} = g^j, \quad j = 1, \ldots, n$$

correspond to the equation

$$\bar{\partial} f = g, \tag{2.2.22}$$

where

$$g = \sum_{j=1}^{n} g^j d\bar{z}^j.$$

We call (2.2.22) the $\bar{\partial}$-*equation on M* ($\bar{\partial}$ is usually pronounced 'dee-bar', or, if there is any confusion, as in the relation between the operators ∂ and d, one often says 'round-dee' for the operator ∂). The $\bar{\partial}$-equation makes sense when $f \in \mathcal{E}^{p,q}$ and $g \in \mathcal{E}^{p,q+1}$, and this family of differential equations plays an extremely important role in the theory of complex manifolds. They are analogous to the equation

$$df = g,$$

which was used to define the de Rham groups of a differentiable manifold. In this spirit, we define the *Dolbeault complex* on a complex manifold M,

$$\mathcal{E}^{p,0}(M) \xrightarrow{\bar{\partial}} \mathcal{E}^{p,1}(M) \xrightarrow{\bar{\partial}} \mathcal{E}^{p,2}(M) \xrightarrow{\bar{\partial}} \cdots \xrightarrow{\bar{\partial}} \mathcal{E}^{p,n}(M). \tag{2.2.23}$$

This is a complex since $\bar{\partial}^2 = 0$, by Proposition 2.2.2. We have a similar complex of smooth forms with the ∂-operator which is less important and which we shall not write down. However, if we define

$$\Omega^p(M) := \ker \bar{\partial} : \mathcal{E}^{p,0} \to \mathcal{E}^{p,1}, \quad \text{for } p = 0, 1, \ldots, n, \tag{2.2.24}$$

then we have the following complex on M:

$$\Omega^0(M) \xrightarrow{\partial} \Omega^1(M) \xrightarrow{\partial} \Omega^2(M) \xrightarrow{\partial} \cdots \xrightarrow{\partial} \Omega^n(M). \qquad (2.2.25)$$

We see that if $\omega \in \Omega^p(M)$, then locally

$$\omega = \sum_{|I|=p} f_I dz^I,$$

and $\bar{\partial}\omega = 0$, so $\bar{\partial}f_I = 0$, for each multiindex I, and hence each of the coefficients of the differential form ω is a holomorphic function. Thus we call $\Omega^p(M)$ the vector space of *holomorphic p-forms on M*, and (2.2.25) is called the *holomorphic de Rham complex on M*. Note that on holomorphic p-forms $d = \partial$, so we could write (2.2.24) also as

$$\Omega^0(M) \xrightarrow{d} \Omega^1(M) \xrightarrow{d} \cdots \xrightarrow{d} \Omega^n(M) \rightarrow$$

if we desire, and hence the name de Rham complex in this case.

We have analogues to the Poincaré lemma for each of these complexes (2.2.23) and (2.2.25). Let

$$P_r = \{z \in \mathbf{C}^n : |z_i| < r, \quad \text{for all } i = 1, \ldots, n\} \text{ be the polydisk of}$$
$$\text{radius } r \text{ in } \mathbf{C}^n.$$

Lemma 2.2.3. *Suppose $r > 0$, and $\varepsilon > 0$.*

    (a) *Let $\omega \in \mathcal{E}^{p,q}(P_{r+\varepsilon})$ satisfy $\bar{\partial}\omega = 0$ on $P_{r+\varepsilon}$; then there exists $\varphi \in \mathcal{E}^{p,q-1}(P_r)$ such that $\bar{\partial}\varphi = \omega$ on P_r.*

    (b) *Let $\omega \in \Omega^p(P_r)$ satisfy $\partial\omega = 0$ on P_r; then there exists $\varphi \in \Omega^{p-1}(P)$ such that $\partial\varphi = \omega$ on P_r.*

Remark: Note that in (a) it was necessary to shrink the polydisk slightly in order to solve the equation. It is also true that one can solve $\bar{\partial}\varphi = \omega$ precisely on P_r as in (b), but this is a much deeper result and the elementary proof which we shall give here does not yield this finer result. For applications to sheaf theory the weaker version given by this lemma is more than sufficient.

Proof: The proof is a simple adaptation of the proof of Lemma 2.2.1. In that lemma we needed to solve the equation

$$\frac{\partial u(x^1, \ldots, x^j, \ldots, x^n)}{\partial x^j} = g^j(x^1, \ldots, x^j, \ldots, x^n)$$

on the rectangle R, and the solution

$$u(x^1, \ldots, x^j, \ldots, x^n) = \int_0^{x^j} g^j(x^1, \ldots, \underset{(j)}{t}, \ldots, x^n) dt$$

was given by the fundamental theorem of calculus. We need to simply formulate the analogous one-variable results for the other two operators. Namely, we need to solve

$$\frac{\partial u(z^1, \ldots, z^j, \ldots, z^n)}{\partial \bar{z}^j} = g^j(z^1, \ldots, z^j, \ldots, z^n) \qquad (2.2.26)$$

on the smaller polydisk P_r, given the data g^j on the larger polydisk $P_{r+\epsilon}$. The solution is given by

$$u(z^1, \ldots, z^j, \ldots, z^n) = \frac{1}{2\pi i} \int_{P_r} \frac{g^j(z^1, \ldots, \overset{(j)}{\zeta}, \ldots, z^n)}{\zeta - z^j} d\zeta \wedge d\bar{\zeta},$$

i.e., in one complex variable

$$\frac{\partial}{\partial \bar{z}} \left(\frac{1}{2\pi i} \int_{|\zeta| < r} \frac{f(\zeta)}{\zeta - z} d\zeta \wedge d\bar{\zeta} \right) = f(z),$$

provided f is smooth on $|\zeta| \leq r$. This follows from the general Cauchy integral formula for smooth (C^∞) functions

$$f(z) = \frac{1}{2\pi i} \int_{|\zeta| = r} \frac{f(\zeta)}{\zeta - z} d\zeta + \frac{1}{2\pi i} \int_{|\zeta| \leq r} \frac{\frac{\partial f}{\partial \bar{\zeta}}(\zeta)}{\zeta - z} d\zeta \wedge d\bar{\zeta},$$

which is an elementary consequence of Green's theorem in the plane.

Similarly, for holomorphic p-forms, we shall need to solve the equation

$$\frac{\partial u(z^1, \ldots, z^j, \ldots, z^n)}{\partial z^j} = g^j(z^1, \ldots, z^j, \ldots, z^n), \qquad (2.2.27)$$

where g^j is holomorphic. We can represent holomorphic functions in a polydisk by convergent power series, by an iteration of the standard one-variable argument. For one variable we simply want to solve

$$\frac{\partial}{\partial z} \left(\sum_{n=0}^\infty a_n z^n \right) = \sum_{n=0}^\infty b_n z^n,$$

and clearly, we can just integrate term by term obtaining $a_{n+1} = (n+1)^{-1}b_n$, a_0 arbitrary, as usual.

To prove part (a) of the lemma, we replace dx^j by $d\bar{z}^j$ for the definition of F^p in the proof of Lemma 2.2.1, and similarly, for part (b), we replace dx^j by dz^j for the definition of F^p. The induction proof proceeds in the same manner, using the solutions to (2.2.26) and (2.2.27) at the appropriate steps in the proof. We leave the details to the reader.   ∎

We now have one last generalization of the de Rham complex. In §2.1. we have introduced vector bundles, and here we have been discussing the complexes of differential forms. Now we want to combine these two notions. The differential forms we have been considering have complex-valued functions as coefficients. Recalling that sections of complex vector bundles are locally vector-valued functions, we want to consider differential forms with sections of given vector bundles as coefficients. Let E and F be smooth complex vector bundles over a differentiable manifold M. Let $x_0 \in M$, and choose frames $e = (e_1, \dots, e_r)$, $f = (f_1, \dots, f_m)$ for the bundles E and F near x_0. Then any section of $E \otimes F$ near x_0 is of the form

$$\sigma = \sum_{i=1}^{r} \sum_{j=1}^{m} \sigma^{ij} e_i \otimes f_j,$$

where σ^{ij} are smooth functions. We can interpret σ in two ways:

(a) $\sigma = \sum_{j=1}^{m} \left(\sum_{i=1}^{r} \sigma^{ij} e_i \right) \otimes f_j = \sum_{j=1}^{m} c^j \otimes f_j;$

$$\text{(2.2.28)}$$

(b) $\sigma = \sum_{i=1}^{r} e_i \otimes \left(\sum_{j=1}^{m} \sigma^{ij} f_i \right) = \sum_{i=1}^{r} e_i \otimes d^i.$

In (a) we have expanded σ with respect to the frame f for F, and the coefficients c^j of σ are sections of E, and in (b) we have expanded σ with respect to the frame e for E, and the coefficients d^i are sections of F.

We can now consider $E \otimes \bigwedge^p(M)$, and we see that locally a section of this bundle is of the form

$$\sigma = \sum c^j \otimes \omega_j,$$

where $\{\omega_1, \ldots, w_N\}$ is a local frame for $E \otimes \bigwedge^p(M)$, and $N = \binom{n}{p}$ is the number of linearly independent p-forms in $\bigwedge^p(M)$ at each point of M. In other words, the trivialization of E given by $e = (e_1, \ldots, e_r)$ defined on $U \subset M$ yields

$$E \otimes \textstyle\bigwedge^p(M)|_U \cong \mathbf{C}^r \otimes \textstyle\bigwedge^p(U)$$

$$\cong \left\{ \begin{pmatrix} \omega^1 \\ \vdots \\ \omega^r \end{pmatrix} : \omega^j \in \textstyle\bigwedge^p(U) \right\}.$$

In other words, sections of $E \otimes \bigwedge^p(M)$ are locally vectors of p-forms, just like sections of E are locally vectors of functions. The transition functions for E give the relation between two different representations of vectors of p-forms. That is, any section $\sigma \in \Gamma^\infty(M, E \otimes \bigwedge^p(M))$ is equivalent to a covering $\{U_\alpha\}$ and vectors of p-forms $\begin{pmatrix} \omega_\alpha^1 \\ \vdots \\ \omega_\alpha^r \end{pmatrix}$ satisfying

$$\begin{pmatrix} \omega_\alpha^1 \\ \vdots \\ \omega_\alpha^r \end{pmatrix} = (g_{\alpha\beta}) \begin{pmatrix} \omega_\beta^1 \\ \vdots \\ \omega_\beta^r \end{pmatrix},$$

and $g_{\alpha\beta} : U_\alpha \cap U_\beta \to \mathrm{GL}(r, \mathbf{C})$ are the transition functions for E. We set

$$\mathcal{E}^p(M, E) := \Gamma^\infty(M, E \otimes \textstyle\bigwedge^p(M)),$$

and call $\mathcal{E}^p(M, E)$ the vector space of *p-forms on M with coefficients in E* (*E*-valued *p*-forms). We would like to extend the exterior derivative operator d to act on *E*-valued *p*-forms so that we could have a complex of the form

$$0 \to \mathcal{E}^0(M, E) \xrightarrow{\tilde{d}} \mathcal{E}^1(M, E) \xrightarrow{\tilde{d}} \cdots \xrightarrow{\tilde{d}} \mathcal{E}^n(M, E).$$

Without additional information, this is generally not possible, and when it is possible, it may be that $(\tilde{d})^2 \neq 0$. As we shall see in Chapter 3, we can extend d to $\tilde{d}$ by introducing a *connection* on E, and the obstruction $\tilde{d}^2$ to having a de Rham type complex turns out to be the curvature of the connection, suitably formulated.

However, if we consider the Dolbeault complex and tensor with a holomorphic vector bundle, there is a natural extension of the $\bar{\partial}$-operator. Let $\omega \in \mathcal{E}^{p,q}(M, E)$, where E is a holomorphic vector bundle over a complex manifold. We want to define $\bar{\partial}\omega$ as a well-defined element of $\mathcal{E}^{p,q+1}(M, E)$. Since $\bar{\partial}$ is to be a differential operator, we shall define it locally. Let $x_0 \in M$, and suppose that U is a trivializing open set for E containing x_0. Thus,

$$E|_U \cong U \times \mathbf{C}^r,$$

and so $\omega|_U$ can be expressed in the form $\begin{pmatrix} \omega^1 \\ \vdots \\ \omega^r \end{pmatrix}$, where $\omega^j \in \mathcal{E}^{p,q}(M)$.

Define

$$\bar{\partial}\omega = \begin{pmatrix} \bar{\partial}\omega^1 \\ \vdots \\ \bar{\partial}\omega^r \end{pmatrix},$$

i.e., we simply take $\bar{\partial}$ of each of the components with respect to the trivialization. Suppose we had taken a different trivialization at the same point (say the V-trivialization where $V \ni x_0$); then we shall have

$$\omega \sim \begin{pmatrix} \widetilde{\omega}^1 \\ \vdots \\ \widetilde{\omega}^r \end{pmatrix},$$

where

$$\begin{pmatrix} \widetilde{\omega}^1 \\ \vdots \\ \widetilde{\omega}^r \end{pmatrix} = g_{UV} \begin{pmatrix} \omega^1 \\ \vdots \\ \omega^r \end{pmatrix},$$

and where g_{UV} is the transition function between the two trivialization representations. But we see that

$$\begin{pmatrix} \bar{\partial}\widetilde{\omega}^1 \\ \vdots \\ \bar{\partial}\widetilde{\omega}^r \end{pmatrix} = \bar{\partial}\left\{ g_{UV} \begin{pmatrix} \omega^1 \\ \vdots \\ \omega^r \end{pmatrix} \right\}$$

$$= \bar{\partial}g_{UV} \begin{pmatrix} \omega^1 \\ \vdots \\ \omega^r \end{pmatrix} + g_{UV} \begin{pmatrix} \bar{\partial}\omega^1 \\ \vdots \\ \bar{\partial}\omega^r \end{pmatrix}$$

$$= g_{UV} \begin{pmatrix} \bar{\partial}\omega^1 \\ \vdots \\ \bar{\partial}\omega^r \end{pmatrix},$$

since g_{UV} is holomorphic. Thus, the vector $\begin{pmatrix} \bar{\partial}\omega^1 \\ \vdots \\ \bar{\partial}\omega^r \end{pmatrix}$ transforms exactly as a section of $E \otimes \bigwedge^{p,q+1}(M)$ should under changes of trivialization for E, and hence defines a section of this bundle. Therefore we have a well-defined complex

$$\mathcal{E}^{p,0}(M,E) \xrightarrow{\bar{\partial}} \mathcal{E}^{p,1}(M,E) \xrightarrow{\bar{\partial}} \cdots \xrightarrow{\bar{\partial}} \mathcal{E}^{p,n}(M,E),$$

and we define the cohomology groups of this complex

$$H^{p,q}_{\text{DOL}}(M,E) := H^q(\mathcal{E}^{p,q}(M,E), \bar{\partial}) \qquad (2.2.29)$$

to be the *Dolbeault cohomology groups of the complex manifold M with coefficients in the vector bundle E.*

The cohomology groups $H^{p,q}(M,E)$ are abstract quotients of solutions of the Cauchy–Riemann equation, and it is remarkable that for many examples they can be quite accurately computed, or at least their dimensions as vector spaces can be determined. We shall not give any examples here, as various examples will be more amenable to sheaf-theoretic techniques to be developed in Chapter 3. If M is compact, one can show from the general theory of elliptic differential equations that these vector spaces are always finite-dimensional. We shall use this fact when we need it, but refer the reader to Wells (1980) for the details of this important fact.

Exercises for Section 2.2

2.2.1. Derive the local exterior differentiation formula (2.2.3) from properties (2.2.2).

2.2.2. Let ω be a p-form on an open subset U of $\mathbf{R}^n$, and let M be a closed submanifold of U of dimension p. Show that $\int_M \omega$ is well-defined and independent of the choice of coordinate charts and partitions of unity used in its definition.

2.2.3. If ω is the closed one-form on $\mathbf{C} - \{0\}$ defined by (2.2.5), and $\tilde{\omega}$ is any other closed one-form defined on $\mathbf{C} - \{0\}$, such that $\tilde{\omega}$ represents a nonzero element of $H^1_{\text{DR}}(\mathbf{C} - \{0\})$, then show that there is a constant c and a function $\varphi \in \mathcal{E}^0(\mathbf{C} - \{0\})$ such that $\omega - c\tilde{\omega} = d\varphi$.

2.2.4. Suppose that ω is a closed differential form defined on S^n, and let $i : S^n \to \mathbf{R}^{n+1}$ be the standard embedding. Moreover, suppose that $\omega = i^*(\tilde{\omega})$, where $\tilde{\omega}$ is a closed form on $\mathbf{R}^{n+1}$. Show that ω is an exact form.

2.2.5. Show that if ω is a d-closed $(1,1)$-form on a complex manifold X, then locally, near any point of X, there exists a function φ such that $\omega = \partial\bar{\partial}\varphi$ (this is often referred to as a Kähler potential for ω).

2.2.6. Let X be an even-dimensional differentiable manifold, and suppose that there is a real-linear mapping $J : T(X) \to T(X)$ such that $J^2 = -\,\mathrm{id}$. We call J an *almost-complex tensor* on X. Show that on X one can associate in a unique manner to a given almost-complex tensor J a complex structure on the real vector space $T_x(X)$, in such a manner that a mapping $L : T(X) \to T(X)$ is complex-linear on the fibers of $T(X)$ if and only if L commutes with J.

2.2.7. An *almost-complex manifold* X is an even-dimensional manifold equipped with an almost-complex tensor J (or equivalently, by Exercise 2.2.6, with a complex structure on each tangent space). Define $T^{1,0}(X)$ and $T^{0,1}(X)$ just as for complex manifolds using the almost-complex tensor J. On any almost-complex manifold the notion of (p,q)-forms and the projection $\pi_{p,q} : \mathcal{E}^{p+q}(X) \to \mathcal{E}^{p,q}(X)$ is well-defined (using the eigenspaces of J), and we can define $\bar{\partial} := \pi_{p,q+1} \circ d$, as before. An almost-complex structure is integrable if $\bar{\partial}^2 = 0$. Show that each complex manifold X induces a unique integrable almost-complex structure such that $T^{1,0}(X)$ becomes the holomorphic tangent bundle. The converse is also true: all integrable almost-complex structures are induced from a complex manifold structure (this is a deep theorem of Newlander and Nirenberg 1957; cf. Hörmander 1973).

2.2.8. A differentiable submanifold M of a complex manifold X is *totally real* if $T(M) \cap JT(M) = 0$, where J is the almost-complex tensor acting on $T(X)$ considered as a real tangent bundle induced by the complex structure of X (intuitively, totally real means that multiplication of a real tangent vector to M by i gives a vector which is no longer tangent to M). Show that the following are totally real submanifolds:

(a) any real curve in a complex manifold X,
(b) $\mathbf{R}^n$ embedded canonically in $\mathbf{C}^n$,
(c) $\{(z^1, \ldots, z^n) \in \mathbf{C}^n : |z^j| = 1, \ j = 1, \ldots, n\}$,

(d) the embedding of S^4 in **M** given by (1.6.13),

(e) the embedding of M in **M** given by (1.6.13).

2.3 Tensor and spinor bundles

In §2.1 we introduced briefly the *tensor algebra* $T(M)$ of a differentiable manifold M. This was defined as the algebra of bundles generated by $T(M)$ and $T^*(M)$ using tensor product as the multiplication operation. There are various ways of describing this tensor algebra so that one can make calculations. The classical 'tensor analysis' and index notation made popular by Einstein's use of it in his description of the field equation of general relativity, is an important descriptive tool. At the other extreme is the attempt by many contemporary mathematicians to 'get rid of indices' by using the language of vector bundles and an invariant notation, which gives a certain elegance and clarity to many tensorial calculations, in particular to the calculus of differential forms. In between these two points of view is the abstract index notation of Penrose (1968b), which combines some of the best features of both the other viewpoints. It is always useful to have several different points of view, and we shall use all three descriptions in this book, letting the context determine which is the most convenient. We shall spend a little time in this section discussing these ideas, but the real learning process will take place in later chapters, when these concepts are used in concrete situations. We shall be discussing both tensor bundles and spinor bundles on a manifold M, and the remarks concerning notations will apply to both. We shall first concentrate on the tensor algebra to get some feel for the system, and then we shall discuss spinor bundles.

Let us start with the definition of a metric or pseudometric on a differentiable manifold M. If $T(M)$ is the tangent bundle, then a pseudometric g is an element

$$g \in \Gamma^\infty(M, \odot^2 T^*(M))$$

(where we recall that '$\odot$' denotes symmetric tensor product) with the property that

$$g_x : T(M) \times T(M) \to \mathbf{R}$$

is a nondegenerate bilinear form for each $x \in M$. We say that g is a *Riemannian metric* (or sometimes simply a *metric*) if this quadratic form is positive-definite. If we choose a frame $f_0 = (e_1, \ldots, e_n)$ for $T(M)$ near $x_0 \in M$, then f_0 induces a frame $f^0 = (e^1, \ldots, e^n)$ for $T^*(M)$ by duality, and we can write

$$g = \sum_{ab} g_{ab} e^a \odot e^b,$$

where $g(f) = (g_{ab})$ is a symmetric nondegenerate matrix, which locally represents g, depending on the choice of frame. If we change frames to $\tilde{f}_0 = (\tilde{e}_1, \ldots, \tilde{e}_n)$, then we have

$$g = \sum_{ab} \tilde{g}_{ab} \tilde{e}^a \odot \tilde{e}^b.$$

Moreover, if $\tilde{f} = f \cdot N$ where N is a $GL(n, \mathbf{R})$-valued function defined on the overlap of the defining domains for the frames, i.e.,

$$\tilde{e}_a = \sum_b e_b N^b_a,$$

then we find that, as matrices,

$$\tilde{g} = N^t g N.$$

Since $g \in \Gamma(M, \odot^2 T^*(M))$, we see that g is a tensor field or simply tensor (of type $(0, 2)$). Then we have the description

$$g \in \Gamma^\infty(M, \odot^2 T^*(M)) \leftrightarrow \{g_{ab}\}, \quad g_{ab} \text{ symmetric}.$$

This is simply a special case of the general principle of representing sections of bundles in terms of local vector-valued functions enunciated in Proposition 2.1.3. Choose a coordinate frame $e_i = \frac{\partial}{\partial x^i}$, $i = 1, \ldots, n$ and write

$$g = ds^2 = \sum_{ab} g_{ab} dx^a dx^b,$$

where $dx^a dx^b$ means $dx^a \odot dx^b$, the symmetric tensor product. The coefficients are the $\{g_{ab}\}$ of classical differential geometry. Now any tensor field

$$s \in \Gamma^\infty(M, T(M) \otimes \cdots \otimes T(M) \otimes T^*(M) \otimes \cdots \otimes T^*(M))$$

can be represented in the form, in terms of a frame,

$$s = \sum s^{ab...d}{}_{rs...t}\, e_a \otimes e_b \otimes \cdots \otimes e_d \otimes e^r \otimes e^s \otimes \cdots \otimes e^t.$$

It is certainly simpler to write

$$s = s^{ab...d}{}_{rs...t}$$

to represent a tensor of this form if the frame quantities are clearly understood. We say that the tensor is of *type* (r, s) if it consists of r factors of $T(M)$ and s factors of $T^*(M)$. This representation then changes by a generalization of the transformation for the tensor of type $(0, 2)$ above when the frames undergo a linear transformation. There are various operations one can perform on tensor fields. One important operation is *contraction*. This is a mapping of (r, s)-tensors to $(r-1, s-1)$-tensors of the sort

$$s = s^{ab...d}{}_{rs...t} \mapsto s^{ab...d}{}_{as...t},$$

where we again use the Einstein convention that

$$s^{ab...d}{}_{as...t} = \sum_a s^{ab...d}{}_{as...t}.$$

This operation is *independent* of the choice of frame, and this depends on the fact that similarity transformations preserve the trace of a matrix.

A second operation on tensor fields depends on having a given metric or pseudometric on the manifold M. If g is a pseudometric on M, then g induces for each point $x \in M$ an isomorphism between $T_x(M)$ and its dual space $T_x^*(M)$. Given this isomorphism, we could cease to distinguish between upper and lower indices in a local representation of a tensor field. However, it is wiser to keep the distinction, as we sometimes vary the metric, and one needs to keep track of the nature of the tensor in question. If $v \in \Gamma^\infty(M, T(M))$ is a vector field on M, then with respect to a frame we have

$$v = \sum_a v^a e_a,$$

and we can write simply

$$v \sim v^a.$$

If g is a pseudometric and g is represented with respect to a frame as g_{ab}, then we can write

$$v_b := v^a g_{ab} = \sum_a v^a g_{ab}.$$

This means that

$$\tilde{v} = \sum_b v_b e^b$$

is a differential one-form and $\tilde{v}$ is well-defined and independent of the choice of frame, as is easy to show. More generally, if we let g^{ab} be the coefficients of the *inverse matrix* for g_{ab}, then we have

$$g^{ab} g_{bc} = \delta^a{}_c,$$

where

$$\delta^a{}_c = \left\{ \begin{matrix} 1, & a = c \\ 0, & a \neq c \end{matrix} \right\}$$

is the Kronecker delta, a well-defined tensor field of type $(1,1)$ on the manifold M.

Thus, we can contract tensor fields and raise and lower indices. The only remaining fundamental operations on tensors is that of co-variant differentiation ∇, which we shall introduce in Chapter 3 when we discuss connections and curvature. Covariant differentiation will be a natural generalization of the derivative of a function. This will depend on additional data on the manifold, such as a metric (or more generally, a *connection* which relates tangent vectors from one point of a manifold to another).

The abstract index notation of Penrose mirrors the index notation derived by looking at coefficients of sections of bundles with respect to a choice of frame. The formal operations, contraction, raising, lowering indices with respect to a metric, and covariant differentiation are almost identical to the calculus of numerically indexed tensorial quantities, and can be manipulated just as smoothly. In fact, in the usual tensor calculus, one rarely works with the actual numerical indices; almost always it is the formal calculations which are used. This is what Penrose has formalized.

Let $V^\bullet$ be a module over a ring R and let $V_\bullet$ be the dual module, i.e., R-linear mappings from $V^\bullet$ into R. The example we want to keep in mind is $V^\bullet = V(M)$, the module of smooth vector fields over the ring $\mathcal{E}(M)$ of smooth functions on M. The dual module

$V_\bullet$ in this case will be simply $\mathcal{E}^1(M)$, the smooth one-forms on M. The tensor algebra of $V^\bullet$ is defined as the direct sum of the tensor products

$$\mathcal{T}(V^\bullet) = \sum_{r,s} \underbrace{V^\bullet \otimes \cdots \otimes V^\bullet}_{r \text{ factors}} \otimes \underbrace{V_\bullet \otimes \cdots \otimes V_\bullet}_{s \text{ factors}},$$

and we want to describe elements of $\mathcal{T}(V^\bullet)$ in a convenient fashion. Let L be an infinite set consisting of distinct symbols;

$$L = \{a, b, c, \ldots, a_0, b_0, \ldots, a_1, b_1, \ldots, a_2, \ldots\}.$$

This will be the set of *labels* for the tensor algebra. Define, for each label $l \in L$, an isomorphic copy of the module $V^\bullet$, and we set, for $l \in L$,

$$V^l := V^\bullet \times \{l\} \subset V^\bullet \times L.$$

If $\xi \in V^\bullet$, then we denote by ξ^l the corresponding element of V^l. Since V^l is a module, the usual module operations in V^l are applicable. Thus we have, for instance, $\xi^l + \eta^l$ is a well-defined sum, and

$$\xi^l + (\eta^l + \zeta^l) = (\xi^l + \eta^l) + \zeta^l,$$

etc. However, the addition $\xi^l + \xi^{\tilde{l}}$, for $l \neq \tilde{l}$, does not make sense. The indexed dual modules are defined the same way:

$$V_l := V_\bullet \times \{l\} \subset V_\bullet \times L.$$

If $\xi^l \in V^l, \eta_l \in V_l$, then we denote by $\xi^l \eta_l$ or $\eta_l \xi^l$ the duality pairing between the elements ξ^l and η_l. In other words, since η_l is a linear mapping

$$\eta_l : V^l \to R,$$

then

$$\xi^l \eta_l := \eta_l \xi^l := \eta_l(\xi^l) \in R.$$

It follows from the definition that the duality pairing is independent of the index, i.e., that

$$\xi^a \eta_a = \xi^b \eta_b = \cdots, \quad a, b, \ldots \in L.$$

We want to generate $\mathcal{T}(V)$ with the elements of $R, V^a, V^b, \ldots, V_a, V_b, \ldots$. To do this, we shall define a product $\xi^a \eta^b \in V^{ab}$, for $\xi^a \in V^a, \eta^b \in V^b$, where $a \neq b$. The product will *not* be defined for

$a = b$. The space V^{ab} will be described below. The product should be commutative, i.e., $\xi^a \eta^b = \eta^b \xi^a$. This product is generated by taking elementary products in the symmetric tensor algebra of the direct sum

$$V^a \oplus V_a \oplus V^b \oplus V_b \oplus \cdots .$$

Let

$$V^{abc...d}_{efg...h} = \{\text{linear combinations of } \xi^a \eta^b \zeta^c \ldots \chi^d \rho_e \sigma_f \tau_g \ldots \nu_h\},$$

for any set of labels $a, b, c, \ldots$ and elements $\xi^a, \eta^b, \ldots$, etc.

In this manner we generate a labled tensor system

$$\{V\} = \{R, V^a, V^b, \ldots, V_a, V_b, \ldots, V^{ab}, \ldots, V^{abc...d}_{efg...h}, \ldots\}.$$

The module $V^{abc...d}_{efg...h}$ is defined by

$$V^{abc...d}_{efg...h} \cong V^{\bullet} \times \cdots \times V^{\bullet} \times V_{\bullet} \times \cdots \times V_{\bullet},$$

where the number of products corresponds to the number of labels. The labeling indicates how many tensor products of each kind are present. There are four basic algebraic operations with indexed quantities: addition, multiplication, index substitution, and contraction. We shall illustrate this with special cases:

$$\sigma^{ab}_e + \tau^{ab}_e \in V^{ab}_e \qquad \text{(addition)},$$
$$\sigma^{ab}_e \cdot \tau^m_{rs} \in V^{abm}_{ers} \qquad \text{(multiplication)},$$
$$\sigma^{ab}_e \mapsto \sigma^{rs}_m \qquad \text{(index substitution)},$$
$$\sigma^{as...t}_{bm...n} \mapsto \sigma^{as...t}_{am...n} \in V^{s...t}_{m...n} \qquad \text{(contraction)}.$$

Since the elements of each of these spaces are generated by products $\xi^a \eta^b \ldots$ etc., (not necessarily uniquely), one needs to define these operations in terms of the generators. We have already seen, for instance, how to contract $\xi^a \eta_b \mapsto \xi^a \eta_a$ by using the duality pairing, and this is simply extended to the more general situation in a natural manner. Index substitution means that any formula in the tensor quantities is invariant under permutations of elements of the labeling set L.

One can choose a frame and represent a tensorial quantity in terms of numerical indices (which we shall denote by *sans serif* font, e.g., g_{ab}). The formal operations will be the same for both the abstract and numerical indices (except, as mentioned before, when we come to covariant differentiation, where certain specific differences arise). We have sketched briefly Penrose's ideas concerning abstract index notation stemming from his original paper (1968b). We refer the reader to the recent treatise by Penrose and Rindler (1984, 1986) which goes into much more detail on the whole subject. In any context in which we use abstract index labels, it is always possible to convert the calculation to numerical index calculations.

At the end of §2.1 we briefly discussed vector bundles associated with principal G-bundles and a representation of the group G. These are then vector bundles whose transition functions belong to matrix representations of the group G. A special case of this is when G is a subgroup of $GL(r, \mathbf{C})$, when the vector bundle has rank r, and the transition functions of the bundle E can be chosen to be in the subgroup G. If this is the case, we say that E is a G-bundle and that the structure group of E can be *reduced* from $GL(r, \mathbf{C})$ to G. More generally, we can talk about reduction of a given structure group G of a bundle to a subgroup H.

When a vector bundle is given, the ability to reduce its structure group often reflects some global property of the base space of the bundle. This will be illustrated in the following examples.

Example 2.3.1 (Tangent bundle structure group reduction): Let M be a differentiable manifold. Then $T(M)$, the tangent bundle to M, is a $G = GL(n, \mathbf{R})$-bundle, where $n = \dim M$. We have the following global structures which correspond to reductions to specific subgroups of G (if H is a topological group then we shall denote by H_0 the connected component containing the identity element):

Reduction to subgroup	Global structure
(a) $O(n)$	Riemannian metric;
(b) $O(p, q)$	Indefinite metric of signature (p, q);
(c) $GL_0(n, \mathbf{R})$	Choice of orientation;
(d) $SL(n, \mathbf{R})$	Choice of volume form;
(e) $GL(n/2, \mathbf{C})$	Almost complex structure.

We shall discuss each of these in turn, but they all hinge on the same idea of being able to find frames adapted to the given structure. What the list means is that you can reduce to the given subgroup if and only if the corresponding global structure exists on the manifold M.

For case (a) we note that any manifold M admits a Riemannian metric. Locally, one can always construct one (just choose a coordinate system and the standard metric there), and then one can use a partition of unity to construct a global one. What is needed here is simply that the sum of two positive-definite matrices is still positive definite. Suppose g is a Riemannian metric on M, then for each point $p \in M$ there is a neighborhood U of p and a frame $f = (e_1, \ldots, e_n)$ for $T(M)|_U$, so that

$$g(e_i, e_j) = \delta_{ij}, \quad i, j = 1, \ldots, n,$$

i.e., f is an *orthonormal frame* with respect to the metric g. One simply applies the Gram–Schmidt orthonormalization process to any frame defined near p. Thus, we can find a covering of M by open sets $\{U_\alpha\}$ with orthonormal frames f_α defined on U_α. The transition functions $\{g_{\alpha\beta}\}$ defined with respect to this family of frames will necessarily be of the form

$$g_{\alpha\beta} : U_\alpha \cap U_\beta \to \mathrm{O}(n),$$

since the induced linear mapping from an orthonormal basis to an orthonormal basis is an orthogonal matrix. Thus we see that if a Riemannian metric is given on M it induces a reduction of the structure group G to $\mathrm{O}(n)$.

Conversely, suppose such a reduction is given. Then we can construct a Riemannian metric on M. Simply let f_α be the frame defined on U_α so that the transition functions $g_{\alpha\beta}$ take values in $\mathrm{O}(n)$. Define the vector fields in the frame f_α to be orthonormal at each point. This is then enough information to determine a Riemannian metric.

Case (b) is similar. The only difference is that it is not necessarily the case that an arbitrary manifold admits an indefinite metric of signature (p, q) (see Beem and Ehrlich 1981, Steenrod 1951). A pseudo-Riemannian manifold has signature (p, q) if its pseudo-metric has p positive eigenvalues and q negative eigenvalues at each point. A pseudo-Riemannian manifold is *Lorentzian* if it has signature $(1, n-1)$ or $(n-1, 1)$. The Lorentzian manifolds of dimension four are the manifolds on which the space-time manifolds of general relativity are modeled. We shall discuss these manifolds in more detail in Chapter 6. A manifold M of dimension n admits a reduction to $\mathrm{O}(p, q)$ if and only if there are q global vector fields on M which are linearly independent at each point (Steenrod 1951). We shall see that the theory of characteristic classes (§3.4) gives topological restrictions for such global vector fields to exist on a given manifold. For instance, for a single vector field $(q = 1)$, the necessary and sufficient condition is that the Euler characteristic of M be zero (see §3.4).

Cases (c) and (d) are related. A volume form exists on a manifold if and only if it is orientable, and a choice of volume form induces a choice of orientation. But, of course, not all manifolds are orientable, so the corresponding reductions are not always possible. Recall that one definition of orientability is that one can choose an atlas of coordinate systems so that the transition matrices have positive determinants for each overlap. But $\mathrm{GL}(n, \mathbf{R})_0$ is precisely the set of matrices with positive determinant, so the correspondence in (c) is essentially the definition of orientability and a choice of orientation. Case (d) is a further reduction, which is always possible if (c) is possible. Namely, recall that a volume form on M is an element

$$\Omega \in \mathcal{E}^n(M),$$

such that $\Omega \neq 0$ at each point of M. Suppose such an Ω is given. For each point $p \in M$ choose a frame $f = (e^1, \ldots, e^n)$ for $T^*(M)$, defined on an open set $U \ni p$ so that

$$e^1 \wedge \cdots \wedge e^n = \Omega|_U. \qquad (2.3.1)$$

This is clearly possible, since any frame $\tilde{f} = (\tilde{e}^1, \ldots, \tilde{e}^n)$ defined near p will have

$$\tilde{e}^1(x) \wedge \cdots \wedge \tilde{e}^n(x) = c(x)\Omega(x),$$

where $c(x) \in \mathbf{R}$ and $c(x) \neq 0$ for $x \in U$. Let

$$e^j(x) = c(x)^{-1/n}\tilde{e}^j(x),$$

and $f = (e^1, \ldots, e^n)$ will satisfy (2.3.1). Now cover M by a family of such adapted frames. Let $g_{\alpha\beta}$ be the corresponding transition functions and we observe that if f_α, f_β are two frames defined on $U_\alpha \cap U_\beta$, then

$$e^1_\alpha \wedge \cdots \wedge e^n_\alpha = \det(g_{\alpha\beta})e^1_\beta \wedge \cdots \wedge e^n_\beta,$$

but since

$$e^1_\alpha \wedge \cdots \wedge e^n_\alpha|_{U_\alpha \cap U_\beta} = \Omega|_{U_\alpha \cap U_\beta},$$
$$e^1_\beta \wedge \cdots \wedge e^n_\beta|_{U_\alpha \cap U_\beta} = \Omega|_{U_\alpha \cap U_\beta},$$

it follows that $\det(g_{\alpha\beta}) = 1$, i.e., $g_{\alpha\beta}$ takes values in $\mathrm{SL}(n, \mathbf{R})$.

Case (e) can be taken as a definition of an almost-complex structure. Alternatively, an almost-complex structure is a (smoothly varying) *choice* of complex-linear structures on the real tangent spaces $T_x(M)$, for each $x \in M$ (see Exercises 2.2.6 and 2.2.7). In particular, M must be even-dimensional. Once again, not all real even-dimensional manifolds admit almost complex structures. We leave the details of case (e) to the exercises. It is similar to the others in that frames adapted to the complex structure in each fiber must be chosen.

Our next example is similar in spirit, but is not restricted to tangent bundles.

Example 2.3.2 ($\mathrm{SL}(2, \mathbf{C})$-**bundles**): Let S be a complex vector bundle of rank two over a manifold M which is reduced to an $\mathrm{SL}(2, \mathbf{C})$-bundle. Just as in Example 2.3.1, we have that a complex bundle $S \to M$ of rank two can be reduced to $\mathrm{SL}(2, \mathbf{C})$ if and only if there exists a nondegenerate skew form

$$\varepsilon \in \Gamma(M, {\textstyle\bigwedge}^2 S).$$

This is completely analogous to the reduction to $\mathrm{SL}(n, \mathbf{R})$ in the case of the tangent bundle. Namely, if ε is given, then choose adapted frames $f = (e_1, e_2)$ for S of the form

$$e_1 \wedge e_2 = \varepsilon \quad \text{(where defined),}$$

and the corresponding transition functions will take values in $\mathrm{SL}(2, \mathbf{C})$. Similarly, if such a reduction is given, define $\varepsilon = e_1 \wedge e_2$ locally by frames which effect the reduction, and ε will be a globally defined nondegenerate skew form on M. Just as in the case of orientability of a manifold, a given rank two complex vector bundle need not be reducible to an $\mathrm{SL}(2, \mathbf{C})$-bundle. We make the observation that reduction of a complex bundle S of rank two to an $\mathrm{SL}(2, \mathbf{C})$-bundle is equivalent to the triviality of the bundle $\bigwedge^2 S$. If M is a complex manifold of dimension n, then we define the *canonical bundle* on M to be

$$K_M := {\textstyle\bigwedge}^n T^*(M).$$

Considering the case of $\dim M = 2$, there are examples of K_M being either trivial or nontrivial, thus showing that such a reduction is not a trivial concept. For instance, if $M = \mathbf{P}_2$, then K_M is not trivial (in fact, $K_{\mathbf{P}_2} \cong H^{-3}$, where $H \to \mathbf{P}_2$ is the hyperplane section bundle of $\mathbf{P}_2$). Moreover, if M is a hypersurface of degree four in $\mathbf{P}_3$ (i.e., the zero set of a homogeneous polynomial of degree four), then K_M is trivial. Both of these are discussed in Griffiths and Harris (1978). In each case let $S = T^*(M)$, and we see that we can reduce to $\mathrm{SL}(2, \mathbf{C})$ or not, depending on whether K_M is trivial.

Let S be a complex bundle of rank two on a manifold M. We want to construct from S in a canonical fashion three more bundles. Let $S_{\mathbf{C}} := S \otimes_{\mathbf{R}} \mathbf{C}$ be the *complexification* of S. On $S_{\mathbf{C}}$ there is a conjugation defined, and we let $\overline{S}$ be the conjugate of S in $S_{\mathbf{C}}$. Thus S and $\overline{S}$ are both rank two complex vector bundles. Let S^* and $\overline{S}^*$ be the (complex-linear) duals of S and $\overline{S}$, respectively. Thus, given S, we generate $\overline{S}$, S^*, and $\overline{S}^*$.

Consider a real four-dimensional manifold M equipped with an $\mathrm{SL}(2, \mathbf{C})$-bundle $S \to M$ such that

$$T(M) \otimes \mathbf{C} \cong S \otimes \overline{S}.$$

Such bundles S are examples of what are called *spinor bundles* generically. Globally, the existence of such bundles is a delicate question, and will be taken up in §3.5 and §4.2 in more detail. We call the elements of S, $\overline{S}$, etc., 'spinors' or 'spinor fields'. This means that any tangent vector can be written as the sum of tensor products of spinors. Classically, spinors were introduced by E. Cartan and Dirac in different contexts. The two-component spinor formalism of Infeld and van der Waerden (1933) provided a convenient notation to describe spinor fields. We refer the reader to the treatise of Penrose and Rindler (1984, 1986), which is dedicated to the subject of spinors and space-time, for a more thorough treatment. We shall develop here and in §3.5 and §4.2 a few of the basic ideas necessary for the presentation of the later topics this book. The canonical local example is Minkowski space M^4. We shall see later in §3.5 global examples of spinor bundles which are locally equivalent to this one.

Example 2.3.3 (Spinors on Minkowski space): Choosing an origin and a basis for M^4, we can identify M^4 with its tangent space at $0, T_0(M^4)$, and with the space of 2×2 Hermitian matrices, as usual (see §1.6). Then let $S = \mathbf{C}^2$ be given by column vectors, and let S^* be the dual space of row vectors. If a vector space V is a tensor product $W \otimes Z$, and we choose bases for W and Z, then we can represent elements of W by column vectors, elements of Z by row vectors, and elements of V as the sum of products of the form (column) × (row). Thus we can express

$$T_0(M^4) \otimes \mathbf{C} \cong M^4 \otimes \mathbf{C}$$
$$\cong M_2(\mathbf{C}) \quad (2 \times 2 \text{ complex matrices}),$$
$$\cong S \otimes S^*.$$

A vector in $T_0(M^4) \otimes \mathbf{C}$ is *null* if it is a single product

$$v = s \times \tilde{s},$$

i.e., a 2×2 matrix v has zero determinant if and only if

$$v = (\text{row vector}) \times (\text{column vector}).$$

The vector v is real and null if

$$v = s \times \bar{s},$$

that is,

$$v = \begin{pmatrix} a & b \\ c & d \end{pmatrix} \text{ is Hermitian, } \det v = 0$$

if and only if

$$\begin{pmatrix} a & b \\ c & d \end{pmatrix} = \begin{pmatrix} s^0 \\ s^1 \end{pmatrix} \overline{(s^0 s^1)},$$
$$= \begin{pmatrix} s^0 \overline{s^0} & s^0 \overline{s^1} \\ s^1 \overline{s^0} & s^1 \overline{s^1} \end{pmatrix}.$$

Thus, in this sense, the spinor s is a 'square root' of the real null vector v.

We want to introduce some useful notation for the local structure of two-component spinor bundles as above. Let $S \to M$ be a spinor bundle. Then we define:

$$S^- := S,$$
$$S^+ := \overline{S},$$
$$S_- := S^*,$$
$$S_+ := \overline{S}^*.$$

Thus a given spinor bundle S generates automatically four associated spinor bundles $S^{\pm}$, $S_{\pm}$. This would, of course, also apply to a general complex vector bundle, if we wished. The fact that S is a spinor bundle means, in particular, as we have seen, that there is a non-degenerate skew form $\varepsilon \in \Gamma(M, \wedge^2 S)$. But, since $S^{\pm}$, $S_{\pm}$ are also spinor bundles, we also have associated nondegenerate skew forms

$$\varepsilon^{\pm} \in \Gamma(M, \wedge^2 S^{\pm}), \qquad \varepsilon_{\pm} \in \Gamma(M, \wedge^2 S_{\pm}).$$

The existence of any one of these skew forms implies the existence of the others. We can use the skew forms $\varepsilon^{\pm}$, $\varepsilon_{\pm}$ to provide vector bundle isomorphisms between a given spinor bundle and its dual:

$$\varepsilon^+ : S_+ \xrightarrow{\cong} S^+,$$

$$\varepsilon^- : S_- \xrightarrow{\cong} S^-,$$

$$\varepsilon_+ : S^+ \xrightarrow{\cong} S_+,$$

$$\varepsilon_- : S^- \xrightarrow{\cong} S_-.$$

Let us introduce a convenient index notation for these spinor bundles, which we can use at either the abstract or numerical level. Consider two independent sets of labels of the form $L = \{A, B, \dots\}$ and $L' = \{A', B', \dots\}$. Define

$$S^A = S^-,$$
$$S_A = S_-,$$
$$S^{A'} = S^+,$$
$$S_{A'} = S_+.$$

We label the tensor products of S^- and its dual by elements from the labelling set L and the tensor products of S^+ and its dual by elements from the 'primed' label set L'. The $\pm$ refers to the $\pm$ in self-dual and anti-self-dual two-forms on a four-dimensional space. Mnemonically, '+' means *with* primed indices, and '−' means *without*. Since these two labelings are independent, the ordering of primed and unprimed indices does not play a role in the calculations. It therefore turns out that any spinor field (section of the spinor algebra generated by S^A, S_A, $S^{A'}$, $S_{A'}$) is of the form

$$\varphi^{ABC\dots A'B'\dots}_{EFG\dots E'F'\dots}$$

or sums of such terms. In particular, if we let $[A \ldots D]$ denote skew-symmetrization of indices and $(A \ldots D)$ denote symmetrization of indices, then we see that $S^{[AB]} \cong \bigwedge^2 S^-$, and hence the skew form ε^- is of the form

$$\varepsilon^{AB} \in \Gamma(M, S^{[AB]}).$$

Generalizing the notation $\mathcal{E}^p$ of differential forms, we can define

$$\varepsilon^{AB} \quad \in \mathcal{E}^{[AB]}\ (M) \cong \Gamma^\infty(M, \bigwedge^2 S^-),$$

$$\varepsilon_{AB} \quad \in \mathcal{E}_{[AB]}\ (M) \cong \Gamma^\infty(M, \bigwedge^2 S_-),$$

$$\varepsilon^{A'B'} \in \mathcal{E}^{[A'B']}(M) \cong \Gamma^\infty(M, \bigwedge^2 S^+),$$

$$\varepsilon_{A'B'} \in \mathcal{E}_{[A'B']}(M) \cong \Gamma^\infty(M, \bigwedge^2 S_+).$$

We have introduced the two types of notation S^A, S^B, etc., and $S^\pm$, etc., since both play an important role in the literature, each is very useful in certain contexts, and it is important to be able to go back and forth. Note that the labels $+$, $-$ correspond to the symbols A', A, and we shall use these notations rather interchangeably in the remainder of the book, using whichever is convenient, depending on the context.

More concretely, if we have a trivialization over $U \subset M$ of these bundles, then the transition function matrices $g \in \mathrm{SL}(2, \mathbf{C})$ act on the vector representations in the following manner:

$$S^-|_U \cong \mathbf{C}^2 \ni v \mapsto g \cdot v \qquad \text{(regular representation)},$$

$$S_-|_U \cong \mathbf{C}^2 \ni v \mapsto {}^t v \cdot {}^t g \qquad \text{(dual representation)},$$

$$S^+|_U \cong \mathbf{C}^2 \ni v \mapsto \bar{g} \cdot v \qquad \text{(conjugate representation)},$$

$$S_+|_U \cong \mathbf{C}^2 \ni v \mapsto {}^t v \cdot {}^t \bar{g} \quad \text{(conjugate dual representation)}.$$

These are distinct representations of $\mathrm{SL}(2, \mathbf{C})$ acting on $\mathbf{C}^2$, and from a group-representation-theoretic point of view, the bundles $S^\pm$, $S_\pm$ are generated from S by changing the representation of $\mathrm{SL}(2, \mathbf{C})$ from the regular representation (left-multiplication by matrices) to the others indicated above.

In terms of numerical indices, if we choose a frame (e_0^A, e_1^A) for $S^A(\cong S^-)$ and let (e_A^0, e_A^1) be the dual frame, then we can express a section of S^A as

$$\sigma = \sigma^0 e_0^A + \sigma^1 e_1^A = \sigma^A e_A^A,$$

using the summation convention. So $\sigma^A, A = 0, 1$, is a numerical indexed quantity which represents σ with respect to the frame (e_0^A, e_1^A). We can choose a frame so that

$$\varepsilon^- = \varepsilon^{AB} \sim \varepsilon^{AB} = \begin{pmatrix} 0 & 1 \\ -1 & 0 \end{pmatrix}.$$

Using the dual frame we shall have

$$\varepsilon_- = \varepsilon_{AB} \sim \varepsilon_{AB} = \begin{pmatrix} 0 & 1 \\ -1 & 0 \end{pmatrix}.$$

We shall call any such frame a *spinor frame*, and shall usually make this choice. This is analogous to choosing an orthonormal frame in Riemannian geometry. The mapping

$$\varepsilon^- : S_- \to S^-$$

is given in terms of numerical indices by

$$\xi_A \mapsto \xi^A := \varepsilon^{AB} \xi_B,$$

i.e.,

$$\xi^0 = \varepsilon^{00} \xi_0 + \varepsilon^{01} \xi_1 = \xi_1,$$
$$\xi^1 = \varepsilon^{10} \xi_0 + \varepsilon^{11} \xi_1 = -\xi_0.$$

In terms of abstract indices, we would write

$$\mathcal{E}_A(M) \ni \xi_A \mapsto \xi^{AB} \xi_B \ni \mathcal{E}^A(M),$$

where the repeated index is contracted as described earlier in this section.

Similarly, the mapping

$$\varepsilon_+ : S^+ \to S_+$$

is given in terms of numerical indices by

$$\xi^A \mapsto \xi_B := \xi^A \varepsilon_{AB},$$

i.e.,

$$\xi_0 = \xi^0 \varepsilon_{00} + \xi^1 \varepsilon_{10} = -\xi^1,$$
$$\xi_1 = \xi^0 \varepsilon_{01} + \xi^1 \varepsilon_{11} = \xi^0.$$

We note that

$$\xi^A \xi_A = \xi^B \xi^A \varepsilon_{AB} = 0,$$

since ε_{AB} is a skew-symmetric form. Concretely, in terms of a spinor frame, we see explicitly that

$$\xi^A \xi_A = \xi^0 \xi_0 + \xi^1 \xi_1 = \xi_1 \xi_0 - \xi_0 \xi_1 = 0.$$

Let T be twistor space equipped with an Hermitian form Φ of signature $(2,2)$. The Lie group $\mathrm{SU}(\mathsf{T}, \Phi) \cong \mathrm{SU}(2,2)$ acts irreducibly on T, but if we consider $\mathrm{SL}(2,\mathbf{C})$ as a subgroup of $\mathrm{SU}(\mathsf{T}, \Phi)$, then $\mathrm{SL}(2,\mathbf{C})$ will act reducibly on T and T will split into a direct sum $\mathsf{T} = S_1 \oplus S_2$, $\dim S_j = 2$, and $\mathrm{SL}(2,\mathbf{C})$ acts irreducibly on S_1 and S_2. These will be spinor spaces, and coordinates for S_1 and S_2 will be spinor coordinates for T. They will not be invariant under the full action of $\mathrm{SU}(2,2)$ (conformal invariance), but for Lorentz-invariant questions, these will be very good coordinates. Explicitly, if we choose coordinates for T,

$$Z^\alpha = (\omega^A, \pi_{A'}), \qquad \omega^A = (\omega^0, \omega^1), \qquad \pi_{A'} = (\pi_{0'}, \pi_{1'}), \qquad (2.3.2)$$

where $\Phi(Z^\alpha) = \Phi(\omega^A, \pi_{A'}) = \omega^A \overline{\pi_{A'}} + \overline{\omega^A} \pi_{A'}$, then Φ will have the matrix form

$$\Phi = \begin{pmatrix} 0 & I_2 \\ I_2 & 0 \end{pmatrix},$$

and if $S^A = \mathrm{span}\{\omega^0, \omega^1\}$, $S_{A'} = \mathrm{span}\{\pi_{0'}, \pi_{1'}\}$, then

$$\mathsf{T} = S^A \oplus S_{A'}$$

is such an $\mathrm{SL}(2,\mathbf{C})$-invariant decomposition of twistor space with respect to the quadratic form Φ. Using these coordinates for T, we can rewrite the local coordinates for the double fibration

$$
\begin{array}{ccc}
 & \mathsf{F} & \\
\mu \swarrow & & \searrow \nu \\
\mathsf{P} & & \mathsf{M}
\end{array}
$$

as

homogeneous coordinates for P : $[\omega^A, \pi_{A'}]$

coordinates for $\mathsf{F}^I \cong \mathsf{M}^I \times \mathsf{P}_1$: $(z^{AA'}, [\pi_{A'}])$ $\qquad (2.3.3)$

coordinates for M^I : $z^{AA'} = \begin{bmatrix} z^{00'} & z^{01'} \\ z^{10'} & z^{11'} \end{bmatrix}$

$$= \frac{1}{\sqrt{2}} \begin{bmatrix} z^0 + z^3 & z^1 - iz^2 \\ z^1 + iz^2 & z^0 - z^3 \end{bmatrix}.$$

The mappings

$$\mathbb{F}^I$$

$$\mu \swarrow \qquad \searrow \nu$$

$$\mathbb{P}^I \qquad \qquad \mathbb{M}^I$$

are given by (see (1.2.11))

$$[iz^{AA'}\pi_{A'}, \pi_{A'}] \longleftarrow (z^{AA'}, [\pi_{A'}]) \longmapsto (z^{AA'}).$$

These are the coordinates we shall use in our calculations in later chapters.

Exercises for Section 2.3

2.3.1. Show that a manifold X of dimension $2n$ admits an almost-complex structure if and only if the structure group of $T(X)$ is reducible to $\mathrm{GL}(n, \mathbf{C})$.

2.3.2. Show that a 2×2 matrix v has $\det v = 0$ if and only if $v = a{\cdot}b$, where a is a 2×1 matrix, and b is a 1×2 matrix.

2.3.3. Show $K_{\mathbf{P}_n} \cong H^{-(n+1)}$.

2.3.4. Show that if $\varepsilon^+ : S_+ \to S^+$ is a nondegenerate skew form, then there exists a nondegenerate skew form $\varepsilon^- : S_- \to S^-$.

2.3.5. Let ξ^A be a spinor field. Show that $\xi^A \xi_A = 0$.

2.3.6. Show that $\mathrm{SL}(2, \mathbf{C})$ acts reducibly on $\mathbb{T}$ as a subgroup of $\mathrm{SU}(2, 2)$, and that $\mathbb{T}$ is the direct sum of two subspaces on each of which $\mathrm{SL}(2, \mathbf{C})$ acts irreducibly.

2.4 Connections and curvature

In this section we shall introduce and study some fundamental properties of the notion of a connection and its associated curvature. A connection is a mathematical object defined on a vector bundle which allows one to manipulate the vectors in the different fibers of the vector bundle in a consistent fashion. The specification of a connection is, in principle, arbitrary, just as one can specify arbitrary metrics on a differentiable manifold. There are also a variety of equivalent formulations of the notion of a connection, each being useful for a certain framework of questions and applications. We shall present a connection first as a generalization of exterior differentiation. Then we shall find local representations and develop the notion of curvature of a connection. Finally, we shall develop some of the alternative formulations which will be useful in the succeeding chapters.

Let $E \to M$ be a complex vector bundle on a differentiable manifold M. As we saw in §2.2 we can define $\mathcal{E}^p(M, E)$, the vector space of p-forms with coefficients in E. We recall that $\mathcal{E}^p(M, E)$ is a module over $C^\infty(M)$, and that there is a well-defined module isomorphism

$$\mathcal{E}(M, E) \otimes \mathcal{E}^p(M) \to \mathcal{E}^p(M, E),$$

which we denote by

$$s \otimes f \mapsto f \cdot s.$$

We define a *connection on E* to be a complex-linear mapping

$$D : \mathcal{E}^0(M, E) \to \mathcal{E}^1(M, E),$$

satisfying

$$D(fs) = df \cdot s + f \cdot Ds. \tag{2.4.1}$$

Let D be a connection on E (we shall see soon that such connections always exist). Let $f = (e_1, \ldots, e_r)$ be a frame for the vector bundle E defined over some open set U of M. If $s \in \mathcal{E}^p(U, E)$, then

$$s = s^\rho e_\rho,$$

where $s^\rho \in \mathcal{E}^p(U)$. We consider the vector $s^\rho = s^\rho(f)$ as a column vector. Similarly, if $F = (\omega^1, \ldots, \omega^n)$ is a frame for $T^*(M)$ over U, then an element s of $\mathcal{E}^p(U, E)$ can be written as

$$s = h_{a_1 \ldots a_p} \omega^{a_1} \wedge \cdots \wedge \omega^{a_p},$$

where $h_{a_1 \ldots a_p} \in \Gamma^\infty(U, E)$. We shall let Greek letters represent indices running from 1 to r, the rank of the bundle; and lower case Latin indices ranging from 1 to n correspond to a frame for the tangent bundle to M. Similarly, upper case Latin letters will denote spinor indices taking values in $\{0, 1\}$. As in the previous section these tensor or spinor indices can be either abstract or concrete. In this section we shall be dealing mainly with numerical indices, as we want to discuss some local matrix formulas. Now given the frame f we see that

$$D(e_\sigma) = A_\sigma^\rho e_\rho,$$

where $A_\sigma^\rho = A_\sigma^\rho(f, D)$ is an $r \times r$ matrix of one-forms which is defined on U. Here the superscript labels rows and the subscript labels columns. We denote this matrix by $A(f)$, suppressing the indices. Thus we find that if $s \in \mathcal{E}^0(M, E)$, then

$$
\begin{aligned}
Ds &= D(s^\rho e_\rho) \\
&= s^\rho De_\rho + ds^\rho \cdot e_\rho \\
&= s^\rho A_\rho^\tau e_\tau + ds^\rho \cdot e_\rho \\
&= (s^\rho A_\rho^\tau + ds^\tau)e_\tau, \\
&= (ds + A \cdot s)^\tau e_\tau,
\end{aligned}
$$

where $(ds + A \cdot s)^\tau$ denotes the components of the column vector $ds(f) + A(f) \cdot s(f)$. We shall call $ds(f) + A(f) \cdot s(f)$ the *local representation* of the connection D. We shall often suppress the dependence on the frame f by writing $Ds = ds + A \cdot s$, where the right-hand side is defined with respect to a frame f. In addition, we shall call $A = A(f) = A_\sigma^\rho(f)$ the *connection one-form* associated with the frame f and the connection D. If we make a change of frame $f \mapsto \tilde{f} = (\tilde{e}_1, \ldots, \tilde{e}_r)$, where $\tilde{f} = f \cdot g$, and $g : U \to \mathrm{GL}(r, \mathbf{C})$ is a smooth change of frame, i.e.,

$$
\tilde{e}_\rho = e_\sigma g_\rho^\sigma,
$$

then we see that

$$
\tilde{A} = g^{-1}Ag + g^{-1}dg. \tag{2.4.2}
$$

In terms of indices (2.4.2) becomes

$$
\tilde{A}_\sigma^\rho = (g^{-1})_\mu^\rho A_\tau^\mu g_\sigma^\tau + (g^{-1})_\tau^\rho dg_\sigma^\tau.
$$

To verify (2.4.2) we write

$$
\begin{aligned}
D(\tilde{e}_\sigma) &= \tilde{A}_\sigma^\rho \tilde{e}_\rho \\
&= \tilde{A}_\sigma^\rho e_\tau g_\rho^\tau \\
&= \tilde{A}_\sigma^\rho g_\rho^\tau e_\tau.
\end{aligned}
$$

But we also have

$$
\begin{aligned}
D(\tilde{e}_\sigma) &= D(e_\rho g_\sigma^\rho) \\
&= (A_\rho^\tau e_\tau)g_\sigma^\rho + e_\rho dg_\sigma^\rho \\
&= (A_\rho^\tau g_\sigma^\rho)e_\tau + dg_\sigma^\rho e_\rho \\
&= A_\rho^\tau g_\sigma^\rho e_\tau + dg_\sigma^\tau e_\tau.
\end{aligned}
$$

Thus

$$\tilde{A}_\sigma^\rho g_\rho^\tau = A_\rho^\tau g_\sigma^\rho + dg_\sigma^\tau,$$

which immediately yields (2.4.2).

How can we extend the connection to act on E-valued differential forms of higher degree? We proceed as in the case of exterior differentiation. There is a unique operator

$$\tilde{D} : \mathcal{E}^p(M, E) \to \mathcal{E}^{p+1}(M, E),$$

satisfying

(a) $\tilde{D}(fs) = df \wedge s + (-1)^{\deg f} f \wedge \tilde{D}(s),$

$\qquad\qquad f \in \mathcal{E}^p(M), \ s \in \mathcal{E}^q(M, E),$ (2.4.3)

(b) $\tilde{D}s = Ds, \quad \text{for } s \in \mathcal{E}^0(M, E).$

As in §2.2, we can find a local formula for $\tilde{D}$. Let us choose a local frame f for E, and let $s \in \mathcal{E}^p(M, E)$. Then, letting $s^\rho = s^\rho(f)$, we find that:

$$\begin{aligned}
\tilde{D}(s^\rho e_\rho) &= ds^\rho \cdot e_\rho + (-1)^p s^\rho \wedge D(e_\rho), \\
&= ds^\rho e_\rho + (-1)^p s^\sigma \wedge A_\sigma^\rho e_\rho, \\
&= (ds^\rho + (-1)^p s^\sigma \wedge A_\sigma^\rho) e_\rho, \\
&= (ds^\rho + A_\sigma^\rho \wedge s^\sigma) e_\rho, \\
&= (ds + A \wedge s)^\rho e_\rho.
\end{aligned}$$

Thus we obtain the local formula

$$\tilde{D}s(f) = ds(f) + A(f) \wedge s(f),$$

or in short

$$\tilde{D}s \sim ds + A \wedge s.$$

This is a well-defined local formula which agrees with the local formula for the case when $s \in \mathcal{E}^0(M, E)$. One can find global formulas, and these are discussed in the exercises, as well as the uniqueness of the extension of D to the full algebra of forms. We shall denote by D the connection differential operator acting on E-valued forms of any degree.

We now define the curvature $F = F_D$ of a connection D on the vector bundle $E \to M$. We consider the sequence of mappings

$$\mathcal{E}^0(M, E) \xrightarrow{D} \mathcal{E}^1(M, E) \xrightarrow{D} \mathcal{E}^2(M, E) \to, \qquad (2.4.4)$$

and we define the *curvature* of the connection D to be

$$F_D := D \circ D = D^2 : \mathcal{E}^0(M, E) \to \mathcal{E}^2(M, E).$$

Thus the curvature of a connection is the obstruction to the generalized de Rham sequence (2.4.4) being a complex (i.e., the composition of two maps being zero in the sequence of maps). If $F_D = 0$, then we say that D is a *flat connection* on M. Since $Ds = ds + A \wedge s$ locally, where $s \in \mathcal{E}^p(M, E)$, we see that,

$$\begin{aligned}
D^2 s &= (d + A\wedge)(d + A\wedge)s \\
&= d^2 s + A \wedge ds + d(A \wedge s) + A \wedge A \wedge s \\
&= A \wedge ds + dA \wedge s - A \wedge ds + A \wedge A \wedge s \\
&= (dA + A \wedge A) \wedge s.
\end{aligned}$$

Thus $Fs = D^2 s$ does not involve derivatives of s, and is a $C^\infty(M)$ homomorphism of $C^\infty(M)$-modules. That is, F is C-linear, and

$$F(fs) = fF(s),$$

for $f \in C^\infty(M)$, $s \in \mathcal{E}^p(M, E)$. Thus we see that F is a mapping

$$F : \mathcal{E}^0(M, E) \to \mathcal{E}^0(M, \textstyle\bigwedge^2 T^* \otimes E)$$

and is a homomorphism of $C^\infty(M)$-modules. It follows that

$$\begin{aligned}
F &\in \mathcal{E}^0(M, \mathrm{Hom}(E, \textstyle\bigwedge^2 T^* \otimes E)) \\
&\cong \mathcal{E}^2(M, \mathrm{Hom}(E, E)).
\end{aligned}$$

We see that any element

$$\Phi \in \mathcal{E}^p(M, \mathrm{Hom}(E, E))$$

can be represented with respect to a frame f for E in the following manner. Since $\mathrm{Hom}(E, E) \cong E \otimes E^*$, an induced frame for $\mathrm{Hom}(E, E)$ is given by $\{e^\alpha \otimes e_\beta^*\}$, and we find that

$$\begin{aligned}
\Phi s &= \Phi(s^\alpha e_\alpha) \\
&= s^\alpha \Phi(e_\alpha) \\
&= s^\alpha \Phi_\alpha^\beta(f) e_\beta \\
&= \Phi_\alpha^\beta(f) s^\alpha(f) e_\beta,
\end{aligned}$$

where $\Phi^\beta_\alpha s^\alpha$ is the matrix product of the matrix of Φ with the column vector s^α. Thus we see that

$$(\Phi s)(f) = \Phi(f)s(f),$$

i.e., the local representation of Φs with respect to the frame f is given by the matrix product $\Phi(f)s(f)$. When there is no confusion we can drop the dependency on f and use matrix algebra for local calculations. If E and F are two vector bundles over M, then we can form new bundles for E and F by direct sum, tensor products, etc., as described in §2.1. If E and F have connections D_E and D_F, then there are naturally induced connections on the new bundles formed from E and F.

For instance, this is easy to see for a direct sum, where we set

$$D_{E \oplus F}(s \oplus t) = D_E(s) \oplus D_F(t).$$

For a tensor product we have the formula

$$D_{E \otimes F}(s \otimes t) = (D_E(s) \otimes t + s \otimes D_F(t)).$$

Both of these are naturally induced connections, but they are not necessarily unique extensions which restrict to the original connections in a natural manner. To obtain unique extensions of connections we must normally demand additional conditions of the extended connection, as we shall see shortly.

We note that $\mathcal{E}^*(M, \mathrm{Hom}(E, E))$ is an associative algebra defined pointwise by the multiplication

$$(\omega \otimes L)(\omega' \otimes L') = \omega \wedge \omega' \otimes L \circ L',$$

for

$$\omega, \omega' \in \bigwedge{}^* T^*(M)_x, \quad \text{and} \quad L, L' \in \mathrm{Hom}(E, E)_x, \quad x \in M,$$

where we use the isomorphism

$$\mathcal{E}^*(M, \mathrm{Hom}(E, E)) \cong \Gamma^\infty(M, \bigwedge{}^* T^*(M) \otimes \mathrm{Hom}(E, E)).$$

We can make $\mathcal{E}^*(M, \mathrm{Hom}(E, E))$ into a Lie algebra by defining

$$[s, t] = st - (-1)^{(\deg s)(\deg t)} ts, \tag{2.4.5}$$

where st and ts are the products in $\mathcal{E}^*(M, \text{Hom}(E, E))$, as above. More generally, we can consider a vector bundle F whose fibers F_x are Lie algebras. For instance,

$$\text{Hom}(E, E)_x \cong \mathfrak{gl}(r, \mathbf{C}),$$

is an example of this. Then the Lie algebra structure of the fibers induces a Lie algebra structure on $\mathcal{E}^*(M, F)$ by setting

$$[\omega \otimes T, \omega' \otimes T'] = \omega \wedge \omega'[T, T'] \tag{2.4.6}$$

for $\omega \otimes T$, $\omega' \otimes T' \in \bigwedge^* T^*(M) \otimes F$. For the example $F = \text{Hom}(E, E)$, these two Lie algebra structures coincide.

If we have a connection D on a vector bundle E over M, then there is a unique induced connection $\widehat{D}$ on $\text{Hom}(E, E)$ with the properties:

$$\widehat{D}([s, t]) = [\widehat{D}s, t] + [s, \widehat{D}t], \tag{2.4.7}$$

$$D(s \wedge e) = \widehat{D}(s) \wedge e + (-1)^{\deg s} s \wedge De, \tag{2.4.8}$$

for $s \in \mathcal{E}^*(M, \text{Hom}(E, E))$, $e \in \mathcal{E}^*(M, E)$. We see that (2.4.7) says that $\widehat{D}$ is a derivation of the Lie algebra $\mathcal{E}^*(M, \text{Hom}(E, E))$, while (2.4.8) says that $\widehat{D}$ satisfies the usual product rule for differential forms extended to this setting, noting that $s \wedge e$ is a well-defined element of $\mathcal{E}^*(M, E)$. In fact, (2.4.8) gives a formula for $\widehat{D}(s) \otimes e$ for all $e \in \mathcal{E}^*(M, E)$, which is sufficient to uniquely determine $\widehat{D}(s)$. If we choose a frame f we find that the local representation for $\widehat{D}$ is given by

$$(\widehat{D}s) = ds(f) + [A(f), s(f)], \tag{2.4.9}$$

where $d + A(f)$ is the local representation for D acting on sections of E. Moreover, $\widehat{D}$ will satisfy (2.4.7), which follows easily from the local formula (2.4.9), for instance. Alternatively, one can define $\widehat{D}$ by the local formula (2.4.9) and see that it gives a well-defined connection on $\text{Hom}(E, E)$ and it will satisfy (2.4.7) and (2.4.8). We shall denote also by D this unique extension of the connection D on E to $\text{Hom}(E, E)$.

We have one fundamental identity due to Bianchi which shows that the curvature form F_D satisfies a special differential equation. Namely we have the *Bianchi identity*

$$DF_D = 0, \tag{2.4.10}$$

where D is any connection on a vector bundle E and F_D is its associated curvature. Since $F_D \in \mathcal{E}^2(M, \operatorname{Hom}(E, E))$, and

$$D : \mathcal{E}^2(M, \operatorname{Hom}(E, E)) \to \mathcal{E}^3(M, \operatorname{Hom}(E, E)),$$

(recalling the extension of D to $\operatorname{Hom}(E, E)$ discussed above), we see that the left-hand side of (2.4.10) is certainly well-defined. The proof is easy if we use the local forms of D and F_D. Namely, choosing a frame we see that

$$
\begin{aligned}
DF_D &= d(dA + A \wedge A) + [A, dA + A \wedge A] \\
&= dA \wedge A - A \wedge dA + A \wedge (dA + A \wedge A) \\
&\quad - (dA + A \wedge A) \wedge A \\
&= 0.
\end{aligned}
$$

Thus we see that (2.4.10) is elementary from this point of view, but it is important, as we shall see later.

Now let X be a vector field on M, and $E \to M$ a vector bundle with connection D. We define

$$D_X : \Gamma^\infty(M, E) \to \Gamma^\infty(M, E)$$

by setting

$$D_X(s) = X \lrcorner D(S),$$

the contraction of the differential form $D(s)$ with the vector field X. We call D_X *covariant differentiation* with respect to the connection D, along the vector field X. This operator is a generalization of the classical directional derivative to this setting. Note that the contraction operation $X\lrcorner$ is a well-defined mapping from E-valued p-forms to E-valued $(p-1)$-forms given by

$$\omega \otimes e \mapsto X \lrcorner \omega \otimes e,$$

where $\omega \in \mathcal{E}^p(M), e \in \Gamma^\infty(M, E)$.

The curvature $F = F_D$ is an $\operatorname{Hom}(E, E)$-valued two-form, i.e.,

$$F \in \mathcal{E}^2(M, \operatorname{Hom}(E, E)).$$

Since it is a two-form, F is a skew-symmetric bilinear form acting on the tangent bundle. Hence if X and Y are vector fields on M, we see that

$$F(X, Y) \in \mathcal{E}^0(M, \operatorname{Hom}(E, E)),$$

i.e.,

$$F(X, Y) : \Gamma^\infty(M, E) \to \Gamma^\infty(M, E).$$

We can now relate the curvature form to the more classical expression involving commutators of covariant differentiation operators. Namely, we have that

$$F(X, Y) = D_X D_Y - D_Y D_X - D_{[X,Y]}. \qquad (2.4.11)$$

We see that the right-hand side of (2.4.11) is *a priori* a differential operator of order two. The equality (2.4.11) shows that all the terms involving derivatives cancel, and one obtains a linear operator involving no derivatives, namely the curvature form F contracted with the vector fields X and Y. We leave the proof of (2.4.11) to the exercises, as we shall not have too much need for this version of the curvature form. We note that the corresponding torsion form is only defined in the special case where $E = T(M)$. It has the form

$$T(X, Y) := D_X Y - D_Y X - [X, Y], \qquad (2.4.12)$$

and we see that for a general vector bundle E which is not the tangent bundle, the terms in the definition of $T(X, Y)$ are not sections of the same bundle.

Let us consider a special case of a frame f defined on an open set U for which $A(f) \equiv 0$. Since $F_D = dA + A \wedge A$ on the domain of definition of the frame, we see that $F_D \equiv 0$ on U. We shall say that a section $s \in \Gamma^\infty(U, E)$ is *covariantly constant* on U if $Ds = 0$ on U. If $f = (e_1, \ldots, e_r)$ is a frame over U where each e_j is a covariantly constant section of E over U (what we shall call a *covariantly constant frame*), then we see that $A(f) = 0$, and hence $F_D = 0$ on this open set U. The converse is true.

Theorem 2.4.1. *Let $E \to M$ be a vector bundle with connection D which satisfies $F_D \equiv 0$. Then each point $x \in M$ has a neighborhood U and a covariantly constant frame f defined over U.*

To prove Theorem 2.4.1 we want to reformulate the notion of connection in terms of principal bundles, and then we can simply apply Frobenius' theorem. At the same time we shall formulate the notion of a connection for a G-bundle where G is any Lie group, and we shall prove Theorem 2.4.1 in this more general setting, as it involves no additional effort.

Let $P \to M$ be the principal bundle associated with a given vector bundle $E \to M$. If E has the structure of a G-bundle, then this means that the fibers of P are isomorphic to the structure group G. Moreover, we let $\mathfrak{g}$ be the Lie algebra of G, and we note that the tangent space to any of the fibers of P at some point of P is isomorphic to $\mathfrak{g}$. Recall that the principal bundle P has a right action by the Lie group G, $p \mapsto p \cdot g$, which we define to be $R_g(p)$. In the case of a $GL(r, \mathbf{C})$-vector bundle E, where a point in P is a frame for E, then this right action is simply

$$f = (e_1, \ldots, e_r) \mapsto (e_\alpha g_1^\alpha, \ldots, e_\alpha g_r^\alpha),$$

where $g_\beta^\alpha \in GL(r, \mathbf{C})$, i.e., right-matrix multiplication of the row vector of sections by g, yielding a new row vector of sections (a new frame).

Consider now on P a differential one-form ω with values in $\mathfrak{g}$; that is to say, ω is a mapping

$$\omega : T(P) \to \mathfrak{g}.$$

Since ω takes values in $\mathfrak{g}$, we can, for a fixed $g \in G$, consider the new differential form $\mathrm{ad}(g^{-1})\omega$ defined by

$$\mathrm{ad}(g^{-1})\omega(X) = g^{-1}\omega(X)g \in \mathfrak{g}, \quad \text{for } X \in T_p(P), \; p \in P.$$

Similarly, we can consider the pullback $R_g^*(\omega)$ of the form ω under the diffeomorphism R_g. Let f be a frame for E, defined over $U \subset M$, and let

$$\sigma_f : U \to P$$

be the section of P defined by f. Now if ω is a $\mathfrak{g}$-valued one-form on P, then we can define

$$A(f) := \sigma_f^*(\omega), \tag{2.4.13}$$

a $\mathfrak{g}$-valued one-form on U, which depends on the frame f.

Lemma 2.4.2. (a) *If ω satisfies*

$$R_g^*\omega = \mathrm{ad}(g^{-1})\omega, \quad \text{for all } g \in G, \tag{2.4.14}$$

then

$$A(f \cdot g) = \mathrm{ad}(g^{-1})A(f) + g^{-1}dg \tag{2.4.15}$$

for all smooth maps $g : U \rightarrow G$.

(b) *If $A(f)$ is a collection of smooth $\mathfrak{g}$-valued one-forms satisfying (2.4.15), then there is a $\mathfrak{g}$-valued one-form ω defined on P satisfying $\sigma_f^*(\omega) = A(f)$, and where ω satisfies (2.4.14).*

Remark: We see that the smooth maps $g; U \rightarrow G$ are simply changes of frame, and we recognize (2.4.15) as being the (necessary and sufficient) condition that $A(f)$ be a connection one-form associated with a connection D on E.

Proof: (a): Suppose that ω is given satisfying (2.4.14). Let g and f be given in G and P, and let g_t, f_t, and $u_t = f_t \cdot g_t$ be curves in G, P, and P, respectively, such that $g_0 = g$, $f_0 = f$, and $u_0 = f \cdot g$. If we differentiate with respect to the parameter t (which we denote by a prime), we obtain

$$u_t' = f_t' \cdot g_t + f_t \cdot g_t'$$

and we see that for $t = 0$,

$$u_0' = f_0' g_0 + f_0 g_0'$$
$$= f_0' g_0 + u_0 g_0^{-1} g_0'.$$

Interpreting these tangent vectors to P at the point u_0, we see that

$$\omega(u_0') = R_g^* \omega(f_0') + g_0^{-1} g_0' \in \mathfrak{g}, \qquad (2.4.16)$$
$$= \mathrm{ad}(g^{-1})\omega(f_0') + g^{-1} g_0'.$$

Considering any change of frame $g : U \rightarrow G$, and any tangent vector $X \in T_x(M)$, and any curve $\gamma_t \subset U$ such that $\gamma' = X$, we can use (2.4.16) to deduce:

$$A(fg) = g^{-1} A(f) g + g^{-1} dg,$$

as desired.

To see that (b) is true, one can construct ω directly from the data $A(f)$, using (2.4.13), and letting its values on vertical tangent vectors in P be zero. It then has to be checked that the ω so defined does not depend on the choice of frame f. Alternatively, by using the parallel transport defined by D one can define the space $H_p \subset T_p(P)$ of *horizontal vectors* with respect to the connection D. These are tangent vectors $\tilde{X}$ to P with the property that $\tilde{X} = d\sigma_f(X)$, $X \in T_x(M)$, and $A(f)(X) = 0$. This defines H_p, and we see that

$$T_p(P) = V_p(P) \otimes H_p(p),$$

the vertical (tangent to the fibers) and horizontal vectors at the point p. The choice of H_p depends on the choice of D and equivalently on the choice of $A(f)$ satisfying (2.4.15). We recall that the tangent space $V_p(P)$ can be identified (via invariant vector fields on P_x) with the Lie algebra $\mathfrak{g}$. For $X \in T_p(P)$ define $\omega(X)$ to be the projection along H_p onto V_p. This is then a linear mapping $T_p(P) \to \mathfrak{g}$, and is hence a well-defined $\mathfrak{g}$-valued one-form. One checks that ω is the desired one-form on P which defines the family of one-forms $A(f)$ on M, and this concludes a sketch of the proof. Further details on this and related matters are found in Kobayashi and Nomizu (1963, 1969), and Sternberg (1965). ∎

Given a connection on E the associated connection form ω on P defines a curvature form Ω defined by

$$\Omega := d\omega + \tfrac{1}{2}[\omega, \omega]. \qquad (2.4.17)$$

If we pull back ω to M by a frame f we see that, for $\mathfrak{g} = \mathfrak{gl}(r, \mathbf{C})$,

$$dA(f) + \tfrac{1}{2}[A(f), A(f)] = dA(f) + \tfrac{1}{2}(A(f) \wedge A(f) + A(f) \wedge A(f))$$
$$= dA(f) + A(f) \wedge A(f),$$

which is the local form of the curvature form $F = D^2$ in this special case. In general, for a $\mathfrak{g}$-valued form we define the curvature by (2.4.17) and with respect to a frame f we write

$$F_D(f) := A(f) + \tfrac{1}{2}[A(f), A(f)], \qquad (2.4.18)$$

where $[\,,\,]$ is the Lie bracket in $\mathfrak{g}$. For the special case of $\mathfrak{g} = \mathfrak{gl}(n, \mathbf{C})$ it does not matter which version we use, but more generally it is important to see the Lie algebra structure explicitly and to let the differential form structure be implicit. As before, the Lie algebra product is extended to Lie algebra-valued forms as in (2.4.6).

Frobenius' theorem asserts that if θ^α, $\alpha = 1, \ldots, k$, is a family of one-forms on a differentiable manifold M, and if

$$d\theta^\alpha = \sum f_{\alpha\beta}\theta^\alpha \wedge \theta^\beta,$$

i.e., $d\theta^\alpha$ is in the ideal generated by $\{\theta^\alpha\}$ in the algebra of forms on M, then there are coordinates

$$(x^1, \ldots, x^k, x^{k+1}, \ldots, x^n)$$

locally on M so that

$$\theta^\alpha = dx^\alpha, \quad \alpha = 1, \ldots, k.$$

See Sternberg (1965) or most standard references in differential geometry where this is proven. It can also be formulated in terms of vector fields and their commutators, but the description above is most convenient for our purposes. We formulate the following important theorem.

Theorem 2.4.3. *Let $E \to M$ be a G-vector bundle with connection D, and let $F = d\omega + \frac{1}{2}[\omega, \omega]$ be its curvature form on P. If $F \equiv 0$, then for each point $x \in M$, there is an open set U and a frame $f : U \to P$ such that $\omega|_{f(U)} \equiv 0$, i.e.,*

$$f^*\omega = A(f) \equiv 0.$$

Remark: The section f will be a covariantly constant frame for the connection D, and this will prove Theorem 2.4.1 in this somewhat more general setting.

Proof: Since the curvature is zero we see that

$$d\omega = -\frac{1}{2}[\omega, \omega].$$

If we let T_α be a basis for $\mathfrak{g}, \alpha = 1, \ldots, \dim \mathfrak{g} = k$, then $\omega = \omega^\alpha T_\alpha$, and

$$d\omega^\alpha = k^\alpha_{\beta\gamma} \omega^\beta \wedge \omega^\gamma,$$

for suitable $k^\alpha_{\beta\gamma} \in \mathbf{C}^\infty(P)$. It follows that $\{\omega^\alpha\}$ satisfies the integrability conditions of Frobenius, and hence there are local coordinates $(x^1, \ldots, x^n, y^1, \ldots, y^k)$ for P so that $\omega^\alpha = dy^\alpha$. We note that $(x^1, \ldots, x^n)$ give coordinates for M, and that the projection $\pi : P \to M$ is of the form (in these coordinates)

$$(x^1, \ldots, x^n, y^1, \ldots, y^k) \mapsto (x^1, \ldots, x^n),$$

and the horizontal spaces H_p are tangent to the n-planes

$$y^\alpha = \text{constant}, \quad \alpha = 1, \ldots, k.$$

The frames $f_y(x) = (x, y), y$ constant, are all covariantly constant frames, and the theorem is proven. ∎

We have seen that there are a number of equivalent concepts of a connection and we want to summarize them here briefly. Let $E \to M$ be a G-vector bundle over M with an associated principal bundle. The following are equivalent notions of a connection on E:

(1) An operator $D : \Gamma^\infty(M, E) \to \Gamma^\infty(M, E \otimes T^*(M))$ such that

$$D(fs) = df \cdot s + fD(s).$$

(2) A family of g-valued connection one-forms $A(f)$ defined for frames f, where

$$A(fg) = g^{-1}A(f)g + g^{-1}dg,$$

for a change of frame $g : U \to G$.

(3) A g-valued one-form ω defined on P satisfying $R_g^*\omega = g^{-1}\omega g$.

(4) A smooth choice of horizontal subspaces which are complementary to the vertical subspaces $V_p(P)$ tangent to the fibers of $P \to M$.

(5) A choice of parallel transport, i.e., an isomorphism

$$\varphi_\gamma : T_p(E) \to T_q(E)$$

which depends (smoothly) on a curve γ joining p to q in M.

Associated with the connection D is the curvature F_D. In (1) it is of the form D^2, for (2) it has the form

$$F_D(f) = dA(f) + A(f) \wedge A(f),$$

while for (3) it has the form $F = d\omega + \frac{1}{2}[\omega, \omega]$. For (4) the curvature is the obstruction to integrability of the distribution of horizontal subspaces H_p, and for (5) it is a measure of the deviation from the identity for φ_γ acting on $T_p(E)$ for γ a closed path on M.

All of these notions are interchangeable, and we have only indicated proofs of some of the equivalences which we are discussing here. See Kobayashi and Nomizu (1963, 1969), and Sternberg (1965), or most contemporary differential geometry books, for further details on these issues. We shall normally refer to a connection as a differential operator, but we shall use the alternate forms when it is convenient.

An *Hermitian metric* $h = \langle \, , \, \rangle_E$ on E is defined fiberwise as an Hermitian inner product on the complex vector space E_x for each

$x \in M$, which varies smoothly with x. That is to say, the function $\varphi(x) = \langle s(x), t(x) \rangle_{E_x}$ is a smooth function on M if s and t are smooth sections of E over M. A connection D is *compatible* with an Hermitian metric if

$$d\langle s, t \rangle = \langle Ds, t \rangle + \langle s, Dt \rangle, \qquad (2.4.19)$$

for all $s, t \in \Gamma^\infty(M, E)$. Here we extend the inner product from sections of E to E-valued differential forms by defining

$$\langle \omega \otimes s, \omega' \otimes s' \rangle = \omega \wedge \omega' \langle s, s' \rangle,$$

for ω, $\omega' \in \mathcal{E}^*(M)$, s, $s' \in \Gamma^\infty(M, E)$ (and then extending by complex-linearity). Thus the extended inner product is a differential-form valued 'inner product', and (2.4.19) expresses the compatibility of the exterior derivative d, the connection D, and the inner product $\langle \, , \, \rangle$.

If we have a real vector bundle $E \to M$ then a *Euclidean metric* on E is a symmetric bilinear positive-definite inner product on each fiber E_x of E which varies smoothly as x varies. In §2.3 we discussed Riemannian and pseudo-Riemannian manifolds. A Riemannian manifold is an example of a Euclidean metric on the tangent bundle .

If a pseudo-Riemannian manifold (M, g) is given, is there a connection D compatible with g? The answer is: yes, there are many such connections. On the other hand, for a given pseudo-Riemannian metric g on a manifold M, there is a unique connection ∇ which is compatible with g (2.4.19) and which has vanishing torsion tensor (2.4.12). This connection is called the *Levi–Cevita connection* of the metric g, and plays an important role in the geometry of Riemannian and pseudo-Riemannian manifolds. The existence and uniqueness of this connection is discussed in more detail in the exercises. We shall use the Levi–Cevita connection to define the *curvature* of a pseudo-Riemannian manifold, although this was originally introduced by Riemann directly in terms of the metric itself. We shall use the notation ∇ to refer to the covariant differentiation on a pseudo-Riemannian manifold with respect to the Levi–Cevita connection, and let D be a generic connection on any vector bundle E. We shall be considering both types of connection together in the theory of Yang–Mills fields on curved space times.

Exercises for Section 2.4

2.4.1. Show that there is a unique extension of the connection D to E-valued forms of higher degree satisfying (2.4.3).

2.4.2. Show that any vector bundle $E \to X$ admits a connection. (Hint: find connections locally, and patch together with a partition of unity.)

2.4.3. Show that if E and F are differentiable vector bundles on a manifold M, and if $\varphi : \Gamma(M, E) \to \Gamma(M, F)$ is a C^∞-module homomorphism, then there exists a homomorphism of vector bundles $L : E \to F$ such that φ is induced by L on the global sections. Give an example of a complex-linear mapping $\varphi : \Gamma(M, E) \to \Gamma(M, E)$ of vector spaces which is *not* induced by such a bundle mapping L.

2.4.4. Prove that a connection D on a vector bundle E induces uniquely a connection $\hat{D}$ on $\text{Hom}(E, E)$ which satisfies (2.4.7) and (2.4.8).

2.4.5. Show that the curvature F_D has the representation given in (2.4.11) in terms of commutators.

2.4.6. Show that there exists a unique connection ∇ on a Riemannian manifold (M, g) such that ∇ is compatible with the metric g (2.4.19) and has vanishing torsion (2.4.12) (this is the Levi–Cevita connection).

2.4.7. An *Hermitian manifold* X is a complex manifold whose tangent bundle is equipped with an Hermitian metric. Show that if h is the local matrix expression for the Hermitian metric on X with respect to a holomorphic frame, then $\theta := h^{-1}\partial h$ is a local connection one-form for a connection D on $T(X)$. Show that $F = \bar{\partial}(h^{-1}\partial h)$ is a local expression for the curvature of this connection (this connection is called the *canonical connection* of the Hermitian manifold X; cf. Wells 1980).

2.4.8. A *Kähler manifold* is an Hermitian manifold whose metric form

$$ds^2 = g_{a\bar{b}} dz^a \otimes d\overline{z^b}$$

has the associated two-form

$$\Omega = \frac{i}{2} g_{a\bar{b}} dz^a \wedge d\overline{z^b},$$

where $d\Omega = 0$. Show that any complex submanifold of a Kähler manifold is Kähler.

2.4.9. Let $[z_0, \ldots, z_n]$ be homogeneous coordinates for $\mathbf{P}_n$, and consider the homogeneous differential form

$$\Omega = \frac{i}{2|z|^4} \left\{ |z|^2 \sum_{a=0}^{n} dz_a \wedge d\bar{z}_a - \sum_{a,b=0}^{n} \bar{z}_a z_b dz_a \wedge d\bar{z}_b \right\}$$

where $|z|^2 = |z_0|^2 + \cdots + |z_n|^2$.

(a) Show that Ω is closed, and induces a closed two-form Ω on $\mathbf{P}_n$.

(b) Show that Ω has a positive-definite coefficient matrix in local coordinates, and defines an Hermitian metric on $\mathbf{P}_n$ (this is the Fubini–Study metric on $\mathbf{P}_n$, and this shows that $\mathbf{P}_n$ is Kähler, and from Exercise 2.4.8 we conclude that any complex submanifold of $\mathbf{P}_n$ is Kähler; cf. Weil 1958, Griffiths and Harris 1978, and Wells 1980, for a further discussion of Kähler manifolds).

3

The algebraic topology of
manifolds and bundles

An important ingredient in modern geometry has been the development of topological methods for the study of analytic problems. The solutions of the various types of field equations which we shall encounter in Part III will necessitate a familiarity with various tools which have been developed in the past several decades. The study of manifolds was initiated by Riemann in his study of Riemann surfaces. This was continued in the late nineteenth century by the research of Betti who developed the first topological invariants which transcended the original Euler characteristic of a polyhedral surface S ($\chi(S) = V - E + F$, where V is the number of vertices, E is the number of edges, and F is the number of faces). After this, the fundamental work of Poincaré at the turn of the century gave the impetus to make topology a full-fledged mathematical discipline, with the study of manifolds being one of the central problems. The notion of topological invariants of topological spaces has played a critical role in these developments. These invariants are numbers or algebraic objects which one associates with geometric objects, and which help to classify them. We shall see in §3.1 a survey of the basic elements of homology and homotopy invariants which will be important for our field-theory investigations. In §3.2 we discuss sheaf theory, a method of encoding local information on a global scale which turns out to be very useful. Associated with sheaf theory is a sheaf cohomology theory which helps us to understand the relation between the solution of local and global problems in a number of contexts. These give calculational methods for a number of topological invariants, as well as invariants of more refined geometric structures, such as complex manifolds. The theory of characteristic classes is described in §3.4. This involves an assignment of topological invariants to bundles in a systematic manner, and the associated topological invariants help determine how far a given bundle is from being trivial, for instance, just as topological invariants of spaces help determine how far a given space is from being a specific standard model (e.g., Euclidean space or a sphere). In §3.5 we study a special type of geometric structure on a manifold, that of a spin structure, which plays an important role in physics. The global existence of such spin structures depends very much on the topological nature of the manifold in question, as we see in this section. Finally, in §3.6 we give an overview of spectral sequences, a mathematical tool developed by Leray in conjunction with his study of hyperbolic differential equations, but which soon developed in the 1950s into a major tool of algebraic topology. It is a calculational device, which allows one to compute certain cohomological invariants in terms of others, under various kinds of favorable conditions. It will be useful in certain aspects of the integral-geometric transforms which come up in the later chapters.

3.1 Homotopy and homology

Algebraic topology describes topological spaces and more specifically manifolds in terms of *topological invariants*. A topological invariant is an algebraic object (number, group, vector space, etc.), which is associated with a topological space and which is preserved under homeomorphisms. So, for instance, the *area* of a plane figure would not be a topological invariant, but the *number* of holes of a plane figure would be invariant under homeomorphisms. A classical example and fundamental prototype of a topological invariant is the *genus* of a two-dimensional compact surface. The genus is the number of 'holes' in the surface (see Figure 3.1.1).

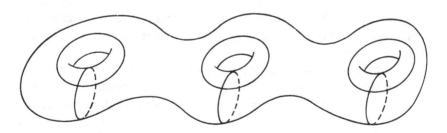

Fig. 3.1.1. Surface M with genus$(M) = 3$, $b_1(M) = 6$.

The word homotopy refers to a continuous deformation from one topological object to another. The word homology refers to a theory in which objects which are homologous are considered equivalent. In the geometric context, two closed curves in a space are homologous if together they form the common boundary of a two-dimensional region (think of an annulus) in the same space. Homotopy groups consist of homotopically equivalent classes of closed curves and higher-dimensional objects, and homology groups consist of homo-logically equivalent classes of closed curves and higher-dimensional objects.

In the late nineteenth century Betti introduced the *Betti numbers* of a manifold of arbitrary dimension, and this was formalized by Poincaré some years later. The Betti numbers measure the 'number of holes' of varying dimensions. They are the number of independent generators of the homology groups we shall study. For instance, for a planar annulus, there will only be one nontrivial generator of the homology group. This class of examples in the plane is studied in detail in the theory of integration of complex-analytic functions (see for instance the classical text of Ahlfors, 1953). In the example shown in Figure 3.1.1 one finds that the first Betti number $b_1(M) = 6$, while genus$(M) = 3$. The six loops sketched in the diagram represent closed paths of nonhomologous curves on the surface M. For oriented surfaces the first Betti number is always twice the genus. For any compact manifold M of dimension n, there are Betti numbers (nonnegative integers)

$$b_0(M), b_1(M), \ldots, b_n(M),$$

which measure the amount of *twisting* or internal holes the surface has (they are well-defined for noncompact manifolds also, but might be infinite). If two surfaces M and M' are homeomorphic, then they have the same Betti numbers (i.e., the Betti numbers are topological invariants). The converse is also true, and this is a fundamental achievement of algebraic topology which was established in the nineteenth century.

A fundamental problem of topology is to find topological invariants which classify the manifold up to homeomorphism type (e.g., by the result mentioned above, any two surfaces with the same number of handles, or which is more precise, the same Betti numbers, are homeomorphic to each other). In geometry the topological invariants are useful measures of the global structures of the manifold, just as the more classical notions of volume and arc length are global descriptions of geometric spaces when these notions make sense.

One defines the Betti numbers as the ranks of certain abelian groups constructed from the manifold which are called *homology* or *cohomology groups*. An example of such numbers is given by the dimensions of the de Rham cohomology groups introduced in §2.2. These dimensions will turn out to agree with the Betti numbers, which have a more geometric definition. So one replaces the association

$$\text{manifold} \;\rightarrow\; \text{finite set of Betti numbers}$$

by

$$\text{manifold} \;\rightarrow\; \text{finite set of abelian groups.}$$

One writes this association as

$$M \rightarrow H_q(M, \mathbf{Z}),$$

where the objects on the right are the *homology groups of M of degree q*, for $q = 0, 1, \ldots, n = \dim M$. One then finds that one can define the Betti numbers to be the ranks of these abelian groups, i.e., the maximal number of independent generators. Namely, if $H_q(M, \mathbf{Z})$ are the homology groups of a manifold M (which we have not yet defined!), then one defines $b_q = b_q(M)$ to be the rank of $H_q(M, \mathbf{Z})$, that is, the number of generators of this abelian group. Before we define the homology groups, we shall give some examples to show the nature of the information which is involved. Here we note that the ring of integers $\mathbf{Z}$ is considered as an infinite cyclic group with one independent generator, which can be chosen to be either 1 or -1.

Example 3.1.1 (Homology groups):

Manifold	*Homology Groups*	*Betti numbers*	
S^1	$H_0(S^1, \mathbf{Z}) \cong \mathbf{Z}$	$b_0 = 1$	
	$H_1(S^1, \mathbf{Z}) \cong \mathbf{Z}$	$b_1 = 1$	
	$H_q(S^1, \mathbf{Z}) = 0$	$b_q = 0,$	$q > 1$
$S^1 \times S^1$	$H_0(S^1 \times S^1, \mathbf{Z}) \cong \mathbf{Z}$	$b_0 = 1$	
	$H_1(S^1 \times S^1, \mathbf{Z}) \cong \mathbf{Z} \times \mathbf{Z}$	$b_1 = 2$	
	$H_2(S^1 \times S^1, \mathbf{Z}) \cong \mathbf{Z}$	$b_2 = 1$	
	$H_q(S^1 \times S^1, \mathbf{Z}) = 0$	$b_q = 0,$	$q > 2$
S^2	$H_0(S^2, \mathbf{Z}) \cong \mathbf{Z}$	$b_0 = 1$	
	$H_1(S^2, \mathbf{Z}) = 0$	$b_1 = 0$	
	$H_2(S^2, \mathbf{Z}) \cong \mathbf{Z}$	$b_2 = 1$	
	$H_q(S^2, \mathbf{Z}) = 0$	$b_q = 0,$	$q > 2$

S^n $H_0(S^n, \mathbf{Z}) \cong \mathbf{Z}$ $b_0 = 1$

                $H_q(S^n, \mathbf{Z}) = 0$ $b_q = 0, \quad q \neq 0, n$

                $H_n(S^n, \mathbf{Z}) \cong \mathbf{Z}$ $b_n = 1$

$\mathbf{P}_n = \mathbf{P}_n(\mathbf{C})$ $H_q(\mathbf{P}_n, \mathbf{Z}) \cong \mathbf{Z}$ $b_q = 1, \quad q = 2k, k \leq n$

                $H_q(\mathbf{P}_n, \mathbf{Z}) = 0$ $b_q = 0, \quad q = 2k + 1,$

                                                 $k \leq n$

                $H_q(\mathbf{P}_n, \mathbf{Z}) = 0$ $b_q = 0, \quad q > 2n$

Intuitively, we think of this table in the following manner: For all of the above $b_0 = 1$, and this means the spaces S^1, $S^1 \times S^1$, etc., are *connected*. When $b_1 = 1$, there is one generator, a copy of the circle. When $b_1 = 2$, then each factor $S^1 \times \{\text{pt}\}$, and $\{\text{pt}\} \times S^1$ is an independent generator of the homology groups in degree one. When each such loop bounds a two-disk in the space (or can be shrunk to a point) as in S_n, $n > 1$, then $b_1 = 0$. Similarly, S^n is a generator of homology in degree n, and there is no other individual homology.

Let us define the (singular) homology of a space. We refer to Greenberg and Harper (1981) for details of the construction. If we let

$$\Delta^k = \{(x^1, \ldots, x^k) \in \mathbf{R}^k : x^i \geq 0, \ i = 1, \ldots, k, \ x^1 + \cdots + x^k \leq 1\},$$

then this is the *standard k-simplex*. Then we let $\{p_0, \ldots, p_k\}$ be the vertices

$$p_0 = (0, 0, \ldots, 0),$$
$$p_1 = (1, 0, \ldots, 0),$$
$$\vdots$$
$$p_k = (0, \ldots, 0, 1)$$

and we see that Δ^k is the *convex span* of the points $\{p_0, \ldots, p_k\}$. Thus, Δ^0 is a point, Δ^1 is an interval, Δ^2 is a triangle, Δ^3 is a three-dimensional solid tetrahedron, etc.. These are the basic 'building blocks' of combinatorial topology described quite explicitly with respect to this coordinate system. We shall envision a space as being decomposed in continuous images of these fundamental objects. For instance, if we have a two-sphere as in Figure 3.1.2, then S^2 is the union of eight triangular regions $R_1, \ldots, R_8$, where each region R_i has a boundary consisting of great circles and has three vertices. This is an example of a *triangulation* of S^2, i.e., a decomposition

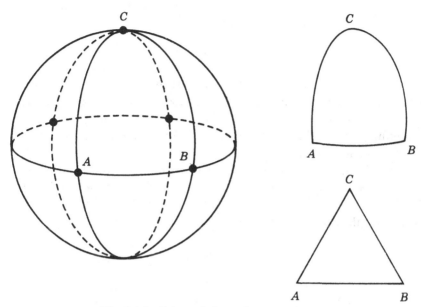

Fig. 3.1.2. Triangulation of a two-sphere.

of S^2 in terms of suitable triangular regions. This is also an example of what we mean by building blocks. Algebraic topology has many techniques for calculating topological invariants in terms of a given decomposition such as this triangulation, but where the result is independent of the specific decomposition used for the calculation. Triangulations using simplices (homeomorphic images of the standard simplices) were the first ones used, starting with Poincaré's, and later other decompositions came into use. In this section we shall encounter a cell decomposition, and in §3.3 we shall see decompositions in terms of open coverings (Čech coverings) used in calculating sheaf cohomology, an extremely useful generalization of the geometric theory we develop in this section.

We shall first discuss homology groups of a given topological space, and then we shall indicate how they can be calculated for cases such as those cited in Example 3.1.1. Cohomology groups of a space will then be defined in terms of duality, and these are very much related to the de Rham cohomology discussed earlier. Finally we shall briefly discuss homotopy groups and some examples, as they will play a role in some of the Yang–Mills theory, to be developed in Part II.

Let M be a topological space; then we shall define a *singular k-simplex* to be a continuous mapping

$$\sigma : \Delta^k \to M.$$

Intuitively, we visualize a mapping such as that given from the triangle to the sphere in Figure 3.1.2. This is a generalization of a path, which is a singular one-simplex. The word singular means that the mapping may be allowed to degenerate (for instance, a triangle may collapse to an interval).

Consider now the set of all singular k-simplices, and suppose we are given a commutative ring R, for instance $R = \mathbf{Z}, \mathbf{Q}, \mathbf{R}$, or $\mathbf{C}$ will be the common cases to be considered. We define $S_k(M, R)$ to be the set of all formal finite sums of the form

$$\sigma = \sum_\mu a^\mu \sigma_\mu,$$

where a^μ is an element of the ring R, and σ_μ is a singular k-simplex. We add such finite formal sums by adding coefficients of the corresponding simplices. A sum is defined to be zero if its coefficients are zero. This makes $S_k(M, R)$ into a module over the ring R, as the module multiplication is defined also in terms of the coefficients, and it is called the module of *singular chains* on M of dimension k. This is a very large infinite-dimensional space, and we want to introduce an equivalence relation on it in order to obtain finite quantities. We want to define the boundary operator

$$\partial : S_k(M, R) \to S_{k-1}(M, R).$$

If we have a simple two-simplex (e.g., Figure 3.1.2 again),

$$\sigma : \Delta^2 \to M,$$

then $\partial\sigma$ will be a singular one-chain which will be the formal sum of the singular one-simplices which are the images of the three sides of the triangle. We need to orient these sides of the triangles so that they close up and form a loop which is the boundary of the original singular two-simplex. The generalization of 'side of a triangle' which we shall use is that of a *face* of a k-simplex. Define

$$F_k^i : \Delta^{k-1} \to \Delta^k$$

by

$$F_k^i := \text{ convex span of } \{p_0, \ldots, \hat{p}_i, \ldots, p_k\},$$

where $\hat{p}_i$ means we omit the point p_i and where $\{p_0, \ldots, p_{k-1}\} \rightarrow$ $\{p_0, \ldots, \hat{p}_i, \ldots, p_k\}$ and their convex combinations are mapped in the natural ordered manner. This is called the ith *face* of the standard k-simplex. For instance if we identify F_k^i with its image in Δ^k for purposes of visualization, then in Figure 3.1.3 the faces of Δ^2 are the three sides of the triangle opposite the points omitted. The higher-dimensional situation is analogous. If σ is a k-simplex, then the ith *face of σ* is defined by

$$\sigma^{(i)} := \sigma \circ F_k^i.$$

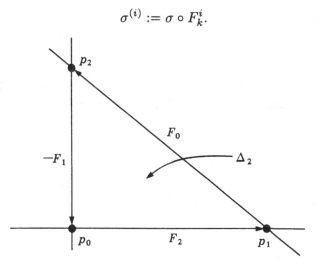

Fig. 3.1.3. The boundary mapping ∂ for a simple simplex.

Thus each k-simplex σ has associated with it $(k+1)$ faces. We now can define the boundary of a k-simplex σ to be the formal sum

$$\partial\sigma = \sum_{i=0}^{k}(-1)^i\sigma^{(i)}.$$

Thus the formal sum only makes sense as a singular chain, which is why this notion was introduced. The boundary one-simplex is an ordered difference of points. This ordering gives an orientation to the one-simplex, and the formal sum in the boundary of a two-simplex consists of three oriented one-simplices which are attached to one another in a loop, as would be desired. In particular, if we try to take the boundary of the boundary of a two-simplex, this looping means that there are no boundary points, and the three segments form a closed path (see Figure 3.1.3). The formal sum, which we have called the boundary, will agree with the classical notion of oriented boundary which occurs in Green's and Gauss' theorems, for instance. We can then extend ∂ to any singular chain by R-linearity, i.e., if

$$\sigma = \sum a^\mu \sigma_\mu,$$

then

$$\partial \sigma := \sum a^\mu \partial \sigma_\mu.$$

It is a fundamental, but not too difficult calculation to verify that

$$\partial^2 := \partial \circ \partial = 0, \tag{3.1.1}$$

which we shall omit. The basic idea is quite clearly illustrated for ∂^2 acting on two-simplices in Figure 3.1.3. We see that

$$\partial F_0 = p_2 - p_1,$$
$$\partial F_1 = p_2 - p_0,$$
$$\partial F_2 = p_1 - p_0,$$

and also

$$\partial \Delta_2 = F_0 - F_1 + F_2,$$

by definition, so

$$\partial^2 \Delta_2 = \partial(F_0 - F_1 + F_2)$$
$$= (p_2 - p_1) - (p_2 - p_0) + (p_1 - p_0)$$
$$= 0.$$

Using (3.1.1) we then have the complex

$$\rightarrow S_q(M, R) \xrightarrow{\partial} S_{q-1}(M, R) \xrightarrow{\partial} S_{q-2}(M, R) \rightarrow .$$

The kernel divided by the image will be our desired homology groups. More specifically, a chain σ is called a *cycle* if $\partial \sigma = 0$, and it is called a *boundary* if it is of the form $\sigma = \partial(\sigma')$ for some σ'. The set of all cycles is denoted by $Z_q(M, R)$, and the set of all boundaries is denoted by $B_q(M, R)$. These are both submodules of $S_q(M, R)$, and in view of (3.1.1) we also have $B_q(M,R) \subset Z_q(M,R)$. Thus we can define the quotient module

$$H_q(M,R) := Z_q(M,R)/B_q(M,R).$$

These are called, for variable q, the *singular homology groups of M with coefficients in R*. We shall simply say homology groups for short. The term 'group' is used much more commonly than 'module', although the module structure is always understood from the coefficient ring.

If we consider the case of $R = \mathbf{R}$, then we can define the *Betti numbers* of M by the formula

$$b_q(M) := \dim H_q(M,\mathbf{R}), \quad q = 0, 1, 2, \ldots.$$

This is the same as the rank of $H_q(M,\mathbf{Z})$, in fact (Exercise 3.1.2),

$$H_q(M,\mathbf{R}) = H_q(M,\mathbf{Z}) \otimes_{\mathbf{Z}} \mathbf{R}.$$

We have now defined all of the quantities appearing in Example 3.1.1. The isomorphisms there are as $\mathbf{Z}$-modules, that is, as abelian groups. The infinite cyclic group $\mathbf{Z}$ has one generator. The assertion that a given homology group, say $H_1(S^1, \mathbf{Z})$, is isomorphic to $\mathbf{Z}$ means that there is a distinguished cycle (in this case the cycle $\sigma(t) = e^{2\pi i t}$, for $0 \leq t \leq 1$), which generates the module in an infinite cyclic fashion. This is a nontrivial assertion, as a little consideration will show. There are many formal properties of these combinatorically-defined homology groups, which allows one to calculate them in special cases, such as the one just mentioned. We can only discuss briefly some of this methodology here and refer to the standard texts in algebraic topology for a systematic development, which will allow one to make such calculations readily. We shall give some indication of why one has the results in Example 3.1.1. In §3.3 we shall learn more about calculating cohomology in the context of sheaf theory, which will give us tools which apply in the cases under consideration here.

In order to calculate homology groups of a space, one needs one more fundamental topological notion, that of homotopy. This is the formalization of the intuitive notion of continuous deformation. Two curves in a planar region are homotopic if one of them can be continuously deformed into the other. If this is always possible, then the planar domain is simply-connected, as we recall from classical complex analysis. Integrals of holomorphic functions over homotopic curves with the same end points have the same value—the classical Cauchy theorem. In another vein, the common example of a torus and a coffee cup being topologically equivalent would be expressed by continuously deforming one into the other via a homotopy. Let us look at homotopy in a more formal manner to make things a little more precise. We shall first look at homotopic maps and use this to talk about homotopic spaces. Let I denote the unit interval $[0, 1]$, and let

$$f_1 : M \to N$$
$$f_2 : M \to N$$

be two continuous mappings of topological spaces. We shall say that f_1 is *homotopic* to f_2 if there is a continuous mapping

$$F : M \times I \to N,$$

such that

$$F(x, 0) = f_1(x),$$
$$F(x, 1) = f_2(x).$$

The mappings

$$f_t(x) := F(x, t)$$

form a one-parameter family of mappings which interpolate continuously between f_1 and f_2. Thus we can speak of paths in N as being homotopic, for instance, by taking the special case of $M = I$.

Let M and N be two topological spaces, then M is *homotopic* to N if there are mappings

$$f : M \to N,$$
$$g : N \to M,$$

so that $f \circ g$ and $g \circ f$ are both homotopic to the identity mapping. We say that f is a homotopy equivalence and write $f : M \overset{\simeq}{\to} N$. If f is a homeomorphism, then taking $g = f^{-1}$, we see that homeomorphic spaces are homotopic, but, as we shall see shortly, homotopic spaces need not be homeomorphic. Suppose that $f : M \to N$ is a continuous mapping, then there is a naturally induced sequence of homomorphisms of the homology groups

$$f_q : H_q(M, R) \to H_q(N, R), \quad q = 0, 1, 2, \ldots. \qquad (3.1.2)$$

Namely, if $\sigma \in S_q(M, R)$, then we define $f_q(\sigma)$ to be $f \circ \sigma$. It will follow that f_q maps cycles to cycles, boundaries to boundaries, and hence induces the required mapping by taking homology classes to classes which defines (3.1.2).

A fundamental result which relates the concept of homotopy and homology is given by the following theorem.

Theorem 3.1.2. *If $f : M \xrightarrow{\simeq} N$ is a homotopy equivalence, then the induced mappings on homology are isomorphisms of R-modules*

$$f_q : H_q(M, R) \xrightarrow{\cong} H_q(N, R), \quad q = 0, 1, 2, \ldots.$$

In particular, homeomorphic spaces have the same (up to isomorphism) homology groups. Thus the homology groups are topological invariants in the sense stated at the beginning of this section. This fundamental result is proven in detail in Greenberg and Harper (1981), Spanier (1966), and other texts on algebraic topology.

The theorem above allows one to calculate homology groups by looking at homotopically equivalent but simpler spaces. For instance, $\mathbf{R}^n$ is homotopic to a single point, and it follows readily that $b_r(\mathbf{R}^n) = 0, r = 1, 2, \ldots$. Note that $\mathbf{R}^n$ is definitely not homeomorphic to a point, and this example illustrates some of the power of the notion of homotopy.

A second type of formal algebraic topological object we shall need to use are the *cohomology groups* of a space. These are defined in terms of the dual space of the module of singular chains. *Singular cochains* on M are R-homomorphisms

$$c : S_q(M, R) \to R.$$

The set $S^q(M, R)$ of singular cochains forms an R-module in a natural manner. The *coboundary* operator

$$\delta : S^q(M, R) \to S^{q+1}(M, R)$$

is defined as the transpose of the operator ∂ with respect to the duality pairing

$$\langle c, \sigma \rangle := c(\sigma), \quad c \in S^q(M, R), \ \sigma \in S_q(M, R).$$

In other words δ is defined by the relation

$$\langle \delta c, \sigma \rangle = \langle c, \partial \sigma \rangle.$$

Let us give an example of a cochain. If ω is a differential form of degree q on a differentiable manifold M, then we can define, for $\sigma \in S_q(M, R)$,

$$c(\sigma) = \int_\sigma \omega = \sum a^\mu \int_{\sigma_\mu} \omega,$$

where $\sigma = \sum a^\mu \sigma_\mu$, σ_μ are singular q-simplices, and

$$\int_{\sigma_\mu} \omega := \int_{\Delta^q} \sigma_\mu^*(\omega).$$

This last definition is only well-defined if σ_μ is a smooth singular q-simplex. To define $c(\sigma)$ for all singular q-chains σ, we would need to approximate continuous chains by differentiable chains and then take a suitable limit. We shall not worry about this nontrivial and subtle point (see, for instance, Godement (1964) for a more thorough discussion). We shall simply remark that integration of a specific differential form over chains is the fundamental example of a cochain. The coboundary operation for such differential forms considered as a cochain is simply exterior differentiation.

We define a *cocycle* to be a cochain c such that $\delta c = 0$, and a *coboundary* c is a cochain of the form $c = \delta c'$, for some c'. It follows readily that $\delta^2 = \delta \circ \delta = 0$, as before, and hence we can define the quotient space

$$H^q(M, R) := Z^q(M, R)/B^q(M, R),$$

letting $Z^q(M, R)$ be the cocycles and $B^q(M, R)$ be the coboundaries. We call $H^q(M, R)$ the *singular cohomology group of M with coefficients in R*. We shall normally simply say the 'cohomology groups' (with a specific coefficient group). We shall refer to $H^q(M, \mathbf{Z})$ as the *integral cohomology groups*, $H^q(M, \mathbf{R})$ as the *real cohomology groups*, and $H^q(M, \mathbf{C})$ as the *complex cohomology groups*, respectively.

If M is a compact manifold, then $H_q(M, \mathbf{R})$ is a finite-dimensional vector space and

$$H^q(M, \mathbf{R}) \cong [H_q(M, \mathbf{R})]',$$

the real-linear dual. Thus the Betti numbers can be calculated using either homology groups or cohomology groups. We now formulate the following important theorem due to de Rham, the proof of which will be discussed in more detail after we introduce sheaf cohomology in §3.3.

Theorem 3.1.3 (de Rham). *Let M be a differentiable manifold, then*

$$H^q(M, \mathbf{C}) \cong H^q_{\text{DR}}(M),$$

where the de Rham cohomology groups (2.2.6) are defined in terms of complex-valued differential forms.

Remark: The same result holds if we change the coefficients of the singular cohomology to $\mathbf{R}$ and use differential forms with real coefficients in the definition of the de Rham group.

This result shows that the calculation of cohomology groups, which reflects only the geometry of the manifold, can be reduced to the study of global differential equations on the manifold. In particular, it formalizes the example of a cochain being given by integration of a differential form ω over singular chains. All cocycles have representatives which are of this form and, moreover, this representative will be a closed differential form. Exterior multiplication induces a product structure on the de Rham cohomology and therefore on the singular cohomology. This product structure, which can also be intrinsically defined for the singular theory, is an important invariant of the topological structure and is useful for calculations.

There are fundamental tools for calculating cohomology (excision, relative cohomology, long exact sequences, etc.), and we shall mention only one here explicitly. Additional tools will appear when we discuss sheaf cohomology, as there are methods which are valid at that level which are quite applicable in the current setting.

Suppose we know the cohomology groups of two spaces M and N; can we calculate the cohomology of the product space $M \times N$? If we consider simple complex coefficients, then we have the following formula due to Künneth:

$$H^q(M \times N, \mathbf{C}) \cong \sum_{r+s=q} H^r(M, \mathbf{C}) \otimes_{\mathbf{C}} H^s(N, \mathbf{C}), \qquad (3.1.3)$$

where the tensor product is the tensor product of complex vector spaces. This is similar in spirit to the product formula (Leibniz' formula) for calculating higher derivatives of products of functions:

$$D^q(fg) = \sum_{r+s=q} \binom{q}{r} (D^r f)(D^s g).$$

We shall not prove the Künneth formula here, but refer to Green-berg and Harper (1981), Godement (1964), or other algebraic topology texts for a proof of this standard result. Just as in the case of the product rule for derivatives of functions, we shall not have to go back to the definition of the derivative in order to be able to use the formula. Applications are discussed in the exercises. Thus, if a space is a product, then its homology is determined by the homology of its factors. If we consider integral homology or cohomology, then the Künneth formula is more delicate, and we shall not write it down here. The complication is due to the possible presence of torsion elements in the **Z**-modules. For instance, as is discussed below, $H_1(\mathbf{P}_2(\mathbf{R}), \mathbf{Z}) \cong \mathbf{Z}_2$, and hence there are nilpotent elements of order two in this integral homology group. These nilpotent elements do not contribute to the Betti numbers $(b_1(\mathbf{P}_2(\mathbf{R})) = 0)$, but are important additional topological invariants in their own right.

Suppose one does not have a product, what can one do then? There are two approaches. One is to generalize the Künneth formula, as we shall see in our discussion of the Leray spectral sequence in §3.6, and a second one is to look for specific decompositions of the space from which one can read off the homology groups. This latter approach will now be discussed.

A *cell complex* is a topological space which is created by successive patchings of cells of higher and higher dimension. By starting with a single point and by adding a cell of one dimension (an interval), we obtain a circle. Similarly, a two-sphere is the attachment of a two-cell to a point (recall the addition of a single point ∞ to the complex plane to obtain the Riemann sphere). An n-sphere is by analogy the attachment of an n-cell to a point. A different type of example is provided by projective space. If we let M be real projective n-space, $\mathbf{P}_n(\mathbf{R})$, then letting H_{n-1} be a projective hyperplane in M and letting $c_n = M - H_{n-1}$, we see that c_n is a cell of dimension n (it is a standard coordinate chart). Now H_{n-1} is a copy of $\mathbf{P}_{n-1}(\mathbf{R})$, and we can consider a hyperplane section H_{n-2} in H_{n-1}, and the difference is a cell c_{n-1} of dimension $n-1$. Continuing this process, we see that we have a sequence of attachments of cells of the form:

$$\mathbf{P}_n(\mathbf{R}) = \{\text{pt}\} \cup c_1 \cup c_2 \cup \cdots \cup c_n.$$

This is a more complicated version than the examples of spheres. The attaching mappings are also important and are implicit in this construction. For complex projective space there is a similar decomposition, but since the coordinate charts are each even-dimensional (since they are complex coordinates) the cells are all even-dimensional, and one has a pattern like this:

$$\mathbf{P}_n(\mathbf{C}) = \{\text{pt}\} \cup c_2 \cup c_4 \cup \cdots \cup c_{2n}.$$

There is an inductive procedure for calculating homology groups for cell decompositions of the above sort, and this is outlined nicely in Greenberg and Harper (1981). The induction is much simpler if there is a skip in dimensions as in the case of complex projective space, and one obtains very nice results, the homology being supported in the same dimensions where the cells appear, namely, every even dimension. Real projective space is much more complicated due to the nonskipping of dimensions, and one has as an example that

$$H_q(\mathbf{P}_2(\mathbf{R}), \mathbf{Z}) \cong \begin{cases} \mathbf{Z}, & q = 0 \\ \mathbf{Z}_2, & q = 1 \\ 0, & q \geq 2 \end{cases}.$$

Here $\mathbf{Z}_2 = \mathbf{Z}/2\mathbf{Z}$, the cyclic group of order two. Thus we have torsion in the homology which is not detectable by Betti numbers and, moreover there is no homology in the top dimension (in this case, two), and this is a reflection of the fact that this manifold is not orientable. In fact, a connected manifold M of dimension n is orientable if and only if $H_n(M, \mathbf{Z}) \cong \mathbf{Z}$, and a choice of generator for this homology group is equivalent to a choice of orientation for the manifold. The fundamental point is that one can read off the homology groups from a very specific geometric decomposition of the space involved. This is possible for two-dimensional surfaces, for the projective spaces, Grassmannian manifolds and flag manifolds studied in Chapter 1, and for the classical Lie groups and their homogeneous spaces, and we shall use this kind of detailed information about these manifolds as we need it, with suitable references. The decompositions in most of these cases are quite analogous to that given for projective space above.

The final topic we need to discuss concerns homotopy groups of a space. These are generally much more difficult to compute than homology or cohomology groups. Just computing the homotopy groups of the standard n-sphere has been a difficult task for algebraic topologists over the past several decades, although a lot is now known about this. Let us first define the homotopy groups for a path-wise connected space M. Fix a point on M and consider the set of all paths

$$\gamma : [0, 1] \to M$$

such that $\gamma(0) = \gamma(1) = p$, i.e., closed paths passing through the point p. Each such path is a *loop* at p, and we denote by $\Omega_{M,p}$ the space of all such loops at p. Composition of two such loops will be again a loop of the same type (after reparametrizing). We say that two loops are *homotopically equivalent* if the maps defining them are homotopic to one another, and we let $\pi_1(M, p)$ be the group of homotopy equivalence classes of loops at p. The group operation is induced by the composition of loops mentioned above, and this group is known as the *fundamental group* of M or as the *first homotopy group* of M (the point p being understood). The point p is referred to as the *base point*, and the fundamental groups for different base points are conjugate to one another. The fundamental group of a space is, in general, a nonabelian group. A space is *simply-connected* if its fundamental group is trivial. There is a subtle relation between $\pi_1(M, p)$ and $H_1(M, \mathbf{Z})$, which we shall not develop here. Both of these groups have representatives which are closed loops on M, and a simply-connected space will have $H_1(M, \mathbf{Z}) = 0$, but the converse is false.

The higher homotopy groups are defined in an inductive manner by using the loop space at p. The loop space can also be made into a nice topological space by introducing a suitable topology on it (the natural topology is the compact-open topology; see Greenberg and Harper 1981), which allows us to define

$$\pi_2(M, p) := \pi_1(\Omega_{M,p}, C),$$

where C is the constant loop. Then we define inductively

$$\pi_n(M, p) := \pi_{n-1}(\Omega_{M,p}, C).$$

The higher homotopy groups turn out to be commutative, and we shall have occasion to refer to one of these groups in connection with the theory of magnetic monopoles. We shall summarize some of the results from Steenrod (1951) which illustrates the complexity of the situation.

Example 3.1.4:

$$\pi_i(S^n) = \begin{cases} 0, & i < n \\ \mathbf{Z}, & i = n \end{cases}$$
$$\pi_i(S^1) = 0, \quad i > 1,$$
$$\pi_3(S^2) \cong \mathbf{Z},$$
$$\pi_i(S^4) \cong \pi_{i-1}(S^3), \quad 2 \le i < 7.$$

This example indicates that homotopy groups are nontrivial in degree above the dimension of the space, which is not the case for the homology or cohomology groups. These examples and others like it depend heavily on the principal bundle theory of orthonormal frame bundles developed in Steenrod's book, and homotopy-theoretic calculations using these bundle constructions. We shall use this example as we need it.

Exercises for Section 3.1

3.1.1. Show that $\partial^2 = 0$ in (3.1.1).

3.1.2. Show that $H_q(M, \mathbf{R}) \cong H_q(M, \mathbf{Z}) \otimes_{\mathbf{Z}} \mathbf{R}$.

3.1.3. Show that if $f : M \to N$ is a continuous mapping, then there is an induced sequence of homomorphisms

$$f_q : H_q(M, R) \to H_q(N, R), \quad q = 1, 2, \dots.$$

Show that if f is a homeomorphism, then each f_q will be an isomorphism.

3.1.4. Calculate the Betti numbers of $S^3 \times S^1$ using the Künneth formula (3.1.3).

3.1.5. Show that if a p-cochain is represented by a differential form ω, then $\delta\omega$ is represented by the exterior derivative $d\omega$. (Hint: use Stokes' theorem.)

3.1.6. Show that $S^1 \times S^1$ and S^2 are not homotopic.

3.2 Sheaf theory

Sheaf theory was introduced in the late 1940s by Jean Leray in his study of hyperbolic differential equations. Sheaf theory and its associated cohomology theory, which we shall discuss in the next section, had little impact on partial differential equations at the time. However, in the early 1950s J. P. Serre applied the ideas to algebraic topology and, soon thereafter, along with H. Cartan, A. Grothendieck, and others, it became an indispensable tool in the disciplines of several complex variables and algebraic geometry. It became an important adjunct to the algebraic topology which had developed in the previous three decades. The major difference is that it could relate to a larger class of geometric problems, not only to the topological ones. The fundamental issue is the interaction of local versus global solutions of certain specific geometric problems. A sheaf incorporates local geometric information, and the cohomology theory allows one to codify the passage from local to global. Let us give one example of this process. Suppose that in the complex plane we are given a discrete set of points $\{p_j\}_{j=1}^{\infty}$, and for each point p_j an integer m_j. Can we find a meromorphic function $f(z)$ which has zeros (or poles) at p_j with multiplicity m_j? Locally, in any relatively compact set U, there are only a finite number of such points, and the function

$$f_U(z) = \prod_{p_j \in U} (z - p_j)^{m_j}$$

is a rational function with the prescribed zeros and poles in U. Globally, however, this process might not converge. The answer to the question above, nevertheless, is yes; and it is a classical theorem of Mittag-Leffler that there is a global meromorphic function with the prescribed zeros and poles. This problem can be recast in terms of sheaf theory and as such admits a generalization and a solution in much more general situations than the complex plane. This is the flavor of 'local versus global solutions' in our discussion above.

In this section we shall introduce sheaves and their fundamental properties along with a number of examples. For more details and information, the reader is referred to Griffiths and Harris (1978), Hartshorne (1977), Hirzebruch (1966), and Wells (1980), which all relate to complex and algebraic geometry, as well as the treatises Bredon (1967) and Godement (1964). We shall attempt to describe the aspects of the theory which we need and which are useful for later calculations.

Let M be a topological space. A *presheaf* $\mathcal{S}$ of abelian groups on M is an assignment

$$U \to \mathcal{S}(U)$$

of an open set U to an abelian group $\mathcal{S}(U)$, with the property that for any two open sets $V \subset U \subset M$, there are *restriction homomorphisms*

$$r_V^U : \mathcal{S}(U) \to \mathcal{S}(V)$$

with the property that for open sets $W \subset V \subset U$,

$$r_V^U \circ r_W^V = r_W^U, \qquad r_U^U = id.$$

We could also talk about a presheaf of vector spaces over $\mathbf{C}$, a presheaf of rings, a presheaf of algebras, etc. If $\mathcal{R}$ is a presheaf of rings (with identity), for instance, then we can talk about a presheaf $\mathcal{M}$ of modules over $\mathcal{R}$, i.e., for each open set $U \subset M, \mathcal{M}(U)$ is a module over the ring $\mathcal{R}(U)$. Thus we see that a presheaf is a family of algebraic objects of the same type, parametrized by the *open sets* of M. Recall that a bundle $B \to M$ is a family of geometric objects of a certain type parametrized by the *points* of the space M.

A *sheaf* will be a presheaf $\mathcal{S}$ satisfying two additional properties:

If $\{U_\alpha\}$ is a family of open sets in M, where $U = \bigcup U_\alpha$, and if $s_\alpha \in \mathcal{S}(U_\alpha)$ and

$$r_{U_\alpha \cap U_\beta}^{U_\alpha}(s_\alpha) = r_{U_\alpha \cap U_\beta}^{U_\beta}(s_\beta),$$

for all α, β, then there exists an $s \in \mathcal{S}(U)$ such that

$$r_{U_\alpha}^U(s) = s_\alpha, \quad \text{for all } \alpha. \tag{3.2.1)(S_1)}$$

If $\{U_\alpha\}$ is a family of open sets in M, with $U = \bigcup U_\alpha$, and if s and $s' \in \mathcal{S}(U)$ satisfy

$$r_{U_\alpha}^U(s) = r_{U_\alpha}^U(s'), \tag{3.2.2)(S_2)}$$

for all α, then $s = s'$.

We refer to the elements of $\mathcal{S}(U)$ as *sections* of the sheaf $\mathcal{S}$ over the open set U. The examples below will illustrate this terminology, as the principal examples consist of sections of vector bundles. Moreover, we shall see that the sections of a sheaf as we have defined it can be considered as 'sections' of an appropriate abstract fibration which will be introduced shortly as a technical tool, and this will justify the terminology.

Axiom S_1 (3.2.1) above asserts that local sections can be pieced together to give global sections, while Axiom S_2 (3.2.2) means that any section is uniquely determined by its local behavior.

Let us give some examples. Let M be a manifold, then define the presheaves

$$\mathcal{E} = \{\mathcal{E}(U) := C^\infty \text{ functions on } U\}, \tag{3.2.3}$$
$$\mathcal{E}^r = \{\mathcal{E}^r(U) := \text{smooth } r\text{-forms on } U\}, \tag{3.2.4}$$
$$\mathcal{V} = \{\mathcal{V}(U) := \text{smooth vector fields on } U\}, \tag{3.2.5}$$
$$\mathcal{E}(E) = \{\mathcal{E}(U, E) := \text{smooth sections of a} \tag{3.2.6}$$
$$\text{vector bundle } E \to M\}.$$

In each case the restriction mappings are the usual restriction mappings, and they all satisfy Axioms S_1 and S_2, and hence are sheaves. Examples of presheaves which do not satisfy the sheaf axioms are given in the exercises.

On a complex manifold M one has similar sheaves. Let $E \to M$ be a holomorphic vector bundle. Then we have the examples:

$$\mathcal{O} = \{\mathcal{O}(U) := \text{holomorphic functions on } U\}, \tag{3.2.7}$$

$$\Omega^p = \{\Omega^p(U) := \text{holomorphic } p\text{-forms on } U\}, \tag{3.2.8}$$

$$\mathcal{E}^{p,q} = \{\mathcal{E}^{p,q}(U) := \text{smooth } (p,q)\text{-forms on } U\}, \tag{3.2.9}$$

$$\mathcal{O}(E) = \{\mathcal{O}(E)(U) := \text{holomorphic sections of}$$
$$E \text{ over } U\}, \tag{3.2.10}$$
$$\Omega^p(E) = \{\Omega^p(U, E) := E\text{-valued holomorphic}$$
$$p\text{-forms on } U\}. \tag{3.2.11}$$

We call $\mathcal{O}_M := \mathcal{O}$ on M the *structure sheaf* of the complex manifold M, and similarly $\mathcal{E}_M$, the sheaf of smooth functions on a differentiable manifold M, is called the structure sheaf of M.

We see that both $\mathcal{O}$ and $\mathcal{E}$ are sheaves of rings, and the examples in (3.2.3)–(3.2.6) are sheaves of modules over $\mathcal{E}$, while the examples (3.2.7)–(3.2.11) are sheaves of modules over $\mathcal{O}$. A final family of examples is given by the *constant sheaves*. Let R be a ring, and X a topological space. Let $R(U)$ be the ring of locally constant continuous functions on U, then this determines a presheaf, and this presheaf is a sheaf (Exercise 3.2.7) which is called the *constant sheaf* R on X. For instance we can let $R = \mathbf{Z}, \mathbf{R}, \mathbf{C}$, etc. These are the most common examples of constant sheaves. We shall not distinguish notationally between the constant sheaf and its corresponding algebraic object. Note that a section $s \in \mathbf{C}(U)$ is the assignment of a complex number s to a component of the open set U, which we can view as a locally constant continuous complex-valued function on U. Thus there is a natural inclusion

$$\mathbf{C}(U) \subset \mathcal{E}(U)$$

on a smooth manifold. This inclusion will be denoted by $\mathbf{C} \to \mathcal{E}$ and is an example of a sheaf morphism.

A *morphism* of sheaves $f : \mathcal{A} \to \mathcal{B}$ is a family of maps

$$f_U : \mathcal{A}(U) \to \mathcal{B}(U),$$

where f_U is an algebraic homomorphism of the algebraic structures on $\mathcal{A}(U)$ and $\mathcal{B}(U)$. The maps should be *compatible* with the restriction mappings in the natural manner, i.e.,

$$
\begin{array}{ccc}
\mathcal{A}(U) & \xrightarrow{\ f_U\ } & \mathcal{B}(U) \\
\downarrow & & \downarrow \\
\mathcal{A}(V) & \xrightarrow{\ f_V\ } & \mathcal{B}(V)
\end{array}
$$

should be commutative, where the vertical mappings are the restriction mappings. For instance, the de Rham complex on a differentiable manifold M of dimension n,

$$0 \to \mathbf{C} \to \mathcal{E}^0 \xrightarrow{d} \mathcal{E}^1 \xrightarrow{d} \cdots \to \mathcal{E}^n \to 0, \qquad (3.2.12)$$

is well-defined at the sheaf level, where the sheaf morphism d is defined by $d_U : \mathcal{E}^p(U) \to \mathcal{E}^{p+1}(U)$ as the usual exterior derivative, and the sheaf morphism $\mathbf{C} \to \mathcal{E}^0$ is just the natural inclusion of constant functions in the smooth functions discussed above.

Let S be a presheaf on M. We want to define the *stalk* of S at any point $x \in M$. The stalk is similar to the fiber of a vector bundle and is the localization of the information in S near the point x. Let

$$\widetilde{S}_x = \bigcup_{U \ni x} S(U) \quad \text{(disjoint union)},$$

and we shall say that two elements in $\widetilde{S}_x$,

$$f \in S(U), \quad U \ni x, \quad \text{and} \quad g \in S(V), \quad V \ni x,$$

are *equivalent* if there is some open set $W \subset (U \cap V)$ such that $x \in W$ and

$$r_W^U(f) = r_W^V(g),$$

i.e., f and g *agree* after restriction to some common domain of definition. We define S_x to be the set of equivalence classes in $\widetilde{S}_x$ with respect to this equivalence relation. We call S_x the *stalk* of S at the point x. We see that S_x inherits the algebraic structure of the presheaf S, i.e., we can add elements in S_x by adding representatives of the equivalence classes. We shall let

$$r_x^U : S(U) \to S_x$$

be the natural restriction mapping to the stalks (cf. Exercise 3.2.2).

We say that a sequence of homomorphisms of sheaves

$$A \xrightarrow{f} B \xrightarrow{g} C$$

is *exact* at B if the induced sequence

$$A_x \xrightarrow{f_x} B_x \xrightarrow{g_x} C_x$$

is exact, i.e., $\ker g_x = \operatorname{im} f_x$, for all $x \in M$. A *short exact sequence* of sheaves is a sequence of the form

$$0 \to A \to B \to C \to 0,$$

which is exact at A, B, and C. Let us give some examples. It follows from the Poincaré Lemma (Lemma 2.2.1) that the de Rham complex of sheaves (3.2.12) is exact at each $\mathcal{E}^p$. Exactness at C is simply the fact that a local smooth function f satisfying $df = 0$ is locally constant. Similarly, on a complex manifold M with a holomorphic vector bundle E, we have the Dolbeault complex

$$0 \to \Omega^p \to \mathcal{E}^{p,0} \overset{\bar{\partial}}{\to} \mathcal{E}^{p,1} \overset{\bar{\partial}}{\to} \cdots \overset{\bar{\partial}}{\to} \mathcal{E}^{p,n} \to 0,$$

and the Dolbeault complex with vector bundle coefficients

$$0 \to \Omega^p(E) \to \mathcal{E}^{p,0}(E) \overset{\bar{\partial}}{\to} \mathcal{E}^{p,1}(E) \overset{\bar{\partial}}{\to} \cdots \overset{\bar{\partial}}{\to} \mathcal{E}^{p,n}(E) \to 0,$$

where $\mathcal{E}^{p,q}(E)$ is the sheaf of smooth (p,q)-forms with coefficients in the vector bundle E. The exactness of these sequences follows from the $\bar{\partial}$-variant of the Poincaré lemma in Lemma 2.2.3. We also have the exact sequence of holomorphic p-forms, noting that $d = \partial$, acting on holomorphic forms,

$$0 \to \mathbf{C} \to \Omega^0 \overset{d}{\to} \Omega^1 \overset{d}{\to} \cdots \to \Omega^n \to 0.$$

Variants on these sequences relative to the twistor manifolds of Chapter 1 will play a role in our development of the Penrose transform. This will be discussed in later chapters.

We need to discuss one technical point which is fundamental to sheaf theory, and that is the process of generating a sheaf from a presheaf $\mathcal{S}$ which might itself not be a sheaf, i.e., it violates either S_1 or S_2. Let

$$\hat{\mathcal{S}} = \bigcup_{x \in M} \mathcal{S}_x$$

be the disjoint union of the stalks. We define $\tilde{\mathcal{S}}(U)$ to be the set of maps of the form $s : U \to \hat{\mathcal{S}}$ with the two properties:

$$s(x) \in \mathcal{S}_x; \qquad (3.2.13)$$

if $s(x)$ has a representative $s \in \mathcal{S}(W)$, $W \ni x$, then for $y \in W$,

$$s(y) = r_y^W(s). \qquad (3.2.14)$$

The presheaf $\tilde{\mathcal{S}}$ is a subpresheaf of the presheaf of all functions

$$\mathcal{S}'(U) = \{s : U \to \hat{\mathcal{S}}\},$$

which is equipped with the natural restriction mappings (restriction as functions). It is clear that $\mathcal{S}'$ satisfies the Axioms S_1 and S_2. The restrictions (3.2.13) and (3.2.14) do not affect this, since (3.2.13) is not affected by the axioms, and (3.2.14) is a local 'continuity property'.

The sheaf $\widetilde{S}$ constructed from the presheaf S is called the *sheaf generated* by S. If S is a sheaf, then one can show that $S \cong \widetilde{S}$, and we can identify the original sheaf with the sheaf $\widetilde{S}$ which is a sheaf of functions of a specific type, and this is a useful point of view. Most sheaves which arise in geometry are functions of some sort or other, i.e., sections of bundles, differential forms, etc.

We shall give an example to clarify this point. Consider the presheaf $\mathcal{B}$ on the complex plane $\mathbb{C}$ defined by

$$\mathcal{B}(U) = \{\text{bounded holomorphic functions on } U\}.$$

As is claimed in the exercises (Exercise 3.2.5), this is not a sheaf, essentially because of the nonlocal nature of the boundedness property. The sheaf generated by $\mathcal{B}$ can be identified with $\mathcal{O}_{\mathbb{C}}$, the sheaf of holomorphic functions on $\mathbb{C}$. This is easy to see since any $f \in \mathcal{O}(U)$ is locally bounded, and hence $\mathcal{B}_x \cong \mathcal{O}_x$, from which it follows that $\widetilde{\mathcal{B}} \cong \mathcal{O}$ (Exercise 3.2.9). The principal application of this construction is to define the quotient sheaf of two sheaves. Let $\mathcal{A}$ be a subsheaf of $\mathcal{B}$ on a topological space M, i.e., $\mathcal{A}(U)$ is a abelian subgroup of $\mathcal{B}(U)$ for all open sets $U \subset M$. Define the presheaf $\mathcal{C}'$ by

$$U \to \mathcal{B}(U)/\mathcal{A}(U) = \mathcal{C}'(U)$$

of abelian groups. The presheaf $\mathcal{C}'$ is not necessarily a sheaf, and we define $\mathcal{C} := \widetilde{\mathcal{C}'}$, the sheaf generated by $\mathcal{C}'$, to be the quotient sheaf of $\mathcal{B}$ by the subsheaf $\mathcal{A}$. We write notationally $\mathcal{C} = \mathcal{B}/\mathcal{A}$, and note that $(\mathcal{B}/\mathcal{A})(U)$ contains $\mathcal{B}(U)/\mathcal{A}(U)$, but in principle is a *larger* abelian group (see Exercise 3.2.10). The fundamental fact is that the sequence

$$0 \to \mathcal{A} \to \mathcal{B} \to \mathcal{B}/\mathcal{A} \to 0 \qquad (3.2.15)$$

is an exact sequence when $\mathcal{A}$ is a subsheaf of $\mathcal{B}$. This is the natural analogue for the corresponding statement for abelian groups of vector spaces, but we had to take special care to make sure that $\mathcal{B}/\mathcal{A}$ was a *sheaf*, before this could be formulated correctly.

If $f : \mathcal{A} \to \mathcal{B}$ is a sheaf morphism, then we define $\ker f$ and $\operatorname{im} f$ to be the sheaves generated by the presheaves $U \to \ker f_U$ and $U \to \operatorname{im} f_U$, respectively. The $\ker f$ presheaf is itself a sheaf, but the presheaf $U \to \operatorname{im} f_U$ need not be a sheaf, hence the necessity to go over to the sheaf generated by it (see Exercise 3.2.10).

We shall close with one important family of short exact sequences. On a complex manifold M, let $\mathcal{O}^*$ be the sheaf of nonvanishing holomorphic functions, constructed as an abelian group under multiplication. Then we define the morphism of sheaves

$$\mathcal{O} \xrightarrow{\exp} \mathcal{O}^*.$$

If $f \in \mathcal{O}(U)$, for U open in M, then $\exp(f)$ is defined to be $e^{2\pi i f}$, which is holomorphic and nonvanishing on U. The kernel of this sheaf morphism is a specific sheaf (the kernel of *any* sheaf morphism is always a sheaf), and it is easy to verify that

$$\ker(\exp) \cong \mathbf{Z},$$

the constant sheaf of integers. Moreover, the sheaf morphism exp is surjective. This follows from the fact that locally on a simply-connected neighborhood of a point one can take a logarithm. If $\exp(f) \equiv 1$ on a connected set U, then f must be an integer on the set U. Since exactness is a local property (i.e., one only has to check exactness at the level of germs), these two remarks suffice to prove that

$$0 \to \mathbf{Z} \to \mathcal{O} \xrightarrow{\exp} \mathcal{O}^* \to 0 \qquad (3.2.16)$$

is a short exact sequence of sheaves on M. We have similar exact sequences if we replace $\mathcal{O}$ by $\mathcal{E}$, $\mathcal{A}$, or simply $\mathcal{C}$, the sheaf of continuous functions on a topological space M.

Exercises for Section 3.2

3.2.1. Let I be a partially ordered set with partial ordering $\geq$. Let S_α be a family of modules indexed by I, with homomorphisms

$$f_\beta^\alpha : S_\alpha \to S_\beta$$

defined if $\alpha \geq \beta$, with the properties that

$$f_\alpha^\alpha = \operatorname{id}, \qquad f_\beta^\alpha \circ f_\gamma^\beta = f_\gamma^\alpha \quad \text{if } \alpha \geq \beta \geq \gamma.$$

Then define the *inductive limit*

$$S = \varprojlim_{\alpha \in I} \operatorname{ind} S_\alpha$$

to be the equivalence classes in the disjoint union

$$\hat{S} = \bigcup_{\alpha \in I} S_\alpha$$

under the equivalence relation:

x_α and $x_\beta \in \hat{S}$ are equivalent if and only if there exists a γ such that $\alpha \geq \gamma$, $\beta \geq \gamma$, and $f_\gamma^\alpha(x_\alpha) = f_\gamma^\beta(x_\beta)$.

Show that:

a) if S_α are modules over a ring R, then S is a module over R.

b) There is a canonical mapping

$$f_\alpha : S_\alpha \to S.$$

3.2.2. Show that if S is a sheaf, then

$$S_x = \varinjlim_{U \ni x} S(U).$$

3.2.3. Give an example of an inductive limit which has the property that

$$\varinjlim_{\alpha \in I} S_\alpha = S_{\alpha_0},$$

for some $\alpha_0 \in I$. (Hint: consider a constant sheaf.) Show that

$$\varinjlim_{x \in U} \mathcal{O}(U)$$

does not have this property, where $\mathcal{O}$ is the sheaf of holomorphic functions on $\mathbf{C}^n$.

3.2.4. Let $X = (0, 1)$, and let G be an abelian group. Define the presheaf of all mappings from subsets of X to G with proper restrictions being zero. Show that this presheaf is not a sheaf.

3.2.5. Define $\mathcal{B}(U)$ to be the bounded holomorphic functions on an open subset U of $\mathbf{C}$. Show that the presheaf $U \to \mathcal{B}(U)$ is not a sheaf.

3.2.6. Let $\mathcal{R}$ be a sheaf of rings; then a *locally free sheaf* of $\mathcal{R}$-modules is a sheaf S with the property that near each point the restriction of the sheaf S to some neighborhood of the point is isomorphic to the direct sum $\bigoplus_1^r \mathcal{R}$, where r is the *rank* of the locally free sheaf. Show that if $\mathcal{E}$ is the sheaf of smooth functions on a differentiable manifold, then, up to isomorphism, locally free sheaves of $\mathcal{E}$-modules of rank r are in one-to-one correspondence with vector bundles of rank r on M.

3.2.7. Show that the presheaf $U \to R(U)$, where $R(U)$ is the ring of continuous R-valued locally-constant functions on U, is a sheaf.

3.2.8. Show that if S is a sheaf then $S \cong \tilde{S}$, where $\tilde{S}$ is the sheaf generated by (the presheaf) S.

3.2.9. Show that if B is the presheaf of bounded holomorphic functions (Exercise 3.2.5), then $\tilde{B} \cong \mathcal{O}$, the sheaf of holomorphic functions on $\mathbf{C}$.

3.1.10. Give an example of a sheaf morphism $f : \mathcal{A} \to \mathcal{B}$, where the presheaf $U \mapsto \text{im } f(U)$ is not a sheaf.

3.3 Sheaf cohomology

Sheaf cohomology gives a tool for relating local to global sections of sheaves. For instance, if we consider any short exact sequence of sheaves on some topological space M,

$$0 \to \mathcal{A} \to \mathcal{B} \to \mathcal{C} \to 0,$$

then let us consider, for some fixed open set U

$$0 \to \mathcal{A}(U) \to \mathcal{B}(U) \to \mathcal{C}(U) \to 0.$$

This sequence is *not* necessarily exact for each such $U \subset M$, as we shall see by an example below. However, it is exact at $\mathcal{A}(U)$ and $\mathcal{B}(U)$, which is not too difficult to verify, using Axioms S_1 and S_2 for a sheaf. Thus the question arises: when is the sheaf morphism

$$\mathcal{B}(U) \to \mathcal{C}(U) \tag{3.3.1}$$

surjective? Sheaf cohomology gives an accurate measure of the failure of surjectivity of the mapping in (3.3.1). For some classes of sheaves, as we shall see, (3.3.1) is always surjective, and there are no obstructions. These sheaves form the basic 'building blocks' for more complicated sheaves without this property, just as a complicated topological space is built out of local cells with trivial topology.

Example 3.3.1: Recall the short exact sequence (3.2.16)

$$0 \to \mathbf{Z} \to \mathcal{O} \to \mathcal{O}^* \to 0$$

on a complex manifold M. Consider the case of $M = \mathbf{C}$, and let U be the annulus $0 < |z| < 1$ in $\mathbf{C}$. Then we see that

$$\mathcal{O}(U) \xrightarrow{\text{exp}} \mathcal{O}^*(U) \qquad (3.3.2)$$

is not surjective, since there is no well-defined branch of the logarithm on U. We shall see that the obstruction to this mapping is precisely $H^1(U, \mathbf{Z})$, which involves the *first* sheaf of the short exact sequence. We observe that this group is zero if U is simply-connected, and this fact is also sufficient for (3.3.2) to be surjective, i.e., that there is a branch of the logarithm. We shall come back to this example later, after we develop the language of sheaf cohomology.

There are two fundamental approaches to developing sheaf cohomology. The first is that of *resolution of sheaves*, where we relate a given sheaf to an associated sequence of sheaves where there are no obstructions to (3.3.1) being surjective. This is an elegant way to derive the fundamental properties of sheaf cohomology, which is found in various books (e.g., Godement 1964, Bredon 1967, Hartshorne 1977, and Wells 1980). However, for certain purposes of calculation, in particular for detailed examples, it is much more convenient to use the Čech theory. This is a combinatorial theory, depending on a covering of the space M in question by open sets on which the restriction of the sheaves behave nicely, e.g., (3.3.1) becomes surjective. Using coverings, one has a calculus of cohomology which is similar to the classical homology and cohomology theory of topological spaces (as was discussed in §3.1). The coefficients of this theory are sections of a sheaf. If the sheaf is a locally constant sheaf such as $\mathbf{Z}$ or $\mathbf{C}$, then the theory will agree with the integral cohomology or complex cohomology of a space. To prove the complete set of properties of sheaf cohomology from the covering point of view (or any point of view, for that matter) requires some work, but one can find this developed quite completely in the treatise of Hirzebruch (1966). We shall formulate sheaf cohomology in terms of Čech theory, indicating the main points of the theory which will be useful for our purposes, leaving some of the details to the books cited above. The relation of the theory to resolutions of sheaves will be brought out later in our discussion of de Rham's theorem and related topics in §3.6.

Let M be a paracompact topological space, and let $\mathfrak{U} = \{U_\alpha\}$ be some open covering of M. A *refinement* of $\mathfrak{U}$ is an open covering $\mathfrak{V} = \{V_\beta\}$ with the property that each V_β from $\mathfrak{V}$ is contained in some U_α from $\mathfrak{U}$. We say that $\mathfrak{U} > \mathfrak{V}$ if $\mathfrak{V}$ is a refinement of $\mathfrak{U}$. The relation '$>$' makes the set of all open covers of M a partially ordered set.

Now let $\mathfrak{U} = \{U_\alpha\}$ be a fixed open covering of M, and let $\mathcal{S}$ be a sheaf on M. A *q-simplex*, σ, is an ordered collection of $q+1$ open sets of the covering $\mathfrak{U}$ with nonempty intersection. If $\sigma = (U_0, \dots, U_q)$, then we let

$$|\sigma| := U_0 \cap \cdots \cap U_q,$$

and this is called the *support* of the simplex σ. A *q-cochain* of $\mathfrak{U}$ with coefficients in $\mathcal{S}$ is a mapping

$$f : \sigma \to \mathcal{S}(|\sigma|).$$

The set of q-cochains will be denoted by $C^q(\mathfrak{U}, \mathcal{S})$, and it is an abelian group under point-wise addition. Since the cochains are parametrized by ordered $(q+1)$-tuples from the indexing set for $\mathfrak{U}$, we can assume, without loss of generality, that the covering is indexed by positive integers, and express a given cochain in the form:

$$f_{i_0 \dots i_q} = f((U_0, \dots, U_q)) \in \mathcal{S}(U_0 \cap \cdots \cap U_q).$$

For instance a zero-cochain is a family of sections of $\mathcal{S}$ of the form

$$f_i \in \mathcal{S}(U_i),$$

and a one-cochain has the form

$$f_{ij} \in \mathcal{S}(U_i \cap U_j).$$

So a cochain is a suitably indexed family of sections of the sheaf defined on a similarly indexed family of intersections of elements of the given open covering.

We define a *coboundary operator*,

$$\delta : C^q(\mathfrak{U}, \mathcal{S}) \to C^{q+1}(\mathfrak{U}, \mathcal{S}),$$

as follows. If $f \in C^q(\mathfrak{U}, \mathcal{S})$, and $\sigma = (U_0, \dots, U_{q+1})$, define

$$\delta f(\sigma) = \sum_{i=0}^{q+1} (-1)^i r_{|\sigma_i|}^{|\sigma|} f(\sigma_i),$$

where $\sigma_i = (U_0, \ldots, U_{i-1}, U_{i+1}, \ldots, U_{q+1})$ and $r_{|\sigma_i|}^{|\sigma|}$ is the sheaf restriction mapping. It is clear that δ is a group homomorphism, and one can check readily that $\delta^2 = 0$ (cf. (3.1.1)). Thus we have a complex of cochains

$$0 \to C^0(\mathfrak{U}, \mathcal{S}) \to \cdots \to C^q(\mathfrak{U}, \mathcal{S}) \xrightarrow{\delta} C^{q+1}(\mathfrak{U}, \mathcal{S}) \to \cdots .$$

Now let $H^q(\mathfrak{U}, \mathcal{S})$ be the cohomology groups of this complex (see (2.2.8)), where $Z^q(\mathfrak{U}, \mathcal{S})$ and $B^q(\mathfrak{U}, \mathcal{S})$ will denote the *cocycles* and the *coboundaries* in this context. This is the *Čech cohomology of the space M with coefficients in the sheaf $\mathcal{S}$ and with respect to the covering* $\mathfrak{U}$. For specific spaces and sheaves this is what we shall work with. First we need to obtain the intrinsic sheaf cohomology which does not depend on a covering. If $\mathfrak{V}$ is a refinement of $\mathfrak{U}$, then there is a natural group homomorphism (induced by the restriction mappings of the sheaf $\mathcal{S}$),

$$h_{\mathfrak{V}}^{\mathfrak{U}} : H^q(\mathfrak{U}, \mathcal{S}) \to H^q(\mathfrak{V}, \mathcal{S}).$$

We can then define the inductive limit of these cohomology groups with respect to the partially ordered set of all coverings (see Exercise 3.2.1):

$$H^q(M, \mathcal{S}) := \lim_{\mathfrak{U}} \text{ind } H^q(\mathfrak{U}, \mathcal{S}).$$

This limit is referred to as the *sheaf cohomology of M of degree q with coefficients in the sheaf $\mathcal{S}$*. By the properties of inductive limits (Exercise 3.2.1), we have a homomorphism

$$H^q(\mathfrak{U}, \mathcal{S}) \to H^q(M, \mathcal{S}).$$

What we would like to do is find coverings for which this homomorphism is an isomorphism. We shall encounter situations where finite coverings will suffice, and also a significant family of examples for which the covering consists of only two open sets!

First, we note that a given cohomology class $c \in H^q(M, \mathcal{S})$ can be represented by a cocycle defined for a specific open covering (by the nature of the inductive limit), but as we vary from cohomology class to cohomology class in $H^q(M, \mathcal{S})$, we shall, in principle, have to vary the covering as we vary the cocyle (cf. Exercise 3.2.3).

For a specific sheaf $\mathcal{S}$ on a specific space M, it is sometimes possible to find a *fixed* covering $\mathfrak{U}$ of M with the property that

$$H^q(\mathfrak{U}, \mathcal{S}) \cong H^q(M, \mathcal{S}),$$

and we call such a covering a *Leray covering* of M with respect to the sheaf $\mathcal{S}$. As remarked above, if we can find a Leray covering (or Leray cover, as it is commonly also known) then we can work with that fixed covering for all calculations. Leray gave a criterion for the existence of such covers.

Lemma 3.3.2 (Leray). *Let $\mathfrak{U}$ be a covering of M with the property that for all simplices σ of the cover*

$$H^q(|\sigma|, \mathcal{S}) = 0, \quad q = 1, 2, \ldots.$$

then

$$H^q(\mathfrak{U}, \mathcal{S}) \cong H^q(M, \mathcal{S}).$$

The proof of this is found in Hirzebruch (1966), for instance. We shall use this criterion to produce several specific Leray covers as we go. We notice here that the intersections on which one must have vanishing cohomology can be small relative to the global extent of the space M. Again we have the notion of building blocks. The small intersections are cohomologically trivial, and we can use them together to build up the global cohomology of the space, just as in a cell complex, for instance.

In order to calculate sheaf cohomology, we shall need to have some of the general properties, which one can derive from the basic definitions involved. These properties will be summarized in the form of a theorem.

Theorem 3.3.3. *Let M be a paracompact Hausdorff space. Then:*
(a) *for any sheaf over M, $H^0(M, \mathcal{S}) = \Gamma(M, \mathcal{S}) \ (= \mathcal{S}(M))$.*
(b) *for any sheaf morphism*

$$h : \mathcal{A} \to \mathcal{B},$$

there is, for each $q \geq 0$, a group homomorphism

$$h^q : H^q(M, \mathcal{A}) \to H^q(M, \mathcal{B}),$$

such that
(1) *$h^0 = h_M : \mathcal{A}(M) \to \mathcal{B}(M)$,*
(2) *h^q is the identity map if h is the identity map, $q \geq 0$,*
(3) *$g^q \circ h^q = (g \circ h)^q$ for all $q \geq 0$, if $g : \mathcal{B} \to \mathcal{C}$ is a second sheaf morphism,*

(c) *for each short exact sequence of sheaves*

$$0 \to \mathcal{A} \to \mathcal{B} \to \mathcal{C} \to 0,$$

there is a group homomorphism

$$\delta^q : H^q(M, \mathcal{C}) \to H^{q+1}(M, \mathcal{A}),$$

for all $q \geq 0$, such that

    (1) *The induced sequence*

$$0 \to H^0(M, \mathcal{A}) \to H^0(M, \mathcal{B}) \to H^0(M, \mathcal{C}) \xrightarrow{\delta^0} H^1(M, \mathcal{A}) \to \cdots$$

$$\to H^q(M, \mathcal{A}) \to H^q(M, \mathcal{B}) \to H^q(M, \mathcal{C}) \xrightarrow{\delta^q} H^{q+1}(M, \mathcal{A}) \to$$

    is exact,

    (2) *A commutative diagram*

$$
\begin{array}{ccccccccc}
0 & \longrightarrow & \mathcal{A} & \longrightarrow & \mathcal{B} & \longrightarrow & \mathcal{C} & \longrightarrow & 0 \\
& & \downarrow & & \downarrow & & \downarrow & & \\
0 & \longrightarrow & \mathcal{A}' & \longrightarrow & \mathcal{B}' & \longrightarrow & \mathcal{C}' & \longrightarrow & 0
\end{array}
$$

induces a commutative diagram

$$
\begin{array}{ccc}
0 \longrightarrow & H^0(M, \mathcal{A}) & \longrightarrow H^0(M, \mathcal{B}) \\
& \downarrow & \downarrow \\
0 \longrightarrow & H^0(M, \mathcal{A}') & \longrightarrow H^0(M, \mathcal{B}')
\end{array}
$$

$$
\begin{array}{ccc}
\longrightarrow & H^0(M, \mathcal{C}) & \longrightarrow H^1(M, \mathcal{A}) \longrightarrow \cdots \\
& \downarrow & \downarrow \\
\longrightarrow & H^0(M, \mathcal{C}') & \longrightarrow H^1(M, \mathcal{A}') \longrightarrow \cdots
\end{array}
$$

Proof: (a) By convention $H^0(M, .) = Z^0(M, .)$, as there are no zero-coboundaries. If $c \in H^0(M, \mathcal{S})$, then c can be represented by a cocycle $f = (f_i)$, where $f_i \in \mathcal{S}(U_i)$ for some covering $\mathfrak{U}$ of M. The condition $\delta f = 0$ means that[1]

$$(\delta f)_{ij} = f_j - f_i = 0 \quad \text{on} \quad U_i \cap U_j. \tag{3.3.3}$$

---

[1] In (3.3.3) we are implicitly using the restriction mappings of the sheaf. More pedantically, one could write

$$r^{U_i}_{U_i \cap U_j}(f_i) = r^{U_j}_{U_i \cap U_j}(f_j),$$

but (3.3.3) means the same thing since sections of a sheaf can be considered as functions with values in the stalks, and the restriction mappings are the usual mappings of restriction of functions, thus justifying the usual function-theoretic notation of (3.3.3).

Thus condition (3.3.3) says that the sections f_i agree on the overlaps and hence must define a global section which agrees with the f_i locally.

(b) This is a straightforward application of the definitions.

(c) This is the crux of the theorem. The mapping δ^q is the *connecting* homomorphism or the *Bockstein* mapping. Our first task is to define the mapping. Suppose that we have a fixed covering $\mathfrak{U}$ of M; then we have a natural commutative diagram:

$$
\begin{array}{ccccccccc}
& & \downarrow & & \downarrow & & \downarrow & & \\
0 & \to & C^{q-1}(\mathfrak{U},\mathcal{A}) & \to & C^{q-1}(\mathfrak{U},\mathcal{B}) & \to & C^{q-1}(\mathfrak{U},\mathcal{C}) & \to & 0 \\
& & \downarrow & & \downarrow & & \downarrow & & \\
0 & \to & C^{q}(\mathfrak{U},\mathcal{A}) & \to & C^{q}(\mathfrak{U},\mathcal{B}) & \overset{\alpha}{\to} & C^{q}(\mathfrak{U},\mathcal{C}) & \to & 0 \\
& & \downarrow & & \downarrow & & \downarrow & & \\
0 & \to & C^{q+1}(\mathfrak{U},\mathcal{A}) & \overset{\beta}{\to} & C^{q+1}(\mathfrak{U},\mathcal{B}) & \overset{\gamma}{\to} & C^{q+1}(\mathfrak{U},\mathcal{C}) & \to & 0 \\
& & \downarrow & & \downarrow & & \downarrow & &
\end{array}
$$

The horizontal complexes are induced from the given exact sequence of sheaves, and they are exact except at the last term (recalling the initial discussion of this section), while the vertical mappings are the coboundary mappings δ (not labeled). The vertical complexes are in principle not exact at any term. Let us choose a given element $c \in H^q(M,\mathcal{C})$. We want to define $\delta^q(c)$ to be an element of $H^{q+1}(M,\mathcal{A})$. Now c can be represented by a cocycle f with respect to an appropriate covering. Although the mapping α may not be surjective, we can choose a refinement (which we shall call $\mathfrak{U}$) of the covering by which f is defined so that $f = \alpha(f')$, for some $f' \in C^q(\mathfrak{U},\mathcal{B})$. This now fixes the covering which we shall work with in the above diagram. Now let $f'' = \delta(f')$. Then, since the diagram is commutative, we have $\gamma(f'') = \delta(\alpha(f')) = \delta(f) = 0$. Since the horizontal sequence at the $(q+1)$th level is exact in the middle it follows that there is some element $f''' \in C^{q+1}(\mathfrak{U},\mathcal{A})$ so that $f'' = \beta(f''')$. One uses the commutativity of the diagram, including the $(q+2)$th level (not shown), to verify that $\delta(f''') = 0$. Thus f''' is a cocycle in $Z^{q+1}(\mathfrak{U},\mathcal{A})$, and hence represents a cohomology class in the direct limit $H^{q+1}(M,\mathcal{A})$. This cohomology class will be denoted by $\delta^q(c)$. This mapping as defined depends on specific choices of f, f', f'', and f'''. One has to check that the various changes in a choice will involve only changing by a coboundary, and that we obtain in this manner a well-defined mapping of cohomology classes

$$
H^q(M,\mathcal{C}) \overset{\delta^q}{\to} H^{q+1}(M,\mathcal{A}).
$$

Once this mapping is defined, one has to check that assertions in (1) and (2) of the theorem are correct. We shall leave these details for the reader; they can also be found in the references cited earlier.

Remark: This set of properties essentially characterizes sheaf cohomology and can be used to give an axiomatic formulation, where the Čech theory gives one of several existence proofs for the theory (see, for instance, Godement 1964, for details). If we had a Leray cover, then the horizontal sequences in the commutative diagram used in the construction of δ^q in the proof of the theorem would be exact, and we would have a formula for the connecting homomorphism valid for all cocycles, without taking a further refinement. We shall be able to use this comment later on.

Let us now compare this cohomology with the singular cohomology developed earlier. If R is a ring, and if we let $\widetilde{R}$ be the sheaf generated by the presheaf which allocates an element of R to each open connected subset of M, then $\widetilde{R}$ is a (locally) constant sheaf as discussed in §3.2 (which we are notationally distinguishing from R, for the moment).

Theorem 3.3.4. *There is a natural isomorphism*

$$H^q(M, R) \cong H^q(M, \widetilde{R}).$$

There are various ways of proving this theorem. For instance, both theories have axiomatic formulations, and this can be used to show that they are identical. One can also use resolutions and the abstract de Rham theorem (see Bredon 1967, Godement 1964, or Wells 1980, for instance). We shall simply identify these two classes of cohomology groups, and consider them all as sheaf cohomology groups, although we can interpret the singular cohomology as such whenever we need to (in particular, for using concrete results such as those listed in Example 3.1.1).

Now we turn to a special family of sheaves which always has trivial cohomology. These will be the *fine* sheaves. Fine sheaves will include the sheaves of smooth functions or smooth differential forms, for instance, but will not include the locally constant sheaves, or the sheaf of holomorphic functions on a complex manifold and its generalizations. Essentially, the fine sheaves are those which admit a suitable partition of unity and cutoff functions. We shall give a formal definition first. A sheaf $\mathcal{S}$ on M is a *fine* sheaf if, for any locally finite covering $\{U_j\}$ of M, there is a family of sheaf homomorphisms

$$\varphi_i : \mathcal{S} \to \mathcal{S},$$

such that

$$
\begin{aligned}
\text{(a)} & \quad \operatorname{supp} \varphi_i(\mathcal{S}) \subset U_i, & (3.3.4)\\
\text{(b)} & \quad \sum_i \varphi_i = \operatorname{id}.
\end{aligned}
$$

It is shown in the exercises that the sheaf of smooth functions $\mathcal{E}$ on a differentiable manifold is a fine sheaf. Moreover, any $\mathcal{E}$-module will also be fine (Exercises 3.3.2, 3.3.3). So, on a complex manifold, for instance, $\mathcal{E}^r, \mathcal{E}^{p,q}$, and $\mathcal{E}^{p,q}(E)$, where $E \to M$ is a complex vector bundle, are all fine sheaves.

Theorem 3.3.5. *Let $\mathcal{S}$ be a fine sheaf on a paracompact Hausdorff space M. Then*

$$H^q(M, \mathcal{S}) = 0, \quad q = 1, 2, \ldots.$$

Proof: We shall write out the proof for the case $q = 1$, since that is simpler to see, and it involves the full set of ideas. Let

$$c \in H^1(M, \mathcal{S}),$$

then we want to show that $c = 0$. The cohomology class c can be represented by a cocycle f with respect to some open covering, which by a refinement, if necessary, we can assume to be locally finite. Therefore f has the form

$$f = (f_{ij}), \quad f_{ij} \in \Gamma(U_i \cap U_j, \mathcal{S}).$$

Moreover $\delta f = 0$, that is to say,

$$(\delta f)_{ijk} = f_{jk} - f_{ik} + f_{ij} = 0 \quad \text{on} \quad U_i \cap U_j \cap U_k. \quad (3.3.5)$$

Now choose a 'partition of unity' $\{\varphi_i\}$ for the sheaf S, coming from the definition of being a fine sheaf. Then we define the one-cochain g by

$$g_i = -\sum_k \varphi_k(f_{ik}),$$

which is well-defined because of the support property of $\{\varphi_i\}$ and the local finiteness of the cover. Then we calculate

$$
\begin{aligned}
(\delta g)_{ij} &= g_j - g_i, \\
&= -\sum_k \varphi_k(f_{jk}) + \sum_k \varphi_k(f_{ik}), \\
&= \sum_k \varphi_k(f_{ik} - f_{jk}),
\end{aligned}
$$

and using (3.3.5) we obtain

$$
\begin{aligned}
(\delta g)_{ij} &= -\sum_k \varphi_k(-f_{ij}) \\
&= f_{ij}.
\end{aligned}
$$

Thus the cocycle f is a coboundary, and hence $c = 0$. The higher order cases can be treated very similarly, and will be omitted. ∎

We now turn to a sequence of examples of calculations of cohomology.

There are basically three classes of sheaves which are of interest to us. The first are the constant sheaves. The sheaf cohomology with respect to constant sheaves coincides with singular cohomology, as remarked before, and reflects the global topology of the space. The second class contains the sheaves of smooth functions $\mathcal{E}$ and their generalizations ($\mathcal{E}^r$, $\mathcal{E}^{p,q}$, and other such locally free $\mathcal{E}$-modules). These are all fine sheaves as we have just seen, and they have vanishing cohomology, which plays a useful role in conjunction with other sheaves where this is not the case. The third class includes the sheaf $\mathcal{O}$ of holomorphic functions on a complex manifold M, and locally free $\mathcal{O}$-modules such as Ω^p, the sheaf of holomorphic p-forms on M, $\Omega^p(E)$, the E-valued holomorphic p-forms, where $E \to M$ is a holomorphic vector bundle over M, and $\mathcal{O}(H^m)$, the sheaf of sections of the mth power of the hyperplane section bundle over $\mathbf{P}_n$, among others. These sheaves, in general, have nontrivial cohomology where they are defined, or, if it is trivial cohomology, it is often a difficult task to verify this, using diverse tools from analysis and geometry. We shall give a sequence of examples to illustrate the subtleties involved.

The theory of such cohomology calculations (as we are choosing to call it) is developed in various books dealing with complex analysis in several variables, and is a subject which has undergone great developments in the past 40 years. The theory splits naturally into two areas. The first is the theory of open complex manifolds (domains of holomorphy, pseudoconvexity, Stein manifolds,) which is developed in the books by Gunning and Rossi (1965), and Hörmander (1973), for instance. The second area deals with the theory of compact complex manifolds (projective algebraic manifolds, Kodaira's classification, Riemann surfaces, Kähler manifolds, Hodge theory), and you will find these topics in Griffiths and Harris (1978), Morrow and Kodaira (1971), and Wells (1980). We shall mention a general result which will limit how much we have to calculate and which will play a role in some of the examples. Let $E \to M$ be a holomorphic vector bundle over a complex manifold M. Recall the Dolbeault groups

$$H^{p,q}_{\mathrm{DOL}}(M, E),$$

which were defined in (2.2.29). These were, like the de Rham groups, $\bar{\partial}$-closed forms modulo $\bar{\partial}$-exact forms, i.e., quotients of solutions of global differential equations on the manifold. We have an analogue to the de Rham theorem which relates sheaf cohomology to such solutions of differential equations and which is due to Dolbeault.

Theorem 3.3.6 (Dolbeault). *There is a natural isomorphism*

$$H^q(M, \Omega^p(E)) \cong H^{p,q}_{\mathrm{DOL}}(M, E),$$

for $p, q = 0, 1, 2, \ldots$.

The proof of this will be given in §3.6. As a consequence, we have the following corollary which shows that cohomology of high degree is always trivial, depending on the dimension.

Corollary 3.3.7. *Let M be an n-dimensional complex manifold. Then, for $p = 0, 1, 2, \ldots$, $H^q(M, \Omega^p(E)) = 0$, for $q > n$.*

This result is a simple consequence of Dolbeault's theorem and the fact that on a complex manifold of dimension n there are no (p, q)-forms for $q > n$.

We now turn to our sequence of examples. We shall give some indication of proof, especially in the first one, in the context of one complex variable, to give some flavor of the subject. The later examples would require much more effort, and there we shall rely on the more detailed reference materials. We shall have occasion to use some of these specific results in our work in Part III on the Penrose transform.

Example 3.3.8 (One complex variable): Let us consider first the complex plane. We have:

$$H^0(\mathbf{C}, \mathcal{O}) \cong \mathcal{O}(\mathbf{C}) = \{\text{entire functions on the plane}\} \qquad (3.3.6)$$
$$H^1(\mathbf{C}, \mathcal{O}) = 0. \qquad (3.3.7)$$

On the other hand, for one-dimensional projective space we have:

$$H^0(\mathbf{P}_1, \mathcal{O}) \cong \mathbf{C}, \qquad (3.3.8)$$
$$H^1(\mathbf{P}_1, \mathcal{O}) = 0, \qquad (3.3.9)$$
$$H^0(\mathbf{P}_1, \Omega^1) = 0, \qquad (3.3.10)$$
$$H^1(\mathbf{P}_1, \Omega^1) \cong \mathbf{C}. \qquad (3.3.11)$$

We shall go through these one by one. First, we note that on $\mathbf{C}$ we have $\Omega^1 \cong \mathcal{O}$, so there is no need to calculate the holomorphic one-form case. On the other hand (3.3.8)–(3.3.11) above show dramatically that Ω^1 is not isomorphic to $\mathcal{O}$ on $\mathbf{P}_1$.

Now (3.3.6) is clear from Theorem 3.3.3(a). To see (3.3.7), let $\mathfrak{U} = \{U_j\}$ be a locally finite covering of $\mathbf{C}$ and let $f = (f_{ij})$ be a cocycle in $Z^1(\mathfrak{U}, \mathcal{O})$. Then we want to find a cochain $g = (g_i), g_i \in \mathcal{O}(U_i)$, so that $\delta g = f$, i.e.,

$$f_{ij} = g_j - g_i \quad \text{on} \quad U_i \cap U_j.$$

Let us first find a C^∞ solution to this problem. Let

$$h_i = -\sum_k \varphi_k f_{ik},$$

where $\{\varphi_i\}$ is a partition of unity relative to $\mathfrak{U}$. Then we see that (as in the proof of Theorem 3.3.5)

$$f_{ij} = h_j - h_i \quad \text{on} \quad U_i \cap U_j.$$

Let
$$\omega_j = \bar{\partial} h_j \quad \text{on} \quad U_j.$$

Then we see that on $U_i \cap U_j$ we have

$$
\begin{aligned}
\omega_i - \omega_j &= \bar{\partial}(h_i - h_j) \\
&= \bar{\partial}(f_{ji}) \\
&= 0.
\end{aligned}
$$

Thus the cochain $\omega = (\omega_i)$ defines a global $(0,1)$-form on $\mathbf{C}$ which we shall also denote by ω. Therefore we have that

$$\omega(z) = k(z) d\bar{z}, \quad k \in C^\infty(\mathbf{C}).$$

Suppose that we can solve the equation

$$\bar{\partial} u = \omega, \tag{3.3.12}$$

for some $u \in C^\infty(\mathbf{C})$, that is to say, the inhomogeneous Cauchy-Riemann equation,

$$\frac{\partial u}{\partial \bar{z}} = k.$$

Then we can set

$$g_j = h_j - u,$$

and we then see that

$$
\begin{aligned}
\bar{\partial} g_j &= \bar{\partial} h_j - \bar{\partial} u \\
&= \omega - \bar{\partial} u \\
&= 0.
\end{aligned}
$$

Moreover, $\delta g = f$, since, on $U_i \cap U_j$,

$$
\begin{aligned}
g_j - g_i &= (h_j - u) - (h_i - u) \\
&= h_j - h_i \\
&= f_{ij}.
\end{aligned}
$$

Let us now show that (3.3.12) is always solvable. First, if we let Δ_n be the disk centered at the origin, of radius n, then we can set

$$u_n(z) = \frac{1}{2\pi i} \int\limits_{\Delta_{n+1}} \frac{k(\zeta) d\zeta \wedge d\bar{\zeta}}{\zeta - z},$$

and we see that this is a solution to (3.3.12) on the closure of Δ_n, which we denote by $\bar{\Delta}_n$. However, the sequence of solutions u_n defined on larger and larger sets may not converge. Consider the infinite series

$$u = \sum_{k=1}^{\infty} (u_k - u_{k-1}), \quad u_0 = 0.$$

Then we have on $\bar{\Delta}_n$

$$u = u_n + \sum_{k=n+1}^{\infty} (u_k - u_{k-1}),$$

since the earlier terms all cancel. If the series were to converge, then we would have

$$\bar{\partial} u = \bar{\partial} u_n + \sum_{k=n+1}^{\infty} (\bar{\partial} u_k - \bar{\partial} u_{k-1})$$
$$= \bar{\partial} u_n + 0$$
$$= \omega,$$

on $\bar{\Delta}_n$, where we used the fact that on $\bar{\Delta}_n$, $\bar{\partial} u_k = \bar{\partial} u_{k-1} = \omega$, for $k > n$. This would then be the desired solution of the equation. However, there is no reason for this to converge, but we can use the Mittag-Leffler trick to force it to converge. Each term $(u_k - u_{k-1})$ is holomorphic on $\bar{\Delta}_{k-1}$, and hence on this same set it can be approximated uniformly by a holomorphic polynomial p_k so that

$$|(u_k - u_{k-1}) - p_k| < 2^{-k} \quad \text{on} \quad \bar{\Delta}_{k-1}. \tag{3.3.13}$$

Namely, the Taylor series for $(u_k - u_{k-1})$ converges uniformly on compact subsets of the domain of definition of this function, and choosing a sufficiently high-order Taylor approximating polynomial will give (3.3.3). Since $\bar{\partial} p_k = 0$ for all k, we see that if we let

$$u = \sum_{k=1}^{\infty} (u_k - u_{k-1} - p_k),$$

then on Δ_n we still have $\bar{\partial} u = \omega$, as before, and this time the series converges. Thus we have solved equation (3.3.12), and consequently $H^1(\mathbf{C}, \mathcal{O}) = 0$.

We carried this out in detail, as it is a very good model of what a vanishing theorem on an open complex manifold involves. The Mittag-Leffler process was used in the proof, and, by similar arguments one can prove that the equation $\bar{\partial}u = \omega$ is solvable on any plane domain in the complex plane. One first solves the equation on compact subsets by a Cauchy integral formula, then one takes a limiting process to get a solution up to the boundary, and one would normally use some variation of the Mittag-Leffler process. Holomorphic functions on the whole domain will replace the polynomials in the above proof (see Hörmander 1973 for more details). Thus we have for any plane domain $U \subset \mathbf{C}$,

$$H^1(U, \mathcal{O}) = 0. \tag{3.3.14}$$

More generally, one has $H^1(M, \mathcal{O}) = 0$ for any one-complex-dimensional connected noncompact manifold (an open Riemann surface), by generalizations of these same arguments.

It follows from this that any nontrivial covering of a one-complex-dimensional manifold is a Leray cover. Namely, all of the supports of the one-cochains will be noncompact one-complex-dimensional manifolds, and hence by the remark following (3.3.14), the criterion of Leray in Lemma 3.3.2 will be satisfied.

Now (3.3.8) is clear. This is simply a consequence of the maximum principle for holomorphic functions. Namely, any global holomorphic function on $\mathbf{P}_1$ (or any compact complex manifold, for that matter) must assume its maximum somewhere, and hence would be locally constant near that maximum. Then by the identity theorem for holomorphic functions, it would be constant everywhere. To compute $H^1(\mathbf{P}_1, \mathcal{O})$, let us use the standard covering (which we shall call $\mathfrak{U}$) by two coordinate charts:

$$U_0 = \{[z^0, z^1] : z^0 \neq 0\},$$
$$U_1 = \{[z^0, z^1] : z^1 \neq 0\}.$$

Then

$$H^1(\mathbf{P}_1, \mathcal{O}) \cong H^1(\mathfrak{U}, \mathcal{O}).$$

But letting $z = z^1/z^0$ be an affine coordinate, we can define, for some cocycle $g_{01} \in Z^1(\mathfrak{U}, \mathcal{O})$,

$$g(z) = \int\limits_{|\zeta|=1} \frac{g_{01}(\zeta)d\zeta}{\zeta - z}, \qquad |z| \neq 1,$$

and we then set

$$g_0(z) = g(z), \quad |z| < 1,$$
$$g_1(z) = g(z), \quad |z| > 1.$$

It follows that $g_0 \in \mathcal{O}(U_0)$, $g_1 \in \mathcal{O}(U_1)$, and

$$g_0 - g_1 = g_{01} \quad \text{on} \quad U_0 \cap U_1,$$

as one can easily check by using standard complex variable arguments. But this says that the cocycle g_{01} is a coboundary. Therefore we conclude that $H^1(\mathfrak{U}, \mathcal{O}) = 0$, and hence that $H^1(\mathbf{P}_1, \mathcal{O}) = 0$. Note that if we did not have the Leray theorem, we could not necessarily conclude from the vanishing of the cohomology for the particular covering the vanishing of the direct limit.

To calculate (3.3.10) we let ω be a global holomorphic one-form on $\mathbf{P}_1$. Using the same coordinate system as above, we see that

$$\omega = f(z)dz, \quad f \in \mathcal{O}(U_0).$$

Expand f in a Taylor series of the form

$$f(z) = \sum_{n=0}^{\infty} a_n z^n.$$

The change of coordinates implies that we should have a similar expansion in the variable $w = 1/z$, the affine coordinate in U_1. But the expansion in w would have to be given by

$$f(z)dz = f(1/w)d(1/w) = \sum_{n=0}^{\infty} a_n w^{-n}(-w^{-2})dw,$$

but there are only negative powers of w in this Laurent expansion, and hence it can be holomorphic in U_1 only if all the coefficients are zero. Thus we conclude that ω is zero, and that is the content of (3.3.10).

Finally we want to compute $H^1(\mathbf{P}_1, \Omega^1)$. Again, using the standard covering, consider

$$g_{01} \in \Omega^1(U_0 \cap U_1),$$

which we express in the affine coordinate of U_0 as

$$g_{01} = f\,dz, \quad \text{for } f \text{ holomorphic in } 0 < |z| < \infty.$$

Here f is possibly singular at 0 and ∞ in contrast to the previous case. Again expand in a Laurent series, and we find that

$$g_{01}(z) = \left(\sum_{n=-\infty}^{-2} a_n z^n\right)dz + a_{-1}z^{-1}dz + \left(\sum_{n=0}^{\infty} a_n z^n\right)dz$$
$$= g_1(z) + a_{-1}z^{-1}dz + g_0(z).$$

Now by changing variables where necessary, we can check that

$$g_0 \in \Omega^1(U_0),$$
$$g_1 \in \Omega^1(U_1),$$

and hence g_{01} is cohomologous to the new cocycle $\tilde{g}_{01} = a_{-1}z^{-1}dz$ defined on $U_0 \cap U_1$. Moreover, it is clear that the cohomology class of $\tilde{\omega}_{01}$ is 0 if and only if a_{-1} is 0. Thus we conclude that $H^1(\mathfrak{U}, \Omega^1) \cong \mathbf{C}$, and since it is a Leray cover, we obtain (3.3.11).

We have carried out this example in some detail, as it illustrates many important features of sheaf cohomology. First we note that in these cases the cohomological assertions were statements about analysis not topology (see the ingredients in the proof). Secondly, it is always true that sheaf cohomology for locally free $\mathcal{O}$-modules on a compact complex manifold has finite-dimensional cohomology. In the open case one can have also infinite-dimensional cohomology (as in (3.3.6)).

Example 3.3.9 ($\mathbf{P}_n$): Recall that $H^m \to \mathbf{P}_n$ is the mth power of the hyperplane section bundle. Let $\mathcal{O}(m) := \mathcal{O}(H^m)$, and more generally if $\mathcal{S}$ is any $\mathcal{O}$-module on $\mathbf{P}_n$ we set $\mathcal{S}(m) := \mathcal{S} \otimes_{\mathcal{O}} \mathcal{O}(H^m)$. Then we have:

$$H^0(\mathbf{P}_n, \mathcal{O}(m)) \cong \mathbf{C}_m[Z^0, \ldots, Z^n], \quad m \geq 0, \qquad (3.3.15)$$
$$H^0(\mathbf{P}_n, \mathcal{O}(m)) = 0, \quad m < 0, \qquad (3.3.16)$$

where the right hand side denotes the complex vector space of homogeneous polynomials of degree m in $n+1$ variables. In addition,

$$H^q(\mathbf{P}_n, \Omega^q) \cong \mathbf{C}, \quad q = 0, 1, \ldots, n, \qquad (3.3.17)$$
$$H^q(\mathbf{P}_n, \Omega^p) = 0, \quad p \neq q, \qquad (3.3.18)$$
$$H^q(\mathbf{P}_n, \Omega^p(m)) = 0, \quad p + q < n, m < 0. \qquad (3.3.19)$$

These are all found in Griffiths and Harris (1978) and Wells (1980). Equation (3.3.15) is an identification of the global sections, and it follows from a multiple Laurent series argument, for instance. Equation (3.3.16) can be proved in the same manner and is also a special case of what is known as the Kodaira vanishing theorem which is used to prove (3.3.19), where we note that we no longer have line bundle coefficients. The Kodaira vanishing theorem uses Hodge theory (representing Dolbeault cohomology classes by harmonic differential forms) and differential geometry and curvature arguments going back originally to Bochner, so it is considerably less elementary. Part of the Hodge theory is the Hodge decomposition which is valid on any Kähler manifold (see Wells 1980; this is also discussed briefly in §3.6). On projective space it takes the form

$$H^r(\mathbf{P}_n, \mathbf{C}) \cong \sum_{p+q=r} H^q(\mathbf{P}_n, \Omega^p). \qquad (3.3.20)$$

One can find a specific differential form ω on $\mathbf{P}_n$ of type $(1, 1)$ with the properties that: (a) $d\omega = 0$, and, (b) if we consider a complex projective line L in $\mathbf{P}_n$ as a two-cycle, which represents a homology class in $H_2(\mathbf{P}_n, \mathbf{Z})$, then

$$\int_L \omega = 1.$$

One can construct this differential form as the Chern form of the hyperplane section bundle in the next section (cf. §3.4 and Wells 1980). Thus this differential form represents a nonzero cohomology class in $H^2(\mathbf{P}_n, \mathbf{Z})$ (if it were a coboundary, its integral over L would necessarily be zero, by Stokes' theorem). The wedge products $\omega^q = \omega \wedge \omega \wedge \cdots \wedge \omega$ (q factors) will also be closed differential forms of type (q, q), and the integral of ω^q over a projective subspace of dimension q (a cycle representing $H_{2q}(\mathbf{P}_n, \mathbf{Z})$) will also be nonzero. It follows that $H^q(\mathbf{P}_n, \Omega^q)$ is nonzero for $q = 0, 1, 2, \ldots, n$. Now we note that the left-hand side of (3.3.20) is either zero- or one-dimensional, and it follows then that (3.3.17) and (3.3.18) must be valid.

Example 3.3.10 (Riemann surface): Let M be a Riemann surface, that is, a compact connected one-complex-dimensional manifold. Let us define

$$g := \dim H^0(M, \Omega^1).$$

The number g is called the *genus* of M. Now the Hodge decompostion mentioned in the previous example is valid on any Riemann surface also, and we find that

$$H^1(M, \mathbf{C}) \cong H^0(M, \Omega^1) \oplus H^1(M, \mathcal{O}).$$

Moreover, the left-hand side is the complexification of the real vector space $H^1(M, \mathbf{R})$, and the two terms in the direct sum on the right-hand side are complex conjugates with respect to this complexification. In particular, it follows that they have the same dimension and that

$$\dim H^1(M, \mathcal{O}) = g = \tfrac{1}{2}b_1(M), \tag{3.3.21}$$

and thus g is a topological invariant. In the case of $M = \mathbf{P}_1$, the genus is zero, as we saw in Example 3.3.8. For more details on this example see Griffiths and Harris (1978).

Example 3.3.11 (Hartogs' phenomenon): We recall that higher degree cohomology for the structure sheaf vanished for any plane domain in $\mathbf{C}$ (Example 3.3.8). However, for more than one complex variable, this is no longer the case. For instance we have

$$H^1(\mathbf{C}^2 - \{0\}, \mathcal{O}) \cong \mathbf{C}. \tag{3.3.22}$$

This follows from a Laurent series argument and is not difficult (see Griffiths and Harris 1978, for instance). A *domain of holomorphy* in $\mathbf{C}^n$ is an open domain D with the property that there exists a function holomorphic on D which is singular at every boundary point of D. All domains in the complex plane are domains of holomorphy, which follows from the Weierstrass theorem, for instance (construct a function in the domain with zeros which have every boundary point as a limit point). On the other hand,

the domain $D = \mathbf{C}^2 - \{0\}$ is not a domain of holomorphy,

$$\tag{3.3.23}$$

as any function homomorphic on D continues analytically to all of $\mathbf{C}^2$, which can again be seen by a Laurent series argument. This example was discovered in 1906 by Hartogs. For the next 50 years there was a lot of intense research which involved characterizing domains of holomorphy and their generalizations. A deep result of Oka asserts that if D is a domain in $\mathbf{C}^2$, then D is a domain of holomorphy if and only if $H^1(D, \mathcal{O}) = 0$. A second important characterization of domains of holomorphy was the local geometric one in terms of pseudoconvexity, which we shall not describe precisely here (see, e.g., Hörmander 1973). However, a domain which is convex is pseudoconvex and so is a domain of holomorphy, while a domain whose boundary is strictly concave near some point will not be pseudoconvex and not a domain of holomorphy.

For instance, if we take a solid three-ball in $\mathbf{C}^2$ and deform it to have a concave dimple as illustrated in Figure 3.3.1, then this is not a domain of holomorphy. Calling this deformed domain D, we see that, by Oka's result above, $H^1(D, \mathcal{O}) \neq 0$. This is a very delicate result. By deforming the domain slightly near one boundary point, we have changed the cohomology from being zero to being nonzero, and moreover, this illustrates vividly that this kind of sheaf cohomology is definitely not topological in nature, which was not so apparent from all of the other examples, all of which had nontrivial topology of some sort. On the other hand, in the case of some of the compact examples, the holomorphic sheaf cohomology did turn out to be topological in nature, as we saw (in the previous example, for instance).

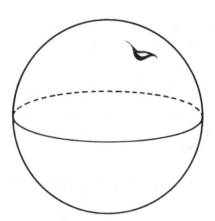

Fig. 3.3.1. Deformed solid three-ball in $\mathbf{C}^2$.

Example 3.3.12 (Classification of line bundles): Consider the exponential sequence on a differentiable manifold M:

$$0 \to \mathbf{Z} \to \mathcal{E} \to \mathcal{E}^* \to 0.$$

Associated with this short exact sequence is the induced long exact sequence in cohomology:

$$\cdots \to H^1(M, \mathcal{E}) \to H^1(M, \mathcal{E}^*) \to H^2(M, \mathbf{Z}) \to H^2(M, \mathcal{E}) \to \cdots .$$
$$(3.3.24)$$

Now we note that

$$H^1(M, \mathcal{E}^*) \cong \{\text{equivalence classes of smooth line bundles on } M\}.$$
$$(3.3.25)$$

This is easy to see in that a cocycle $g = (g_{ij}) \in Z^1(\mathfrak{U}, \mathcal{E}^*)$, for some covering $\mathfrak{U}$ of M, can be interpreted as the set of transition functions for a differentiable line bundle for the same covering. Moreover, the cocycle condition (written multiplicatively for this sheaf) is

$$g_{ij} g_{jk} g_{ki} = 1,$$

and is the necessary compatibility conditions for the (g_{ij}) to be transition functions for a line bundle. This correspondence for a specific covering yields the isomorphism (3.3.25) above.

Now, since $\mathcal{E}$ is a fine sheaf, we have that (Theorem 3.3.3)

$$H^1(M, \mathcal{E}) = H^2(M, \mathcal{E}) = 0,$$

and hence we obtain from (3.3.24) that

$$0 \to H^1(M, \mathcal{E}^*) \xrightarrow{\delta} H^2(M, \mathbf{Z}) \to 0$$

is exact, from which it follows that δ is an isomorphism. Thus we see that the nonequivalent line bundles on M are characterized by the topology of the base space. If $H^2(M, \mathbf{Z}) = 0$ (e.g., $M = S^3$), then there are no nontrivial line bundles on M. On the other hand, for S^2, for instance, we find that

$$H^1(S^2, \mathcal{E}^*) \cong H^2(S^2, \mathbf{Z}) \cong \mathbf{Z},$$

so there is a countable family of inequivalent line bundles on S^2. We shall see in the next section that this family coincides with the equivalence classes of the powers of the hyperplane section bundle on $\mathbf{P}_1$ ($\cong S^2$).

We can attempt to classify holomorphic line bundles on a complex manifold M in the same manner. One also has the isomorphism

$$H^1(M, \mathcal{O}^*) \cong \{\text{equivalence classes of holomorphic line bundles on } M\}$$
$$(3.3.26)$$

just as in the differentiable case. If we now look at the exponential sequence in this context,

$$0 \to \mathbf{Z} \to \mathcal{O} \to \mathcal{O}^* \to 0,$$

and its associated long exact cohomology sequence

$$\cdots \to H^1(M, \mathcal{O}) \to H^1(M, \mathcal{O}^*) \to H^2(M, \mathbf{Z}) \to H^2(M, \mathcal{O}) \to \cdots,$$
$$(3.3.27)$$

we see that things are not as simple as in the differentiable case. For $\mathbf{P}_n$, however, it is quite the same thing, since we know that

$$H^1(\mathbf{P}_n, \mathcal{O}) = H^2(\mathbf{P}_n, \mathcal{O}) = 0,$$

from (3.3.18). Thus we find that

$$H^1(\mathbf{P}_n, \mathcal{O}^*) \xrightarrow[\cong]{\delta} H^2(\mathbf{P}_n, \mathbf{Z}) \cong \mathbf{Z}. \qquad (3.3.28)$$

Now we know from (3.3.15), for instance, that the hyperplane section bundle H on projective space is nontrivial, thus in the isomorphism (3.3.27) $\delta(\text{trivial line bundle}) = 0$, and hence $\delta(H) = k \neq 0$. Moreover, it follows from the fact that δ is a group homomorphism that $\delta(H^m) = mk$. We shall see in the next section that $k = 1$, and it will then follow that the hyperplane section bundle is a generator for the abelian group $H^1(\mathbf{P}_n, \mathcal{O}^*)$.

In the other direction, if we have a Riemann surface of nonzero genus, then we see that $H^1(M, \mathcal{O}) \neq 0$ and the classification of the holomorphic line bundles is a more delicate matter, and not just a matter of the topology (see Griffiths and Harris 1978 for more details on this issue).

This concludes our tour through a representative set of examples of sheaf cohomology. We have seen that the long exact sequence on cohomology is a means for calculating cohomology, in that if one knows some cohomology groups from some source or other, then other groups can be determined in some fashion or other from the relations in the long exact sequence. The methods to be introduced in §3.6 involving spectral sequences will be a similar set of ideas. These will be a somewhat complicated set of relations between families of related cohomology groups, and if one has enough information about some of the groups occurring, one can deduce information about the others. So, we can summarize by saying that sheaf cohomology involves two types of ingredients: the first is computing certain explicit ones by analysis, topology, or geometry, and the second one is to use relationships between various cohomology groups (long exact sequences, and later spectral sequences) to deduce information about unknown groups from known ones. We shall be using both types of methodology in our work in Part III.

Exercises for Section 3.3

3.3.1. Show that the sheaf of differentiable functions on a differentiable manifold is a fine sheaf.

3.3.2. Show that if $E \to X$ is a differentiable vector bundle, then $\mathcal{E}(E)$, the sheaf of differentiable sections of E is a fine sheaf.

3.3.3. Show that if $\mathcal{R}$ is a fine sheaf of rings, then any sheaf of $\mathcal{R}$-modules is a fine sheaf.

3.3.4. Show that if

$$0 \to \mathcal{A} \to \mathcal{B} \to \mathcal{C} \to 0$$

is an exact sequence of sheaves on X, and $\mathcal{A}$ and $\mathcal{B}$ are fine then

$$H^q(X, \mathcal{C}) = 0, \quad q \geq 1.$$

Is $\mathcal{C}$ necessarily a fine sheaf?

3.3.5. Show that $H^n(\mathbf{P}_n, \mathcal{O}(-n-1)) \cong \mathbf{C}$.

3.4 Characteristic classes

In §2.1 we studied the geometry of vector bundles, and in §3.1 we studied topological invariants of manifolds. In this section we want to show how the global topology of a manifold determines to what extent vector bundles over a manifold are trivial. If we restrict our attention to differentiable vector bundles for the time being, then it will follow from the results we develop here that any vector bundle over a cell is (up to isomorphism) the trivial bundle. In other words, one needs a nontrivial topology of the base space (e.g., a two-sphere) in order to have nontrivial vector bundles.

In the last section we used sheaf cohomology to construct a mapping

$$H^1(M, \mathcal{E}^*) \xrightarrow{\delta} H^2(M, \mathbf{Z}).$$

This is an example of a characteristic class. As we saw, elements of $H^1(M, \mathcal{E}^*)$ can be identified with complex line bundles on M, and the mapping above can be interpreted as:

$$\{\text{line bundles on } M\} \rightarrow H^2(M, \mathbf{Z}).$$

This characteristic class is a cohomology class which is characteristic of the bundle (sometimes it is so characteristic, that it characterizes the bundle, as in the case above). More generally a characteristic class is a rule which assigns a topological invariant (usually an element of some cohomology group or homotopy group) to a given vector bundle on a manifold. The first characteristic class was the Euler characteristic discovered by Euler in the eighteenth century. This was a topological invariant assigned to a manifold (in contemporary language, to the tangent bundle of the manifold). Euler expressed his 'Euler number' in terms of a triangulation of a two-manifold. Namely, if V is the number of vertices, E is the number of edges, and F is the number of faces of a triangulated two-manifold M, then the *Euler number* of M is defined to be

$$\chi(M) := V - E + F.$$

This number is the best known example of a topological invariant. For the two-sphere in Figure 3.1.2 we see that $V = 6$, $E = 12$, $F = 8$, so

$$\chi(S^2) = 6 - 12 + 8 = 2.$$

This number is independent of the triangulation. A famous theorem of the nineteenth century is that one can find a smooth global nonvanishing vector field on M if and only if $\chi(M) = 0$. Here we see the tangent bundle coming in. Note that if we have two global vector fields on M which are linearly independent at each point, then this is a global frame for $T(M)$, and hence $T(M) \cong M \times \mathbf{R}^2$, and thus is trivial. So, having no global vector fields implies that $T(M)$ is nontrivial in a strong sense. Characteristic classes in general give restrictions for vector bundles to admit such global frames, or global partial frames (e.g., k linearly independent vector fields on a manifold of dimension n, for $k < n$).

Characteristic classes in their modern form were first formulated by Stiefel and Whitney in the 1930s, using a combinatorial description of the underlying manifolds. In the 1940s Chern used differential-geometric methods to define characteristic classes of complex vector bundles, and at the same time, a theory of characteristic classes more general than that of Stiefel and Whitney was developed by Pontrjagin. It was later discovered how to express the Pontrjagin classes in terms of the Chern classes, and hence in terms of local differential-geometric invariants. We shall concentrate on describing Chern's characteristic classes, which are now called the Chern classes of a complex vector bundle.

Suppose that E is a complex differentiable vector bundle of rank n on a manifold M. Then the Chern classes will be an assignment

$$E \to c_j(E) \in H^{2j}(M, \mathbf{Z}),$$

for $j = 0, 1, \ldots, n$, with certain special properties which help measure how nontrivial such a vector bundle is. Chern's idea was to first introduce an Hermitian metric on E, then to calculate its curvature, which is locally a matrix of two-forms on M, and then to take certain invariant polynomials of these matrix entries of two-forms, obtaining a sequence of differential forms of degree $0, 2, \ldots, 2n$. These forms will turn out to be closed, and will then define elements of the de Rham group of M, which in turn are independent of the original metric. These are the Chern classes, and they turn out to be elements of the integral cohomology group (that is, their integrals over integral cycles are always integers). This brief outline will be carried out in somewhat more detail as the theory allows us to consider classes of bundles over various types of manifolds. For complex line bundles the Chern class $c_1(E)$ turns out to coincide with $\delta(E)$ mentioned earlier (this is verified in Wells 1980, for instance).

First we want to discuss de Rham cohomology. We recall that de Rham's Theorem (Theorem 3.1.3) asserts that, assuming real coefficients,

$$H^r_{\mathrm{DR}}(M) \cong H^r(M, \mathbf{R}).$$

This isomorphism is more explicitly induced by the mapping

$$\mathcal{E}^r(M) \to S^r(M, \mathbf{R})$$

$$\omega \mapsto \int_\gamma \omega,$$

for $\gamma \in H_r(M, \mathbf{R})$. Here we are representing sheaf cohomology with constant coefficients by means of singular cohomology. There are natural homomorphisms of singular cohomology (or sheaf cohomology, if we wish)

$$H^r(M, \mathbf{Z}) \overset{r}{\to} H^r(M, \mathbf{R}) \overset{c}{\to} H^r(M, \mathbf{C}). \qquad (3.4.1)$$

The mapping r might not be injective, since there may be nilpotent cohomology elements on M. On the other hand, c is injective, and is induced by the injection $\mathbf{R} \to \mathbf{C}$. A de Rham class is *integral* if it is in the image of r in (3.4.1). A closed differential form ω of degree r defines an integral de Rham class if and only if

$$\int_\gamma \omega \in \mathbf{Z}, \quad \text{for all } \gamma \in Z_q(M, \mathbf{Z})$$

(to be precise we consider only differentiable cycles γ, but this is sufficient). It will turn out that the Chern classes are always integral de Rham classes in this sense.

Let p be a homogeneous polynomial of degree k defined on the vector space of $r \times r$ matrices ($\mathfrak{gl}(r, \mathbf{C})$). There is a uniquely determined k-linear function $\tilde{p}$ defined on $\mathfrak{gl}(r, \mathbf{C})$ such that $p(X) = \tilde{p}(X, \ldots, X)$, called the *polarization* of p. This is described in detail in Exercise 3.4.1, but for instance for degree two we have

$$\tilde{p}(X, Y) = -\tfrac{1}{2}(p(X) + p(Y) - p(X + Y)).$$

Such a polynomial p is called *invariant* if

$$p(gXg^{-1}) = p(X), \quad \text{for all } g \in \mathrm{GL}(r, \mathbf{C}).$$

Clearly, if p is invariant, then its associated multilinear function $\tilde{p}$ is also invariant in the same sense.

If $E \to M$ is a complex vector bundle, then let D be a connection on E, and let $F = F_D$ be the curvature of D. If p is an invariant polynomial of degree k, then $p(F)$ is a well-defined differential form of degree $2k$ on M. To see this, we choose a frame f so that $F(f)$ is an $r \times r$ matrix of two-forms (F_β^α) with respect to the frame f. Now the polynomial is well-defined on the matrix entries since they are commuting two-forms, thus yielding a $2k$-form. If we consider a change of frame $f \mapsto fg$, for $g \in \mathrm{GL}(r, \mathbf{C})$, we have that

$$F(fg) = gF(f)g^{-1}.$$

Since p is invariant under such similarity transformations, we see that $p(F(f)) = p(F(fg))$, and hence is a well-defined $2k$-form on M, which is independent of the choice of frame and which we denote by $p(F)$. We have the following important theorem due to Weil.

Theorem 3.4.1 (Weil). *Let $E \to M$ be a complex vector bundle of rank r, and let p be an invariant polynomial on $\mathfrak{gl}(r, \mathbf{C})$ of degree k. Then:*

    (a) *If D is a connection on E, $p(F_D)$ is a closed $2k$-form.*

    (b) *If D_1 and D_2 are two different connections on E, then there is a $(2k - 1)$-form χ defined on M such that*

$$p(F_{D_1}) - p(F_{D_2}) = d\chi.$$

Corollary 3.4.2. $p(F_D)$ *defines a de Rham cohomology class which is independent of the connection D.*

The proof of the fundamental result can be found in various references and will not be repeated here (see, e.g., Chern 1967, Griffiths and Harris 1978, and Wells 1980). The basic tools used in the proof are the Bianchi identity $dF = [\theta, F]$, where θ is the connection one-form of D, the polarization of the invariant polynomial p, the differentiation formula for a product, and the fundamental theorem of calculus. It is a very elegant proof, and the reader is advised to see how it is carried out.

We can use this theorem to define the Chern classes. If A is an $r \times r$ matrix, then the expression

$$\det(I + tA),$$

for a fixed value of the parameter t, is an invariant polynomial which is not homogeneous. It has a decomposition

$$\det(I + tA) = \sum_{k=0}^{r} p_k(A) t^k,$$

and each polynomial $p_k(A)$ is homogeneous of degree k and is invariant. We note that $p_1(A) = \operatorname{tr}(A)$ and $p_r(A) = \det(A)$. The intermediate polynomials can be expressed in terms of traces and determinants of the minors of A (see Griffiths and Harris 1978).

If E is a complex vector bundle, and if D is a connection on E, then we define

$$c_k(E, D) := p_k\left(\frac{i}{2\pi}F_D\right)$$

The form $c_k(E, D)$ is, by Theorem 3.4.1, a closed $2k$-form, which is called the *kth Chern form* of E with respect to the connection D. The normalizing factor of $i/2\pi$ is inserted to ensure that the resulting cohomology classes will be integral. We then define

$$c_k(E) := \{\text{de Rham class of } c_k(E, D)\} \in H^{2k}_{\text{DR}}(M). \qquad (3.4.2)$$

This is the kth Chern class of E and is, by Theorem 3.4.1, independent of the choice of connection D. We define the *total Chern form* depending on a connection D to be

$$c(E, D) := \det\left(I + \frac{i}{2\pi}F_D\right),$$

and we denote its de Rham class, which we shall call the *total Chern class*, by $c(E)$, and we have the decomposition

$$c(E) = c_0(E) + c_1(E) + \cdots + c_r(E) \in H^*(M, \mathbf{C}).$$

It is not too difficult to show that the differential forms which appear in the Chern form can be chosen to have real coefficients (Exercise 3.4.2); however, it is a much deeper fact that the integrals of these cohomology classes over integral cycles will always be integers (see Chern 1967, Griffiths and Harris 1978, or Wells 1980, for a proof of this). If M is a manifold, and if $c \in H^k(M, \mathbf{R})$, then we define $c[\gamma]$ to be the evaluation of the cohomology class c on the homology class $[\gamma]$, defined by the k-cycle γ. If we represent c by a closed k-form ω, then we would have

$$c[\gamma] = \int_\gamma \omega.$$

In particular, the manifold M itself is an n-cycle in $H_n(M, \mathbf{Z})$, and if c is a cohomology class of degree n represented by a differential form ω, then we could write

$$c[M] = \int_M \omega.$$

Theorem 3.4.3 (Chern). *The Chern classes have the following properties:*

(a) $c_0(E) = 1$, *for any complex vector bundle E.*
(b) *If $H \to \mathbf{P}_1$ is the hyperplane section bundle then*

$$c_1(H)[\mathbf{P}_1] = 1.$$

(c) *If $f : M \to N$ is a differentiable mapping of manifolds, and if E is a complex vector bundle over N, then*

$$c(f^*E) = f^*c(E),$$

where f^*E is the pullback vector bundle under the mapping[2] f, and $f^*(c(E))$ is the pullback of the cohomology class under the mapping f.
(d) *If $E = E_1 \oplus E_2$ is a direct sum of complex vector bundles, then*

$$c(E) = c(E_1)c(E_2),$$

where the product is in the cohomology ring $H^*(M, \mathbf{R})$.

These four properties characterize the Chern classes, and there are approximately eight independent constructions of Chern classes which have these properties. It is not very difficult to verify the properties using the definition we have given here, and we shall leave the details to the references cited above. As an example, we shall indicate how property (d) is proven. Choose a connection D on E which is the direct sum of two connections in the sense that their connection one-forms have the form

$$A = \begin{pmatrix} A_1 & 0 \\ 0 & A_2 \end{pmatrix}.$$

The curvature will have the form

$$F = \begin{pmatrix} F_1 & 0 \\ 0 & F_2 \end{pmatrix}.$$

---

[2] See Exercise 3.4.3.

It then follows that

$$c(E, D) = \det\left(I + \frac{i}{2\pi}F\right)$$

$$= \det\begin{pmatrix} I_1 + \dfrac{i}{2\pi}F_1 & 0 \\ 0 & I_2 + \dfrac{i}{2\pi}F_2 \end{pmatrix}$$

$$= \det\left(I_1 + \frac{i}{2\pi}F_1\right)\det\left(I_2 + \frac{i}{2\pi}F_2\right)$$

$$= c(E_1, D_1)c(E_2, D_2),$$

as desired.

We see immediately that if E is trivial, then $c_j(E) = 0$, for $j > 0$ (choose a flat connection). Moreover, if $E = G \oplus T^l$, where T^l is a trivial bundle of rank l, then

$$c(E) = c(G)c(T^l) = c(G)\cdot 1.$$

But $c_k(G) = 0$, for $k > \mathrm{rank}\, G = r - l$, so $c_k(E) = 0$, for $k > r - l$.

A final important result we shall mention is the Chern version of a Gauss–Bonnet formula. If we define the *Euler characteristic* of a manifold M of dimension $2n$ to be the alternating sum of the Betti numbers

$$\chi(M) = \sum_{k=0}^{2n}(-1)^k b_k(M),$$

then we have the formula

$$\int_M c_n(T(M)) = \chi(M).$$

This can be found in Chern (1967) or Griffiths and Harris (1978).

We shall end this section with one special case which will play a role in the theory of instantons later. Let M be a compact orientable four-manifold, and let $E \to M$ be an SU(2)-bundle over M. Then we have following formula.

Proposition 3.4.4. *If D is an SU(2)-connection on E, and $F = F_D$ is the curvature two-form of E with respect to D, then*

$$c_2(E)[M] = \frac{1}{8\pi^2}\int_M \mathrm{tr}(F \wedge F) \in \mathbf{Z}.$$

Proof: To check this formula, we can choose an SU(2)-frame for E and represent F with respect to this frame. Note that the trace of $F \wedge F$ which appears in the formula will be independent of the choice of frame. Thus,

$$F = \begin{pmatrix} F_1^1 & F_1^2 \\ F_2^1 & F_2^2 \end{pmatrix}.$$

We note that the connection one-form and curvature two-form are $\mathfrak{su}(2)$-valued. Thus we have

$$\begin{aligned} F_1^1 &= -F_2^2, \\ F_\alpha^\beta &= -\overline{F}_\beta^\alpha, \end{aligned} \tag{3.4.3}$$

as differential forms. Now we can calculate

$$\begin{aligned} c_2(E, D) &= \left(\frac{i}{2\pi}\right)^2 \det \begin{pmatrix} F_1^1 & F_1^2 \\ F_2^1 & F_2^2 \end{pmatrix} \\ &= -\frac{1}{4\pi^2} \left(F_1^1 \wedge F_2^2 - F_2^1 \wedge F_1^2 \right). \end{aligned}$$

On the other hand, we can calculate

$$\operatorname{tr}(F \wedge F) = \operatorname{tr} \left(\begin{pmatrix} F_1^1 & F_1^2 \\ F_2^1 & F_2^2 \end{pmatrix} \wedge \begin{pmatrix} F_1^1 & F_1^2 \\ F_2^1 & F_2^2 \end{pmatrix} \right)$$

Using (3.4.3) above and recalling that two-forms commute, we find that

$$\operatorname{tr}(F \wedge F) = -2 \left(F_1^1 \wedge F_2^2 - F_2^1 \wedge F_1^2 \right).$$

Hence we obtain

$$c_2(E, D) = \frac{1}{8\pi^2} \operatorname{tr}(F \wedge F),$$

and this yields the desired result. ∎

At the beginning of this section we gave the example

$$H^1(M, \mathcal{E}^*) \xrightarrow{\delta} H^2(M, \mathbf{Z})$$

as an example of a characteristic class mapping. In this case we have the line bundle (isomorphism class) $L \mapsto \delta(L) = c_1(L)$, where $c_1(L)$ is the first Chern class of the line bundle L. See, for instance Wells

(1980), for a proof of this fact. Thus the sheaf cohomology gives a means of definition using the exact sequence

$$0 \to \mathbf{Z} \to \mathcal{E} \to \mathcal{E}^* \to 0.$$

This approach does not adapt readily for higher rank vector bundles, due to the fact that sheaf cohomology does not make sense for sheaves of nonabelian groups. However, some aspects of the sheaf cohomology theory do generalize, and they can be very useful, as we shall soon see.

We can consider sheaves of nonabelian groups in the same manner as abelian groups. There is no problem here; the difficulty lies in the cohomology theory. Let G be a Lie group, and let M be a manifold. Define G_c to be the sheaf defined by

$$U \mapsto \{f : U \to G, \quad f \text{ continuous}\}.$$

We see that G_c is clearly a sheaf, and we can define G_d to be the sheaf of differentiable mappings into G. In the case that $G = \mathbf{C}^*$, then $G_d = \mathcal{E}^*$, the sheaf of nonvanishing smooth functions on M, which we used above. If $\mathcal{F}$ is such a sheaf, then we can define $H^0(M, \mathcal{F}) = \Gamma(M, \mathcal{F})$. We can define the *cohomology set*

$$H^1(M, \mathcal{F})$$

with a distinguished element 1, where $\mathcal{F}$ is a sheaf of nonabelian groups, but it will not be a group in general. To do this, we define $H^1(\mathfrak{U}, \mathcal{F})$ for a covering $\mathfrak{U}$ in the same manner as the Čech definition of $H^1(M, \mathcal{F})$ for a sheaf of abelian groups, just writing everything multiplicatively. Namely, if $g_{\alpha\beta} \in \Gamma(U_\alpha \cap U_\beta, \mathcal{F})$, then

$$H^1(\mathfrak{U}, \mathcal{F}) = \frac{\{g_{\alpha\beta} : g_{\alpha\beta} g_{\beta\gamma} g_{\gamma\alpha} = \mathrm{id}\}}{\{g_{\alpha\beta} : g_{\alpha\beta} = g_\alpha g_\beta^{-1}\}}$$

If we take a direct limit over coverings, we obtain

$$H^1(M, \mathcal{F}) = \varinjlim_{\mathfrak{U}} \mathrm{ind}\, H^1(\mathfrak{U}, \mathcal{F}).$$

Now $g_{\alpha\beta} = 1$ on $U_\alpha \cap U_\beta$ defines the distinguished element 1 of $H^1(M, \mathcal{F})$.
   If

$$1 \to \mathcal{A} \to \mathcal{B} \to \mathcal{C} \to 1$$

is a short exact sequence of sheaves of multiplicative abelian groups, there is an exact sequence of mappings (Exercise 3.4.5)

$$1 \to H^0(M, \mathcal{A}) \to H^0(M, \mathcal{B}) \to H^0(M, \mathcal{C})$$
$$\to H^1(M, \mathcal{A}) \to H^1(M, \mathcal{B}) \to H^1(M, \mathcal{C}). \qquad (3.4.4)$$

If $\mathcal{A}$ is abelian, it continues to

$$\cdots \to H^1(M, \mathcal{C}) \to H^2(M, \mathcal{A}),$$

and if $\mathcal{A}$ and $\mathcal{B}$ are abelian, it continues to

$$\cdots \to H^2(M, \mathcal{A}) \to H^2(M, \mathcal{B}).$$

Exactness means that the preimage of 1 agrees with the image of the previous term. The construction of the mapping is just as before. We shall get a chance to use this sequence in the next section, when we have the sequence, writing $\mathbf{Z}_2 = \{\pm 1\}$

$$1 \to \mathbf{Z}_2 \to \text{Spin}(n) \to \text{SO}(n) \to 1,$$

where $\text{Spin}(n)$ will be a covering group of $\text{SO}(n)$ (in fact a two-to-one covering of $\text{SO}(n)$). We shall then use this cohomology theory to help discuss which manifolds admit spin structures.

Exercises for Section 3.4

3.4.1. Define

$$\widetilde{\varphi}(A_1, \ldots, A_k) = \frac{(-1)^k}{k!} \sum_{j=1}^{k} \sum_{i_1 < \cdots < i_j} (-1)^j \varphi(A_{i_1} + \cdots + A_{i_j}),$$

where φ is a homogeneous polynomial of degree k defined on the complex vector space of $r \times r$ matrices. Show that $\widetilde{\varphi}$ is k-linear and invariant under the action of $\text{GL}(r, \mathbf{C})$ if φ is, and that it has the property that $\widetilde{\varphi}(A, \ldots, A) = \varphi(A)$.

3.4.2. Show that the Chern forms on a complex manifold M can be chosen to represent cohomology classes in $H^*(M, \mathbf{R})$.

3.4.3. If $E \to X$ is a vector bundle and $f : Y \to X$ is a differentiable mapping, then the pullback bundle f^*E is a vector bundle $\widetilde{E} \to Y$ such that

$$
\begin{array}{ccc}
\widetilde{E} & \longrightarrow & E \\
\downarrow & & \downarrow \\
Y & \longrightarrow & X
\end{array}
$$

is a commutative bundle mapping. Show that f^*E exists, and is unique up to a bundle isomorphism.

3.4.4. Verify part (c) of Theorem 3.4.3.

3.4.5. Verify (3.4.4).

3.5 Clifford algebras and spin bundles

In §2.3 we discussed a local theory of spin bundles on Minkowski space and, in particular, the local relation between spinors and vectors, the fundamental relation being of the form $v^a \sim v^{AA'}$, where A, A' are spinor indices. This is the relation between four-vectors and 2×2 matrices (see Example 2.3.3). The global theory is more complicated and quite important. In this section we shall use the principal bundle point of view to develop a general theory of spin bundles on real and complex manifolds, using the cohomology theory of §3.3 as a principal tool. We shall use Clifford algebras to describe the Spin and Spin^c groups, which are Lie groups related to orthogonal groups in a natural manner. The basic idea is illustrated by the mapping (described below)

$$
\mathrm{Spin}(4) \xrightarrow{\rho} \mathrm{SO}(4), \tag{3.5.1}
$$

where Spin(4) is a specific Lie group described by the Clifford algebra of E^4, and ρ is a two-to-one mapping. A principal SO(4)-bundle F on a space X generates a tensor algebra of vector bundles on X by representations of the group SO(4). If $\dim X = 4$, then these are *tensor bundles* on X, with a specific Riemannian metric induced from a Riemannian metric on the tangent bundle $T(X) \to X$, given by the principal bundle F. If there is a lifting of the principal bundle of orthonormal frames $F \to X$ (with fiber SO(4)) to a Spin(4)-principal bundle $\widetilde{F}$ with fiber Spin(4) covering the fibers of F by (3.5.1), then associated with $\widetilde{F}$ is a family of vector bundles, induced from representations of the group Spin(4). Some of these representations arise from representations of SO(4), and some do not. The family of bundles will be called spin bundles, each spin bundle being associated with a specific representation of Spin(4). Just as covariant differentiation acts on sections of the tensor bundles, we shall see that there is an associated *Dirac operator* which will act on sections of the spin bundles. In §4.2 we shall see the Dirac operator locally in the original physical context in which it was discovered, and in a fashion used by physicists today. We shall concentrate here only on the global structure of the bundles, and in Chapter 7 we shall see, in an invariant form, the natural differentiation operators which act on sections of these bundles, including the classical Dirac operators.

Let V be a real vector space equipped with an inner product $\langle\ ,\ \rangle$, defined by a nondegenerate quadratic form Q of signature (p,q). Let $\mathcal{T}(V)$ be the tensor algebra of V and consider the two-sided ideal $\mathcal{J}$ in $\mathcal{T}(V)$ generated by $x \otimes x + Q(x)$. The quotient space

$$\mathrm{Cl}(V) := \mathrm{Cl}(V, Q) := \mathcal{T}(V)/\mathcal{J}$$

is the *Clifford algebra* of the vector space V equipped with the quadratic form Q. The product induced by the tensor product in $\mathcal{T}(V)$ is known as *Clifford multiplication* or the *Clifford product* and is denoted by $x \cdot y$ for $x, y \in \mathrm{Cl}(V)$. We shall summarize here some of the main properties of Clifford algebras which are important to us in a development of spin bundles. We refer to Harvey (1989) and Lawson and Michelson (1988) for a more detailed analysis of these topics. The classical reference, which contains most of what we need, is Atiyah, Bott and Shapiro (1964). The Clifford algebras of various dimensions and signatures have been classified, and their structure is periodic of period eight with respect to the dimension of V. For instance, if we let $\mathrm{Cl}(n) = \mathrm{Cl}(V)$ where V has dimension n, and Q is positive-definite, we have

$$\mathrm{Cl}(1) \cong \mathbf{C}, \qquad \dim_{\mathbf{R}} = 2,$$
$$\mathrm{Cl}(2) \cong \mathbf{H}, \qquad \dim_{\mathbf{R}} = 2^2,$$
$$\mathrm{Cl}(3) \cong \mathbf{H} \otimes \mathbf{H}, \qquad \dim_{\mathbf{R}} = 2^3,$$

etc. The dimension of $\mathrm{Cl}(V)$ is 2^n if $\dim V = n$. A basis for $\mathrm{Cl}(V)$ is given by the scalar 1 and the products

$$e_{i_1} \cdot e_{i_2} \cdot \ \cdots \ \cdot e_{i_n}, \quad i_1 < \cdots < i_n,$$

where $\{e_1, \ldots, e_n\}$ is an orthonormal basis for V. Moreover the products satisfy

$$e_i \cdot e_j + e_j \cdot e_i = 0 \quad i \neq j,$$
$$e_i \cdot e_i = -2\langle e_i, e_i \rangle \quad i = 1, \ldots, n.$$

As a vector space $\mathrm{Cl}(V)$ is isomorphic to $\bigwedge^*(V)$, the Grassmann algebra, with

$$e_{i_1} \cdot \cdots \cdot e_{i_n} \mapsto e_{i_1} \wedge \cdots \wedge e_{i_n},$$

but this isomorphism does not preserve the product structure.

There are two natural involutions on $\mathrm{Cl}(V)$. The first, denoted by $\alpha : \mathrm{Cl}(V) \to \mathrm{Cl}(V)$, is induced by the involution $x \mapsto -x$ defined on V, which extends to an automorphism of $\mathrm{Cl}(V)$. The eigenspace of α with eigenvalue $+1$ consists of the *even elements* of $\mathrm{Cl}(V)$ and the eigenspace of α of eigenvalue -1 consists of the *odd elements* of $\mathrm{Cl}(V)$ (corresponding to an even or odd number of products in a typical term of a representation of an element of V in terms of a basis).

The second involution is a mapping $x \mapsto x^t$, induced on generators by

$$(e_{i_1} \cdot \cdots \cdot e_{i_p})^t = e_{i_p} \cdot \cdots \cdot e_{i_1},$$

where e_i are basis elements of V. In addition, we define $x \mapsto \bar{x}$, a third involution of $\mathrm{Cl}(V)$, by $\bar{x} = \alpha(x^t)$. For $\mathrm{Cl}(1)$ and $\mathrm{Cl}(2)$, $\bar{x}$ can be identified easily with complex and quaternionic conjugation, respectively, as we see below. The generator 1 of the scalars in $\mathrm{Cl}(V)$ induces an embedding $\mathbf{R} \to \mathrm{Cl}(V)$, and we see that $x \cdot \bar{x} \in \mathbf{R}$ under this embedding. We define $\mathrm{Cl}^*(V)$ to be the group of invertible elements of $\mathrm{Cl}(V)$, and we define the *Clifford group* $\Gamma(V)$ to be the subgroup of $\mathrm{Cl}^*(V)$ defined by

$$\Gamma(V) := \{x \in \mathrm{Cl}^*(V) : y \in V \Rightarrow \alpha(x)yx^{-1} \in V\},$$

where V is embedded in $\mathrm{Cl}(V)$ in the natural manner. One can show that the mapping $\rho(x) : V \to V$ given by $\rho(x)y = \alpha(x)yx^{-1}$ is an isometry of V with respect to the quadratic form Q (Exercise 3.5.1). The mapping $x \mapsto \|x\| := x\bar{x}$, is the *square-norm mapping*. It defines a homomorphism of groups

$$\mathrm{Cl}^*(V) \xrightarrow{\;\|\;\|\;} \mathbf{R}^*,$$

and thus any element v in $\text{Cl}^*(V)$ has $\|v\| \neq 0$, and this is a necessary and sufficient condition.

We thus have an exact sequence

$$1 \to \mathbf{R}^* \to \Gamma(V) \to O(V) \to 1,$$

where $O(V)$ are the automorphisms of (V, Q), i.e., orthogonal transformations of V with respect to the quadratic form Q. Let $\text{Pin}(V) := \{x \in \Gamma(V) : \|x\| = 1\}$. We have, letting $\mathbf{Z}_2 = \pm 1$,

$$1 \to \mathbf{Z}_2 \to \text{Pin}(V) \xrightarrow{\rho} O(V) \to 1;$$

and if we let $\text{Spin}(V) := \rho^{-1}(SO(V))$, then we have

$$1 \to \mathbf{Z}_2 \to \text{Spin}(V) \xrightarrow{\rho} SO(V) \to 1.$$

For the case where $\dim V = 2$ and Q is positive-definite, we have

$$\text{Cl}(V) \cong_{\mathbf{R}} \text{span}\{1, e_1, e_2, e_1 \cdot e_2\},$$
$$e_1^2 = e_2^2 = (e_1 \cdot e_2)^2 = -1,$$

and thus,

$$\text{Cl}(V) \cong \mathbf{H}, \quad \text{with } \mathbf{i} = e_1, \mathbf{j} = e_2, \mathbf{k} = e_1 \cdot e_2.$$

If

$$v = x^0 + x^1 e_1 + x^2 e_2 + x^3 e_1 \cdot e_2 \in V,$$

then

$$\bar{v} = x^0 - x^1 e_1 - x^2 e_2 + x^3 e_2 \cdot e_1$$
$$= x^0 - x^1 e_1 - x^2 e_2 - x^3 e_1 \cdot e_2,$$

the usual quaternionic conjugation.

If we let G_0 be the identity component of a Lie group G, then we have the following fundamental diagram relating these various groups, which summarizes the above discussion.

$$
\begin{array}{ccccccccc}
1 & \to & \mathbf{R}^* & \to & \Gamma(V) & \to & O(V) & \to & 1 \\
  &   & \uparrow &   & \uparrow &   & \uparrow &   & \\
1 & \to & \mathbf{Z}_2 & \to & \text{Pin}(V) & \to & O(V) & \to & 1 \\
  &   & \uparrow &   & \uparrow &   & \uparrow &   & \\
1 & \to & \mathbf{Z}_2 & \to & \text{Spin}(V) & \to & SO(V) & \to & 1 \\
  &   & \uparrow &   & \uparrow &   & \uparrow &   & \\
1 & \to & \mathbf{Z}_2 & \to & \text{Spin}(V)_0 & \to & SO(V)_0 & \to & 1
\end{array}
\qquad (3.5.2)
$$

This is all proven in the references cited above; see Harvey (1989), in particular, for the indefinite case. For the case $\dim V > 1$, $\mathrm{Spin}(V)$ is a simply-connected two-to-one covering group of $\mathrm{SO}(V)$. If $V = M^4$, with signature $(1,3)$, then we saw in §1.6 that

$$1 \to \mathbf{Z}_2 \to \mathrm{SL}(2, \mathbf{C}) \to \mathrm{SO}(1,3)_0 \to 1, \qquad (3.5.3)$$

and hence we can identify $\mathrm{SL}(2, \mathbf{C}) \cong \mathrm{Spin}(1,3)_0$.

Suppose we have a Lorentzian manifold X which is time-oriented and oriented. That is, the orthonormal frame bundle F for X is an $\mathrm{O}(1, n-1)$-bundle which can be reduced to $\mathrm{SO}(1, n-1)_0$. Then the manifold M has *spin structure* if and only if there is a two-to-one covering $\widetilde{F} \overset{\tilde{\times}\sigma}{\to} F$ of the principal bundle F, where $\widetilde{F}$ is a principal bundle with structure group $\mathrm{Spin}(1, n-1)_0$, and

$$
\begin{array}{ccccc}
\mathrm{Spin}(1, n-1)_0 & \to & \widetilde{F} & \to & X \\
\downarrow \rho & & \downarrow \sigma & & \downarrow \mathrm{id} \\
\mathrm{SO}(1, n-1)_0 & \to & F & \to & X
\end{array}
\qquad (3.5.4)
$$

is a commutative diagram of principal bundles. If X is an oriented Riemannian manifold, then X has a spin structure if and only if there is a two-to-one covering $\widetilde{F} \overset{\tilde{\times}\sigma}{\to} F$ of the $\mathrm{SO}(n)$-frame bundle F for X, where $\widetilde{F}$ is a $\mathrm{Spin}(n)$-bundle, and there is a commutative diagram

$$
\begin{array}{ccccc}
\mathrm{Spin}(n) & \to & \widetilde{F} & \to & X \\
\downarrow & & \downarrow & & \downarrow \mathrm{id} \\
\mathrm{SO}(n) & \to & F & \to & X
\end{array}
\qquad (3.5.5)
$$

We shall refer to the diagrams (3.5.4) and (3.5.5) as the *spin structure* of the manifold X. More generally, one considers a $\mathrm{Spin}(p,q)_0$- or $\mathrm{Spin}(p,q)$-structure on an $\mathrm{SO}(p,q)_0$-manifold or on an $\mathrm{SO}(p,q)$-manifold.

We shall refer to $(\widetilde{F}, \sigma)$ in (3.5.4) and (3.5.5) as the spin structure for simplicity, the diagrams being understood. There is a natural notion of *equivalence* of spin structures, i.e., an isomorphism of principal bundles $f : \widetilde{F} \to \widetilde{F}'$ so that the diagrams with $(\widetilde{F}, \tilde{\times}\sigma)$ and $(\widetilde{F}', \tilde{\times}\sigma')$ of the form (3.5.4) and (3.5.5) commute with the mapping f. Existence of a spin structure involves two things:

(1) the existence of a metric on X of the correct signature, and a reduction of the frame bundle to either $\mathrm{SO}(p,q)$ or $\mathrm{SO}(p,q)_0$ as the case may be;
and

(2) topological conditions on the manifold X, as we shall see below.

More generally, if G is a Lie group, and we have a two-to-one covering group $\widetilde{G}$ with short exact sequence

$$1 \to \mathbf{Z}_2 \to \widetilde{G} \to G \to 1,$$

then a principal G-bundle ξ has a spin structure $(\tilde{\xi}, \sigma)$, if $\tilde{\xi}$ is a $\widetilde{G}$-bundle which is a two-to-one covering manifold of ξ with covering mapping σ, and which satisfies the analogue of (3.5.4).

Let us consider a differentiable manifold X which admits a reduction of its frame bundle to G, where $G = SO(p,q)$ or $SO(p,q)_0$. Let $\widetilde{G}$ be the associated spin group, as in the above discussion. We want to consider the cohomology groups $H^q(X, G_d)$ and $H^q(X, \widetilde{G}_d)$, where we recall from §3.4 that G_d is the sheaf of differentiable mappings from X to G, which is a sheaf of nonabelian groups if G is nonabelian. We have the sequence of Lie groups

$$1 \to \mathbf{Z}_2 \to \widetilde{G} \to G \to 1,$$

which yields the short exact sequence of sheaves

$$1 \to \mathbf{Z}_2 \to \widetilde{G}_d \overset{\rho}{\to} G_d \to 1,$$

and the (truncated) long exact sequence (3.4.4)

$$
\begin{aligned}
1 \quad &\to \quad H^0(X, \mathbf{Z}_2) \quad \to \quad H^0(X, \widetilde{G}_d) \overset{\rho_0}{\to} H^0(X, G_d) \overset{\delta_1}{\to} \\
&\to \quad H^1(X, \mathbf{Z}_2) \quad \to \quad H^1(X, \widetilde{G}_d) \overset{\rho_1}{\to} H^1(X, G_d) \overset{\delta_2}{\to} \\
&\quad H^2(X, \mathbf{Z}_2).
\end{aligned}
$$

$$(3.5.6)$$

Here δ_1 and δ_2 are the connecting homomorphisms, and ρ_0 and ρ_1 are the mappings induced by ρ. We see that if $\xi \in H^1(X, G_d)$, then ξ is a G-principal bundle, and if $\tilde{\xi} \in H^1(X, \widetilde{G}_d)$, then $\tilde{\xi}$ is a $\widetilde{G}$-principal bundle. If $\tilde{\xi}$ is given, then $\rho_1(\tilde{\xi}) = \xi$ is a well-defined G-bundle, and there is an induced mapping $\tilde{\xi} \overset{\times \sigma}{\to} \xi$. Thus, any $\tilde{\xi}$ determines a spin structure for $\rho_1(\tilde{\xi})$. Conversely, if ξ is a given G-bundle, and $\delta_2(\xi) \in H^2(X, \mathbf{Z}_2)$ vanishes, then $\xi = \rho_1(\tilde{\xi})$, for some $\tilde{\xi}$, i.e., ξ admits a spin structure. In fact, ξ admits a spin structure if and only if $\delta_2(\xi) = 0$, as we see.

How many distinct spin structures are there for a given ξ? As Milnor points out (1963), one can have inequivalent spin structures

$(\tilde{\xi}, \tilde{\times}\sigma)$, $(\tilde{\xi}', \sigma')$ for a given G-bundle ξ, where the principal $\widetilde{G}$-bundles $\tilde{\xi}$, $\tilde{\xi}'$ are *equivalent* as $\widetilde{G}$-bundles. In other words, one has the same bundle $\tilde{\xi}$, but inequivalent mappings $\tilde{\xi} \overset{\tilde{\times}\sigma}{\to} \xi$, $\tilde{\xi} \overset{\sigma'}{\to} \xi$. A theorem of Milnor's (1963) shows that the number of inequivalent spin structures for a given $\xi \in H^1(X, G_d)$ is measured by $H^1(X, \mathbf{Z}_2)$. This is proved at the end of §3.6 (Theorem 3.6.7), and we shall assume the result here. We note that in the sequence (3.5.6) the term $H^1(X, \mathbf{Z}_2)$ is an obstruction to the surjectivity of ρ_0. If X is not simply-connected, ρ_0 is not necessarily surjective, as one can see by the example

$$1 \to \mathbf{Z}_2 \to \mathrm{Spin}(2) \overset{\rho}{\to} \mathrm{SO}(2) \to 1,$$
$$\parallel \qquad\qquad \parallel$$
$$S^1 \qquad\qquad S^1$$

where $\rho(e^{i\theta}) = e^{2i\theta}$. If we let $X = S^1 \times \mathbf{R}$, then the mapping

$$\gamma(e^{i\theta}, x) = e^{i\theta}$$

has no lifting to $\tilde{\gamma} : S^1 \times \mathbf{R} \to \mathrm{Spin}(2)$ for which $\rho \circ \tilde{\gamma} = \gamma$, as one easily checks.

So, in general, the number of inequivalent spin bundles is measured by

$$\ker \rho_1 \cong H^1(X, \mathbf{Z}_2)/\delta_1 H^0(X, G_d).$$

In the example above of $X = S^1 \times \mathbf{R}$, we see that in the sequence (3.5.6), ρ_0 is not surjective, so $\delta_1 \neq 0$, which implies in this case that $\ker \rho_1 = 1$. Thus there is a unique spin bundle $\tilde{\xi}$ associated with a given $\xi \cong H^1(S^1 \times \mathbf{R}, \mathrm{SO}(2))$. But $H^1(S^1 \times \mathbf{R}, \mathbf{Z}_2) \cong \mathbf{Z}_2$, so there are two distinct spin structures for the given ξ, by the theorem of Milnor (Theorem 3.6.7).

We now summarize these results in the following theorem for the special case of frame bundles, which are of the most importance here.

Theorem 3.5.1. *A pseudo-Riemannian structure $F \in H^1(X, G_d)$ has a spin structure $\widetilde{F} \overset{\tilde{\times}\sigma}{\to} F$, where $\widetilde{F} \in H^1(X, \widetilde{G}_d)$, if and only if $\delta_2(F) \in H^2(X, \mathbf{Z}_2)$ vanishes. Moreover, the inequivalent spin structures associated with F are in one-to-one correspondence with $H^1(X, \mathbf{Z}_2)$.*

Remark: The topological invariant $\delta_2(F) \in H^2(X, \mathbf{Z}_2)$ can be identified with the *second Stiefel–Whitney class* of the principal bundle F. For other characterizations of this characteristic class see Hirzebruch (1966), for instance.

Corollary 3.5.2. *Compactified Minkowski space M has two distinct spin structures, and the four-sphere S^4 has a unique spin structure.*

Proof: Compactified Minkowski space M is orientable and time-orientable, hence admits a Lorentz metric of signature $(1,3)$ which is reducible to $SO(1,3)_0$ (see Exercise 3.5.3). Since $H^2(S^1 \times S^3, \mathbf{Z}_2) = 0$, there exist spin structures for any $F \in H^1(S^1 \times S^3, (SO(1,3)_0)_d)$. Moreover, since $H^1(S^1 \times S^3, \mathbf{Z}_2) \cong \mathbf{Z}_2$, it follows that there are precisely two such structures. Similarly, S^4 is orientable and admits $SO(4)$-frame bundles. Moreover, $H^2(S^4, \mathbf{Z}_2) = 0$, so each such frame bundle $F \in H^1(S^4, SO(4))$ has a spin structure with spin-frame bundle $\widetilde{F} \in H^1(S^4, \mathrm{Spin}(4)_d)$, and this spin structure is unique, since $H^1(S^4, \mathbf{Z}_2) = 0$. $\blacksquare$

We now want to introduce the notion of Spin^c-bundles. Consider the Clifford algebra $\mathrm{Cl}(V)$ as above, and let $\mathrm{Cl}(V)^c := \mathrm{Cl}(V) \otimes \mathbf{C}$ be its complexification. Then we can consider $(\mathrm{Cl}(V)^c)^*$, the invertible elements, and extend the involutions by $\alpha(x \otimes z) = \alpha(x) \otimes z, (x \otimes z)^t = x^t \otimes \bar{z}$ to the complexification. Then the generalized Clifford group $\Gamma^c(V)$ is defined by

$$\Gamma^c(V) := \{x \in (\mathrm{Cl}(V)^c)^* : \alpha(x)yx^{-1} \in V \text{ if } y \in V\}.$$

Note that we are *not* complexifying V here, so we obtain a mapping $\Gamma^c(V) \to O(V)$, and the kernel is now larger than $\mathbf{Z}_2$. If we define $\mathrm{Pin}^c(V) := \{x \in \Gamma^c(V) : \|x\| = 1\}$, then we have the following proposition whose proof we leave as an exercise.

Proposition 3.5.3. *The following sequence is exact:*

$$1 \to U(1) \to \mathrm{Pin}^c(V) \xrightarrow{\rho} O(V) \to 1.$$

As a corollary we obtain the following result.

Corollary 3.5.4. *Let $\mathbf{Z}_2$ act on $\mathrm{Pin}(V)$ and $U(1)$ as ± 1, then*

$$\mathrm{Pin}(V) \times_{\mathbf{Z}_2} U(1) \cong \mathrm{Pin}^c(V).$$

Define $\mathrm{Spin}^c(V) := \{x \in \mathrm{Pin}^c(V) : \det \rho(x) = 1\}$; then we have from Corollary 3.5.4

$$\mathrm{Spin}^c(V) \cong \mathrm{Spin}(V) \times_{\mathbf{Z}_2} U(1). \tag{3.5.7}$$

A given $U(n)$-bundle on a manifold X is an $SO(2n)$-bundle under the natural embedding $U(n) \to SO(2n)$. However, this mapping may not lift to $\mathrm{Spin}(2n)$. So, the existence of a complex structure on a bundle of rank $2n$ does not necessarily yield a Spin-bundle. However, it does yield a Spin^c-structure, a less restrictive requirement. This is indicated in the following diagram. There is a mapping ℓ so that

$$
\begin{array}{ccc}
U(n) & \xrightarrow{\ \ell\ } & \mathrm{Spin}(2n) \times U(1) \\
\downarrow & & \downarrow \\
SO(2n) & \longleftarrow & \mathrm{Spin}^c(2n)
\end{array}
$$

The mapping ℓ is given by the following expression, where we choose an orthonormal basis $e_1, \ldots, e_{2n}$ for $\mathbf{R}^{2n}$, so that $T \in U(n)$ has the form

$$
T = \begin{pmatrix} e^{i\theta_1} & & 0 \\ & \ddots & \\ 0 & & e^{i\theta_n} \end{pmatrix},
$$

and we let

$$
\ell(T) = \prod_{j=1}^{n} (\cos(\theta_j/2) + \sin(\theta_j/2)(e_{2j-1} \cdot e_{2j}) \times \exp\left(\frac{i\Sigma\theta_j}{2}\right)).
$$

Thus any $U(n)$-frame bundle on a $2n$-manifold induces a $\mathrm{Spin}^c(2n)$-structure on X. In certain cases, we shall be able to see that the $\mathrm{Spin}^c(2n)$-structure reduces to a $\mathrm{Spin}(2n)$-structure on certain real submanifolds of X.

What is the obstruction to reducing from a Spin^c-structure to a Spin-structure on a given manifold? Consider the exact sequence

$$
1 \to \mathbf{Z}_2 \to \mathrm{Spin}(2n) \to \mathrm{Spin}^c(2n) \to U(1) \to 1, \qquad (3.5.8)
$$

which follows from the above description of Spin^c in terms of Spin and $U(1)$ (3.5.7). This reduces to two short exact sequences

$$
1 \to \mathbf{Z}_2 \to \mathrm{Spin}(2n) \to \mathrm{Spin}(2n)/\mathbf{Z}_2 \to 1,
$$
$$
1 \to \mathrm{Spin}(2n)/\mathbf{Z}_2 \to \mathrm{Spin}^c(2n) \to U(1) \to 1.
$$

We thus have the cohomology exact sequences

$$
H^1(X, \mathrm{Spin}(2n)_d) \to H^1(X, \mathrm{Spin}(2n)_d/\mathbf{Z}_2) \to H^2(X, \mathbf{Z}_2),
$$
$$
\qquad (3.5.9)
$$
$$
H^1(X, \mathrm{Spin}(2n)_d/\mathbf{Z}_2) \to H^1(X, \mathrm{Spin}^c(2n)_d) \to H^1(X, U(1)_d).
$$
$$
\qquad (3.5.10)
$$

But

$$H^1(X, U(1)_d) \cong H^2(X, \mathbf{Z}), \qquad (3.5.11)$$

as groups. This follows from the fact that $H^1(X, U(1)_d) \cong H^1(X, \mathcal{E}^*)$ in the differentiable category (by choosing an Hermitian metric in the principal bundle defined by an element of $H^1(X, \mathcal{E}^*)$) and by the Chern class isomorphism (see Example 3.3.12)

$$H^1(X, \mathcal{E}^*) \cong H^2(X, \mathbf{Z}).$$

Thus we have the following proposition.

Proposition 3.5.5. *If $H^2(X, \mathbf{Z}) = 0$, then any* $\mathrm{Spin}^c(n)$*-bundle can be reduced to a* $\mathrm{Spin}(n)$*-bundle.*

Corollary 3.5.6. *Let $S \to \mathsf{M}$ be any* $\mathrm{GL}(2, \mathbf{C})$*-bundle. Then S reduces to a* $\mathrm{Spin}(1, 3)$*-bundle on M and to a* $\mathrm{Spin}(4)$*-bundle on S^4.*

Proof: Let P be a principal $\mathrm{GL}(2, \mathbf{C})$-bundle on M. Then by a choice of Hermitian metric on P, P reduces to a $U(2)$-bundle on M. The $U(2)$-structure on P makes P into a $\mathrm{Spin}^c(4)$-bundle on M and on any subset of M. Restricting P to $S^4 \subset \mathsf{M}$, makes P into a $\mathrm{Spin}^c(4)$-bundle over S^4. But $H^2(S^4, \mathbf{Z}) = 0$, so by Proposition 3.5.5, P reduces to a $\mathrm{Spin}(4)$-bundle on S^4, and hence is a $\mathrm{Spin}(4)$-structure for the associated Riemannian $\mathrm{SO}(4)$-bundle induced by the projection $\mathrm{Spin}(4) \to \mathrm{SO}(4)$.

The reduction of P to M is different. The $\mathrm{GL}(2, \mathbf{C})$-bundle P is not reducible to an $\mathrm{SL}(2, \mathbf{C})$-bundle on M, since $\det P$ may be a nontrivial $\mathrm{GL}(1, \mathbf{C})$-bundle on M. For instance, if P is the principal bundle associated with $U :=$ universal bundle of $\mathsf{M} \cong \mathbf{G}(2, 4, \mathbf{C})$, then $\bigwedge^2 U$ is not a trivial bundle (see Exercise 3.5.4). On the other hand, if we restrict to M, $\det P$ is the principal bundle of a complex line bundle $L \to M$. But $H^1(M, \mathcal{E}^*) \cong H^2(M, \mathbf{Z}) = 0$, and hence $\det P$ is trivial on M. Thus (see Example 2.3.2) P is reducible to an $\mathrm{SL}(2, \mathbf{C})$-bundle on M, and hence P is a $\mathrm{Spin}(1, 3)_0$-bundle on M. ∎

∎

Exercises for Section 3.5

3.5.1. Show that the mapping $\rho : V \to V$ given by $\rho(x)y = \alpha(x)yx^{-1}$ is an isometry of V with respect to the quadratic function Q.

3.5.2. Show that $x \mapsto \|x\| := x\bar{x}$ is a homomorphism of groups

$$s : \mathrm{Cl}^*(V) \to \mathbf{R}^*,$$

where $s(x) = \|x\|$.

3.5.3. A Lorentzian manifold of signature $(1,3)$ is time-orientable if there is a vector field X which has norm-squared $|X|^2 > 0$ at all points of the manifold M (i.e., it has a time-like vector field). Show that a time-oriented and oriented Lorentz manifold has a structure group $O(1,3)$ which is reducible to $SO(1,3)_0$.

3.5.4. Show that $\bigwedge^2 U \to \mathbf{G}(2,4,\mathbf{C})$ is nontrivial, where U is the universal bundle over this Grassmannian manifold. (Hint: calculate the Chern class with respect to a natural Hermitian metric on U; see Chern 1967 or Wells 1980.)

3.6 Spectral sequences

A spectral sequence is a sequence of two-dimensional arrays (possibly infinite) of abelian groups $E_r, r = 1, 2, \ldots$. The abelian groups of the array are labeled

$$\{E_r^{pq}\}, \quad p, q \in \mathbf{Z}.$$

We shall call a specific group E_r^{pq} an *element* of the spectral sequence, and we shall refer to r as the *order* of a given element. Each array E_r is equipped with a *differential* d_r, whereby d_r is a group homomorphism which maps the array to itself in a specific fashion, and it satisfies $d_r^2 = 0$. Namely, d_r will map elements to elements in the following manner:

$$d_r : E_r^{pq} \to E_r^{p+r,q-r+1}. \tag{3.6.1}$$

We shall only be concerned with spectral sequences which have zero elements for p or q negative. The first two orders of a spectral sequence are illustrated for positive p and q in (3.6.2) and (3.6.3) below.

$$
\begin{array}{ccccccccc}
\vdots & & \vdots & & \vdots & & & & \vdots \\
E_1^{0q} & \to & E_1^{1q} & \to & E_1^{2q} & \to & \cdots & \to & E_1^{pq} & \to \\
\vdots & & \vdots & & \vdots & & & & \vdots \\
E_1^{01} & \to & E_1^{11} & \to & E_1^{21} & \to & \cdots & \to & E_1^{p1} & \to \\
E_1^{00} & \to & E_1^{10} & \to & E_1^{20} & \to & \cdots & \to & E_1^{p0} & \to
\end{array}
\tag{3.6.2}
$$

In the diagram above, the horizontal arrows represent the mappings

$$d_1 : E_1^{pq} \to E_1^{p+1,q}.$$

For $r = 2$ we have

$$
\begin{array}{ccccccc}
\vdots & \vdots & \vdots & & \vdots & \\
E_2^{0q} & E_2^{1q} & E_2^{2q} & \cdots & E_2^{pq} & \cdots \\
\vdots & \vdots & \vdots & & \vdots & \\
E_2^{01} & E_2^{11} & E_2^{21} & \cdots & E_2^{p1} & \cdots \\
E_2^{00} & E_2^{10} & E_2^{20} & \cdots & E_2^{p0} & \cdots
\end{array}
\tag{3.6.3}
$$

In (3.6.3) we have $d_2 : E_2^{pq} \to E_2^{p+2,q-1}$, which we recognize as the familiar knight's move from chess (two over and one down).

The next ingredient in a spectral sequence is that the arrays are linked cohomologically from one order to the next. Namely, if we let $H(E_r)$ be the cohomology of the differential module (E_r, d_r), then one must have

$$H(E_r) \cong E_{r+1}.$$

In other words, each element E_{r+1}^{pq} of the array E_{r+1} is isomorphic to

$$H(E_r)^{pq} := \frac{\ker d_r : E_r^{pq} \to E_r^{p+r,q-r+1}}{\operatorname{im} d_r : E_r^{p-r,q+r-1} \to E_r^{pq}}.$$

Thus we see that each term of the array is inductively determined from the previous one with its differential.

The spectral sequence will have convergence properties, and there will be a well-defined limit

$$E_\infty^{pq} = \lim_{r \to \infty} \operatorname{ind} E_r^{pq},$$

which exists in a natural manner (the inductive limit). A special case of such convergence, which we shall encounter most often, is when

$$E_r^{pq} \cong E_{r_0}^{pq}, \quad r \geq r_0,$$

and so we would write

$$E_\infty^{pq} = E_{r_0}^{pq}.$$

This summarizes the basic aspects of a spectral sequence.

Spectral sequences as described above are derived from basic geometric or algebraic data. Namely, if we start with a complex K with a differential d, then we can consider the cohomology of this complex $H(K)$. If we give additional structure to the complex in the form of a *decreasing filtration*, that is, a sequence of submodules $K_p \subset K$ satisfying

$$\cdots \supset K_p \supset K_{p+1} \supset \cdots ,$$

then we can determine a spectral sequence E_r^{pq}, where the limit of the spectral sequence can be identified with the cohomology $H(K)$ in a certain manner, and all of the arrays of the spectral sequence can be considered to be successive approximations to the cohomology $H(K)$. If the spectral sequence degenerates to low order, and we can calculate the low order terms, then we have a tool for calculating the cohomology of the original complex. That is the most common use of the spectral sequence. We shall give a sequence of examples in which a complex is given, and where we determine the low order terms (either E_1 or E_2) and the nature of the limiting cohomology.

All of the spectral sequences we are interested in will start from a *bicomplex* or *bigraded module* of the form

$$K = \sum_{p,q} K^{pq},$$

with two differentials of degree one,

$$d' : K^{pq} \to K^{p+1,q},$$
$$d'' : K^{pq} \to K^{p,q+1},$$

satisfying $(d')^2 = 0$, $(d'')^2 = 0$, and $d'd'' + d''d' = 0$. Given this bicomplex, we form the *standard complex* from it in the following manner. We set

$$K^r = \sum_{p+q=r} K^{pq},$$

and

$$d = d' + d'';$$

then we see that $d^2 = 0$, and (K^*, d) is a complex of the usual form

$$\cdots \xrightarrow{d} K^r \xrightarrow{d} K^{r+1} \xrightarrow{d} \cdots .$$

As an example of such a bicomplex and its associated standard complex we let M be a complex manifold, and let

$$K^{pq} := \mathcal{E}^{p,q}(M),$$
$$d' = \partial, \quad d'' = \bar{\partial}.$$

Then we see that

$$K^r = \mathcal{E}^r(M),$$

and d is ordinary exterior differentiation.

Now we want to consider two natural filtrations on such a bicomplex K. We shall set

$$'K_r = \sum_q \sum_{p \geq r} K^{pq}, \tag{3.6.4}$$

$$''K_r = \sum_p \sum_{q \geq r} K^{pq}. \tag{3.6.5}$$

Then it is clear that both (3.6.4) and 3.6.5) are filtrations of the bicomplex K. These are called the *first* and *second* filtrations of the bicomplex K, respectively. For the time being, let us consider a complex $K = \sum_n K^n$ with a differential d and equipped with a filtration K_p. We want to derive a spectral sequence from this data. We shall then apply this to the special case of a complex coming from a bicomplex with one of the two filtrations (3.6.4) and (3.6.5), but the notation is simpler if we forget about the bicomplex for a little while. Let $H(K) = (H^m(K))$ be the (graded) cohomology of the complex K. We suppose that the filtration K_p satisfies the following properties:

$$dK_p \subset K_p, \tag{3.6.6}$$

$$K_p = \sum_q K_p \cap K^q, \tag{3.6.7}$$

$$K_p \cap K^n = 0, \quad \text{for } p > n. \tag{3.6.8}$$

If we had a bicomplex with $K^{pq} = 0$ for p or q negative, then all three of these properties will be satisfied for the first and second filtrations, for instance. And this is the only kind of bicomplex we shall be considering.

Given a complex K with a filtration $\{K_p\}$ satisfying properties (3.6.6)–(3.6.8), we can now define a spectral sequence which will converge in an appropriate sense to the cohomology $H(K)$. Let

$$Z_r^p = \{x \in K_p : dx \in K_{p+r}\}.$$

Define

$$E_r^p = Z_r^p / (dZ_{r-1}^{p-r+1} + Z_{r-1}^{p+1}),$$

$$E_r = \sum_p E_r^p,$$

which is a graded module, and defines the rth term of the spectral sequence. We now have the first proposition.

Proposition 3.6.1. *The differential d induces a differential of degree r acting on the module E_r, i.e.,*

$$d_r : E_r^p \to E_r^{p+r},$$

and satisfies $d_r^2 = 0$. Moreover, the cohomology groups of the complex (E_r, d_r) satisfy

$$H^k(E_r) = E_{r+1}^k.$$

Proof: First we note that $dZ_r^p \subset Z_r^{p+r}$. Namely, if $x \in Z_r^p$, then by definition, $dx \in K_{p+r}$, and $d(dx) = 0 \in K_{p+2r}$, and hence $dx \in Z_r^{p+r}$. Similarly, d maps $dZ_{r-1}^{p-r+1} + Z_{r-1}^{p+1}$ onto dZ_{r-1}^{p+1}. Thus there is an induced mapping of the quotient spaces, which we denote by d_r:

$$Z_r^p/(dZ_{r-1}^{p-r+1} + Z_{r-1}^{p+1}) \xrightarrow{d_r} Z_r^{p+r}/(dZ_{r-1}^{p+1} + Z_{r-1}^{p+r+1}).$$

This is the required mapping $d_r : E_r^p \to E_r^{p+r}$.

A cocycle of degree p in E_r will be represented by an element $x \in Z_r^p$ satisfying $dx \in dZ_{r-1}^{p+1} + Z_{r-1}^{p+r+1}$, i.e., $dx = dy + z$, where $y \in Z_{r-1}^{p+1}, z \in Z_{r-1}^{p+r+1}$. Now let $u = x - y$, then $du = dx - dy = z$, so $u \in K_p$, and $du \in K_{p+r+1}$, that is, $u \in Z_{r+1}^p$. It follows that

$$x = u + y \in Z_{r+1}^p + Z_{r-1}^{p+1}.$$

Thus, the cocycles of degree p in E_r have the form

$$Z^p(E_r) = (Z_{r+1}^p + Z_{r-1}^{p+1})/(dZ_{r-1}^{p-r+1} + Z_{r-1}^{p+1}).$$

The coboundaries in E_r of degree p, which we denote by $B^p(E_r)$, are represented by elements of dZ_r^{p-r}. This module contains dZ_{r-1}^{p-r+1}. It follows then, using the same argument as above, that

$$B^p(E_r) = (dZ_r^{p-r} + Z_{r-1}^{p+1})/(dZ_{r-1}^{p-r+1} + Z_{r-1}^{p+1}).$$

Therefore, we see that

$$H^p(E_r) = (Z_{r+1}^p + Z_{r-1}^{p+1})/(dZ_r^{p-r} + Z_{r-1}^{p+1}).$$
$$= Z_{r+1}^p / (Z_{r+1}^p \cap (dZ_r^{p-r} + Z_{r-1}^{p+1})).$$

But we note that

$$Z^p_{r+1} \supset dZ^{p-r}_r, \qquad Z^p_{r+1} \cap Z^{p+1}_{r-1} = Z^{p+1}_r,$$

from which we can write

$$H^p(E_r) = Z^p_{r+1}/(dZ^{p-r}_r + Z^{p+1}_r) = E^p_{r+1}.$$

■

We now want to define the limit of the spectral sequence, as r tends to ∞. First we set

$$B^p_r = K_p \cap dK_{p-r},$$

and we can then write

$$E^p_r = Z^p_r/(B^p_{r-1} + Z^{p+1}_{r-1}).$$

We then define

$$K_\infty = 0, \qquad K_{-\infty} = K,$$
$$Z^p_\infty = \{x \in K_p : dx = 0\} = \text{ cocycles in } K_p,$$
$$B^p_\infty = K_p \cap dK_{-\infty} = \text{ coboundaries of } K_p \text{ in } K,$$

and let

$$E^p_\infty = Z^p_\infty/(B^p_\infty + Z^{p+1}_\infty),$$
$$E_\infty = \sum E^p_\infty.$$

With the assumptions we have made above about the filtration one can show (Exercise 3.6.3) that there is a natural inductive limit which can be identified with these groups 'at ∞', i.e.,

$$\lim_{r \to \infty} \text{ind } E^p_r = E^p_\infty.$$

We now turn to the bigrading of this spectral sequence. For this, we now need to use the information (which we have not really used till now) that the original complex K was graded. We define

$$Z^{pq}_r = Z^p_r \cap K^{p+q}, \qquad B^{pq}_r = B^p_r \cap K^{p+q},$$
$$E^{pq}_r = Z^{pq}_r/(B^{pq}_{r-1} + Z^{p+1,q-1}_{r-1}),$$

and it follows that

$$d_r : E^{pq}_r \to E^{p+r,q-r+1}_r,$$

as described in the opening paragraphs of this section.

Now we have the spectral sequence, and we have defined the limit E_∞, which is itself a graded module. We want to relate it to the cohomology of the original complex K. If we let $H(K)$ be the graded cohomology module of the original differential d acting on K, then we let $H(K)_p$ be the image of $H(K_p)$ in $H(K)$, and we then have a decreasing filtration of $H(K)$,

$$\cdots H(K)_p \supset H(K)_{p+1} \supset \cdots .$$

Whenever we have a filtered module $L = \{L_r\}$ of the type we have been encountering, we can form the *associated graded module*:

$$\mathrm{Gr}(L) := \sum_r (L_r/L_{r+1}).$$

So, in particular, we can consider the graded module $\mathrm{Gr}(H(K))$.

Theorem 3.6.2. *There is a canonical isomorphism of graded modules*

$$E_\infty \cong \mathrm{Gr}(H(K));$$

more specifically,

$$E_\infty^{pq} \cong H^{p+q}(K)_p/H^{p+q}(K)_{p+1}.$$

Proof: There is a surjective homomorphism $Z_\infty^p \to H(K)_p$, since Z_∞^p consists of precisely the cocycles in K_p. Now an element $z \in Z_\infty^p$ which maps to $H(K)_{p+1}$ must be cohomologous to an element x in K_{p+1}, that is,

$$z = x + dy, \quad x \in Z_\infty^{p+1} + B_\infty^p, \quad y \in K_p.$$

This proves the first assertion in the theorem. The second part of the theorem follows immediately taking the bigrading into account. ∎

Let us consider the case where $H^{p+q}(K)$ are finite-dimensional vector spaces and $K_p = K$ for $p < 0$. Then we see that

$$H^r(K) = H^r(K)_0 \supset H^r(K)_1 \supset H^r(K)_2 \supset \cdots \supset 0.$$

Since we can find supplementary subspaces in a finite-dimensional vector space, we have a vector space direct sum decomposition

$$H^r(K) \cong (H^r(K)_0/H^r(K)_1) \oplus (H^r(K)_1/H^r(K)_2) \oplus \cdots$$
$$\cong E_\infty^{0r} \oplus E_\infty^{1,r-1} \oplus \cdots .$$

Thus we obtain

$$H^r(K) \cong \sum_{p+q=r} E_\infty^{pq}. \tag{3.6.9}$$

This will be a most useful corollary, as we shall see. The decomposition given in (3.6.9) is not usually canonical, but it gives very nice information about the dimensions of the vector spaces involved.

Now, given a bicomplex K, there are two filtrations $'K_p$ (3.6.4) and $''K_p$ (3.6.5). To each of these filtrations there corresponds a spectral sequence $'E_r^{pq}$ and $''E_r^{pq}$, respectively, which both converge to $H(K)$ in the sense given by Theorem 3.6.2. The *grading* in the limit depends on the filtrations chosen. However, in (3.6.9) the left-hand side does not depend on the filtration, and we shall be able to exploit this fact later, as we shall see in some of the examples.

Example 3.6.3 (Fröhlicher): Let M be a complex manifold, and let K be the bicomplex

$$K = \sum_{p,q} \mathcal{E}^{p,q}(M),$$
$$d' = \partial, \quad d'' = \bar{\partial},$$

and let $\sum_r \mathcal{E}^r(M)$ be the standard complex associated with K, where

$$d = \partial + \bar{\partial}.$$

We can equip K with the first filtration $'K_p$ given by (3.6.4). This determines the spectral sequence $E_r^{pq} := {}'E_r^{pq}$. We want to calculate E_1^{pq}. First we note that $K^{pq} = 0$, for p or $q > n = \dim_{\mathbb{C}} M$, and for p or q negative. It follows that $K_p = 0$ for $p > n$, and hence E_r^{pq} is nontrivial only for $0 \leq p, q \leq n$. In particular, it follows that $d_r = 0$ for $r > n$, and that $E_\infty^{pq} \cong E_n^{pq}$, so the spectral sequence degenerates at least to order n.

Now we want to calculate E_1^{pq}. First we can express any homogeneous differential form of degree k in the form

$$\omega^k = \omega^{0,k} + \cdots + \omega^{k,0},$$

then

$$Z_1^{pq} = \{\omega^{p+q} = \omega^{p,q} + \omega^{p+1,q-1} + \cdots + \omega^{p+q,0} :$$
$$d\omega^{p+q} = \eta^{p+1,q} + \cdots + \eta^{p+q+1,0}\}.$$

In other words, since $d\omega^{p,q} = \partial\omega^{p,q} + \bar{\partial}\omega^{p,q}$, we require that $\bar{\partial}\omega^{p,q} = 0$. Thus we find that

$$Z_1^{pq} = \{\omega = \omega^{p,q} + \cdots + \omega^{p+q,0} : \bar{\partial}\omega^{p,q} = 0\}.$$

Now

$$
\begin{aligned}
B_1^{pq} &= B_1 \cap K^{p+q} \\
&= K_p \cap dK_{p-1} \cap K^{p+q} \\
&= \{\omega^{p+q} = \omega^{p,q} + \cdots + \omega^{p+q,0} : \omega^{p+q} = d\eta^{p+q-1}, \text{ with} \\
&\quad \eta^{p+q-1} = \eta^{p-1,q} + \cdots + \eta^{p+q-1,0}, \text{ and } \bar{\partial}\eta^{p-1,q} = 0\}.
\end{aligned}
$$

Note that all terms of the form $\omega = \omega^{r,s}$ with $r + s = p + q$, $r \geq p + 1$ are in $Z_0^{p+1,q-1}$. Therefore, we see that

$$
\begin{aligned}
E_1^{pq} &= Z_1^{pq}/(B_1^{pq} + Z_0^{p+1,q-1}) \\
&\cong \{\omega^{p,q} : \bar{\partial}\omega^{p,q} = 0\}/\{\omega^{p,q} = \bar{\partial}\eta^{p,q-1}\} \\
&= H_{\text{DOL}}^{p,q}(M).
\end{aligned}
$$

What is the mapping d_1 in this case? We see that d induces the natural mapping

$$\partial : H_{\text{DOL}}^{p,q}(M) \rightarrow H_{\text{DOL}}^{p+1,q}(M),$$

where we use the fact that $\partial\bar{\partial} + \bar{\partial}\partial = 0$. This can easily be identified with d_1 (Exercise 3.6.5).

Thus we have a spectral sequence with limit being

$$H^r(K) = H^r(\mathcal{E}^*(M)) = H_{\text{DR}}^r(M),$$

the de Rham cohomology of M, while

$$E_1^{pq} = H_{\text{DOL}}^{p,q}(M).$$

We write this mnemonically as

$$E_1^{pq} = H_{\text{DOL}}^{p,q}(M) \Rightarrow H_{\text{DR}}^r.$$

If the spectral sequence degenerates to first order, i.e., if it has the property that

$$E_1^{pq} \cong E_2^{pq} \cong \cdots \cong E_\infty^{pq},$$

then we would have, by (3.6.9),

$$H_{\mathrm{DR}}^r(M) \cong \sum_{p+q=r} E_1^{pq}$$

$$\cong \sum_{p+q=r} H_{\mathrm{DOL}}^{p,q}(M). \tag{3.6.10}$$

This is, in fact, the case for compact Kähler manifolds, and (3.6.10) is called the Hodge decomposition for compact Kähler manifolds. We shall briefly discuss Kähler manifolds and why the spectral sequence degenerates to first order.

A Kähler manifold is a complex manifold with an Hermitian metric $h = h_{ab}dz^a \otimes d\bar{z}^b$, satisfying $d(h_{ab}dz^a \wedge d\bar{z}^b) = 0$ (see Weil 1958, Wells 1980). Complex projective space and any of its submanifolds are Kähler manifolds (see Exercises 2.4.8 and 2.4.9). If M is a compact Kähler manifold, then we let $\overline{\square} = \bar{\partial}\bar{\partial}^* + \bar{\partial}^*\bar{\partial}$, and $\square = \partial\partial^* + \partial^*\partial$, where $\bar{\partial}^*$ and ∂^* are the L^2 adjoints of the operators $\bar{\partial}$ and ∂, respectively, with respect to the inner product

$$(\alpha, \beta) = \int_M \alpha \wedge *\bar{\beta}, \quad \alpha, \beta \in \mathcal{E}^{p,q}(M).$$

On a Kähler manifold one has the identity $\square = \overline{\square}$, which follows from the local differential geometry of the Kähler manifold (see Weil 1958, Wells 1980). Then we set

$$\mathcal{H}_{\overline{\square}}^{p,q}(m) = \{\omega^{p,q} : \overline{\square}\,\omega^{p,q} = 0\}.$$

These are the $\overline{\square}$-*harmonic* (p,q)-*forms*. A fundamental theorem of Hodge (see Wells 1980) gives us an isomorphism

$$\mathcal{H}_{\overline{\square}}^{p,q}(M) \cong H_{\mathrm{DOL}}^{p,q}(M).$$

Now consider $\partial\omega^{p,q}$, where $\omega^{p,q}$ is harmonic. Then, by the equality of the two second order (elliptic) operators indicated above, we have that $\overline{\square}(\partial\omega^{p,q}) = \square(\partial\omega^{p,q}) = 0$. Now the Hodge decomposition for $\square$-harmonic forms has the following unique decomposition (true for any elliptic differential operator on a compact manifold, see Wells 1980):

$$\omega = \{\text{harmonic part of } \omega\} + \{\partial(\eta)\} + \{\partial^*(\tau)\}, \quad \text{for some } \eta \text{ and } \tau.$$

Thus, it follows that $\partial \omega^{p,q}$ is harmonic if and only if it is zero. Therefore, we see that $d_1 = 0$, the spectral sequence degenerates at the first order, and the Hodge decomposition for Kähler manifolds follows.

This could have been proven directly by Hodge theory, and usually is, but we give it here as an example of the nontrivial degeneration of a spectral sequence and its consequences. On the other hand, it follows from the spectral sequence, that for *any* compact complex manifold

$$\sum_{r=0}^{2n}(-1)^r b_r = \sum_{p,q=1}^{n}(-1)^{p+q}h^{p,q},$$

where

$$h^{p,q} := \dim H_{\text{DOL}}^{p,q}(M)$$

are the *Hodge numbers* of the complex manifold M. Alternating sums of dimensions of this type are preserved from one term to the next of the spectral sequence, and from this one derives the result (see Fröhlicher 1955). There is not really any simpler way to see this nontrivial relationship between the Betti numbers and the Hodge numbers of a complex manifold. On the other hand, on a Kähler manifold more is true, namely,

$$b_r = \sum_{p+q=r} h^{p,q},$$

which is a stronger result and follows from (3.6.10) and immediately yields the result of Fröhlicher, in this case.

A *graded sheaf* $\mathcal{S}^*$ is a sequence of sheaves $\{\mathcal{S}^n\}, n \in \mathbf{Z}$. A *differential sheaf* is a graded sheaf with a differential d of degree $+1$ (i.e., d is a homomorphism of sheaves of the form $d : \mathcal{S}^n \to \mathcal{S}^{n+1}$, which satisfies $d^2 = 0$). For instance, the graded sheaf of differential forms $\mathcal{E}^*$ on a differentiable manifold M, with $\mathcal{E}^m = 0$ for $m < 0$, $m > \dim M$, is a differential sheaf, where d is exterior differentiation (this is the canonical model).

A *resolution* of a sheaf $\mathcal{S}$ is an exact sequence of sheaves

$$0 \to \mathcal{S} \to \mathcal{A}^0 \to \mathcal{A}^1 \to \cdots \to \mathcal{A}^n \to \cdots.$$

For instance, as was discussed in §3.2, the de Rham complex

$$0 \to \mathbf{C} \to \mathcal{E}^0 \to \mathcal{E}^1 \to \cdots \to \mathcal{E}^n \to 0$$

is exact at each term of the sequence, and hence is a resolution of the constant sheaf $\mathbf{C}$ on a differentiable manifold.

There is a *canonical resolution* of any sheaf S, defined in the following manner. Recall that

$$\widehat{S} = \bigcup_{x \in M} S_x$$

is the union of the stalks, and let $\pi : \widehat{S} \to M$ be the natural projection. Let $C(S)$ be the sheaf defined by the presheaf (which in fact is a sheaf)

$$U \to \{f : U \to \widehat{S} \quad \text{where } \pi \circ f = \text{id}\},$$

(cf. §3.2 where $C(S)$ was denoted by S'). There is a natural injection $S \to C(S)$, but there are many sections of $C(S)$ which violate the continuity property (3.2.14), and hence S does not map onto $C(S)$. Define inductively,

$$C^0(S) := C(S),$$
$$C'(S) := C(S)/S,$$
$$C^1(S) := C(C'(S)),$$
$$C^2(S) := C(C'(C^1(S))),$$
$$\text{etc.}$$

We then obtain a sequence of sheaves

$$0 \to S \to C^0(S) \to C^1(S) \to \cdots \to C^n(S) \to \cdots, \qquad (3.6.11)$$

which, one can easily see, is exact. This is referred to as the *canonical resolution* of C. We now denote by

$$C^p(M, S) := \Gamma(M, C^p(S)),$$

the global sections of the sheaf $C^p(S)$ over M. Thus we have an induced sequence of global sections (not necessarily exact!).

$$0 \to \Gamma(M, S) \to C^0(M, S) \to C^1(M, S) \to \cdots \to C^n(M, S) \ldots.$$

We see that $C^*(M, S)$ is a complex of abelian groups, and we let $H^q(C^*(M, S))$ be the cohomology of this complex. Note that the cohomology groups depend only on the sheaf S and the space M; there have been no other choices made.

Theorem 3.6.4. *There is an isomorphism*

$$H^q(M, \mathcal{S}) \cong H^q(C^*(M, \mathcal{S})).$$

There are several ways to prove this, and we shall not carry any of them out here. One way is to check that $H^q(C^*(M, \mathcal{S}))$ satisfies the axioms of sheaf cohomology (essentially all of the ingredients in Theorem 3.3.3), and then use the uniqueness of the sheaf cohomology. Alternatively, there is a spectral sequence relating Čech cohomology to cohomology defined in this manner, which one can use to deduce the above isomorphism (see Godement 1964, for either of these proofs). Defining cohomology by means of the canonical resolution of a sheaf is the alternate definition of sheaf cohomology, alluded to in §3.3. It is quite impractical for calculating specific cohomology groups, as we did by using Čech cohomology in that section. However, it is an extremely useful definition from the point of view of certain theoretical calculations, as we shall see in the next examples.

Example 3.6.5 (Spectral sequence of a differential sheaf):
   Let

$$0 \rightarrow \mathcal{S} \rightarrow \mathcal{A}^0 \rightarrow \mathcal{A}^1 \rightarrow \cdots \rightarrow \mathcal{A}^n \rightarrow \cdots \tag{3.6.12}$$

be a complex of sheaves and consider the canonical resolution of each term of this sequence

$$
\begin{array}{ccccccccc}
\vdots & & \vdots & & \vdots & & \vdots & & \\
C^1(\mathcal{S}) & \rightarrow & C^1(\mathcal{A}^0) & \rightarrow & C^1(\mathcal{A}^1) & \rightarrow & C^1(\mathcal{A}^2) & \rightarrow & \\
\uparrow & & \uparrow & & \uparrow & & \uparrow & & \\
C^0(\mathcal{S}) & \rightarrow & C^0(\mathcal{A}^0) & \rightarrow & C^0(\mathcal{A}^1) & \rightarrow & C^0(\mathcal{A}^2) & \rightarrow & \\
\uparrow & & \uparrow & & \uparrow & & \uparrow & & \\
\mathcal{S} & \rightarrow & \mathcal{A}^0 & \rightarrow & \mathcal{A}^1 & \rightarrow & \mathcal{A}^2 & \rightarrow & \cdots
\end{array}
\tag{3.6.13}
$$

where the horizontal maps are induced from the mappings of (3.6.12). Then, taking global sections of the bicomplex of sheaves in (3.6.13), we obtain a bicomplex

$$K = K^{**} = C^*(M, \mathcal{A}^*) = \{C^p(M, \mathcal{A}^q)\}.$$

We have mappings

$$d' : C^p(M, \mathcal{A}^q) \rightarrow C^{p+1}(M, \mathcal{A}^q)$$
$$d'' : C^p(M, \mathcal{A}^q) \rightarrow C^p(M, \mathcal{A}^{q+1})$$

induced from (3.6.13). We choose d' and d'' to be the same as the induced maps up to a change in sign so that $d'd'' + d''d' = 0$. We then let

$$K^n = \sum_{p+q=n} C^p(M, \mathcal{A}^q)$$

be the associated standard complex and let $d = d' + d''$ as before.

If we choose the first filtration $'K_p$ of this bicomplex K, then we have its associated spectral sequence $\{'E_r^{pq}\}$. We now claim that

$$'E_2^{pq} = H^p(M, \mathcal{H}^q(\mathcal{A}^*)), \tag{3.6.14}$$

where $\mathcal{H}^q(\mathcal{A}^*)$ is the derived sheaf of the differential sheaf $\mathcal{A}^*$ defined by

$$\mathcal{Z}^n(\mathcal{A}^*) := \ker(\mathcal{A}^n \to \mathcal{A}^{n+1}),$$
$$\mathcal{B}^n(\mathcal{A}^*) := \operatorname{im}(\mathcal{A}^{n-1} \to \mathcal{A}^n),$$
$$\mathcal{H}^n(\mathcal{A}^*) := \mathcal{Z}^n(\mathcal{A}^*)/\mathcal{B}^n(\mathcal{A}^*).$$

In addition we have, using now the second filtration, a second spectral sequence whose first and second order terms we can identify as

$$''E_1^{pq} = H^q(M, \mathcal{A}^p),$$
$$''E_2^{pq} = H^p(H^q(M, \mathcal{A}^*)). \tag{3.6.15}$$

Here $H^q(M, \mathcal{A}^*)$ is a complex of abelian groups, and $H^p(H^q(M, \mathcal{A}^*))$ is the cohomology of that complex. The verification of (3.6.14) and (3.6.15) is not too difficult and will be omitted (Exercise 3.6.6).

Now suppose we have that (3.6.12) is a resolution of the sheaf $\mathcal{S}$. Then $\mathcal{H}^q(\mathcal{A}^*) = 0$ for $q > 0$, while $\mathcal{H}^0(\mathcal{A}^*) = \mathcal{S}$. Thus we find that

$$'E_2^{p0} = H^p(M, \mathcal{S}),$$
$$'E_2^{pq} = 0, \quad q > 0.$$

It follows that the array $'E_2^{pq}$ has the following form

$$
\begin{array}{cccc}
0 & 0 & 0 & \cdots \\
0 & 0 & 0 & \cdots \\
0 & 0 & 0 & \cdots \\
H^0(M, \mathcal{S}) & H^1(M, \mathcal{S}) & H^2(M, \mathcal{S}) & \cdots
\end{array}
$$

It is clear from this array that $d_2 = 0$ and, moreover that we have $'E_2^{pq} = 'E_\infty^{pq}$. It then follows from Theorem 3.6.2 that (see Exercise 3.6.4)

$$H^p(K) = 'E_2^{p0} = H^p(M, \mathcal{S}). \tag{3.6.16}$$

Similarly, suppose now that instead of being a resolution, the differential sheaf in (3.6.12) is *acyclic*. That is to say that $H^q(M, \mathcal{A}^p) = 0$, for $q \geq 1, p \geq 0$. This will be the case if $\mathcal{A}^*$ consists of fine sheaves, for instance. Then we see that

$$''E_2^{p0} = H^p(\Gamma(M, \mathcal{A}^*))$$
$$''E_2^{pq} = 0, \quad q > 0.$$

Thus, this spectral sequence for the acyclic differential sheaf degenerates also to second order and the array $''E_2^{pq}$ has the form

$$
\begin{array}{cccc}
0 & 0 & 0 & \cdots \\
0 & 0 & 0 & \cdots \\
0 & 0 & 0 & \cdots \\
H^0(\Gamma(M, \mathcal{A}^*)) & H^1(\Gamma(M, \mathcal{A}^*)) & H^2(\Gamma(M, \mathcal{A}^*)) & \cdots
\end{array}
$$

and we obtain as before from Exercise 3.6.4 that

$$H^p(K) = ''E_2^{p0} = H^p(\Gamma(M, \mathcal{A}^*)). \tag{3.6.17}$$

If the differential sheaf in (3.6.12) is both a resolution and acyclic at the same time, then we find from (3.6.16) and (3.6.17) that

$$H^q(M, \mathcal{S}) \cong H^q(\Gamma(M, \mathcal{A}^*)). \tag{3.6.18}$$

The isomorphism in (3.6.18) is known as the *abstract de Rham theorem*. If we apply this to the de Rham resolution

$$0 \to \mathbf{C} \to \mathcal{E}^0 \to \mathcal{E}^1 \to \cdots \to \mathcal{E}^n \to 0$$

on a differentiable manifold M, which we note is an acyclic resolution, since these are fine sheaves, we obtain from (3.6.18)

$$H^q(M, \mathbf{C}) = H^q(\Gamma(M, \mathcal{E}^*)) = H^q_{\mathrm{DR}}(M),$$

which is simply de Rham's theorem (Theorem 3.1.3, here we have identified sheaf cohomology with singular cohomology; see Godement 1964 or Wells 1980 for a more complete description of de Rham's theorem, including its specific relation to singular cohomology and where the isomorphism is induced by integration of differential forms over cycles).

In a similar manner, we can consider the Dolbeault resolution on a complex manifold M with a holomorphic vector bundle $E \to M$,

$$0 \to \Omega^p(E) \to \mathcal{E}^{p,0}(E) \to \mathcal{E}^{p,1}(E) \to \cdots \to \mathcal{E}^{p,n}(E) \to 0.$$

This is an acyclic resolution of the sheaf $\Omega^p(E)$. It is acyclic since the sheaves involved are fine sheaves, and it is a resolution because of the $\bar{\partial}$-Poincaré lemma. Thus, we obtain from (3.6.18)

$$H^q(M, \Omega^p(E)) \cong H^q(\Gamma(M, \mathcal{E}^{p,*}(E)))$$
$$= H^{p,q}_{\text{DOL}}(M, E).$$

This is Dolbeault's theorem (Theorem 3.3.6).

We shall also need the spectral sequence in the case of a resolution which is *not* necessarily acyclic, e.g.,

$$0 \to \mathbf{C} \to \Omega^0 \to \Omega^1 \to \cdots \to \Omega^n \to 0$$

on a complex manifold is not acyclic in general (recall that $H^1(\mathbf{P}_1, \Omega^1) \neq 0$, for instance). Thus, the spectral sequence $''E_r^{pq}$ does not necessarily degenerate at the second order. Variations of this will come up in the Penrose transform for massless fields in Chapter 7. In addition we shall encounter acyclic differential sheaves which are not resolutions.

Let us now summarize two principal conclusions which we can draw from the above discussions. If (3.6.12) is an acyclic differential sheaf, then we have the spectral sequence

$$E_2^{pq} := {}'E_2^{pq} = H^p(M, \mathcal{H}^q(\mathcal{A}^*)) \Rightarrow H^r(\Gamma(M, \mathcal{A}^*)) = H^r(K).$$
$$(3.6.19)$$

On the other hand, if the differential sheaf $\mathcal{A}^*$ in (3.6.12) is a resolution which is not necessarily acyclic, then we have the spectral sequence

$$E_2^{pq} := {}''E_2^{pq} = H^p(H^q(M, \mathcal{A}^*)) \Rightarrow H^r(M, \mathcal{S}) = H^r(K).$$
$$(3.6.20)$$

We shall have occasion to use both of these spectral sequences later.

Example 3.6.6 (Leray spectral sequence): Let $S \xrightarrow{f} B$ be a proper surjective mapping of topological spaces, where we recall that proper means $f^{-1}(K)$ is compact if K is. Let $\mathcal{S}$ be a sheaf on S. We define, for q a nonnegative integer, $f_*^q S$ to be a sheaf on B defined by the presheaf

$$U \to H^q(f^{-1}(U), \mathcal{E}).$$

The sheaves $f_*^q S$ are called the *direct image sheaves* of S with respect to the mapping f. We see that the stalks of the direct image sheaves have the form

$$(f_*^q S)_x = \varinjlim_{x \in U} H^q(f^{-1}(U), S).$$

Thus, we see that the stalk at x can be interpreted as the cohomology along the fiber of the mapping f at x which is $f^{-1}(x)$.

There is a spectral sequence

$$E_2^{pq} = H^p(B, f_*^q S) \Rightarrow H^r(S, S). \qquad (3.6.21)$$

This means, as we have seen in Example 3.6.5, that there is a filtered differential module K with $H^r(K) = H^r(S, S)$, in which the second order term of the associated spectral sequence can be identified as $E_2^{pq} = H^p(B, f_*^q S)$. So, in particular, if the spectral sequence degenerates for some reason at the second order, then we obtain a relationship between the cohomology of the base B, the cohomology of the fibers, and the cohomology of the total space S. To derive the spectral sequence, consider the canonical resolution $C^*(S)$ and let

$$\mathcal{A}^* = f_*^0(C^*(S)).$$

This is a differential sheaf on B which turns out to be acyclic (but not necessarily exact; see Exercise 3.6.7). We use the spectral sequence of this differential sheaf (3.6.19)

$$E_2^{pq} = H^p(B, \mathcal{H}^q(\mathcal{A}^*)) \Rightarrow H^r(\Gamma(B, \mathcal{A}^*)).$$

But we see that

$$\Gamma(B, \mathcal{A}^*) = \Gamma(f^{-1}(B), C^*(\mathcal{A})) = C^*(S, S).$$

Thus, using Theorem 3.6.4, we find that $H^r(\Gamma(B, \mathcal{A}^*)) = H^r(S, S)$. On the other hand, $\mathcal{H}^q(\mathcal{A}^*)$ is generated by the presheaf

$$
\begin{aligned}
U \to H^q(\mathcal{A}^*(U)) &= H^q(\Gamma(f^{-1}(U), C^*(S))) \\
&= H^q(C^*(f^{-1}(U), S)) \\
&= H^q(f^{-1}(U), S).
\end{aligned}
$$

Thus we obtain

$$E_2^{pq} = H^p(B, f_*^q(\mathcal{S})) \Rightarrow H^r(S, \mathcal{S}).$$

We shall now give one application of this spectral sequence, and others in Chapter 7. Recall from §3.4 that a spin structure on a principal G-bundle ξ on a differentiable manifold M, where $G = SO(n)$, for instance, was a pair $(\tilde{\xi}, \sigma)$ where $\tilde{\xi}$ is a two-to-one covering of ξ by the mapping σ and $\tilde{\xi}$ was a Spin(n)-bundle. We saw in Theorem 3.5.1 that $(\tilde{\xi}, \sigma)$ exists if and only if $\delta_2(\xi) \in H^2(M, \mathbf{Z}_2)$ vanishes. We now want to show that the distinct spin structures $(\tilde{\xi}, \sigma)$ associated to a G-bundle ξ are classified by $H^1(M, \mathbf{Z}_2)$. The proof we are giving is due to Milnor (1963). First we shall give an alternate definition of spin structure. Let ξ have the form:

$$G \to P \xrightarrow{\pi} M,$$

where P is the total space of the fiber bundle, π is the projection, and G is isomorphic to the fibers. A *spin structure* on ξ is a cohomology class $c \in H^1(P, \mathbf{Z}_2)$ whose restriction to the fibers of π generates $H^1(\text{fiber}, \mathbf{Z}_2)$, for all fibers of π. It is not difficult to see that this is equivalent to the previous definition of spin structure. Namely, if $\tilde{P} \xrightarrow{\sigma} P$ is the two-to-one covering, then this defines a cohomology class $c \in H^1(P, \mathbf{Z}_2)$ by considering $\tilde{P} \xrightarrow{\sigma} P$ as a $\mathbf{Z}_2$-principal bundle over P, and the transition functions for this principal bundle for a covering $\{U_a\}$ are a cocycle of the form $c_{\alpha\beta} : U_\alpha \cap U_\beta \to \mathbf{Z}_2$, which defines an element $c \in H^1(P, \mathbf{Z}_2)$. Conversely, any element of $H^1(P, \mathbf{Z}_2)$ will define a $\mathbf{Z}_2$-principal bundle of the form

$$\mathbf{Z}_2 \to \tilde{P} \to P.$$

The requirement that $c|_{\text{fibers}}$ generate $H^1(\text{fiber}, \mathbf{Z}_2)$ will imply that the preimage of the fiber will be isomorphic to $\tilde{G}$, since the two-to-one spin covering groups are unique.

To classify the spin structure we need to classify cohomology classes of this type. Let us use the Leray spectral sequence of the fibration $P \xrightarrow{\pi} M$ for the constant sheaf $\mathbf{Z}_2$ on P. We have, by (3.6.21),

$$E_2^{p,q} = H^p(M, \pi_*^q(\mathbf{Z}_2)) \Rightarrow H^r(P, \mathbf{Z}_2),$$

Now

$$\pi_*^p(\mathbf{Z}_2)_x = \varinjlim_{U \ni x} H^q(\pi^{-1}(U), \mathbf{Z}_2),$$

and for U topologically trivial and sufficiently small, $\pi^{-1}(U) \cong U \times G$, and by Künneth's formula (see, e.g., Greenberg and Harper 1981),

$$H^q(U \times G, \mathbf{Z}_2) = H^0(U, \mathbf{Z}_2) \otimes H^q(G, \mathbf{Z}_2)$$
$$= \mathbf{Z}_2 \otimes H^q(G, \mathbf{Z}_2).$$

Thus $\pi_*^q(\mathbf{Z}_2)$ is a constant sheaf on M of the form $\mathbf{Z}_2 \otimes H^1(G, \mathbf{Z}_2)$. Therefore, the spectral sequence at the E_2 level has the form

$$\begin{array}{ccc} E_2^{01} & E_2^{11} & \cdots \\ E_2^{00} & E_2^{10} & E_2^{20} \end{array}$$

Now

$$\begin{aligned} E_2^{01} &= H^0(M, \mathbf{Z}_2 \otimes H^1(G, \mathbf{Z}_2)) \\ &\cong H^0(M, \mathbf{Z}_2) \otimes H^1(G, \mathbf{Z}_2) \\ &\cong \mathbf{Z}_2 \otimes H^1(G, \mathbf{Z}_2) \\ &\cong H^1(G, \mathbf{Z}_2). \end{aligned}$$

Also

$$\begin{aligned} E_\infty^{10} = E_2^{10} &= H^1(M, \mathbf{Z}_2 \otimes H^0(G, \mathbf{Z}_2)) \\ &\cong H^1(M, \mathbf{Z}_2) \otimes H^0(G, \mathbf{Z}_2) \\ &\cong H^1(M, \mathbf{Z}_2) \otimes \mathbf{Z}_2 \\ &\cong H^1(M, \mathbf{Z}_2). \end{aligned}$$

Similarly, $E_2^{20} \cong H^2(M, \mathbf{Z}_2)$, and we have the spectral sequence mapping $d_2 : E_2^{01} \to E_2^{20}$. Thus we see that (since $d_3 = 0$ on E_3^{01}) there is a mapping,

$$H^1(G, \mathbf{Z}_2) \xrightarrow{d_2} H^2(M, \mathbf{Z}_2),$$

so that we have an exact sequence

$$0 \to E_\infty^{01} \to H^1(G, \mathbf{Z}_2) \xrightarrow{d_2} H^2(M, \mathbf{Z}_2). \tag{3.6.22}$$

Moreover, since in the standard filtration of this bicomplex (see Example 3.6.6), we have $K_2 \cap K^1 = 0$, it follows that

$$H^1(K) = H^1(P, \mathbf{Z}_2)$$

and

$$H^1(K) = H^1(K)_0 \supset H^1(K)_1 \supset H^1(K)_2 = 0.$$

Therefore, by Theorem 3.6.2,

$$E_\infty^{01} \cong H^1(K)/H^1(K)_1,$$
$$E_\infty^{10} \cong H^1(K)_1,$$

and hence we have the short exact sequence

$$0 \to E_\infty^{10} \to H^1(P, \mathbf{Z}_2) \to E_\infty^{01} \to 0. \qquad (3.6.23)$$

Using the fact that $E_\infty^{10} \cong H^1(M, \mathbf{Z}_2)$, and by splicing the two sequences (3.6.22) and (3.6.23) together, we obtain the exact sequence

$$0 \to H^1(M, \mathbf{Z}_2) \to H^1(P, \mathbf{Z}_2) \xrightarrow{\alpha},$$
$$H^1(G, \mathbf{Z}_2) \to H^2(M, \mathbf{Z}_2). \qquad (3.6.24)$$

If $c \in H^1(P, \mathbf{Z}_2)$ satisfies $\alpha(c) \neq 0$ in $H^1(G, \mathbf{Z}_2)$ (writing the groups additively here), then c is a spin structure. Here α represents the restriction of the cohomology class c to the typical fiber of P, which is isomorphic to G. The inequivalent spin structures are measured by $\ker \alpha \cong H^1(M, \mathbf{Z}_2)$, as desired. Thus we have proved the following theorem.

Theorem 3.6.7. *Let $P \to M$ be an* $\mathrm{SO}(n)$*-bundle on the differentiable manifold M which admits a spin structure $\widetilde{P} \xrightarrow{\sigma} P$, then the inequivalent spin structures in P are parametrized by $H^1(M, \mathbf{Z}_2)$.*

The proof and the theorem are also valid for other orthogonal groups $\mathrm{SO}(p, q)$, $\mathrm{SO}(p, q)_0$ and we have only formulated it in this one case for simplicity. ∎

Exercises for Section 3.6

3.6.1. Let K^{pq} be a bicomplex which vanishes for p or q negative, and let K_r be one of the two standard filtrations on K^{**}. Show that there is a canonical homomorphism

$$E_r^{p0} \to H^p(K).$$

3.6.2. Let $0 \to S \to A^*$ be a resolution of S on X, then use Exercise 3.6.1 to show that there is a canonical homomorphism

$$H^p(\Gamma(X, A^*)) \to H^p(X, S).$$

3.6.3. Show that

$$\lim_a \mathrm{ind}\, E_r^p = E_\infty^p$$

(see Exercise 3.2.1 for the definition of inductive limit).

3.6.4. Let $\{E_r^{pq}\}$ be a spectral sequence with the property that, for some r_0,

$$E_{r_0}^{pq} = 0, \quad q \neq q_0.$$

Show that the spectral sequence degenerates at order r_0, and moreover,

$$H^{p+q_0} \cong E_{r_0}^{pq_0}.$$

Show that one has a similar result if

$$E_{r_0}^{pq} = 0, \quad p \neq p_0.$$

3.6.5. Show that

$$\partial : H_{\mathrm{DOL}}^{p,q}(M) \to H_{\mathrm{DOL}}^{p+1,q}(M)$$

can be identified with d_1 in the Fröhlicher spectral sequence (Example 3.6.3).

3.6.6. Show that for the spectral sequence of a differential sheaf (3.6.12), the first and second order terms are given by (3.6.15).

3.6.7. Give an example to show that the sheaf $A^* = f_*^0(C^*(S))$ used in Example 3.6.6 is not necessarily an exact differential sheaf.

Part II

Classical field theory

4

Linear field theories

This chapter aims at providing, for mathematicians who are unacquainted with such matters, some essential background on the subject of linear field theories in flat space-time. These include, for example, the wave equation, Maxwell's equations, and the Dirac equation. The first two of these are dealt with in §4.1. The Dirac equation requires the use of spinors, which are introduced in §4.2. We shall mainly use a two-component spinor notation, but since most of the physics literature uses four-component (Dirac) spinors, we shall give the translation between these two alternative descriptions. §4.2 also introduces the idea of massless fields of arbitrary helicity.

Each of these theories arises from an action principle, or variational principle. This is discussed in §4.3. Finally, we come in §4.4 to the key ideas of Poincaré invariance and conformal invariance.

4.1 The wave equation and Maxwell's equations

The basic space of special relativity is *Minkowski space-time* M^4. This is the manifold $\mathbf{R}^4$ equipped with a flat, pseudo-Riemannian metric of signature $+---$. Thus if (x^0, x^1, x^2, x^3) are the standard coordinates on $\mathbf{R}^4$, the line-element takes the form

$$ds^2 = (dx^0)^2 - (dx^1)^2 - (dx^2)^2 - (dx^3)^2.$$

Think of x^0 as 'time' and x^1, x^2, x^3 as 'space-coordinates'.

Throughout the rest of this book we shall use the abstract index notation of Penrose, which was discussed in §2.3, and about which more details can be found in Penrose and Rindler (1984). Recall that a vector field on M^4 is denoted by a symbol such as v^a, with a lower case italic Roman superscript serving to remind us that the object we are dealing with is indeed a vector-field, that is, a contravariant tensor of rank one. Similarly, a one-form in M^4 is denoted by a symbol like ω_a; and the natural pairing between v^a and ω_a is denoted $v^a\omega_a$. The tensor product $v^a \otimes w^b$ of two vectors is simply written $v^a w^b$, or, equivalently, $w^b v^a$. The abstract indices keep track of the order of the two factors. Thus the tensor product of the two vectors in the opposite order is $w^a v^b$, which is to be distinguished from $w^b v^a$.

The metric of M^4 is denoted η_{ab}, and it is symmetric, a fact which we express as $\eta_{ba} = \eta_{ab}$. So the 'length-squared' of a vector v^a is $\eta_{ab} v^a v^b$. Indices may be lowered and raised by using η_{ab} and its inverse η^{ab}, respectively. So if v^a is a vector, then $v_a = \eta_{ab} v^b$ is the corresponding one-form; and if ω_a is a one-form, then $\omega^a = \eta^{ab} \omega_b$ is the corresponding vector. The product $\eta_{ab} \eta^{ac} = \delta_b^c$ is, of course, the identity endomorphism on the space of tangent vectors, or on the space of one-forms. It is the abstract index version of the classical Kronecker delta function.

Let us move on now to discuss some linear field equations. Denote by $\Box$ the d'Alembertian on M^4, that is, the second-order differential operator

$$\Box = \eta^{ab} \nabla_a \nabla_b,$$

where ∇_a is the covariant derivative operator on M^4. In terms of the standard coordinates, therefore,

$$\Box = \frac{\partial^2}{(\partial x^0)^2} - \frac{\partial^2}{(\partial x^1)^2} - \frac{\partial^2}{(\partial x^2)^2} - \frac{\partial^2}{(\partial x^3)^2}.$$

A *massless scalar field* is a real-valued function φ on M^4 satisfying the *wave equation*

$$\Box \varphi = 0.$$

A *scalar field of mass m*, where $m > 0$, is a solution φ of the *Klein–Gordon equation*

$$(\Box + m^2)\varphi = 0.$$

(In the physics literature, this equation is sometimes written with $(mc/\hbar)^2$ instead of m^2, where c denotes the speed of light and $\hbar$ denotes Planck's constant. We shall omit c and $\hbar$, which is tantamount to choosing units of measurement such that $c = \hbar = 1$.)

In general, a section φ of a bundle over M^4 is said to be a *field of mass m* if it is an eigenfunction of $\Box$ with eigenvalue $-m^2$. The field is said to be massless in the case where $m = 0$. The motivation for this terminology comes from quantum theory and special relativity.

Let us turn now to Maxwell's theory of electromagnetic fields. A *Maxwell field* is a two-form F_{ab} on M^4 (so F_{ab} is skew-symmetric: $F_{ab} = -F_{ba}$), satisfying

$$\nabla_{[a} F_{bc]} = 0.$$

The square brackets denote skew-symmetrization. In other words,

$$G_{[ab]} = \tfrac{1}{2}(G_{ab} - G_{ba}),$$
$$G_{[abc]} = \tfrac{1}{6}(G_{abc} + G_{bca} + G_{cab} - G_{bac} - G_{acb} - G_{cba}),$$

and so forth. In index-free notation, therefore, we can simply write $dF = 0$, where d is the exterior derivative; so F_{ab} is a closed two-form.

The one-form $J_a = \nabla^b F_{ab}$ is called the *four-current density* of the field, In index-free notation, this may be written as $J = -3 *d *F$, where $*$ is the Hodge star-operator on M^4. This star-operator is defined by using a volume element (four-form) ε_{abcd} on M^4, and so it involves choosing an orientation on M^4. In terms of the standard coordinates, this volume element is $24dx^0 \wedge dx^1 \wedge dx^2 \wedge dx^3$. The action of 'star' on a two-form F_{ab} is defined by

$$*F_{ab} = \tfrac{1}{2}\varepsilon_{abcd}F^{cd}$$

and its action on a three-form G_{abc} by

$$*G_a = \tfrac{1}{6}\varepsilon_{abcd}G^{bcd}.$$

In general, the 'star' of a p-form is a $(4 - p)$-form (see Exercise 4.1.1 for a discussion of the general properties of the star-operator).

To prove the above statement concerning the index-free version of J, we use the identity

$$\varepsilon_{abcd}\varepsilon^{cdef} = -2(\delta_a^e\delta_b^f - \delta_a^f\delta_b^e). \tag{4.1.1}$$

For then we have

$$
\begin{aligned}
-3(*d *F)_a &= -\tfrac{1}{4}\varepsilon_{abcd}\nabla^b\varepsilon^{cdef}F_{ef} \\
&= \tfrac{1}{2}(\delta_a^e\delta_b^f - \delta_a^f\delta_b^e)\nabla^b F_{ef} \\
&= \nabla^b F_{ab},
\end{aligned}
$$

as required.

A Maxwell field F_{ab} is said to be *source-free* if $J_a = 0$, that is, if $\nabla^a F_{ab} = 0$. In index-free notation, this is $d *F = 0$: F is co-closed. If F_{ab} is a source-free Maxwell field, then

$$
\begin{aligned}
0 &= \nabla^a\nabla_{[a}F_{bc]} \\
&= \tfrac{1}{3}\Box F_{bc};
\end{aligned}
$$

in other words, F_{ab} is a *massless field*. This corresponds to the photon, the quantum of the electromagnetic field, being a particle with zero rest-mass.

One often sees Maxwell's equations written in a way which separates space and time. Let t^a denote a vector of unit length: for example, the vector with components $(1, 0, 0, 0)$. So t^a is a unit vector in the time-direction. Then we call $E_a = t^b F_{ab}$ the *electric field* and $B_a = t^b *F_{ab}$ the *magnetic field*. These one-forms E_a and B_a are spatial: since both are annihilated by t^a, they 'live' in a hyperplane orthogonal to t^a, and this hyperplane is Euclidean three-space. In terms of E and B, the source-free Maxwell equations are

$$\frac{\partial B}{\partial t} + \operatorname{curl} E = 0$$
$$\frac{\partial E}{\partial t} - \operatorname{curl} B = 0 \tag{4.1.2}$$
$$\operatorname{div} E = 0,$$
$$\operatorname{div} B = 0,$$

where $\frac{\partial}{\partial t} = t^a \nabla_a$ is the 'time-derivative'.

Another fact that we deduce from the identity (4.1.1) is that $** = -1$ when acting on two-forms. (The minus sign occurs because spacetime has the Lorentzian signature $+ - - -$; in a positive-definite four-space, we would have $** = +1$, as we shall see in the next chapter.) So it makes sense to speak of the eigenspaces of $*$ with eigenvalues $\pm i$. A two-form ω satisfying $*\omega = i\omega$ is said to be *self-dual*, while one satisfying $*\omega = -i\omega$ is said to be *anti-self-dual*. Any two-form F can be decomposed as

$$F = F^+ + F^-,$$

where

$$F^+ = \tfrac{1}{2}(F - i *F)$$

is the self-dual part of F, and

$$F^- = \tfrac{1}{2}(F + i *F)$$

is its anti-self-dual part. In other words, the vector space of all two-forms is the direct sum of the space of self-dual two-forms and the space of anti-self-dual two-forms.

The source-free Maxwell equations are clearly equivalent to

$$dF^+ = 0$$

which in turn is equivalent to

$$dF^- = 0,$$

since F^+ and F^- are complex-valued forms which are complex conjugates of each other.

Exercises for Section 4.1

4.1.1. Prove the identity (4.1.1) and also

$$\varepsilon_{abcd}\varepsilon^{abcd} = -24$$
$$\varepsilon_{abcd}\varepsilon^{abce} = -6\delta_d^e$$
$$\varepsilon_{abcd}\varepsilon^{aefg} = -6\delta_{[b}^e\delta_c^f\delta_{d]}^g.$$

The 'star' of a two-form and of a three-form were defined in the text; the star of a one-form H_a is the three-form

$$*H_{abc} = \varepsilon_{abcd}H^d.$$

Prove that $** = (-1)^{p+1}$ when acting on p-forms.

4.1.2. Derive, from the source-free Maxwell equations $dF = 0 = d *F$, the equations (4.1.2) for E and B.

4.1.3. If φ is a function of $r^2 = x_a x^a$, then show that it satisfies the wave equation $\Box\varphi = 0$ if and only if it has the form $\varphi = a + br^{-2}$, where a and b are constants.

4.2 Spinors and spinor fields

The particles (or physical fields) that occur in nature are divided into two main types: the bosons and the fermions. Bosons are characterized by the fact that they can be described by tensor fields on space time M^4. For example, the photon is a boson, since the electromagnetic field is described by a two-form on M^4. On the other hand, fermions are not describable by tensors: for them we need *spinor* fields.

Spinors were introduced geometrically in Part I of this book. In this section we shall give an independent (and consistent!) definition. Some group-theoretic motivation appears in §2.3 and §3.5. For more details on the subject of spinors, the reader may consult Penrose and Rindler (1984).

The (abstract) spinor index is a capital Roman letter, either primed $(A', B', \dots)$ or unprimed $(A, B, \dots)$. (In some books, a dot is used instead of a prime, that is, $\dot{A}$ instead of A'.) The basic spin-space is a two-dimensional complex vector space equipped with a symplectic form, i.e., a skew-symmetric complex bilinear form. An element of this space is written with an unprimed superscript, like ψ^A. The symplectic form is denoted ε_{AB}. So $\varepsilon_{AB} = -\varepsilon_{BA}$, and if ψ^A and φ^B are spinors, then $\varepsilon_{AB}\psi^A\varphi^B$ is a complex number (the action of the bilinear form ε_{AB} on the pair of vectors ψ^A and φ^B). We let ψ_A denote an element of the dual spin-space, and the dual symplectic form is ε^{AB}, so $\varepsilon_{AB}\varepsilon^{CB} = \varepsilon_A{}^C$ is the identity endomorphism on spin-space, and on its dual. The forms ε_{AB} and ε^{AB} provide a natural isomorphism between spin-space and its dual; in other words, they are used to raise and lower indices. But since the εs are skew-symmetric, we have to specify exactly how this is to be done: our convention for raising and lowering is

$$\varepsilon^{AB}\psi_B = \psi^A,$$
$$\psi^B\varepsilon_{BA} = \psi_A.$$

In components, for example, we might have

$$\begin{pmatrix} \varepsilon_{00} & \varepsilon_{01} \\ \varepsilon_{10} & \varepsilon_{11} \end{pmatrix} = \begin{pmatrix} 0 & 1 \\ -1 & 0 \end{pmatrix},$$

so that $\psi^0 = \psi_1$ and $\psi^1 = -\psi_0$.

All this structure has a 'primed' version. In particular, $\varepsilon_{A'B'}$ and $\varepsilon^{A'B'}$ are the symplectic forms on primed spin-space and its dual, respectively. There is an anti-isomorphism from primed to unprimed spin-space which is denoted $\psi^A \mapsto \bar{\psi}^{A'}$ and called *complex conjugation*. It is required to preserve the symplectic structure, so $\bar{\varepsilon}_{A'B'} = \varepsilon_{A'B'}$ and so forth. In components, we could, for example, take $\bar{\psi}^{0'} = \overline{\psi^0}$ and $\bar{\psi}^{1'} = \overline{\psi^1}$.

Mixed spinors such as $\psi^A{}_{BA'}$ are elements of the appropriate tensor product of the basic spin-spaces (see §2.3). The rules for abstract indices dictate that we observe the ordering of primed indices, whether subscripts or superscripts; and also the ordering of unprimed indices. But the relative ordering between a primed and unprimed index is irrelevant. For example,

$$\psi^A{}_{BA'} \equiv \psi^A{}_{A'B} \equiv \psi_{A'}{}^A{}_B \neq \psi_{A'B}{}^A.$$

The two-dimensionality of spin-space implies that 'skew spinors are pure traces'. For example, if ξ_{AB} is any spinor, then

$$\xi_{AB} - \xi_{BA} = \varepsilon_{AB}\xi_C{}^C.$$

We come now to the link between spinors on the one hand, and vectors and tensors in M^4 on the other. The idea is to set up an isomorphism between vectors v^a, and spinors $v^{AA'}$ satisfying the reality condition $\bar{v}^{AA'} = v^{AA'}$. This is achieved by the mixed spinor-tensor $\sigma_a{}^{AA'}$, which is required to satisfy

(a) $\qquad \bar{\sigma}_a{}^{AA'} = \sigma_a{}^{AA'},$

(b) $\qquad \sigma_a{}^{AA'}\sigma^b{}_{AA'} = \delta_a^b,$

(c) $\qquad \sigma_a{}^{AA'}\sigma^a{}_{BB'} = \varepsilon_B{}^A\varepsilon_{B'}{}^{A'},$ $\qquad\qquad$ (4.2.1)

(d) $\qquad \sigma_{[a}{}^{AA'}\sigma_{b]A}{}^{B'} = -\tfrac{1}{2}i\varepsilon_{abcd}\sigma^{cAA'}\sigma^d{}_A{}^{B'},$

where, as usual, indices are raised and lowered with η or ε. So the spinor equivalent of a vector v^a is

$$v^{AA'} = \sigma_a{}^{AA'}v^a,$$

and the vector equivalent or a spinor $v^{AA'}$ is

$$v^a = \sigma^a{}_{AA'}v^{AA'}.$$

(This v^a will be real provided $v^{AA'}$ is real in the sense mentioned above: $\bar{v}^{AA'} = v^{AA'}$.) Similarly, a one-form ω_a has a spinor equivalent $\omega_{AA'}$, and so forth. The spinor equivalent of the metric η_{ab} is evidently

$$\eta_{ab}\sigma^a{}_{AA'}\sigma^b{}_{BB'} = \varepsilon_{AB}\varepsilon_{A'B'}.$$

It follows that the operation of raising and lowering indices commutes with the passage from vectors to spinors.

In components, think of $v^{AA'}$ as a 2 × 2 matrix, with A labeling rows and A' labeling columns. The reality condition on $v^{AA'}$ means this matrix is Hermitian. One possible choice of $\sigma_a{}^{AA'}$ is expressed by

$$\begin{pmatrix} v^{00'} & v^{01'} \\ v^{10'} & v^{11'} \end{pmatrix} = \frac{1}{\sqrt{2}} \begin{pmatrix} v^0 + v^3 & v^1 - iv^2 \\ v^1 + iv^2 & v^0 - v^3 \end{pmatrix}. \qquad (4.2.2)$$

The role of the condition (4.2.1d) on $\sigma_a{}^{AA'}$ is to relate the choice between primed and unprimed spinor indices to the choice of orientation on M^4. We shall be using this in a moment. If we choose a frame for the spin-space, we obtain a coordinate system for M^4 which we write as (see (2.3.3))

$$x^{AA'} = \frac{1}{\sqrt{2}} \begin{pmatrix} x^0 + x^3 & x^1 - ix^2 \\ x^1 + ix^2 & x^0 - x^3 \end{pmatrix}. \qquad (4.2.2')$$

This is a *spinor description* of the 2 × 2 matrix coordinates used for M^I in Chapter 1 (see (1.2.8)).

In practice, it is safe to omit σ, and to write expressions such as $v^a = v^{AA'}$ and $\omega_b = \omega_{BB'}$. From now on we shall adopt this convention, and identify a lower-case letter with the pair of capital versions (one primed and one unprimed) of the *same* letter.

Let us now translate the source-free Maxwell equations into spinor form. The two-form F_{ab} can be written

$$\begin{aligned}
F_{AA'BB'} &= \tfrac{1}{2}(F_{AA'BB'} - F_{BB'AA'}) \\
&= \tfrac{1}{2}(F_{AA'BB'} - F_{AB'BA'} + F_{AB'BA'} - F_{BB'AA'}) \\
&= \tfrac{1}{2}F_{AC'B}{}^{C'}\varepsilon_{A'B'} + \tfrac{1}{2}F_{CB'}{}^{C}{}_{A'}\varepsilon_{AB} \\
&= \varphi_{AB}\varepsilon_{A'B'} + \varphi_{A'B'}\varepsilon_{AB},
\end{aligned}$$

$$(4.2.3)$$

where

$$\varphi_{AB} = \tfrac{1}{2}F_{AC'B}{}^{C'},$$
$$\varphi_{A'B'} = \tfrac{1}{2}F_{CB'}{}^{C}{}_{A'}.$$

Now φ_{AB} is symmetric in AB, since

$$\begin{aligned}
\varphi_{[AB]} &= \tfrac{1}{4}\varepsilon_{AB}F_{CC'}{}^{CC'}, \\
&= \tfrac{1}{2}\varepsilon_{AB}\eta^{cd}F_{cd}, \\
&= 0,
\end{aligned}$$

since the trace of a skew-symmetric form is zero. Also, $\varphi_{A'B'}$ is symmetric. In fact, φ_{AB} and $\varphi_{A'B'}$ are complex conjugates of each other:

$$\overline{\varphi}_{A'B'} = \varphi_{A'B'},$$

as follows immediately from the reality condition

$$\overline{F}_{AA'BB'} = F_{AA'BB'}$$

on F_{ab}.

We now claim that $\varphi_{A'B'}\varepsilon_{AB}$ is the self-dual part F_{ab}^+ of F_{ab}. Indeed, if we put $G_{ab} = \varphi_{A'B'}\varepsilon_{AB}$, then condition (4.2.1d) implies that

$$
\begin{aligned}
*G_{ab} &= \tfrac{1}{2}\varphi_{A'B'}\varepsilon_{AB}\sigma^{cAA'}\sigma^{dBB'}\varepsilon_{abcd} \\
&= -i\varphi_{A'B'}\sigma_{[a}{}^{AA'}\sigma_{b]A}{}^{B'} \\
&= i\varphi_{A'B'}\varepsilon_{AB}\sigma_a{}^{AA'}\sigma_b{}^{BB'} \\
&= iG_{ab}.
\end{aligned}
$$

Similarly, $\varphi_{AB}\varepsilon_{A'B'}$ is the anti-self dual part of F_{ab}. So the source-free Maxwell equations $\nabla^a F_{ab}^+ = 0$ become simply

$$\nabla^{AA'}\varphi_{AB} = 0$$

or, equivalently, the conjugate equation

$$\nabla^{AA'}\varphi_{A'B'} = 0.$$

Note, incidentally, that we have extended the domain of definition of the space-time connection $\nabla_a = \nabla_{AA'}$ so that it acts on spinor as well as tensor fields. This is done in such a way that the derivative of an ε vanishes: $\nabla_{AA'}\varepsilon_{BC} = 0$, and so forth. In other words the symplectic form of the spin-space should be covariantly constant with respect to the Levi–Cevita connection on the space M^4.

A totally symmetric spinor field $\varphi_{A'B'...D'}$ on M^4, with n primed indices, and satisfying

$$\nabla^{AA'}\varphi_{A'B'...D'} = 0,$$

is a *massless field of helicity* $\tfrac{1}{2}n$. Similarly, a totally symmetric spinor field $\varphi_{AB...D}$ on M^4, with n unprimed indices, and satisfying

$$\nabla^{AA'}\varphi_{AB...D} = 0, \tag{4.2.4}$$

is said to be a *massless field of helicity* $-\frac{1}{2}n$. One often refers to fields of positive helicity as being *right-handed* and fields of negative helicity as being *left-handed* (cf. Penrose and Rindler 1984). Thus the anti-self-dual part of a source-free Maxwell field is a left-handed massless field of helicity -1, for example.

We now show that these fields are indeed massless according to the definition presented in the previous section. First, note that the part of $\nabla^B_{A'} \nabla^{AA'}$ which is symmetric in BA vanishes, since it corresponds to the anti-self-dual part of the commutator $[\nabla_a, \nabla_b]$ and this vanishes in flat space-time M^4. Thus $\nabla^B_{A'} \nabla^{AA'}$ is skew in BA, whence

$$\nabla^B_{A'} \nabla^{AA'} = \tfrac{1}{2}\varepsilon^{BA}\Box.$$

Applying $\nabla^E_{A'}$ to equation (4.2.4) then yields

$$\Box \varphi_{EB...D} = 0$$

as required.

Particles such as electrons are fermions, and so require spinors for their description. However, they have a nonzero rest-mass, so are not described by the massless field equations given above. A particle of spin $\frac{1}{2}$ and mass $m > 0$, such as the electron, is described in terms of a pair of spinor fields $(\psi^{A'}, \varphi_A)$ satisfying the *Dirac equations*

$$\nabla_{AA'}\psi^{A'} = -\frac{im}{\sqrt{2}}\varphi_A,$$

$$\nabla^{AA'}\varphi_A = -\frac{im}{\sqrt{2}}\psi^{A'}.$$

If we apply $\nabla^A_{B'}$ to the first of these equations and use the second equation, we obtain

$$\nabla^A_{B'} \nabla_{AA'}\psi^{A'} = -\frac{im}{\sqrt{2}}\nabla^A_{B'}\varphi_A = \tfrac{1}{2}\Box \varepsilon_{A'B'}\psi^{A'}$$

$$= -\tfrac{1}{2}m^2\psi_{B'} \qquad = \tfrac{1}{2}\Box\psi_{B'}.$$

Thus $(\Box + m^2)\psi^{A'} = 0$. Similarly, $(\Box + m^2)\varphi_A = 0$, so this is indeed a field of mass m, as defined in the previous section. In the limit when $m \to 0$, we get a pair of massless fields of helicity $\pm\frac{1}{2}$. These are often called *Weyl neutrino fields*.

In the physics literature, one normally encounters a four-component spinor notation, in which the two-component spinors $\psi^{A'}$ and φ_A are 'lumped together'. We shall end the section by describing this notation.

A *Dirac spinor* is a pair $\Psi = (\psi^{A'}, \psi_A)$ of spinors of the type indicated by their indices. (These spinors $\psi^{A'}$ and ψ^A are not necessarily complex conjugates of each other.) There is a natural inner product $\langle \, , \, \rangle$ on the space of Dirac spinors, defined by

$$\langle \Psi, \Phi \rangle = \bar{\psi}_{A'}\varphi^{A'} + \bar{\psi}^A\varphi_A,$$

where $\Psi = (\psi^{A'}, \psi_A)$ and $\Phi = (\varphi^{A'}, \varphi_A)$. It is Hermitian, but *not* positive-definite. The standard physics notation for $\langle \Psi, \Phi \rangle$ is $\overline{\Psi}\Phi$.

We now define *Clifford multiplication* on Dirac spinors. A vector v^a defines a map $\not{v}$ from Dirac spin-space to itself, according to

$$\not{v}\Psi = \sqrt{2}(v^{AA'}\psi_A \, , \, v_{AA'}\psi^{A'}).$$

Note first that

$$\langle \not{v}\Psi, \Phi \rangle = \sqrt{2}(v_{AA'}\bar{\psi}^A\varphi^{A'} + v^{AA'}\bar{\psi}_{A'}\varphi_A)$$
$$= \langle \Psi, \not{v}\Phi \rangle,$$

so $\not{v}$ is an Hermitian operator. Secondly,

$$(\not{v}\not{w} + \not{w}\not{v})\Psi =$$
$$2(v^{AA'}w_{AB'}\psi^{B'} + w^{AA'}v_{AB'}\psi^{B'}, \, v_{AA'}w^{BA'}\psi_B + w_{AA'}v^{BA'}\psi_B)$$
$$= 2(v^a w_a)\Psi,$$

so the *Clifford relations*

$$\not{v}\not{w} + \not{w}\not{v} = 2\eta_{ab}v^a w^b$$

are satisfied.

In the physics notation, Ψ is represented as a column four-vector, and Clifford multiplication is defined in terms of one-form-valued 4×4 matrices γ_a, such that

$$\not{v}\Psi = v^a \gamma_a \Psi.$$

The Clifford relations then read

$$\gamma_a \gamma_b + \gamma_b \gamma_a = 2\eta_{ab}I,$$

where I is the identity 4×4 matrix.

In this Dirac spinor notation, the Dirac equation takes a particularly simple form, namely

$$(\nabla\!\!\!/ + im)\Psi = 0.$$

We end with one final bit of physics notation. Define an operator γ_5 on Dirac spinors by saying that γ_5 maps $\Psi = (\psi^{A'}, \psi_A)$ to $\gamma_5\Psi = (\psi^{A'}, -\psi_A)$. Then clearly $\frac{1}{2}(1+\gamma_5)$ and $\frac{1}{2}(1-\gamma_5)$ are operators which project out the primed and unprimed spinor parts of Ψ, respectively.

Exercises for Section 4.2

4.2.1. Show that $\xi_A\mu^A = 0$ if and only if ξ^A and μ^A are proportional.

4.2.2. Show that a complex vector v^a is null (i.e., $v_a v^a = 0$) if and only if it has the form $v^{AA'} = \lambda^A \xi^{A'}$ for some spinors λ^A and $\xi^{A'}$.

4.2.3. The tensor $T_{ab} = T_{AA'BB'}$ is symmetric (i.e., $T_{ab} = T_{ba}$). Show that its trace-reversed version $\tilde{T}_{ab} = T_{ab} - \frac{1}{2}T_c{}^c \eta_{ab}$ has the spinor form

$$\tilde{T}_{AA'BB'} = T_{BA'AB'} = T_{AB'BA'}.$$

4.2.4. Show that any totally symmetric spinor $\varphi_{AB\ldots D}$ can be 'factorized':

$$\varphi_{AB\ldots D} = \alpha_{(A}\beta_B \ldots \delta_{D)}$$

for some spinors $\alpha_A, \ldots, \delta_D$ (the parentheses denote symmetrization).

4.2.5. The Maxwell spinor field φ_{AB} is said to be *null* if $\varphi_{AB}\varphi^{AB} = 0$. Show that this is equivalent to

$$F_{ab}F^{ab} = *F_{ab} *F^{ab} = *F_{ab}F^{ab} = 0.$$

Show that a null φ_{AB} has the form $\alpha_A\alpha_B$ for some spinor field α_A, and that the source-free Maxwell equations imply $\alpha^A\alpha^B\nabla_{AA'}\alpha_B = 0$.

4.3 The action principle and interactions

It is often useful to regard field equations (such as those mentioned in the previous section) as arising from a variation principle, namely *Hamilton's Principle of Stationary Action*. In particular, this viewpoint is used when quantizing the theory (see, for example, Bjorken and Drell 1965, Glimm and Jaffe 1981). This section is devoted to a description of the action principle as it applies to the linear field theories in which we are interested, and to interactions between these linear fields.

Let ψ_Ξ denote a generic collection of fields on M^4. Thus for each point $x \in M^4$, $\psi_\Xi(x)$ takes its value in a vector space V which is a direct sum of spinor and tensor spaces. In other words, ψ_Ξ is a section of the trivial vector bundle over M^4 with fiber V. A *Lagrangian* is a smooth map $\mathcal{L}$ from $(V \otimes T^*(M^4)) \times V$ to $\mathbf{R}$. So we can write

$$\mathcal{L} = \mathcal{L}(\psi_{\Xi a}, \psi_\Xi).$$

Usually $\mathcal{L}$ is quadratic or linear in $\psi_{\Xi a}$. If Ω is a region in M^4, then the *action* of the field $\psi_\Xi(x)$ in Ω is

$$\mathcal{A} = \int_\Omega \mathcal{L}(\nabla_a \psi_\Xi(x), \psi_\Xi(x)) d^4 x$$

(provided this integral exists). Here $d^4 x$ denotes the volume element on M^4; we regard M^4 as coming equipped with an orientation. So $\mathcal{A}$ is a real-valued functional on the space of smooth field configurations in Ω.

The Principle of Stationary Action states that the action should be stationary with respect to variations of the field ψ_Ξ which vanish on the boundary of Ω. Some or all of this boundary may be 'at infinity', in which case we require that the field variations fall off sufficiently fast at infinity. As is well-known (see, for example, Hawking and Ellis 1973, p. 65), the action is stationary for all Ω if and only if the *Euler–Lagrange equations* hold:

$$\frac{\partial \mathcal{L}}{\partial \psi_\Xi} - \nabla_a \frac{\partial \mathcal{L}}{\partial(\nabla_a \psi_\Xi)} = 0.$$

These are therefore the field equations that $\psi_\Xi(x)$ is required to satisfy.

Example 4.3.1 (The scalar field): The Lagrangian for a free (that is, noninteracting) scalar field $\varphi(x)$ is

$$\mathcal{L}_{KG} = \tfrac{1}{2}(\nabla^a \varphi)(\nabla_a \varphi) - \tfrac{1}{2}m^2 \varphi^2.$$

Substituting this into the Euler–Lagrange equations yields the Klein–Gordon equation $(\Box + m^2)\varphi = 0$ (or the wave equation, if $m = 0$).

Example 4.3.2 (The Dirac field): Here ψ_Ξ takes values in Dirac spin-space, which is a complex vector space. When dealing with complex fields, the rule is that the fields ψ_Ξ and their complex conjugates $\bar\psi_\Xi$ are to be treated as independent arguments of $\mathcal{L}$. The Dirac Lagrangian is

$$\mathcal{L}_D = \mathrm{Re}\langle \Psi, (i\slashed{\nabla} - m)\Psi \rangle.$$

The Euler–Lagrange equations obtained by varying $\overline\Psi$ are

$$(i\slashed{\nabla} - m)\Psi = 0;$$

that is, the Dirac equation. Varying Ψ leads to the adjoint equation. All this can, of course, be written in terms of two-component spinors; but the Dirac-spinor notation is neater, except when $m = 0$. The Lagrangian for a massless field of helicity $\tfrac{1}{2}$ is

$$\mathcal{L}_W = \frac{i}{\sqrt{2}}(\bar\psi_{A'}\nabla^{AA'}\psi_A - \psi_A \nabla^{AA'}\bar\psi_{A'}),$$

and this leads to the Weyl neutrino equation

$$\nabla^{AA'}\psi_A = 0$$

and its complex conjugate.

Example 4.3.3 (The Maxwell field): Here we restrict *a priori* to field configurations F_{ab} satisfying $\nabla_{[a}F_{bc]} = 0$. Thus F_{ab} is a closed two-form, and therefore exact: there exists a one-form A_a, called the *potential*, such that

$$F_{ab} = 2\nabla_{[a}A_{b]}.$$

We are assuming for the time being that the region Ω we are working in is topologically trivial, so that there is no obstruction to the existence of A_a. The Maxwell Lagrangian is

$$\mathcal{L}_M(A_b, \nabla_a A_b) = -\tfrac{1}{4}F_{ab}F^{ab}$$
$$= -(\nabla_{[a}A_{b]})(\nabla^{[a}A^{b]})$$

which indeed yields the source-free Maxwell equations $\nabla_a F^{ab} = 0$.

Example 4.3.4 (The Yukawa interaction): This is our first example of an interaction. In this case, the Klein–Gordon field and the Dirac field are involved. What we do is to add to the 'free' Lagrangian $\mathcal{L}_{KG} + \mathcal{L}_D$ an interaction term $\mathcal{L}_Y = K\langle \Psi, \Psi \rangle \varphi$, where K is a real constant. The Euler–Lagrange equations arising from $\mathcal{L} = \mathcal{L}_{KG} + \mathcal{L}_D + \mathcal{L}_Y$ are easily seen to be

$$(i\slashed{\nabla} - m_D + K\varphi)\Psi = 0,$$
$$(\Box + m_{KG}^2)\varphi = K\langle \Psi, \Psi \rangle.$$

We are allowing the Klein–Gordon and Dirac fields to have different masses, namely m_{KG} and m_D, respectively. Note that the coupled field equations are semi-linear rather than linear. In other words, the term involving the highest-order derivative in each equation (in this case, $i\slashed{\nabla}\Psi$ and $\Box\varphi$) is linear in the field variable; but there are lower-order terms (in this case, $K\varphi\Psi$ and $K\langle \Psi, \Psi \rangle$) which are nonlinear. In computing the mass of each field, according to the definition given in §4.1, one ignores such nonlinear terms; that is, one uses the linearized version of the equations.

Example 4.3.5 (The Maxwell–Dirac field): The standard interaction between the Maxwell and Dirac fields is described by adding the term $\mathcal{L}_{MD} = \langle \Psi, \slashed{A}\Psi \rangle$ to the 'free' Lagrangian. So the full Lagrangian is $\mathcal{L} = \mathcal{L}_M + \mathcal{L}_D + \mathcal{L}_{MD}$, and the corresponding Euler–Lagrange equations are

$$(i\slashed{D} - m)\Psi = 0,$$
$$\nabla^b F_{ab} = \langle \Psi, \gamma_a \Psi \rangle, \tag{4.3.1}$$

where $D_a = \nabla_a - iA_a$. Thus the Maxwell field has a source, namely the four-current-density

$$J_a = \langle \Psi, \gamma_a \Psi \rangle$$
$$= \sqrt{2}(\bar{\psi}_A \psi_{A'} + \psi_A \bar{\psi}_{A'})$$

where, as usual, $\Psi = (\psi^{A'}, \psi_A)$. For example, Ψ might represent an electron field, and since the electron is a charged particle, it acts as an electromagnetic source.

The operator D_a should really be thought of as a connection on a U(1)-principal bundle over M^4. This will be discussed in detail in the next chapter. For the moment, let us introduce the important concept of *gauge invariance*.

The potential A_a is not uniquely determined by the field F_{ab}: the freedom in A_a is clearly

$$A_a \mapsto A_a + \nabla_a \Lambda, \qquad (4.3.2)$$

where Λ is a real-valued function on M^4. Notice that the Maxwell Lagrangian $\mathcal{L}_M$ is *gauge-invariant*, that is, invariant under the *gauge transformation* (4.3.2). We want the Maxwell–Dirac Lagrangian $\mathcal{L} = \mathcal{L}_M + \mathcal{L}_D + \mathcal{L}_{MD}$ to be gauge-invariant as well. This requirement of gauge-invariance is applied by physicists as a general principle. The way to achieve it in the present case is to specify that the Dirac field Ψ should transform as

$$\Psi \mapsto e^{i\Lambda} \Psi$$

under a gauge transformation. It is simple to verify that $\mathcal{L}$ is then indeed gauge-invariant. As an independent check, observe that the Maxwell–Dirac equations (4.3.1) are also gauge-invariant, and in particular that $D_a \Psi$ transforms as

$$D_a \Psi \mapsto e^{i\Lambda} D_a \Psi.$$

Note also that $\mathcal{L}_D + \mathcal{L}_{MD}$ is equal to the Dirac Lagrangian $\mathcal{L}_D$ with ∇_a replaced by D_a.

Example 4.3.6 (The charged scalar field): Here we want to couple a Klein–Gordon field φ to a Maxwell field F_{ab}. To achieve gauge invariance, we have to specify that φ transforms as $\varphi \mapsto \exp(i\Lambda)\varphi$ under gauge transformations. But then it clearly no longer makes sense to require that φ be real-valued: we have to allow it to be complex-valued. The appropriate Lagrangian is

$$\mathcal{L}_{CKG} = \tfrac{1}{2} \overline{(D^a \varphi)} (D_a \varphi) - \tfrac{1}{2} m^2 \overline{\varphi} \varphi,$$

which reduces to the Klein–Gordon Lagrangian $\mathcal{L}_{KG}$ if $A_a = 0$ and $\overline{\varphi} = \varphi$. It is clear that $\mathcal{L}_{CKG}$ is gauge-invariant. The full Lagrangian $\mathcal{L} = \mathcal{L}_M + \mathcal{L}_{CKG}$ gives the field equations

$$D_a D^a \varphi = 0,$$
$$\nabla^b F_{ab} = \tfrac{1}{2} i (\overline{\varphi} D_a \varphi - \varphi \overline{D_a \varphi}).$$

So φ represents a charged scalar field, which acts as a source for the Maxwell field.

Thus far in this section, nothing has been said about higher-spin or higher-helicity fields, such as massless fields of helicity $\geq \frac{3}{2}$. Such fields have traditionally been less important in physics than low-spin ones (however, in supersymmetric theories these higher spin fields play a significant role). In addition, problems arise when we try to introduce interactions. To illustrate this, let us attempt to couple a Maxwell field to a massless field of helicity ≥ 1.

Note first that the commutator $[D_a, D_b]$ equals $-iF_{ab}$ (in fact, F_{ab} represents the curvature of the connection D_a, as we shall see in the next chapter). The anti-self-dual part of the commutator is

$$\tfrac{1}{2}(D^B_{A'}D^{AA'} + D^A_{A'}D^{BA'}) = -i\varphi^{AB}.$$

Now the most obvious way of modifying the field equation

$$\nabla^{AA'}\psi_{AB\cdots E} = 0$$

is to replace ∇_a by D_a, so that we have

$$D^{AA'}\psi_{AB\cdots E} = 0. \tag{4.3.3}$$

But acting on this equation with $D^B_{A'}$ gives

$$\varphi^{AB}\psi_{AB\cdots E} = 0,$$

which is an *algebraic* restriction on the fields. This is physically unacceptable, since field equations are meant to be *hyperbolic* systems, for which the initial-value problem is well-posed for arbitrary initial data. The field equations in all of the examples in this section have this property, but equation (4.3.3) does not. In general, the problem of finding 'good' equations for interacting high-spin fields is a difficult (if not impossible) one. See the discussion at the end of §8.1, and also Penrose and Rindler (1984).

Exercises for Section 4.3

4.3.1. Show that one can choose the potential A_a for a Maxwell field in such a way that Maxwell's equations imply $\Box A_a = -J_a$.

4.3.2. The Lagrangian is, in effect, a function of the metric η^{ab}. For example,

$$\mathcal{L}_{KG} = \tfrac{1}{2}\eta^{ab}(\nabla_a\varphi)(\nabla_b\varphi) - \tfrac{1}{2}m^2\varphi^2.$$

The *energy-momentum tensor* T_{ab} is defined by

$$T_{ab} = 2\frac{\partial\mathcal{L}}{\partial\eta^{ab}} - \mathcal{L}\eta_{ab}.$$

Show that for the Klein–Gordon field, we have

$$T_{ab} = (\nabla_a\varphi)(\nabla_b\varphi) - \tfrac{1}{2}[(\nabla_c\varphi)(\nabla^c\varphi) - m^2\varphi^2]\eta_{ab}$$

and that $\nabla^a T_{ab} = 0$ if the Klein–Gordon equation is satisfied.

4.3.3. Show that for a source-free Maxwell field, the energy-momentum tensor is

$$T_{ab} = -F_{ac}F_b{}^c + \tfrac{1}{4}F_{cd}F^{cd}\eta_{ab}$$

and $\nabla^a T_{ab} = 0$.

4.4 Poincaré and conformal invariance

In this section we shall discuss the isometries and the conformal mappings of Minkowski space-time, and the question of how field theories behave under these transformations. Some of these matters were discussed in Part I of this book, from a somewhat different point of view. General references on the subject include Penrose (1968b), Penrose and MacCallum (1973), Kuiper (1949), Penrose and Rindler (1984, 1986).

The Lie group of isometries of M^4 is called the *Poincaré group*. If we represent each point in M^4 by its position vector x^a with respect to some (arbitrarily chosen) origin O, then the mappings in the Poincaré group have the form

$$x^a \mapsto \Lambda_b{}^a x^b + \xi^a$$

where ξ^a and $\Lambda_b{}^a$ are constant, and $\Lambda_b{}^a$ satisfies

$$\eta_{ac}\Lambda_b{}^a\Lambda_d{}^c = \eta_{bd},$$

so that the (pseudo-)metric η_{ab} is preserved. The transformation $x^a \mapsto x^a + \xi^a$ is called a *translation*, while $x^a \mapsto \Lambda_b{}^a x^b$ is called a

Lorentz transformation. The Lie group of Lorentz transformations is called the *Lorentz group* and denoted $O(1,3)$. The translations depend on four parameters, and the Lorentz transformations on six parameters, so the Poincaré group is ten-dimensional.

Now $O(1,3)$ has four disconnected components, since each of time-inversion and space-reflection is a Lorentz transformation which is disconnected from the identity. (If coordinates are chosen so that the line-element is $ds^2 = (dx^0)^2 - (dx^1)^2 - (dx^2)^2 - (dx^3)^2$, then $x^0 \mapsto -x^0, x^j \mapsto x^j$ for $j = 1, 2, 3$ is a time-inversion; while $x^0 \mapsto x^0$, $x^j \mapsto -x^j$ for $j = 1, 2, 3$ is a space-reflection.) Let $O(1,3)_0$ denote that component of $O(1,3)$ which contains the identity.

This group $O(1,3)_0$ is connected (of course), but not simply-connected. In fact, it is doubly-connected: $\pi_1(O(1,3)_0) = \mathbf{Z}_2$, essentially because it contains the doubly-connected group $SO(3)$. The universal cover (in this case, a double cover) of $O(1,3)_0$ is $SL(2, \mathbf{C})$, the group of unimodular 2×2 complex matrices. This is why spinors arise: a two-component spinor transforms according to the fundamental representation of $SL(2, \mathbf{C})$. The symplectic form on spin-space is preserved because of the unimodularity.

We can describe the covering map $SL(2, \mathbf{C}) \rightarrow O(1,3)_0$ by using the vector-spinor correspondence developed earlier. If $\Lambda_A{}^B \in SL(2, \mathbf{C})$, then the covering map is

$$\Lambda_A{}^B \mapsto \Lambda_a{}^b = \Lambda_A{}^B \bar{\Lambda}_{A'}{}^{B'}.$$

This is clearly a two-to-one mapping, since $\Lambda_A{}^B$ and $-\Lambda_A{}^B$ are mapped to the same $\Lambda_a{}^b$.

We move on now to the subject of conformal transformations. A *conformal rescaling* of the metric η_{ab} is a transformation

$$\eta_{ab} \mapsto \hat{\eta}_{ab} = \Omega^2 \eta_{ab},$$

where $\Omega = \Omega(x)$ is a smooth, positive scalar function on M^4. In general, the new metric $\hat{\eta}_{ab}$ is not flat, although it is (by definition) conformally flat. But for certain Ωs, $\hat{\eta}_{ab}$ *is* flat. For example, if Ω is constant, or if $\Omega(x) = (x^a x_a)^{-1}$, then $\hat{\eta}_{ab}$ is flat (see Chapter 6). The latter of these Ωs is singular on $\{x^a : x^a x_a = 0\}$, the *null cone* of the origin O, so this null cone must be excised from the space.

If M_1 and M_2 are two pseudo-Riemannian manifolds, then a diffeomorphism $f : M_1 \rightarrow M_2$ is called a *conformal mapping* if the metric induced on M_2 by f from the metric on M_1, is a conformal rescaling of the given metric on M_2. In particular, therefore, if f is an isometry, then f is a conformal mapping.

Example 4.4.1: Take $M_1 = M_2 = M^4$ (Minkowski space-time), and $f : x^a \mapsto kx^a$, where k is a positive constant. Then the induced metric $\hat{\eta}_{ab}$ is related to the original metric η_{ab} by $\eta_{ab} = k^2\hat{\eta}_{ab}$. So f is a conformal mapping. It is called a *dilation*.

Example 4.4.2: Take

$$M_1 = M^4 - \text{(null cone of } p^a)$$
$$= \{x^a : (p^a - x^a)(p_a - x_a) \neq 0\}$$

and

$$M_2 = M^4 - \text{(null cone of } O).$$

Define f by

$$f : x^a \mapsto \frac{p^a - x^a}{(p_b - x_b)(p^b - x^b)}.$$

Then f is a diffeomorphism, and the induced metric $\hat{\eta}_{ab}$ on M_2 is related to the original metric η_{ab} on M_2 by $\hat{\eta}_{ab} = \Omega^2\eta_{ab}$, where $\Omega(x) = (x^a x_a)^{-1}$. So f is a conformal mapping; it is called an *inversion*. To understand it properly, one should compactify M^4 by adjoining to M^4 a 'null cone at infinity', denoted I. So one gets a compact space M containing M^4; for details, see Chapter 1, Kuiper (1949), Penrose and MacCallum (1973), and Penrose and Rindler (1986). M does not admit a flat metric, but it does admit one which is conformally flat. Moreover, the inversion map f extends to a conformal mapping from M to itself. This map sends the null cone of p^a to I, while I is sent to the null cone of the origin O.

The Lie group of conformal mappings from M to itself is called the *conformal group* and denoted $C(1,3)$. It is 15-dimensional, being generated by the Poincaré transformations (ten parameters), the dilations (one parameter), and the inversions (four parameters). The Poincaré transformations and the dilations leave I invariant, whereas the inversions move I. The precise metric on M is unimportant; all that matters is its *conformal structure*, or null-cone structure.

We turn now to the question of how field theories behave under Poincaré and conformal transformations. Since all the field theories mentioned in the previous three sections involve naturally-defined differential operators acting on spinor or tensor fields, these theories are all Poincaré-invariant. Strictly speaking, the theories involving spinors are invariant not under the Poincaré group, but rather under its universal cover, the identity-connected component of which is generated by translations and $SL(2, \mathbf{C})$ transformations. For example, if $\Lambda_A{}^B$ is an element of $SL(2, \mathbf{C})$, and $x^{AA'} \mapsto \Lambda_B{}^A \bar{\Lambda}_{B'}{}^{A'} x^{BB'}$, denoted by $x \mapsto f(x)$, is the corresponding Lorentz transformation, then a spinor field φ_A is transformed to $\widehat{\varphi}_A$, where

$$\widehat{\varphi}_A(x) = \Lambda_A{}^B \varphi_B(f^{-1}(x)),$$

and the Weyl neutrino equation $\nabla^{AA'} \varphi_A = 0$ is preserved under the transformation.

When it comes to conformal transformations, there are two possible meanings for the statement that a field theory is conformally invariant. First, one might demand that it be possible to attach a conformal weight to each field in the theory, in such a way that the field equations are invariant under arbitrary conformal rescalings

$$\eta_{ab} \mapsto \Omega^2(x)\eta_{ab}$$

of the metric. A scalar, tensor or spinor field φ is said to have *conformal weight* k if it transforms as $\varphi \mapsto \Omega^k \varphi$ under such a rescaling. Since $\Omega^2 \eta_{ab}$ is not flat, in general, this takes us into the realm of curved space-time. We shall deal with this subject in Chapter 6, and we shall see that, for example, massless free fields of any helicity are invariant under conformal rescalings (except for helicity 0, the wave equation, where a slight modification is required). One thing one can see immediately is that theories involving mass are *not* conformally invariant. For under a dilation $\eta_{ab} \mapsto k^2 \eta_{ab}$, the d'Alembertian transforms as $\square \mapsto k^{-2}\square$, so the equation $(\square + m^2)\varphi = 0$ is not invariant unless $m = 0$. To be conformally invariant, a field theory has to involve only massless fields.

The second possible definition of conformal invariance is to demand that the field theory be invariant under the conformal group $C(1,3)$. If a theory is Poincaré-invariant, and also invariant under conformal rescalings (or even just under those rescalings which leave the metric flat), then it is conformally invariant in this second sense. For the Poincaré transformations become conformal transformations according to any other conformally rescaled flat metric; and the conformal transformations obtained in this way, together with the Poincaré transformations, generate the whole conformal group (as discussed in Chapter 2, cf. Penrose and MacCallum 1973).

So the massless free field theories are also conformally invariant in this second sense. This involves extending the fields from M^4 to its compactification M. Some subtleties arise with this: the fields must be thought of as sections of certain nontrivial spinor bundles over M (as discussed in Chapter 1), and these must have conformal weights attached to them (see Chapter 7).

Exercises for Section 4.4

4.4.1. Show that the inverse of a Lorentz transformation $\Lambda_b{}^a$ is given by

$$(\Lambda^{-1})_b{}^a = \Lambda^a{}_b$$

and check that the set of Poincaré transformations does indeed form a group.

4.4.2. Show that the proper conformal group $C(1,3)_0$ is doubly-covered by $SO(2,4)$, and that this, in turn, is doubly-covered by the group $SU(2,2)$.

5

Gauge theory

In the previous chapter we encountered simple gauge theories such as Maxwell–Dirac theory. The key idea in all gauge theories is that of *gauge invariance*. The word 'gauge' is really a misnomer: it arose as a result of an unsuccessful attempt by H. Weyl to unify general relativity and Maxwell theory. This involved tampering with the local length scale in space-time, that is, by allowing local conformal rescalings of the metric; hence the word 'gauge', in the sense of 'measurement scale'. A more accurate term would be 'phase invariance', since the basic transformation in Maxwell theory is $\varphi \mapsto e^{i\Lambda}\varphi$.

The field equations that we deal with in this book are *classical* (that is, nonquantum); they describe the propagation of classical fields through space-time. This provides, at best, an approximation to physical reality: for greater accuracy one has to go to the corresponding quantum field theory, in which the various fields become operator-valued (they act on a suitable Hilbert space). For example, classical Maxwell–Dirac theory predicts the spectrum of the hydrogen atom; and this prediction is pretty close to, but certainly different from, what is observed in experiments. The corresponding quantum field theory, which is called quantum electrodynamics, makes predictions about (say) the spectrum of hydrogen which have unprecedented accuracy (theory agrees with experiment to something like one part in 10^{10}). However, the passage from the classical to the quantum field theory is difficult and as yet not well understood.

Gauge theories seem to be particularly amenable to quantization (in physics jargon, they lead to quantum field theories which are renormalizable, and which therefore make sense). For this and other reasons, gauge theories are now being used to model not only electromagnetism, but also two of the other forces of nature, namely the weak and the strong nuclear forces. The remaining force is that of gravity, where no quantum version as yet exists. The accepted classical model of gravity is Einstein's theory of general relativity; see Chapter 6.

Maxwell theory is said to be a gauge theory with gauge group U(1), because the phase factor $e^{i\Lambda}$ takes values in U(1). General gauge theories are obtained by replacing U(1) with more general, non-abelian Lie groups. This process was begun by Yang and Mills in 1954. It was subsequently realized that the underlying mathematical structure of gauge theories is exactly that of connections on fiber bundles. This correspondence becomes especially useful when one considers global questions, as we shall see in §§5.2 and 5.3. These sections serve as an introduction to two specific types of gauge fields, namely instantons and magnetic poles. But first, §5.1 describes what gauge theories are, and provides a dictionary between the physics and the mathematics terminology.

5.1 The essentials of gauge theory

We want to generalize Maxwell theory by replacing its gauge group U(1) by a nonabelian gauge group. Usually this group is taken to be compact and semi-simple, or to be the product of several such groups and copies of U(1). Particularly common in physical applications are the special unitary groups $SU(n)$.

The general set-up is as follows. Let G be a Lie group (the gauge group) and let $\mathfrak{g}$ denote its Lie algebra. A *gauge potential* on Minkowski space-time M^4 is a map A_a from M^4 to $\mathfrak{g} \otimes T^*(M^4)$. In other words, A_a is a $\mathfrak{g}$-valued one-form on M^4.

Any (scalar, tensor or spinor) field φ which is to be coupled to the gauge field, must belong to some representation of G. For example, if G is $SU(n)$, then φ might be in the fundamental representation, and so be represented as a column n-vector; or φ might be in the adjoint representation, and so be represented as an $n \times n$ trace-free anti-Hermitian matrix. The potential A_a gives rise to a differential operator D_a acting on such fields φ. If φ is in the fundamental representation, then

$$D_a\varphi = \nabla_a\varphi + A_a\varphi,$$

with A_a acting by matrix multiplication. If φ is in the adjoint representation, then

$$D_a\varphi = \nabla_a\varphi + [A_a, \varphi],$$

where $[\ ,\ ]$ is the matrix commutator (or, equivalently, the Lie algebra 'multiplication').

The commutator $[D_a, D_b]$ is given by

$$[D_a, D_b]\varphi = F_{ab}\varphi \qquad \text{(fundamental repr.)}$$
$$[D_a, D_b]\varphi = [F_{ab}, \varphi] \qquad \text{(adjoint repr.)},$$

where

$$F_{ab} = \nabla_a A_b - \nabla_b A_a + [A_a, A_b].$$

This F_{ab} is a $\mathfrak{g}$-valued two-form on M^4, and is called the *gauge field*. A *gauge transformation* involves a map $g : M^4 \to G$ and is given by

$$A_a \mapsto g^{-1} A_a g + g^{-1} \nabla_a g,$$
$$\varphi \mapsto g^{-1}\varphi, \qquad \text{(fundamental repr.)},$$
$$\varphi \mapsto g^{-1}\varphi g. \qquad \text{(adjoint repr.)}$$

It follows that

$$F_{ab} \mapsto g^{-1} F_{ab} g,$$
$$D_a\varphi \mapsto g^{-1} D_a\varphi, \qquad \text{(fundamental repr.)},$$
$$D_a\varphi \mapsto g^{-1}(D_a\varphi)g. \qquad \text{(adjoint repr.)}.$$

In the special case when $G = U(1)$, all of this reduces to the Maxwell theory discussed in the previous chapter, except that the A_a and F_{ab} there differ from those here by a factor of i.

Mathematically, what all this amounts to is that we have a trivial vector bundle V over M^4 with structure group G. The operator D_a defines a connection on V, and its curvature is F_{ab}. A gauge transformation is just an automorphism of the G-bundle V. See Chapter 2 for more details of all this. (Sometimes gauge theory is described in terms of principal bundles rather than vector bundles. The choice between these is largely a matter of taste.)

Let us move on now to the question of field equations. The Lagrangian for gauge fields is a natural generalization of that for Maxwell fields, namely the following. Let $(\ ,\ )$ denote the natural bilinear form (Killing form) on $\mathfrak{g}$. For example, if $G = SU(n)$ or $U(n)$, and $A, B \in \mathfrak{g}$ are anti-Hermitian $n \times n$ matrices, then

$$(A, B) = -\operatorname{tr}(AB),$$

where $\operatorname{tr}(AB)$ denotes the trace of the matrix AB. From now on, assume that $(\ ,\ )$ is an Hermitian, positive-definite form on $\mathfrak{g}$, as is for example the case if G is a product of copies of $SU(n)$ and $U(1)$. Regarding F_{ab} as a function of A_a and $\nabla_a A_b$, we say that

$$\mathcal{L}_{YM} = \mathcal{L}_{YM}(\nabla_a A_b, A_a) = -\tfrac{1}{4}(F_{ab}, F^{ab}) \qquad (5.1.1)$$

is the Lagrangian for gauge fields. It is called the *Yang–Mills Lagrangian*. We note immediately that it is gauge-invariant, since the Killing form is invariant under the adjoint action of G on $\mathfrak{g}$.

The gauge field F_{ab} automatically satisfies the Bianchi identity

$$D_{[a} F_{bc]} = 0;$$

whereas the Euler–Lagrange equations obtained from $\mathcal{L}_{YM}$ are

$$D^a F_{ab} = 0,$$

which are called the *Yang-Mills equations*. (Recall that $D_a F_{bc}$ is

$$D_a F_{bc} = \nabla_a F_{bc} + [A_a, F_{bc}],$$

since F_{ab} belongs to the adjoint representation.) The Yang–Mills equations generalize the source-free Maxwell equations.

Since F_{ab} is a two-form (with values in $\mathfrak{g}$), we can repeat the analysis of §4.1, and decompose F_{ab} into its self-dual and anti-self-dual parts

$$F_{ab}^{\pm} = \tfrac{1}{2}(F_{ab} \mp i *F_{ab}).$$

Note that F_{ab}^+ and F_{ab}^- are two-forms with values in $\mathfrak{g}_{\mathbb{C}}$ (the complexification of $\mathfrak{g}$) rather than $\mathfrak{g}$. For example, if $\mathfrak{g}$ is the Lie algebra of $SU(n)$, then $\mathfrak{g}_{\mathbb{C}}$ is the Lie algebra of $SL(n, \mathbb{C})$.

An important class of gauge fields are those which are self-dual:

$$*F_{ab} = i F_{ab}$$

or anti-self-dual:

$$*F_{ab} = -i F_{ab}.$$

In either of these cases, the Yang–Mills equations are automatically satisfied, since the Bianchi identity says that

$$D^a *F_{ab} = 0.$$

The self-duality equation $*F = iF$ (or the anti-self-duality equation) is a first-order semi-linear system of equations for the gauge potential A_a. However, the factor of i (or $-i$) means that for 'real' gauge groups such as $SU(n)$, there are no self-dual gauge fields on Minkowski space-time, except the trivial field $F_{ab} = 0$. For if F_{ab} is an anti-Hermitian matrix, then so is $*F_{ab}$, and so $*F_{ab} = \pm i F_{ab}$ implies $F_{ab} = 0$. There are two ways out of this impasse. First, we could allow 'complex' gauge groups such as $SL(n, \mathbf{C})$ or $GL(n, \mathbf{C})$: then there are many self-dual solutions on Minkowski space-time. Or, alternatively, we could reformulate the whole theory on a positive-definite base space, rather than on one with Lorentzian signature such as Minkowski space-time. This idea is pursued in §5.2.

It is clear that the Yang–Mills equations and the self-duality equations are Poincaré-invariant. Furthermore, and very importantly, they are conformally invariant; in fact, invariant under arbitrary conformal rescalings of the space-time metric. (Both A_a and F_{ab} are taken to have conformal weight zero.) This will be shown in Chapter 6. Note in particular that the $*$-operator acting on two-forms is conformally invariant. This is a special feature that occurs only in dimension four, the dimension of space-time.

Let us now look at a couple of examples of interactions. A gauge field couples to other fields in a natural way, generalizing the Maxwell examples of the previous chapter.

Example 5.1.1 (Yang–Mills–Higgs field): A Yang–Mills field coupled to a scalar field φ of mass m, in the adjoint representation, is referred to as a Yang–Mills–Higgs field, and the scalar field is referred to as the Higgs field in this context. So $\varphi = \varphi(x)$ is a map from M^4 to $\mathfrak{g}$. The appropriate Lagrangian is

$$\mathcal{L} = \mathcal{L}_{\text{YM}} - \tfrac{1}{2} \operatorname{tr}(D^a \varphi)(D_a \varphi) + \tfrac{1}{2} m^2 \operatorname{tr} \varphi^2,$$

which is clearly gauge-invariant. The corresponding Euler–Lagrange equations are

$$D_a D^a \varphi + m^2 \varphi = 0,$$
$$D^b F_{ab} = -[\varphi, D_a \varphi].$$

Example 5.1.2 (The Yang–Mills–Dirac field): This requires some additional machinery. Let $\{T_\alpha\}_{\alpha=1,2,\dots,p}$ be a basis for the Lie algebra $\mathfrak{g}$. Think of each T_α as an $n \times n$ matrix. For example, if $G = SU(2)$, then $p = 3$, and we may take the T_α to be the three Pauli matrices multiplied by i:

$$T_1 = i \begin{bmatrix} 0 & 1 \\ 1 & 0 \end{bmatrix},$$

$$T_2 = i \begin{bmatrix} 0 & -i \\ i & 0 \end{bmatrix},$$

$$T_3 = i \begin{bmatrix} 1 & 0 \\ 0 & -1 \end{bmatrix}.$$

Denote by $g_{\alpha\beta} = (T_\alpha, T_\beta) = -\operatorname{tr}(T_\alpha T_\beta)$ the metric (Killing form) on $\mathfrak{g}$.

Now let Ψ denote a Dirac spinor field in the fundamental representation of G. Think of Ψ as a column n-vector of Dirac spinors. Its conjugate $\overline{\Psi}$ is a row n-vector of dual Dirac spinors. And $\overline{\Psi}\Psi = \langle \Psi, \Psi \rangle$ is an $n \times n$ matrix of numbers, whose trace is the norm-squared of Ψ. In other words, the field Ψ takes values in $V \otimes_{\mathbb{C}} D$, where D denotes the (four-complex-dimensional) space of Dirac spinors, and V denotes the (n-complex-dimensional) representation space of G. Now V comes equipped with a G-invariant Hermitian form, while $\langle \ , \ \rangle$ is an Hermitian form on D. So there is a natural Hermitian form on $V \otimes_{\mathbb{C}} D$, and this is $\operatorname{tr}\langle \Psi, \Psi \rangle$.

The Yang–Mills–Dirac Lagrangian is

$$\mathcal{L} = \mathcal{L}_{\mathrm{YM}} + \operatorname{Re} \operatorname{tr} \langle \Psi, i \slashed{D} \Psi \rangle$$

and it leads to the Euler–Lagrange equations

$$\begin{aligned} \slashed{D}\Psi &= 0, \\ D^b F_{ab} &= i g^{\alpha\beta}\{\operatorname{tr}\langle \Psi, \gamma_a T_\beta \Psi\rangle\}T_\alpha, \end{aligned} \qquad (5.1.2)$$

where $g^{\alpha\beta}$ is the inverse of $g_{\alpha\beta}$; that is, $g^{\alpha\gamma}g_{\alpha\beta} = \delta^\gamma_\beta$. The right-hand side of (5.1.2) is independent of the choice of basis $\{T_\alpha\}$, since $\mathcal{L}$ is. But if we tried to write it in a basis-free, manifestly invariant form, then that form would depend on precisely what gauge group G we were dealing with.

In both of these examples, the field equation for Ψ or φ is what it would have been for a free (noninteracting) field, except that ∇_a is replaced by D_a. Geometrically, each of Ψ and φ is a section of a vector bundle over M^4, and D_a is a connection on this bundle;

∇_a is the flat connection. Each field Ψ or φ generates a four-current density J_a, a one-form on M^4 with values in $\mathfrak{g}$; it acts as a source for the gauge field F_{ab}.

Finally, here is a summary of the correspondence between gauge theory and fiber bundle theory.

Gauge Theory	*Fiber Bundles*
gauge group	structure group
gauge potential	connection one-form
gauge field	curvature two-form
gauge	(local) trivialization of bundle
gauge transformation	(local) automorphism of bundle.

Minkowski space-time M^4 being topologically trivial, we have not as yet made any distinction between local and global features. But in the next section this distinction will become important. Incidentally, most physics books use the terms 'local gauge transformation' and 'global gauge transformation', but these have nothing to do with 'local' and 'global' in the mathematical sense. They refer, rather, to whether the gauge-change matrix g, or $e^{i\Lambda}$ in the Maxwell case, depends on the position coordinates x^a. If it does, the gauge transformation is called 'local'; if not, it is 'global'. All our gauge transformations are local in this sense, and we have just called them 'gauge transformations'.

Exercises for Section 5.1

5.1.1. Show that if $D_a\varphi = \nabla_a\varphi + A_a\varphi$ is required to transform like $D_a\varphi \mapsto g^{-1}D_a\varphi$ when $\varphi \mapsto g^{-1}\varphi$, then the transformation law for A_a is uniquely determined.

5.1.2. Show that the gauge field F_{ab} necessarily satisfies the Bianchi identity $D_{[a}F_{bc]} = 0$.

5.1.3. In Examples 5.1.1 and 5.1.2 derive the field equations from the given Lagrangians.

5.2 Yang–Mills instantons

This section is an introduction to the subject of Yang–Mills instantons (namely, finite-action solutions of the Yang–Mills equations on Euclidean four-space E^4), and a list of mathematical results concerning them. It is a subject to which twistor theory applies very effectively, as we shall see in Chapter 8. The reason instantons arise in quantum field theory is summarized in the following discussion.

The standard approach to the quantization of gauge theory (that is, passing from a classical gauge theory to the corresponding quantum field theory) is that of the Feynman path integral. This involves formal functional integrals such as

$$\int_{\Phi(\Omega)} \mathcal{P}[\varphi] \exp(i\mathcal{A}[\varphi])\mathcal{D}[\varphi],$$

where φ denotes the field(s) of the theory, $\Phi(\Omega)$ the space of all such fields φ on the region Ω of space-time (with φ satisfying suitable boundary and regularity conditions), $\mathcal{A}[\varphi]$ the action of φ on Ω, $\mathcal{P}[\varphi]$ a polynomial in φ, and $\mathcal{D}[\varphi]$ a suitable measure on $\Phi[\Omega]$. This space $\Phi(\Omega)$ is infinite-dimensional, and it is very difficult, if not impossible, to make sense of this integral. The strategy is to analytically continue from Minkowski space-time to Euclidean four-space E^4 (with its standard positive-definite metric). Physicists call this a 'Wick rotation'. The integral becomes

$$\int_{\Phi(\Omega)} \mathcal{P}[\varphi] \exp(-\mathcal{A}[\varphi])\mathcal{D}[\varphi],$$

where Ω is now a region of E^4. In many cases of interest, $\mathcal{A}[\varphi]$ is a positive-definite functional of φ (on E^4), and the exponential factor makes the integral manageable. This sort of integral was studied by Wiener in connection with the theory of Brownian motion.

In order to analyze and estimate such an integral, the first step is to find the critical points of $\mathcal{A}[\varphi]$ in $\Phi(E^4)$ (that is, the solutions of the Yang–Mills equations), since the first-order functional variation of $\mathcal{A}[\varphi]$ vanishes for such fields. In particular, we want the local minima of $\mathcal{A}[\varphi]$ in $\Phi(E^4)$, so that we can carry out a perturbation analysis around these minima. Perturbations around a minimum correspond to quantum-mechanical fluctuations around classical solutions. In practice, most calculations in quantum field theory are done in this way, using perturbation theory.

Our first task, therefore, is to reformulate gauge theory in the Euclidean space E^4. This is easily done: one simply replaces the Minkowski metric η_{ab} by its Euclidean counterpart

$$\delta_{ab} = \text{diag}(1, 1, 1, 1).$$

An object such as v^a is now a vector in E^4, and indices are lowered and raised with δ_{ab} and its inverse δ^{ab}. The rules for spinors change slightly, but we do not have to concern ourselves with this at the moment. The Lagrangian (5.1.1) of a gauge field F_{ab} on E^4 is now a negative-definite functional of F_{ab}, so we change its sign and write

$$\mathcal{L}_{\text{YM}} = \tfrac{1}{4}(F_{ab}, F^{ab})$$

for gauge fields on E^4. The use of the same symbol $\mathcal{L}_{\text{YM}}$ should not cause confusion, as it is always clear from the context whether one is working in a space with Lorentzian signature or with positive-definite signature. In the Lorentzian case, one inserts the minus sign in order to ensure that the energy of the field (which is related to the spatial integral of $\mathcal{L}$) is positive.

In E^4, therefore, the action $\mathcal{A}$ of a gauge field is

$$\mathcal{A} = \int_{E^4} \mathcal{L}_{\text{YM}}(\nabla_a A_b(x), A_a(x)) d^4 x$$

$$= \tfrac{1}{4} \int_{E^4} (F_{ab}, F^{ab}) d^4 x,$$

and it is a positive-definite functional of the field. Clearly it is just the natural L^2-norm (squared) of F_{ab}:

$$\mathcal{A} = \tfrac{1}{4} \|F\|^2.$$

This action is manifestly invariant under the group of orientation-preserving rigid motions of E^4, the analogue of the Poincaré group. Furthermore, it is conformally invariant: under a conformal rescaling of the metric, (F_{ab}, F^{ab}) and $d^4 x$ scale with weights -4 and 4, respectively, so $\mathcal{A}$ does not change.

Now we are only interested in fields with finite action (bearing in mind the exponential factor $\exp(-\mathcal{A})$ in the functional integral).

Finite action means, as we shall see presently, that the gauge field
extends to the one-point compactification of E^4, namely S^4, the four-
sphere equipped with its standard metric. The sphere S^4 is obtained
by adding a 'point at infinity', denoted ∞, to E^4. The stereographic
projection $f_\infty : S^4 - \{\infty\} \to E^4$ is a conformal mapping.

But S^4 is topologically nontrivial, so we have to begin to worry
about global matters, and to use the language of vector bundles and
connections. The data specifying the gauge field on S^4 consists of a
vector bundle E on S^4, and a connection D_a on E. For example,
if the gauge group G is $\mathrm{SU}(n)$, then E is a $\mathbf{C}^n$-bundle, and is not
necessarily trivial. If we project E and its connection to E^4 via f_∞,
then we get a trivial bundle on E^4, and a connection on it of the
form $D_a = \nabla_a + A_a$, where A_a is a smooth g-valued one-form on
E^4. In general (in fact, if E is nontrivial) the pullback $f_\infty^* A_a$ of A_a
to $S^4 - \infty$ does not extend over the point ∞, that is, it does not
extend to the whole of S^4. To describe things near ∞, we have to
project stereographically from some other point (say zero) in S^4, via
the conformal mapping

$$f_0 : S^4 - \{0\} \to E^4.$$

This gives a trivial bundle on E^4 and a connection on it of the form
$D_a = \nabla_a + \hat{A}_a$. Now the pullback $f_0^* \hat{A}_a$ of $\hat{A}_a$ is defined everywhere
on S^4 except at zero. The two pullbacks $f_\infty^* A_a$ and $f_0^* \hat{A}_a$ differ by a
gauge transformation on $S^4 - \{0, \infty\}$.

Let ω denote the four-form $(F_{ab}, F^{ab})d^4x$ obtained from the gauge
potential A_a, and $\hat{\omega}$ that obtained from $\hat{A}_a$. By gauge invariance and
conformal invariance, it is clear that $f_\infty^* \omega = f_0^* \hat{\omega}$ on $S^4 - \{0, \infty\}$,
and therefore that $f_\infty^* \omega$ extends smoothly to the whole of S^4. The
action $\mathcal{A}$ is just the integral of this form over S^4, and is finite, because
S^4 is compact. We say that the connection on E satisfies the Yang–
Mills equations if its stereographic projection A_a does. This is a
reasonable and consistent definition, bearing in mind that the Yang–
Mills equations are gauge-invariant and conformally invariant.

We see, therefore, that a Yang–Mills solution on S^4 gives rise to
a finite-action Yang–Mills solution on E^4. The following theorem
shows that *every* finite-action solution arises in this way.

Theorem 5.2.1 (Uhlenbeck 1982). *Let A_a be a finite-action solution
of the Yang–Mills equations on E^4. Then there exists a bundle E*

over S^4, with a connection which is mapped (under the stereographic projection f_∞) to $D_a = \nabla_a + A_a$ on E^4.

We are therefore led to study G-bundles on S^4, and connections on them, satisfying the Yang–Mills equations. Such objects are called *instantons*. We note that if G is simple, then the G-bundles over S^4 are classified by an integer. For example, if G is $SU(n)$, then this integer is the second Chern number of the bundle E, given by (Proposition 3.4.4)

$$c_2(E) := \frac{1}{8\pi^2} \int_{S^4} \operatorname{tr}(F \wedge F)$$

$$= -\frac{1}{8\pi^2} \int_{E^4} (F_{ab}, *F^{ab}) d^4x.$$

The number $k = -c_2(E)$ is called the *topological charge* or *instanton number* of the gauge field. For other gauge groups, there is a similar formula, with perhaps a change in the overall numerical factor (Atiyah, Hitchin and Singer 1978).

The Hodge star operator on E^4 is defined just as in §4.1. But because of the different signature of the metric, the identity (4.1.1) must be replaced by

$$\varepsilon_{abcd}\varepsilon^{cdef} = 2(\delta_a^e \delta_b^f - \delta_a^f \delta_b^e);$$

that is, there is a change of sign. Consequently, we have $** = 1$ when $*$ acts on two-forms, not $** = -1$ as previously. So now the eigenspaces of $*$ have eigenvalues ± 1. A two-form ω_{ab} is said to be *self-dual* if $*\omega_{ab} = \omega_{ab}$, and *anti-self-dual* if $*\omega_{ab} = -\omega_{ab}$. Any two-form ω_{ab} may be decomposed into its self-dual and anti-self-dual parts:

$$\omega_{ab} = \omega_{ab}^+ + \omega_{ab}^-,$$

where

$$\omega_{ab}^\pm = \tfrac{1}{2}(\omega_{ab} \pm *\omega_{ab}).$$

The self-dual and anti-self-dual parts are mutually orthogonal:

$$\omega_{ab}^+ \omega^{-ab} = 0.$$

Let us now substitute the decomposition $F_{ab} = F_{ab}^+ + F_{ab}^-$ of a gauge field into the expressions for the action and the topological charge k. This gives

$$4\mathcal{A} = \|F\|^2$$
$$= \|F^+ + F^-\|^2$$
$$= \|F^+\|^2 + \|F^-\|^2,$$

$$8\pi^2 k = \int (F_{ab}, *F^{ab}) d^4 x$$
$$= \int (F_{ab}^+ + F_{ab}^-, F^{+ab} - F^{-ab}) d^4 x$$
$$= \|F^+\|^2 - \|F^-\|^2.$$

Thus for fields of topological charge k, the action is bounded below by $2\pi^2|k|$; and this lower bound is attained if and only if

$$F_{ab}^+ = 0, \quad k < 0,$$
$$F_{ab} = 0, \quad k = 0,$$
$$F_{ab}^- = 0, \quad k > 0.$$

So the self-dual and anti-self-dual fields correspond to the absolute minima of the action in each topological class. These fields are, of course, solutions of the Yang–Mills equations (being critical points of the action functional), so they are instantons. Sometimes the term 'instanton' is taken to refer only to finite-action self-dual or anti-self-dual fields; sometimes, even more restrictively, it is applied only to the $k = 1$ self-dual solutions, in which case a $k = -1$ anti-self-dual field is an 'anti-instanton', and $k > 1$ self-dual fields are 'multi-instantons'. But for us, an instanton is any finite-action critical point of $\mathcal{A}$.

The question immediately arises as to whether there are any instantons that are *not* self-dual or anti-self-dual. In other words, are there any critical points of $\mathcal{A}$ other than the absolute minima? The answer to this question is not known, but some partial results exist, and we shall now describe a few of them (see note at the end of §5.2).

If G is SO(4), then there *is* a Yang–Mills solution which is neither self-dual nor anti-self-dual, namely the tangent bundle of S^4 with its canonical Riemannian connection (Bourguignon and Lawson 1981). This happens because the group SO(4) holds a rather special position, being the structure group of the tangent bundle. In addition, SO(4) is not simple.

We can ask if there are local minima of the action which are not self-dual or anti-self-dual. For this we need the appropriate generalization of the second derivative test for local minima in this setting. We say that a connection D_a on S^4 is *weakly stable* if

$$\frac{d^2}{dt^2} \mathcal{A}(t)|_{t=0} \geq 0$$

for all smooth one-parameter families of connections $D_a(t)$ with $|t| < \varepsilon$ and $D_a(0) = D_a$. Here $\mathcal{A}(t)$ is the action of $D_a(t)$. We have the following theorem of Bourguignon and Lawson (1981).

Theorem 5.2.2. *Any weakly stable instanton with group* SU(2), SU(3), *or* U(2) *is either self-dual or anti-self-dual.*

So there are no local minima which are not absolute minima: any other critical points must be 'saddle points'. In a somewhat different direction one can show that self-dual instantons are isolated. That is, there are no non-self-dual instantons nearby. More precisely, we have the following theorem of Min-Oo (1982).

Theorem 5.2.3. *Given any gauge group* G, *there exists an* $\varepsilon > 0$ *such that if* F_{ab} *is an instanton field and* $\|F^-\| < \varepsilon$, *then* $F_{ab}^- = 0$. *The same statement holds with* F_{ab}^+ *in place of* F_{ab}^-.

There are several other theorems along these lines (Bourguignon and Lawson 1981; Flume 1978; Daniel, Mitter and Viallet 1978; Dodziuk and Min-Oo 1982). In a more specialized direction, if we impose O(3) symmetry on the field (axial symmetry about a line in E^4), and take the gauge group to be SU(2), then the only instantons are either self-dual or anti-self-dual (Taubes 1980; Jaffe and Taubes 1980).

For the remainder of this section, we shall restrict our attention to self-dual instantons (anti-self-dual instantons are just the mirror images of self-dual ones, and are obtained by reversing the orientation on E^4). We shall see below that for every positive integer k, there exists a self-dual instanton with topological charge k. Using this, the Atiyah–Singer index theorem, and a few other ingredients, one can prove the following result.

Theorem 5.2.4 (Atiyah, Hitchin and Singer 1978). *The space of charge* k *self-dual instantons with gauge group* SU(2) *is a real manifold of dimension* $8k - 3$.

In this result, the gauge freedom is assumed to have been removed, so 'space of instantons' means 'space of instantons modulo gauge equivalence'. In geometry this space is called the space of moduli. The theorem of Atiyah, Hitchin and Singer (1978) is actually much more general, and allows other gauge groups and base spaces other than S^4. See also (Schwarz 1977; Jackiw and Rebbi 1977; Brown, Carlitz and Lee 1977; Bernard, Christ, Guth and Weinberg 1977), where all or part of the result was proved independently.

Let us now demonstrate, as promised, the existence of an SU(2)-instanton of charge k, for each $k > 0$. The first of these to be discovered was for $k = 1$ (Belavin et al. 1975). (It was at first called a 'pseudoparticle' rather than an 'instanton'.) Subsequently this $k = 1$ solution was generalized to $k > 1$ solutions (Witten 1977; Corrigan and Fairlie 1977; Jackiw, Nohl and Rebbi 1977), namely the following.

The solutions are obtained from the ansatz

$$A_a = i\sigma_{ab}\nabla^b \log \varphi \tag{5.2.1}$$

where φ is a real-valued function on E^4, and $\sigma_{ab} = -\sigma_{ba}$ is a g-valued two-form defined by

$$\sigma_{01} = -\sigma_{23} = -\tfrac{1}{2}\sigma_1 := -\tfrac{1}{2}\begin{bmatrix} 0 & 1 \\ 1 & 0 \end{bmatrix},$$

$$\sigma_{02} = \sigma_{13} = -\tfrac{1}{2}\sigma_2 := -\tfrac{1}{2}\begin{bmatrix} 0 & -i \\ i & 0 \end{bmatrix},$$

$$\sigma_{03} = -\sigma_{12} = -\tfrac{1}{2}\sigma_3 := -\tfrac{1}{2}\begin{bmatrix} 1 & 0 \\ 0 & -1 \end{bmatrix},$$

with respect to the standard basis for E^4. The σ_a for $a = 1, 2, 3$, are the Pauli matrices which generate g, the Lie algebra of SU(2). A short calculation shows that σ_{ab} is anti-self-dual:

$$*\sigma_{ab} = \tfrac{1}{2}\varepsilon_{ab}{}^{cd}\sigma_{cd} = -\sigma_{ab},$$

the orientation being defined by $\varepsilon_{0123} = 1$. Thus what σ_{ab} does is select out the space $\bigwedge^2_-$ of anti-self-dual two-forms from the space $\bigwedge = \bigwedge^2_+ \oplus \bigwedge^2_-$ of all two-forms, and project it onto g ($\bigwedge^2_-$ and g are isomorphic as vector spaces, both being three-dimensional). Or, alternatively, $\bigwedge^2$ is naturally isomorphic to the Lie algebra $\mathfrak{so}(4)$ of $SO(4)$, which decomposes as

$$\mathfrak{so}(4) = \mathfrak{su}(2) \oplus \mathfrak{su}(2);$$

and σ_{ab} projects $\mathfrak{so}(4)$ onto one of these $\mathfrak{su}(2)$ factors.

If we now substitute the ansatz (5.2.1) into the self-duality equation $*F_{ab} = F_{ab}$, we obtain the Laplace equation

$$\Delta\varphi = 0$$

where $\Delta = \delta^{ab}\nabla_a\nabla_b$. The ansatz therefore linearizes the (nonlinear) self-duality condition. So one now has to look for harmonic functions φ which lead to gauge fields that are smooth on E^4 and have finite action. The functions φ which have these properties are sums of elementary solutions

$$\varphi(x) = \sum_{\alpha=0}^{k} \frac{\lambda_\alpha}{(x - x_\alpha)^2}, \tag{5.2.2}$$

where the λ_α are positive real parameters, the x_α are distinct points in E^4, and

$$(x - x_\alpha)^2 = \delta_{ab}(x^a - x_\alpha^a)(x^b - x_\alpha^b)$$

is the Euclidean distance, squared, from the point x to the point x_α. A limiting case of this is allowed, where one point, say x_0, is at infinity, so that

$$\varphi(x) = 1 + \sum_{\alpha=1}^{k} \frac{\lambda_\alpha}{(x - x_\alpha)^2}.$$

Although φ has singularities, the gauge field that it gives rise to is smooth on E^4 (the gauge potential is smooth on E^4, after one has applied a gauge transformation to the A_a of the ansatz). And a straightforward calculation reveals that the solution has finite action, and topological charge equal to k (Jackiw, Nohl and Rebbi 1977).

Thus we have exhibited a self-dual $SU(2)$-instanton for each positive value of the topological charge. It is unnecessary to give all the details of the calculation here, as these solutions will be derived again in Chapter 8, using twistor theory. Yet another description, of a more geometrical character, appears in §7 of Atiyah, Hitchin and Singer (1978).

The solution clearly depends on $5k + 4$ parameters, consisting of the choice of the $k + 1$ points x_α in E^4 or S^4 (four parameters each) and the $k + 1$ constants λ_α, minus an overall scale factor (which disappears when we substitute φ into the ansatz). For $k = 1, 2$ the solutions possess symmetries and so are not all independent: the correct count is 5 for $k = 1$ and 13 for $k = 2$. The following table compares the parameter count for the ansatz solutions to the total, $8k - 3$.

k	1	2	3	4	$\cdots$	k	$\cdots$
Ansatz	5	13	19	24	$\cdots$	$5k + 4$	$\cdots$
Total	5	13	21	29	$\cdots$	$8k - 3$	$\cdots$

For $k = 1$, the ansatz gives all the self-dual instantons (Atiyah, Hitchin and Singer 1978, §9). But for $k \geq 3$ this clearly cannot be so: there are more solutions than can be accounted for by the ansatz. We shall return to this matter in Chapter 8.

The $k = 1$ solution is the original 'pseudoparticle' solution discovered by Belavin et al. (1975). It has a high degree of symmetry, admitting $SO(5)$ as a group of symmetries (Jackiw, Nohl and Rebbi 1977). The corresponding bundle over S^4 is, as a principal bundle, the Hopf bundle $S^7 \to S^4$ with fiber $S^3 \cong SU(2)$ (Trautman 1977).

This is not the place to discuss the importance of instantons in physics. To mathematicians, the problem was (and still is) a nice application of various techniques in geometry. And more recently it has been found that a study of the self-dual instantons on four-dimensional Riemannian manifolds reveals previously unknown facts about the structure of such manifolds (Donaldson 1983).

Note added in proof: Sibner, Sibner and Uhlenbeck (1989) have shown that there are critical points of the action $\mathcal{A}$ (p. 274), which are not absolute minima.

Exercises for Section 5.2

5.2.1. Compute the curvature of the potential 5.2.1, and check that the self-duality condition is equivalent to $\Delta\varphi = 0$.

5.2.2. The harmonic function $\varphi = 1 + x^{-2}$ determines a one-instanton solution. Check, by explicit integration, that the second Chern number of the corresponding bundle is -1.

5.3 Magnetic poles

One of the striking applications of twistor theory is in the problem of describing a certain class of solutions in nonabelian gauge theory related to *magnetic poles*. This will be discussed in §8.4. The present section is an introduction to the subject of magnetic poles (often called monopoles). More details may be found in Goddard and Olive (1978) (a review of the subject); Jaffe and Taubes (1980) (analytical results). Our interest is in the nonabelian case (in fact we shall concentrate on the gauge group SU(2)), but let us first set the scene by discussing the U(1) case, namely electromagnetism.

The source-free Maxwell equations for the two-form (Maxwell field) F, namely $dF = d *F = 0$, clearly have a duality symmetry, i.e., a symmetry under $F \mapsto *F$. This corresponds to a symmetry between the electric and magnetic fields E and B (see §4.1). However, a Maxwell field with source satisfies

$$dF = 0, \qquad *d *F = -\tfrac{1}{3}J,$$

where J is the (electric) four-current density; so the duality symmetry is no longer present. To put this another way, the standard Maxwell theory says that *magnetic* charges and currents do not exist. One could relax this restriction, and allow magnetic charges, by replacing $dF = 0$ with

$$*dF = -\tfrac{1}{3}J_{(m)},$$

where $J_{(m)}$ is a 'magnetic' four-current density. However, if we want to maintain the geometrical interpretation of F as the curvature of a connection on a line-bundle, then we need $dF = 0$. The way out is to allow dF to be nonzero only in some subset of space-time, and to think of the line-bundle as being defined over the complement of this subset. As this complementary region may have nontrivial topology, there is the possibility that the line-bundle could be nontrivial, and this is exactly what happens in the case of U(1) magnetic poles. Such poles are called Dirac magnetic poles, since the whole subject really began with Dirac's (1931) paper.

Consider an isolated magnetic charge in the usual three-space $\mathbf{R}^3$. That is to say we have a magnetic field $B^i_{(m)}$ which is a vector field on $\mathbf{R}^3$, and $\nabla_i B^i_{(m)} = 0$ (i.e., div $B_{(m)} = 0$) outside some compact set U in $\mathbf{R}^3$. (Here the index i is an abstract three-dimensional index, so a symbol ξ^i denotes a vector in $\mathbf{R}^3$.) The scalar $\rho_{(m)} = \nabla_i B^i_{(m)}$ is the magnetic charge density, and its integral over U is the magnetic charge Q. Now outside U, the two-form

$$F_{ij} = \varepsilon_{ijk} B^k_{(m)}$$

is closed, and we wish to think of it as the curvature of a U(1)-bundle over $\mathbf{R}^3 - U$. Since $\mathbf{R}^3 - U$ is topologically $S^2 \times \mathbf{R}$, such U(1)-bundles are labeled by an integer n, the (first) Chern number. And the magnetic charge Q is equal to $2\pi n$ (since, by Stokes' theorem, Q equals the integral of F over a two-sphere surrounding the source region U, and F is 2π times the Chern class: see §3.4). So magnetic charge is 'quantized': if it occurs at all, it can only occur in integer multiples of a certain basic quantity.

Physicists have searched for magnetic poles for many years, but so far there has been no confirmed sighting. The matter has become more important in recent times, because some of the models which unify electromagnetism with other forces actually *predict* the occurrence of such poles. This happens, for example, in 'grand unified theories' involving gauge groups such as SU(5). In these cases, the magnetic poles are of a 'nonabelian' type, and we shall now discuss an example of such nonabelian poles.

We take the gauge group to be SU(2), and the ingredients of the model to be a gauge field F_{ab} and a scalar field φ in the adjoint representation. The Lagrangian is

$$\mathcal{L} = \tfrac{1}{4}\operatorname{tr}(F_{ab}F^{ab}) - \tfrac{1}{2}\operatorname{tr}(D_a\varphi)(D^a\varphi) - \tfrac{1}{4}\lambda(|\varphi|^2 - 1)^2 \quad (5.3.1)$$

(cf. Example 5.1.1). Here $\lambda \geq 0$ is a constant, and $|\varphi|^2 = -\tfrac{1}{2}\operatorname{tr}\varphi^2$. The third term in (5.3.1) is called the Higgs potential (its presence means that φ interacts nonlinearly with itself).

We are interested in solutions of the Euler–Lagrange equations following from (5.3.1). More specifically, let us look for solutions which are static and purely magnetic. That is to say, the fields A_a and φ are taken to be functions only of the spatial variables $x^i = (x^1, x^2, x^3)$, and not of time x^0; and the electric part F_{0a} of the gauge field vanishes. In effect, this reduces us to a gauge theory on $\mathbf{R}^3$ (space rather than space-time): we take A_0 to be zero and A_i to be a gauge potential on $\mathbf{R}^3$, with curvature F_{ij}. The relevant functional density is minus what remains of the Lagrangian, namely the Hamiltonian

$$\mathcal{H} = -\tfrac{1}{4}\operatorname{tr}(F_{ij}F^{ij}) - \tfrac{1}{2}\operatorname{tr}(D_i\varphi)(D^i\varphi) + \tfrac{1}{4}\lambda(|\varphi|^2 - 1)^2. \quad (5.3.2)$$

The *energy E* is the integral of (5.3.2) over $\mathbf{R}^3$. It is a positive-definite functional of the fields (recall that the trace 'tr' is negative-definite, and the first two minus signs in (5.3.2) compensate for this). We are interested in the critical points of this energy functional.

Clearly one of these critical points is an absolute minimum, namely $A_i = 0$ and $\varphi = \varphi_0$, where φ_0 is a constant with $|\varphi_0| = 1$. This is the 'ground state' of the system. The important thing to note is that it is not unique, since it involves the choice of φ_0; and such a choice breaks the gauge symmetry from SU(2) to U(1). This happens because a gauge transformation $g : \mathbf{R}^3 \to \mathrm{SU}(2)$ acts on φ according to $\varphi \mapsto g^{-1}\varphi g$; and we need $g^{-1}\varphi_0 g = \varphi_0$ so as to preserve the ground state, which restricts g to a U(1) subgroup of SU(2). This is an example of 'spontaneous symmetry breaking'.

More generally, finite energy requires that $|\varphi| \to 1$ as $r \to \infty$ in $\mathbf{R}^3$. In other words, the φ-field at infinity,

$$\hat{\varphi}(\hat{x}^i) := \lim_{\xi \to \infty} \varphi(\xi \hat{x}^i)$$

satisfies $|\hat{\varphi}| = 1$. Here $\hat{x}^i = r^{-1}x^i$, where $r^2 = x_i x^i$. That is, $\hat{x}^i$ parametrizes points on the unit sphere in $\mathbf{R}^3$. So $\hat{\varphi}$ is a mapping from the unit sphere in $\mathbf{R}^3$ to the unit sphere in the Lie algebra $\mathfrak{su}(2) \cong \mathbf{R}^3$. Such a map is classified up to homotopy by its degree n. We shall use the following formula for n (see Chapter II, Proposition 3.7 in Jaffe and Taubes 1980),

$$\tfrac{1}{2} \operatorname{tr} \int B_j D^j \varphi \, d^3x = 4\pi n, \tag{5.3.3}$$

where $B_i = \tfrac{1}{2}\varepsilon_{ijk}F^{jk}$.

The appearance of the integer n means that the configuration space C (the domain of E) is a disjoint union:

$$C = \bigcup_{n \in \mathbf{Z}} C_n.$$

Here C_n consists of those finite-energy configurations (A_i, φ) where $\hat{\varphi}$ has degree n. We claim that E is bounded below by $8\pi|n|$ on C_n. This is seen by expanding the inequality

$$-\operatorname{tr}(B_i \pm D_i\varphi)(B^i \pm D^i\varphi) \geq 0,$$

which leads to

$$\mathcal{H} \geq \tfrac{1}{4}\lambda(|\varphi|^2 - 1)^2 \pm \operatorname{tr}(B_i D^i \varphi),$$

the integral of which is

$$E \geq \int \tfrac{1}{4}\lambda(|\varphi|^2 - 1)^2 d^3x \pm 8\pi n,$$

with equality if and only if $B_i = \mp D_i \varphi$. So in particular $E \geq 8\pi|n|$, where we choose the upper or lower sign according to whether $n \geq 0$ or $n < 0$. The integer n is called the *topological charge* or *monopole number* of the configuration, and a solution (critical point) with charge n is called an *n-monopole solution*.

It should be emphasized that the situation here is different from that of Dirac poles discussed earlier, in that the fields here are defined (and smooth) on the whole of $\mathbf{R}^3$, whereas a Dirac pole magnetic field is only defined outside of a compact set. But there is a connection between the two models. Namely, the gauge group SU(2) is 'broken down' to U(1) at infinity in $\mathbf{R}^3$, by the field $\widehat{\varphi}$; and this U(1) is to be thought of as the gauge group of Maxwell theory. A nonabelian *n*-monopole field looks like a Dirac *n*-pole when viewed from a distance: the only part of the gauge field that survives at infinity is that part which is 'in the direction of' $\widehat{\varphi}$, namely

$$B_i^{(m)} = \tfrac{1}{4}\operatorname{tr}(B_i\widehat{\varphi})$$

Think of $B_i^{(m)}$ as a U(1) magnetic field, and let us compute its magnetic charge Q by integrating $B_i^{(m)}$ over the 'sphere at infinity' in $\mathbf{R}^3$. This gives

$$
\begin{aligned}
Q &= \int B_i^{(m)} d^2 x^i \\
&= \tfrac{1}{4}\int \operatorname{tr}(B_i\varphi) d^2 x^i \\
&= \tfrac{1}{4}\int \nabla_i \operatorname{tr}(B_i\varphi) d^3 x \quad \text{(Stokes' theorem)} \\
&= \tfrac{1}{4}\int \operatorname{tr}(B_i D^i \varphi) d^3 x \\
&= 2\pi n,
\end{aligned}
$$

as claimed. Note that in the penultimate step above we used $D_i B^i = 0$, which is the Bianchi identity in this context.

In 1974, 't Hooft and Polyakov pointed out that there exists a one-monopole solution. This was seen by imposing spherical symmetry in $\mathbf{R}^3$, so that the fields are (in effect) functions only of the radial distance r, and then using an existence theorem. But (at least for nonzero values of the parameter λ), the solution cannot be expressed in terms of elementary functions. Of greater mathematical interest is the case $\lambda = 0$, which is sometimes known as the *Prasad–Sommerfield limit*. From now on, let us restrict to the case $\lambda = 0$.

It is known that there exist critical points of E which are 'saddle-points' (i.e., not local minima) (Taubes 1982). But our interest here is in minimum-energy solutions, which are clearly those satisfying (for $n > 0$)

$$B_i = -D_i \varphi. \qquad (5.3.4)$$

These are called the *Bogomolny equations*. They are a set of coupled nonlinear first-order partial differential equations for the fields A_i and φ. For a charge-n solution of (5.3.4), the energy E saturates its lower bound $8\pi n$, and so it represents a stable n-monopole configuration.

It is worth pointing out that for such solutions, one can read off the monopole number n from the asymptotic behavior of $|\varphi|$. Indeed, $|\varphi|$ must behave like

$$|\varphi| = 1 - n/r + O(r^{-2}) \quad \text{as} \quad r \to \infty. \qquad (5.3.5)$$

Let us check that this agrees with (5.3.3). For solutions of (5.3.4), we have

$$\operatorname{tr} \int B_j D^j \varphi \, d^3 x = - \int \operatorname{tr}(D_j \varphi)(D^j \varphi) d^3 x.$$

Now $D_j D^j \varphi = 0$ from (5.3.4) and the Bianchi identity $D_i B^i = 0$, so

$$-\operatorname{tr}(D_j \varphi)(D^j \varphi) = \Delta |\varphi|^2,$$

where $\Delta = \nabla_j \nabla^j$ is the Laplacian on $\mathbf{R}^3$. Thus

$$\tfrac{1}{2} \operatorname{tr} \int B_j D^j \varphi \, d^3 x = \tfrac{1}{2} \int \Delta |\varphi|^2 d^3 x = 4\pi n,$$

as required, using Stokes' theorem and the expression (5.3.5) for $|\varphi|$. We note that in (5.3.5) the symbol $O(r^{-2})$ should be taken to mean a function f such that

$$r^2 \int_{S_r} \frac{\partial f}{\partial r} d\omega \to 0 \quad \text{as} \quad r \to \infty,$$

where S_r is the sphere of radius r in $\mathbf{R}^3$. This is what is needed for the Stokes' theorem argument to work.

Let M_n denote the space of smooth, finite-energy, charge-n solutions of (5.3.4), for $n > 0$. Then we know from an existence theorem of Taubes (1981) that M_n is nonempty; and an index-theory argument was used by Weinberg (1979) to show that M_n is $(4n - 1)$-dimensional. An $n = 1$ solution had already been exhibited by Prasad and Sommerfield (1975): it has a simple form involving hyperbolic functions of r (see Exercise 5.3.1).

Much more is now known about these solutions, mainly through the twistor description. We shall take up this story in §8.4.

The results mentioned here have been extended from SU(2) to other gauge groups. Leznov and Saveliev (1980) and Ganoulis, Goddard and Olive (1982) have studied spherically symmetric solutions of (5.3.4), and their relation to the 'Toda molecule' equations. And there are generalizations of the existence theorem (Taubes 1981) and the parameter-count result (Weinberg 1980, 1982) mentioned above.

Exercise for Section 5.3

5.3.1. Take φ and A_i to have the form

$$\varphi = f(r)x^a \sigma_a$$
$$A_i = g(r)\varepsilon_{ijk}x^j \sigma^k,$$

where σ_a denotes the Pauli matrices. Substitute these expressions into the Bogomolny equations (5.3.4), and find a smooth solution satisfying the correct boundary conditions. This is the Prasad–Sommerfield one-monopole solution.

Notes for Chapter 5

This chapter has dealt with classical gauge fields. As mentioned at the beginning, one has to work with the corresponding quantum field theories, if one wants to obtain the most accurate agreement with experiment. However, quantum field theory is a subject fraught with difficulties and inconsistencies. For a rigorous discussion of some simple quantum field theories, and of the functional integrals mentioned in §5.2, see Glimm and Jaffe (1981).

No quantum gauge-field theory is known to exist in a mathematically rigorous sense, and in fact it is generally believed that the simplest of them, namely quantum electrodynamics (QED), probably does *not* exist. Despite this, however, quantum gauge theories such as QED have impressive predictive power. Details of the quantum gauge-field description of the electromagnetic and nuclear forces may be found in, for example, Cheng and Li (1984); Huang (1982); Rajaraman (1982). These books also contain more comprehensive discussions about the use of instantons and the possible relevance of magnetic poles (for example, in grand unified theories).

6

General relativity

The idea of Einstein's theory of general relativity is that gravity is due to the curvature of four-dimensional space-time. The reader may find a concise introduction to pseudo-Riemannian geometry, and its application in general relativity, in (say) the first few chapters of Hawking and Ellis (1973). Here, we shall merely list some basic definitions and results. §6.1 describes how the field theories of Chapter 4 may be extended to curved space-time, and discusses the use of spinors and the effect of conformal rescalings of the space-time metric. For a comprehensive analysis of these topics, the reader is referred to the two volumes of Penrose and Rindler (1984, 1986). In §6.2, we discuss 'self-dual space-times' and review some of what is known about them, in preparation for the twistor treatment in Chapter 9.

6.1   Space-time, spinors, and Einstein's equations

Space-time is a smooth manifold $\mathcal{M}$, of real dimension four, equipped with a *metric* g_{ab} (a smooth symmetric covariant two-tensor field, with signature $+ - - -$ at every point of $\mathcal{M}$). Associated with the metric there is a unique torsion-free connection ∇_a, the *Levi–Civita connection*, satisfying $\nabla_a g_{bc} = 0$. The *Riemann curvature tensor* $R_{abc}{}^d$ is defined by

$$(\nabla_a \nabla_b - \nabla_b \nabla_a)V^d = R_{abc}{}^d V^c, \qquad (6.1.1)$$

where V^a is an arbitrary vector field. The *Ricci tensor* is

$$R_{ab} = R_{acb}{}^c. \qquad (6.1.2)$$

We say that g_{ab} is a *vacuum metric* if $R_{ab} = 0$ (these are *Einstein's vacuum equations*), and that g_{ab} is an *Einstein metric* if the trace-free part of the Ricci tensor vanishes, i.e., if

$$R_{ab} = \tfrac{1}{4}R g_{ab}, \qquad (6.1.3)$$

where $R = g^{ab}R_{ab}$ is the *scalar curvature*. If (6.1.3) holds, then it follows from the Bianchi identities that R is a constant; the quantity $\frac{1}{4}R$ is (in these circumstances) called the *cosmological constant*.

We are, as usual, using the 'abstract index' convention of Penrose (1968b) and Penrose and Rindler (1984), which is discussed in §2.3. So the symbol V^a, for example, denotes a vector, while $\nabla_a f$ denotes a one-form (the exterior derivative of the scalar f). Indices are lowered and raised with the metric tensor g_{ab} and its inverse g^{ab}. Spinor indices $A, B, \ldots, A', B', \ldots$ are also abstract.

The idea of spinors, and the discussion of §4.2, are easily extended to curved space-time $(\mathcal{M}, g_{ab})$. One has to impose some restrictions on space-time (see Penrose and Rindler 1984, §1.5), namely that it be time-orientable and space-orientable, and that the manifold $\mathcal{M}$ admit a spin-structure (this last is a condition on the topology of $\mathcal{M}$; see §3.5). As in flat space-time, there is a correspondence between tangent vectors V^a, and spinors $V^{AA'}$ satisfying $V^{AA'} = \overline{V}^{AA'}$. The metric g_{ab} corresponds to the product of two epsilons:

$$g_{ab} = \varepsilon_{AB}\varepsilon_{A'B'}.$$

The covariant derivative ∇_a (or $\nabla_{AA'}$) is extended so that one is able to differentiate spinor fields. This is done in such a way that

$$\nabla_a \varepsilon_{BC} = 0, \qquad \nabla_a \varepsilon_{B'C'} = 0.$$

We can use spinors to exhibit the irreducible pieces of the curvature tensor R_{abcd} (irreducible, that is, under the action of the Lorentz group). These are

(i) $C^-_{abcd} = \Psi_{ABCD}\varepsilon_{A'B'}\varepsilon_{C'D'}$, the anti-self-dual part of the Weyl tensor; the spinor Ψ_{ABCD} is totally symmetric in its four indices;

(ii) $C^+_{abcd} = \overline{\Psi}_{A'B'C'D'}\varepsilon_{AB}\varepsilon_{CD}$, the self-dual part of the Weyl tensor, and the complex conjugate of (i);

(iii) $\Phi_{ab} = \Phi_{AA'BB'}$, the trace-free part of the Ricci tensor, with $\Phi_{AA'BB'}$ being symmetric both in AB and in $A'B'$; and

(iv) the scalar curvature R. We shall use the symbol $\Lambda = \frac{1}{24}R$.

The sum of (i) and (ii) gives the Weyl tensor, or conformal curvature tensor

$$C_{abcd} = C^+_{abcd} + C^-_{abcd},$$

which is related to the Riemann tensor and Ricci tensor by

$$C_{ab}{}^{cd} = R_{ab}{}^{cd} - 2R_{[a}{}^{[c}\delta_{b]}{}^{d]} + \tfrac{1}{3}R\delta_{[a}{}^{[c}\delta_{b]}{}^{d]},$$

and which is trace-free: $C_{acb}{}^{c} = 0$.

The commutator expression (6.1.1) has a spinor version, as follows. One can decompose the commutator $[\nabla_a, \nabla_b]$ into its self-dual and anti-self-dual pieces

$$[\nabla_a, \nabla_b] = \varepsilon_{AB}\square_{A'B'} + \varepsilon_{A'B'}\square_{AB},$$

where

$$\square_{AB} = \nabla_{A'(A}\nabla_{B)}^{A'}, \qquad \square_{A'B'} = \nabla_{A(A'}\nabla_{B')}^{A}.$$

(The parentheses denote symmetrization over the indices they enclose.) Then $\square_{AB}$ and $\square_{A'B'}$ can be applied either to primed spinor fields or to unprimed ones. The resulting identities are (see Penrose and Rindler 1984, §4.9).

$$\square_{AB}\xi^C = \Psi_{ABD}{}^{C}\xi^D - 2\Lambda\xi_{(A}\varepsilon_{B)}{}^{C}, \tag{6.1.4}$$

$$\square_{A'B'}\xi^{C'} = \overline{\Psi}_{A'B'D'}{}^{C'}\xi^{D'} - 2\Lambda\xi_{(A'}\varepsilon_{B')}{}^{C'}, \tag{6.1.5}$$

$$\square_{AB}\xi^{B'} = \Phi_{ABA'}{}^{B'}\xi^{A'}, \tag{6.1.6}$$

$$\square_{A'B'}\xi^{B} = \Phi_{A'B'A}{}^{B}\xi^{A}. \tag{6.1.7}$$

The field theories in flat space-time, discussed in Chapters 4 and 5, can be extended to curved space-time by simply replacing the flat metric and connection with their curved versions. However, complications can arise in particular cases. Let us examine some examples. First, the wave equation: we let $\square$ denote the 'curved' d'Alembertian

$$\square = g^{ab}\nabla_a\nabla_b,$$

and we take as our 'curved-space' wave equation the following:

$$(\square + \tfrac{1}{6}R)\varphi = 0. \tag{6.1.8}$$

This, of course, reduces to the flat-space wave equation in Minkowski space-time (where $R = 0$). The reason for the $\tfrac{1}{6}R$ term is that it ensures that (6.1.8) is invariant under arbitrary conformal rescalings of the metric g_{ab} (as we shall see later in this section).

Secondly, the equation for a massless field of helicity $-\frac{1}{2}n$, namely

$$\nabla^{AA'}\varphi_{AB...D} = 0, \qquad (6.1.9)$$

could be adopted as it stands (where ∇_a now denotes the 'curved' connection). However, if $n \geq 3$, i.e., if $\varphi_{AB...D}$ has more than two indices, then (6.1.9) is not really a satisfactory equation. To see why, operate on (6.1.9) with $\nabla^B{}_{A'}$ and use (6.1.4):

$$
\begin{aligned}
0 &= \nabla^B{}_{A'}\nabla^{AA'}\varphi_{ABC...E} \\
&= \nabla^{(B}{}_{A'}\nabla^{A)A'}\varphi_{ABC...E} \quad (\varphi_{A...} \text{ being symmetric}) \\
&= \Box^{AB}\varphi_{ABC...E} \\
&= (n-2)\Psi_{ABP(C}\varphi^{ABP}{}_{...E)},
\end{aligned}
\qquad (6.1.10)
$$

after a bit of algebra (see Penrose and Rindler 1984, §5.8). If $n > 2$, and Ψ_{ABCD} is nonzero, then the algebraic consistency condition (6.1.10) effectively renders the equations (6.1.9) useless as physical field equations (cf. the discussion at the end of §4.3). For $n = -1$ (the Weyl neutrino equation $\nabla^{AA'}\varphi_A = 0$) or $n = -2$ (the Maxwell equation $\nabla^{AA'}\varphi_{AB} = 0$) there is no problem: these equations are perfectly acceptable in curved space-time.

 Thirdly, consider gauge theory. Here, note that the expression for the field in terms of the potential, namely

$$F_{ab} = \nabla_a A_b - \nabla_b A_a + [A_a, A_b],$$

is *independent* of the choice of connection ∇_a (and hence of the metric). But the Yang–Mills equations

$$D^a F_{ab} := g^{ac} D_c F_{ab} = 0$$

depend on the metric. And so do the self-duality equations $*F_{ab} = iF_{ab}$, since $*$ is defined (in terms of a volume element ε_{abcd} and the metric g_{ab}) by

$$*F_{ab} := \tfrac{1}{2}\varepsilon_{abcd}g^{ce}g^{df}F_{ef}.$$

We consider next the effect of a conformal rescaling of the metric

$$g_{ab} \mapsto \widehat{g}_{ab} = \Omega^2 g_{ab}, \qquad (6.1.11)$$

where Ω is a smooth, positive scalar field on $\mathcal{M}$. As a result of the change (6.1.11) in the metric, the connection changes, and hence also the curvature. Some of the relevant formulae are given in the next two paragraphs (see Penrose and MacCallum 1972; Penrose and Rindler 1984, §5.6; Penrose and Rindler 1986, §6.8).

Let $\widehat{\nabla}_a$ denote the connection determined by $\widehat{g}_{ab}$. Acting on a scalar φ, we have $\widehat{\nabla}_a \varphi = \nabla_a \varphi$. When $\widehat{\nabla}_a$ acts on spinors, we get

$$\widehat{\nabla}_{AA'} \xi_B = \nabla_{AA'} \xi_B - T_{BA'} \xi_A, \tag{6.1.12}$$

$$\widehat{\nabla}_{AA'} \eta_{B'} = \nabla_{AA'} \eta_{B'} - T_{AB'} \eta_{A'}, \tag{6.1.13}$$

$$\widehat{\nabla}_{AA'} \xi^B = \nabla_{AA'} \xi^B + \varepsilon_A{}^B T_{CA'} \xi^C, \tag{6.1.14}$$

$$\widehat{\nabla}_{AA'} \eta^{B'} = \nabla_{AA'} \eta^{B'} + \varepsilon_{A'}{}^{B'} T_{AC'} \eta^{C'}, \tag{6.1.15}$$

where $T_a = \Omega^{-1} \nabla_a \Omega$. When $\widehat{\nabla}_a$ acts on higher-valence spinors (and hence on vectors, tensors, etc.), one treats each index in turn according to the above scheme. In particular, when $\widehat{\nabla}$ acts on a covector V_a or on a two-form ω_{ab}, we have

$$\widehat{\nabla}_a V_b = \nabla_a V_b - Q_{ab}{}^c V_c, \tag{6.1.16}$$

$$\widehat{\nabla}_a \omega_{bc} = \nabla_a \omega_{bc} - Q_{ab}{}^d \omega_{dc} - Q_{ac}{}^d \omega_{bd}, \tag{6.1.17}$$

where

$$Q_{ab}{}^c = 2T_{(a} \delta_{b)}{}^c - g_{ab} T^c. \tag{6.1.18}$$

The change in curvature is given by the following expressions:

$$\widehat{\Psi}_{ABCD} = \Psi_{ABCD}, \qquad \widehat{\overline{\Psi}}_{A'B'C'D'} = \overline{\Psi}_{A'B'C'D'}, \tag{6.1.19}$$

$$\widehat{\Phi}_{ABA'B'} = \Phi_{ABA'B'} - \nabla_{A(A'} T_{B')B} + T_{A(A'} T_{B')B}, \tag{6.1.20}$$

$$\Omega^2 \widehat{R} = R + 6 \nabla_a T^a + 6 T_a T^a \tag{6.1.21}$$

(see Penrose and Rindler 1986, §6.8). It follows from (6.1.19) that the Weyl tensor is *conformally invariant*: $\widehat{C}_{abc}{}^d = C_{abc}{}^d$. The space-time is *conformally flat* (i.e., there exists a conformal factor Ω such that the curvature $\widehat{R}_{abc}{}^d$ of $\widehat{g}_{ab}$ is zero) if and only if $C_{abc}{}^d$ vanishes.

If we begin in flat space-time, and put $\Omega = (x_a x^a)^{-1}$, then the new metric $\Omega^2 \eta_{ab}$ is still flat, since

$$T_a = -2\Omega x_a,$$

$$\nabla_a T_b = -2\Omega(\eta_{ab} - 2\Omega x_a x_b),$$

and the appropriate bits of (6.1.20) and (6.1.21) vanish:

$$\nabla_{A(A'}T_{B')B} - T_{A(A'}T_{B')B} = 0,$$
$$\nabla_a T^a + T_a T^a = 0.$$

All of the above rescalings $\nabla \mapsto \widehat{\nabla}, R \mapsto \widehat{R}$, etc., have been *induced* by the assumed change of scale (6.1.11) of the metric g_{ab}. We can also consider objects which do not depend on the metric as such, but which do rescale under the metric rescalings. Namely consider scalar functions φ defined on $\mathcal{M} \times \mathbf{R}^+$, which satisfy $\varphi(x, \Omega) = \Omega^k \varphi(x, 1)$, for any $\Omega \geq 0$. We call such a function a *conformal density of weight* k. We could interpret the rescaling used earlier on this product space in a similar manner. Note that any scalar field φ on $\mathcal{M}$ defines a conformal density by setting $\varphi(x, \Omega) = \Omega^k \varphi(x, 1)$, where we define $\varphi(x, 1) := \varphi(x)$. Thus if we consider a scalar field to be a conformal density of weight k in this sense, then we can write $\varphi \mapsto \widehat{\varphi} = \Omega^k \varphi$, by analogy with the previous rescalings. See Penrose and Rindler (1984) for a more detailed discussion of this concept of rescaling. One says that field equations are *conformally invariant* if after rescaling the metric, and with the field quantities chosen to be conformal densities of suitable weights, the equations remain invariant.

. In §7.1 we shall discuss conformal weights from a different but related point of view. There we shall be concerned with sections of bundles where the bundles themselves have conformal weights in the sense described here.

Now let φ be a conformal density of weight -1, i.e., $\widehat{\varphi} = \Omega^{-1}\varphi$. Then

$$\widehat{\nabla}_a \widehat{\varphi} = \nabla_a(\Omega^{-1}\varphi)$$
$$= \Omega^{-1}\nabla_a \varphi - \Omega^{-1}\varphi T_a,$$

and so

$$\widehat{\Box}\widehat{\varphi} := \widehat{g}^{ab}\widehat{\nabla}_a \widehat{\nabla}_b \widehat{\varphi}$$
$$= \Omega^{-3}[\Box\varphi - \tfrac{1}{6}\varphi(\Omega^2 \widehat{R} - R)],$$

after a straightforward calculation involving (6.1.16) and (6.1.21). Thus we get

$$(\widehat{\Box} + \tfrac{1}{6}\widehat{R})\widehat{\varphi} = \Omega^{-3}(\Box + \tfrac{1}{6}R)\varphi; \qquad (6.1.22)$$

so the equation $(\Box + \tfrac{1}{6}R)\varphi = 0$ is *conformally invariant*, provided we specify that φ is a conformal density of weight -1. Similarly, one easily deduces from (6.1.12) and (6.1.13) that the massless free-field equations of arbitrary helicity

$$\nabla^{AA'}\varphi_{AB...D} = 0, \qquad \nabla^{AA'}\varphi_{A'B'...D'} = 0$$

are each conformally invariant, provided the fields are taken to be conformal densities of weight -1 (see Penrose and Rindler 1984, §5.7).

Finally, we come to gauge theory. The gauge potential A_a (and therefore also the gauge field F_{ab}) are taken to have conformal weight zero (i.e., they are invariant under rescaling). Then it follows from (6.1.17) that the Yang–Mills equations $D^a F_{ab} = 0$ are conformally invariant. And the (anti)-self-duality equations $*F_{ab} = \pm i F_{ab}$ are invariant as well; for this, one uses the fact that the volume four-form transforms as

$$\widehat{\varepsilon}_{abcd} = \Omega^4 \varepsilon_{abcd}. \qquad (6.1.23)$$

To put this another way, the Hodge star operator $*$, when acting on two-forms in a four-dimensional space, is conformally invariant.

Exercises for Section 6.1

6.1.1. The spinorial version of the curvature tensor R_{abcd} is

$$\begin{aligned}
R_{abcd} = &\ \Psi_{ABCD}\varepsilon_{A'B'}\varepsilon_{C'D'} + \overline{\Psi}_{A'B'C'D'}\varepsilon_{AB}\varepsilon_{CD} \\
&+ \Phi_{ABC'D'}\varepsilon_{A'B'}\varepsilon_{CD} + \Phi_{A'B'CD}\varepsilon_{AB}\varepsilon_{C'D'} \\
&+ 2\Lambda(\varepsilon_{AC}\varepsilon_{BD}\varepsilon_{A'B'}\varepsilon_{C'D'} + \varepsilon_{AB}\varepsilon_{CD}\varepsilon_{A'D'}\varepsilon_{B'C'}).
\end{aligned}$$

Use this to derive equations (6.1.4)–(6.1.7) from equation (6.1.1).

6.1.2. The Bianchi identity is

$$\nabla_{[a}R_{bc]d}{}^e = 0.$$

Use this to show that if the Einstein condition (6.1.3) holds, then the scalar curvature R must be a constant. Show that in vacuum, the Bianchi identity is equivalent to

$$\nabla^{AA'}\Psi_{ABCD} = 0.$$

6.1.3. Show that if ξ_A satisfies $\nabla_{AA'}\xi_B = 0$, then $\Psi_{ABCD}\xi^D = 0$; while if ξ_A satisfies

$$\xi^A\xi^B\nabla_{AA'}\xi_B = 0,$$

then
$$\Psi_{ABCD}\xi^A\xi^B\xi^C\xi^D = 0.$$

6.1.4. Show that the spinor

$$P_{ABA'B'} = \Phi_{ABA'B'} - \Lambda\varepsilon_{AB}\varepsilon_{A'B'}$$

corresponds to the tensor

$$P_{ab} = \tfrac{1}{12}Rg_{ab} - \tfrac{1}{2}R_{ab},$$

and that under a conformal transformation (6.1.11), it transforms as

$$\widehat{P}_{ABA'B'} = P_{ABA'B'} - \nabla_{AA'}T_{BB'} + T_{AB'}T_{BA'}.$$

6.1.5. Show that the massless free-field equations (for nonzero helicity) are conformally invariant.

6.2 Self-duality and gravitational instantons

When one tries to extend twistor theory to curved space-time, one discovers that a natural extension exists for space-times whose Weyl tensor C_{abcd} is either self-dual or anti-self-dual. This section is a brief review of anti-self-dual spaces (the self-dual case is, of course, entirely similar, the only difference between the two being the choice of orientation on the space-time).

In a space-time with Lorentzian signature $+ - - -$, the self-dual and anti-self-dual parts of the Weyl tensor (namely C^+_{abcd} and C^-_{abcd}, or equivalently $\bar{\Psi}_{A'B'C'D'}$ and Ψ_{ABCD}) are complex conjugates of each other. So an anti-self-dual space-time (one with $C^+_{abcd} = 0$) is necessarily conformally flat (with $C_{abcd} = 0$). However, this restriction does not apply to positive-definite four-spaces, or to 'complex' space-times, and so it is in these two contexts that we shall discuss anti-self-duality.

By a 'complex space-time' we mean a four-dimensional complex manifold $\mathcal{M}$, equipped with a *holomorphic* metric g_{ab}. In other words: with respect to a holomorphic coordinate basis $x^a = (x^0, x^1, x^2, x^3)$, the metric g_{ab} is a 4×4 matrix of holomorphic functions of x^a, and its determinant is nowhere-vanishing. Analogous to the real case, g_{ab} determines a unique holomorphic connection ∇_a, and hence a holomorphic curvature tensor R_{abcd}. The difference now is that the Ricci tensor (which before was real) becomes complex-valued; and the self-dual and anti-self-dual parts of the Weyl tensor (which before were complex-valued, but conjugate to each other) become *independent* holomorphic tensors. To emphasize this, we use the symbol $\widetilde{\Psi}_{A'B'C'D'}$ (instead of $\overline{\Psi}_{A'B'C'D'}$) for the self-dual Weyl spinor.

Since Ψ_{ABCD} and $\widetilde{\Psi}_{A'B'C'D'}$ are now independent of each other, it is possible for one of them to be zero without the other being zero as well. Complex space-times satisfying

$$\widetilde{\Psi}_{A'B'C'D'} = 0, \qquad R_{ab} = 0 \qquad (6.2.1)$$

are said to be *right-flat* or *anti-self-dual vacuum* spaces. More generally, solutions of

$$\widetilde{\Psi}_{A'B'C'D'} = 0, \qquad R_{ab} = \tfrac{1}{4}Rg_{ab} \qquad (6.2.2)$$

are *anti-self-dual Einstein* spaces. We shall see in Chapter 9 how twistor theory provides a construction for solving (6.2.1) or (6.2.2). The problem can also be tackled more directly: see, for example, Boyer et al. (1980).

Complex self-dual (or anti-self-dual) spaces have at least two possible applications: in the theory of $\mathcal{H}$-space (Ko et al. 1981) which relates to the asymptotic structure of asymptotically flat space-times; and in quantum gravity (Penrose 1976). But real (positive-definite, Riemannian) spaces are classically of greater interest to differential geometers, and are also relevant to quantum gravity, via the 'path-integral' approach (Hawking 1979). So let us now turn to the positive-definite case.

The spaces of interest are complete four-dimensional Riemannian manifolds $(\mathcal{M}, g_{ab})$, which are either Einstein or vacuum, and which satisfy one of the following three boundary conditions.

(i) $\mathcal{M}$ is compact without boundary.

(ii) $(\mathcal{M}, g_{ab})$ is asymptotically locally Euclidean (ALE): in other words, it has an 'infinity' which is like that of E^4, but factored by a discrete subgroup Γ of SO(4) acting freely on S^3 (the 'infinity' of E^4). See Gibbons and Hawking (1978); Eguchi and Hanson (1978); Eguchi, Gilkey and Hanson (1980).

(iii) $(\mathcal{M}, g_{ab})$ is asymptotically locally flat (ALF): this means that it has an infinity which is asymptotically Euclidean in a three-dimensional sense, with the fourth direction being periodic.

The infinity, instead of being S^3/Γ as in (ii), has the topology of an S^1-bundle over S^2. See Hawking (1977); Gibbons and Perry (1980).

We refer to such spaces generically as *gravitational instantons*. There are examples of instantons for which the Weyl tensor is neither self-dual nor anti-self-dual: for example, the 'Riemannian Schwarzschild' space (Gibbons and Hawking 1977) which is asymptotically locally flat; and the compact space $S^2 \times S^2$ with its standard metric. But our interest is in anti-self-dual spaces, and so we shall restrict attention to these.

Examples of compact anti-self-dual instantons include the following. The simplest are the four-sphere S^4 and the four-torus T^4 with their standard metrics: they are conformally flat, and (respectively) Einstein and vacuum. Then there is the complex projective space $\mathbf{P}_2(\mathbf{C})$ with its standard (Fubini–Study) metric (see below): thought of as a four-dimensional Riemannian manifold, this is an anti-self-dual Einstein space (Eguchi and Freund 1976; Gibbons and Pope 1978). Finally, there are the K3 metrics, which must exist, but are not known explicitly (Yau 1978): the K3 space is the only compact simply-connected four-dimensional manifold which admits an anti-self-dual vacuum metric.

The Fubini–Study metric on $\mathbf{P}_2$ may be described as follows. If (Z^1, Z^2, Z^3) are projective coordinates on $\mathbf{P}_2$, let ζ^α denote the affine coordinates

$$\zeta^1 = Z^1/Z^3, \qquad \zeta^2 = Z^2/Z^3$$

on the coordinate chart $Z^3 \neq 0$. Then the metric is given by the 'Kähler' expression

$$ds^2 = \frac{\partial^2 K}{\partial \zeta^\alpha \partial \bar{\zeta}^\beta} d\zeta^\alpha d\bar{\zeta}^\beta \tag{6.2.3a}$$

(summed over α and β), where

$$K = q^{-1} \log[1 + q(|\zeta^1|^2 + |\zeta^2|^2)]. \tag{6.2.3b}$$

Here q is a positive constant. The metric (6.2.3) has anti-self-dual Weyl tensor and is an Einstein metric with scalar curvature $R = -24q$.

Some simple examples of ALE and ALF spaces are drawn from the family of metrics

$$ds^2 = V^{-1}(d\tau - \omega_i dx^i)^2 + V dx_i dx^i, \tag{6.2.4}$$

where the index i ranges over 1, 2, 3 (so τ, x^1, x^2, x^3 are the four coordinates, and $dx_i dx^i$ denotes the three-dimensional Euclidean metric). The scalar V and the one-form $\omega_i dx^i$ are functions only of x^i (and not of τ), and are required to satisfy the linear equation

$$\nabla_i V = \tfrac{1}{2}\varepsilon_{ijk}\nabla^j \omega^k \tag{6.2.5}$$

(in words, $\operatorname{grad} V = \operatorname{curl}\omega$). Provided equation (6.2.5) holds, the metric (6.2.4) is an anti-self-dual vacuum solution. We shall demonstrate in Chapter 9 how (6.2.4) can be derived as an example of the twistor construction.

We next have to look for particular solutions of (6.2.5) which are such that the space-time whose metric is given locally by (6.2.4), has the desired global properties. Such solutions are not hard to find. Observe that (6.2.5) requires V to be a solution of the three-dimensional Laplace equation $\nabla^i\nabla_i V = 0$. We take V to be a sum of elementary solutions:

$$V = \varepsilon + \sum_{\alpha=1}^{n} \|x - x_\alpha\|^{-1}, \tag{6.2.6}$$

where $x_\alpha = (x_1, \ldots, x_n)$ are n distinct points in $\mathbf{R}^3$, and $\|\cdot\|$ denotes the three dimensional Euclidean norm

$$\|y\|^2 = y_i y^i.$$

The quantity ε is a constant, which we may take to be either 1 or 0. Once we specify V, equation (6.2.5) determines ω_i, up to the freedom $\omega_i \mapsto \omega_i + \nabla_i\lambda$; but this change in ω_i can be absorbed by a transformation $\tau \mapsto \tau + \lambda$ of the coordinate τ. So V determines the metric uniquely.

If $\varepsilon = 1$, the metrics defined by (6.2.6) are asymptotically locally flat, and are known as the multi-Taub-NUT metrics. The coordinate τ has to be interpreted as a periodic coordinate. For example, in the case $n = 1$ (Hawking 1977), τ is periodic with period 4π; the whole space has the topology of $\mathbf{R}^4$; taking $x_1 = 0$, the hypersurfaces of constant $\|x\|$ have the topology S^3. Note that S^3 is an S^1-bundle over S^2 (the Hopf fibration): the coordinate τ is a parameter along the Hopf circles on S^3.

On the other hand, $\varepsilon = 0$ gives spaces which are asymptotically locally Euclidean (Gibbons and Hawking 1978). The 'infinity' is the lens space $L(n, 1)$ of S^3, obtained by identifying points of S^3 (the 'boundary' of $\mathbf{C}^2$, on which z^1, z^2 are complex coordinates) by

$$(z^1, z^2) \sim (e^{2\pi i/n} z^1, e^{2\pi i/n} z^2).$$

In other words, S^3 is factored by a cyclic group of order n. If $n = 1$, then the metric determined by (6.2.6) is just the flat Euclidean metric E^4. If $n = 2$, then we get the metric of Eguchi and Hanson (1978); the space it lives on is in fact the holomorphic cotangent bundle of $\mathbf{P}_1(\mathbf{C})$, thought of as a real four-dimensional manifold; its infinity is $\mathbf{P}_3(\mathbf{R}) = S^3/\mathbf{Z}_2$. There are other ALE spaces analogous to these $\varepsilon = 0$ examples, whose infinities involve factoring by other subgroups of SO(3). These have arisen out of the twistor construction: see Chapter 9, Hitchin (1979) and Eguchi, Gilkey and Hanson (1980).

Exercise for Section 6.2

6.2.1. Let h_{ab} be a symmetric, trace-free tensor field in complexified flat space-time (so its spinor version $h_{ABA'B'}$ is symmetric in both primed and unprimed indices). Suppose h_{ab} satisfies the equation

$$\nabla^{AC'} h_{ABA'B'} = 0.$$

Let g_{ab} denote the 'linearized' metric $\eta_{ab} + \lambda h_{ab}$, where η_{ab} is the flat metric and λ is an infinitesimal parameter (i.e., neglect terms of order λ^2). Show that g_{ab} is right-flat, with

$$\Psi_{ABCD} = \tfrac{1}{2}\lambda \nabla_{AA'} \nabla_{BB'} h_{CD}{}^{A'B'}.$$

Part III

The Penrose transform

7

Massless free fields

In Chapter 4 we discussed the massless field equations on real Minkowski space M^4

$$\nabla^{AA'} \varphi_{A'...D'} = 0$$
$$\nabla^{AA'} \varphi_{A...D} = 0$$

for spinor fields $\varphi_{A'...D'}$ and $\varphi_{A...D}$ of helicity s and $-s$, for $s > 0$, respectively. Each component of these massless field equations satisfies the wave equation

$$\Box \varphi = 0,$$

which is the special case of a massless field of helicity zero.

In this chapter we want to show how to solve these equations using the methods of twistor geometry. The massless field equations above have constant coefficients and thus it is natural to look for holomorphic solutions of these equations on $M^I = M^4 \otimes C$, the complexification of affine Minkowski space. We shall find all holomorphic solutions of these equations on suitable open subsets of M^I, including all open convex subsets, for instance. The equations on real Minkowski space are hyperbolic, and so the solutions need not be smooth in general. We shall also find all hyperfunction solutions of these equations, realizing them as boundary values of solutions from the complexified domain.

The representation of the solution will be in terms of sheaf cohomology on the twistor space P associated with the conformal compactification of M^I, which we denoted by M in Chapter 1. There is an explicit isomorphism, which is called the Penrose transform, for U, a convex open subset of M, of the form

$$\mathcal{P} : H^1(\widehat{U}, \mathcal{O}(-2s - 2)) \xrightarrow{\cong} \{\text{holomorphic massless fields}$$
$$\text{of helicity } s \text{ on } U\}.$$

This will be elaborated and proved in §§7.1 and 7.2. In §7.1 we study the local aspects of the problem. We pull back the hyperplane section sheaf from P to F, incorporate it into a natural relative de Rham complex on F (relative to the fibration F $\xrightarrow{\mu}$ P) and then relate these sheaves to spinor sheaves on M. The massless field equations on M arise from exterior differentiation along the μ-fibration in a natural manner. In §7.2 we globalize using sheaf cohomology and spectral sequences and prove the Penrose transform isomorphism alluded to above. In §7.3 we shall look at explicit examples of integral formulas, and in §7.4 we shall indicate how to find hyperfunction solutions of the massless field equations on real Minkowski space. In §7.3 examples of this Penrose transform will be given to show how the left-hand side can be considered as free data generated by functions of homogeneity $-2s - 2$ on $P \cong \mathbf{P}_3(\mathbf{C})$. The Penrose transform $\mathcal{P}$ is an integral transform which can be considered both abstractly and concretely. The concrete point of view is contained in the original integral formulas of Penrose (see, e.g., Penrose and MacCallum 1973 for a survey of this point of view at the time of its development). The cohomology represents an understanding of the amount of 'gauge freedom' in the integrand of a given integral formula. In addition, just as we have seen in the examples in §3.3, sheaf cohomology can be represented quite explicitly in terms of a Čech covering, and in some cases, in particular for $P - I = M^I$, one can represent sheaf cohomology by means of a covering consisting of *two* open sets. The data in the overlap open set in this case can be chosen freely, as there are no compatibility conditions. Examples of this are seen in §7.3 and in the explicit construction of anti-self-dual gauge fields, instantons, and monopoles as described in Chapter 8.

The massless field equations are conformally invariant and so are well-defined on M, the compactification of M^I. This is provided that the spinor fields and the differential operators are extended over the 'points at ∞' in a suitable manner. We shall proceed by deriving, from the twistor geometry, the spinor bundles extended over infinity (these will be conformally weighted spinor bundles), the differential operators in an invariant form, and the solutions to the resulting equations. All of this will arise naturally from the twistor geometry, using only the double fibration

$$
\begin{array}{ccc}
 & P & \\
\mu \swarrow & & \searrow \nu \\
P & & M
\end{array}
$$

and the hyperplane section bundle $H \to \mathsf{P}$. When we are done, we shall *identify* the resulting differential operators and equations with the classical equations on M^I by choosing standard local coordinates on M.

7.1 Holomorphic solutions of the massless field equations

We recall the fundamental twistor correspondence from Chapter 1 given by the double fibration (1.2.4)

$$
\begin{array}{ccc}
 & \mathsf{F} & \\
{}^{\mu}\swarrow & & \searrow^{\nu} \\
\mathsf{P} & & \mathsf{M}
\end{array}
\qquad (7.1.1)
$$

Let U be an open subset of M and let

$$
U' := \nu^{-1}(U)
$$
$$
\widehat{U} := \mu \circ \nu^{-1}(U)
$$

(the latter being the inverse correspondence of U as discussed in §1.2). We then have the derived diagram from (7.1.1)

$$
\begin{array}{ccc}
 & U' & \\
{}^{\mu}\swarrow & & \searrow^{\nu} \\
\widehat{U} & & U,
\end{array}
$$

where both U' and $\widehat{U}$ are 'derived sets' of the set U. The Penrose transform will be a mapping of the form (for $|s| > 0$)

$$
\mathcal{P} : H^1(\widehat{U}, \mathcal{O}(-2s - 2)) \to \ker[\nabla_s : \Gamma(U, E_s) \to \Gamma(U, E'_s)]
$$

where E_s, E'_s are suitable holomorphic vector bundles defined on M, and ∇_s is a first-order linear differential operator. The mapping $\mathcal{P}$ will turn out to be an isomorphism. Our first task is to construct the bundles E_s, E'_s on M and the differential operator ∇_s. We shall construct E_s and E'_s as direct images under ν of pullbacks under μ of powers of the hyperplane section bundle on P. The operator ∇_s will be the direct image under ν of a relative de Rham exterior differential operator on the space F, 'relative' being relative to the fibration μ.

As before, we let

$$\mathcal{O}_{\mathbf{P}}(m) = \mathcal{O}_{\mathbf{P}}(H^m), \quad m \in \mathbf{Z},$$

where $H \to \mathbf{P}$ is the hyperplane section bundle on $\mathbf{P}$. If $\mathcal{F}$ is any sheaf of $\mathcal{O}_{\mathbf{P}}$-modules on $\mathbf{P}$ (e.g., $\mathcal{O}_{\mathbf{P}}(m)$ or the holomorphic sections of a holomorphic vector bundle on $\mathbf{P}$), then we can define two different pullback sheaves of the sheaf $\mathcal{F}$ with respect to the mapping μ. The first is the *topological pullback*, $\mu^{-1}(\mathcal{F})$, defined on $\mathbf{F}$ by the presheaf

$$W \to \mathcal{F}(\mu(W)), \quad W \text{ open in } \mathbf{F},$$

noting that μ is an open mapping since it is surjective and of maximal rank (cf. (1.2.11)). Thus the stalks of $\mu^{-1}(\mathcal{F})$ have the form

$$\mu^{-1}(\mathcal{F})_x = \mathcal{F}_{\mu(x)}, \quad x \in \mathbf{F}.$$

Suppose $\mathcal{F}$ is a sheaf of functions (e.g., sections of a bundle). If we consider any section σ of $\mu^{-1}(\mathcal{F})$ on an open set $W \subset \mathbf{F}$, then σ is *locally constant* along the fibers of μ intersected with W.

A second pullback is defined by the fact that $\mathcal{O}_{\mathbf{F}}$ is, in a natural manner, an $\mathcal{O}_{\mathbf{P}}$-module (defined by pullbacks of functions, see Exercise 7.1.1), and we define

$$\mu^* \mathcal{F} := \mu^{-1}(\mathcal{F}) \otimes_{\mathcal{O}_{\mathbf{P}}} \mathcal{O}_{\mathbf{F}}$$

to be the *analytic pullback* of $\mathcal{F}$ (or the 'module pullback'). We see now that sections of $\mu^* \mathcal{F}$ are no longer constant on the fibers of μ. In fact, if $\mathcal{F} = \mathcal{O}_{\mathbf{P}}(E)$, the holomorphic sections of a holomorphic vector bundle $E \to \mathbf{P}$, then one has the identity (Exercise 7.1.3),

$$\mu^* \mathcal{O}_{\mathbf{P}}(E) = \mathcal{O}_{\mathbf{F}}(\mu^* E),$$

where the right-hand side is the sheaf of the holomorphic sections on $\mathbf{F}$ of the *pullback bundle* $\mu^* E$ on $\mathbf{F}$, defined in Exercise 7.1.2. This makes it abundantly clear that the sections of $\mu^* \mathcal{O}_{\mathbf{P}}(E)$ are not constant on the fibers.

On $\mathbf{F}$, then, we have two pullback sheaves and an exact sequence of the form

$$0 \to \mu^{-1} \mathcal{O}_{\mathbf{P}}(\mathcal{F}) \to \mu^* \mathcal{O}_{\mathbf{P}}(\mathcal{F}),$$

i.e., every section of $\mu^{-1} \mathcal{O}_{\mathbf{P}}(\mathcal{F})$ is naturally a section of $\mu^* \mathcal{O}_{\mathbf{P}}(\mathcal{F})$. This sequence is the beginning of a relative de Rham sequence on $\mathbf{F}$, which we shall now describe.

Define
$$\Omega^1_\mu := \Omega^1_F / \mu^* \Omega^1_P$$

This is the sheaf of *relative one-forms* on F ('relative' meaning relative to the fibration F $\xrightarrow{\mu}$ P; we should say for accuracy 'μ-relative', but there will only be one set of relative forms in this context). Since μ is surjective of maximal rank (see (1.2.11)) we can choose local coordinates $(z, w) \in \mathbf{C}^3 \times \mathbf{C}^2$ for F and $z \in \mathbf{C}^3$ for P so that $\mu(z, w) = z$ (later in this section we choose such coordinates explicitly). Then in these coordinates one can see that equivalence classes in Ω^1_μ can be represented by local differential forms of the form

$$\sum f_i(z, w) dw^i,$$

and this representation is unique. Thus Ω^1_μ consists of the one-forms tangent to the fibers of μ; there are no dz^i components (with respect to these coordinates). The coefficients f_i depend holomorphically on both z and w. If

$$\pi_\mu : \Omega^1_F \to \Omega^1_\mu$$

is the quotient mapping, then there is an induced relative exterior differential mapping (see Figure 7.1.1)

$$d_\mu := \pi_\mu \circ d : \mathcal{O}_F \to \Omega^1_\mu.$$

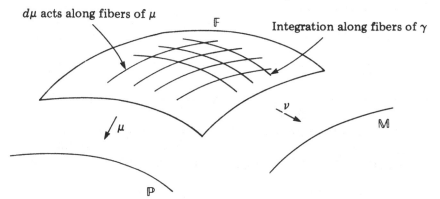

Fig. 7.1.1. The basic geometry of the Penrose transform.

Similarly, we define the sheaf Ω^2_μ of *relative two-forms* by setting

$$\Omega^2_\mu := \Omega^2_F / \mu^* \Omega^1_P \wedge \Omega^1_F.$$

Locally, with respect to the coordinates above, a representative relative two-form has the unique form

$$f(z,w)dw^1 \wedge dw^2,$$

where $f(z,w)$ is a holomorphic function of z and w. We define

$$d_\mu : \Omega^1_\mu \to \Omega^2_\mu$$

by the composition of the two mappings

$$\Omega^1_F \xrightarrow{d} \Omega^2_F \to \Omega^2_\mu$$

and by noting that

$$d(\mu^*\Omega^1_P) \subset \mu^*\Omega^1_P \wedge \Omega^1_F.$$

We now have the following lemma.

Lemma 7.1.1. *The sequence of sheaves*

$$0 \longrightarrow \mu^{-1}\mathcal{O}_P \longrightarrow \mathcal{O}_F \xrightarrow{d_\mu} \Omega^1_\mu \xrightarrow{d_\mu} \Omega^2_\mu \longrightarrow 0 \qquad (7.1.2)$$

is exact.

Proof: We want to show that the sequence of stalks

$$0 \longrightarrow \mu^{-1}\mathcal{O}_{P,x} \longrightarrow \mathcal{O}_{F,x} \xrightarrow{d_\mu} \Omega^1_{\mu,x} \xrightarrow{d_\mu} \Omega^2_{\mu,x} \longrightarrow 0$$

is exact for each $x \in F$. Let x be fixed and choose coordinates for F centered at x of the form (z,w), $z \in \mathbb{C}^3$, $w \in \mathbb{C}^2$, where z is a local coordinate system for P centered at $\mu(x)$ and such that the mapping μ has the form $\mu(z,w) = z$. Then we see that in these coordinates a local section of $\mu^{-1}\mathcal{O}_P$ is a function of the form

$$f(z,w) = g(z).$$

Consider a local section of $\mathcal{O}_F$, i.e., a holomorphic function $f(z,w)$ defined near $(0,0)$. Suppose $d_\mu f = 0$ near $(0,0)$, then

$$df \in \mu^*\Omega^1_P,$$

and moreover,

$$df = \sum \frac{\partial f}{\partial z^i} dz^i + \frac{\partial f}{\partial w^i} dw^i.$$

Any differential form w in $\mu^* \Omega_{\mathbf{P}}^1$ is of the form

$$w = \sum_{i=1}^{3} a_i dz^i, \quad a_i \in \mathcal{O}_{\mathbf{F}}.$$

Therefore $df \in \mu^* \Omega_{\mathbf{P}}^1$ implies that

$$\frac{\partial f}{\partial w^i} = 0, \quad i = 1, 2.$$

Thus f is independent of the variables w^1, w^2 near 0, and is hence a section of $\mu^{-1} \mathcal{O}_{\mathbf{P}}$ near $(0,0)$. This shows that (7.1.2) is exact at $\mathcal{O}_{\mathbf{F},x}$. The arguments for Ω_{μ}^1 and Ω_{μ}^2 are quite similar, and are left as an exercise. ∎

We call (7.1.2) the *relative de Rham complex* on $\mathbf{F}$. Now let $E \rightarrow \mathbf{P}$ be a holomorphic vector bundle on $\mathbf{P}$. Then $\mu^{-1} \mathcal{O}_{\mathbf{P}}(E)$ is an $\mathcal{O}_{\mathbf{P}}$-module induced by the mapping $\mathbf{F} \xrightarrow{\mu} \mathbf{P}$. We note that since d_μ annihilates the transition functions of $\mu^* E$, we have the exact sequence

$$0 \longrightarrow \mu^{-1} \mathcal{O}_{\mathbf{P}} \otimes \mu^{-1} \mathcal{O}_{\mathbf{P}}(E) \longrightarrow \mathcal{O}_{\mathbf{F}} \otimes \mu^{-1} \mathcal{O}_{\mathbf{P}}(E)$$
$$\xrightarrow{d_\mu} \Omega_{\mu}^1 \otimes \mu^{-1} \mathcal{O}_{\mathbf{P}}(E) \longrightarrow \cdots .$$

But we see that

$$\mu^{-1} \mathcal{O}_{\mathbf{P}} \otimes \mu^{-1} \mathcal{O}_{\mathbf{P}}(E) = \mu^{-1} \mathcal{O}_{\mathbf{P}}(E),$$
$$\mathcal{O}_{\mathbf{F}} \otimes \mu^{-1} \mathcal{O}_{\mathbf{P}}(E) = \mu^* \mathcal{O}_{\mathbf{P}}(E),$$
$$= \mathcal{O}_{\mathbf{F}}(\mu^* E),$$

and

$$\Omega_{\mu}^j \otimes \mu^{-1} \mathcal{O}_{\mathbf{P}}(E) = \Omega_{\mu}^j \otimes_{\mathcal{O}_{\mathbf{F}}} \mathcal{O}_{\mathbf{F}}(\mu^* E).$$

Letting

$$\Omega_{\mu}^0(E) := \mathcal{O}(\mu^* E),$$
$$\Omega_{\mu}^p(E) := \Omega_{\mu}^p \otimes_{\mathcal{O}_{\mathbf{F}}} \mathcal{O}(\mu^* E), \quad p \geq 1,$$

we obtain

$$0 \to \mu^{-1} \mathcal{O}_{\mathbf{P}}(E) \to \Omega^0_\mu(E) \to \Omega^1_\mu(E) \to \Omega^2_\mu(E) \to 0. \qquad (7.1.3)$$

For the special case that $E = H^n$, the nth power of the hyperplane section bundle, we shall write

$$0 \to \mu^{-1} \mathcal{O}_{\mathbf{P}}(n) \to \Omega^0_\mu(n) \to \Omega^1_\mu(n) \to \Omega^2_\mu(n) \to 0. \qquad (7.1.4)$$

Now we want to look at certain sheaves on $\mathbf{M}$ which we shall want to pull back to $\mathbb{F}$ in order to compare with the sheaves in (7.1.4). We shall want to use spinor sheaves on $\mathbf{M}$ and convert differential forms to spinorial forms. Let S_+ be the universal bundle over $\mathbf{M} = \mathbf{G}_2(2, \mathbf{T})$ defined by saying that $S_{+,x}$ is the complex-two-plane in $\mathbf{T}$ defining the point $x \in \mathbf{M}$ (see Example 2.1.4). We see that S_+ is embedded into the trivial bundle $\mathbf{M} \times \mathbf{T}$ since each fiber $S_{+,x}$ of S_+ is a subspace of $x \times \mathbf{T}$. Let S^- be the complementary bundle to S_+, i.e.,

$$0 \to S_+ \to \mathbf{M} \times \mathbf{T} \to S^- \to 0$$

is an exact sequence of bundles. Then define

$$S^+ := (S_+)^*, \qquad S_- := (S^-)^*.$$

These four bundles $S^\pm$, $S_\pm$ are all $GL(2, \mathbf{C})$-bundles naturally defined over $\mathbf{M}$. The tensor algebra generated by these bundles is the *spinor algebra* on $\mathbf{M}$. We saw in §3.5 how $GL(2, \mathbf{C})$-bundles relate to spin structures and spin^c-structures on $\mathbf{M}$ and its natural real submanifolds M and S^4. These bundles are themselves not spin bundles in the usual sense, since the determinant of these bundles is not trivial on $\mathbf{M}$, i.e., they do not reduce to $SL(2, \mathbf{C})$-bundles on $\mathbf{M}$. If we introduce conformal weights, then these bundles can be manipulated in the same manner as spin bundles and we shall proceed to do so. First we need to relate the geometry of these bundles to the tangent bundle geometry of $\mathbf{M}$. We recall the following two fundamental facts about the tangent bundle of the Grassmannian (Exercise 2.1.7). Let $\mathbf{G}_{k,n} = \mathbf{G}_{k,n}(\mathbf{C})$, and $U \to \mathbf{G}_{k,n}$ be its universal bundle with Q being its complementary bundle, i.e.,

$$0 \to U \to \mathbf{G}_{k,n} \times \mathbf{C}^n \to Q \to 0.$$

Then we have

$$T(\mathbf{G}_{k,n}) \cong \mathrm{Hom}(U, Q) \cong Q \otimes U^*, \qquad (7.1.5)$$

$$\det T^*(\mathbf{G}_{k,n}) \cong [\det U]^n. \qquad (7.1.6)$$

From (7.1.5) we have immediately the proposition

Proposition 7.1.2. $T(\mathbf{M}) \cong S^- \otimes S^+$.

Proof: We see that

$$\begin{aligned}
S^- \otimes S^+ &\cong S^- \otimes (S_+)^*, \\
&\cong \operatorname{Hom}(S_+, S^-), \\
&\cong T(\mathbf{M}),
\end{aligned}$$

using the standard algebraic fact that $\operatorname{Hom}(A, B) \cong B \otimes A^*$. ∎

We want to introduce the notion of *conformal weights* on **M**. On a differentiable manifold X of dimension n we let $K = \det T^*(X)$ be the canonical bundle. If $K^{1/n}$ makes sense as a line bundle, i.e., there is a line bundle L such that $L^n = K$, and if E is a vector bundle over X, then we say that a section s has *conformal weight* m if

$$s \in \Gamma(E \otimes (K^{1/n})^m).$$

Thus s is an ordinary section of E twisted by the mth power of $L = K^{1/n}$. The concept of conformal weight makes good sense for any manifold, as there are alternative definitions (see Hitchin 1980, for instance). On a complex manifold X we shall use the holomorphic canonical bundle $K = \det T^*(X)$ to define *holomorphic conformal weights*, although we shall drop the adjective holomorphic in context. If we have a submanifold $Y \subset X$ where X is a complexification of Y, e.g., $S^4 \subset \mathbf{M}$ or $M \subset \mathbf{M}$, then $K_Y \otimes \mathbf{C} \cong K_X|_Y$, so the notion of holomorphic conformal weights on X will restrict nicely to real conformal weights on Y. This is important when we want to restrict holomorphic solutions of field equations on open subsets of **M** to S^4 or M.

On **M** we have by (7.1.6) that

$$K_\mathbf{M} = \det T^*(\mathbf{M}) \cong [\det S_+]^4. \tag{7.1.7}$$

Now we also have

$$0 \to S_- \to \mathbf{M} \times \mathbf{T}^* \to S^+ \to 0.$$

If we use an Hermitian metric on **T** to identify $\mathbf{T} \cong \mathbf{T}^*$ as complex vector spaces, then we see that powers of $\det S_-$ will also yield conformal weights on **M**. We want to distinguish between these two conformal weights; although from the standpoint of complex manifolds they can be identified, from the point of view of twistor geometry and the invariants they should be distinguished. Let us define, for any complex vector bundle $E \to \mathbf{M}$,

$$\mathcal{O}_{\mathbf{M}}(E)[m] := \mathcal{O}_{\mathbf{M}}(E \otimes (\det S_-)^m),$$
$$\mathcal{O}_{\mathbf{M}}(E)[m]' := \mathcal{O}_{\mathbf{M}}(E \otimes (\det S_+)^m).$$

These are analogues to twisting a vector bundle with a power of the hyperplane section bundle in projective space $\mathbf{P}_n$ which we denote by $\mathcal{F}(m)$, if $\mathcal{F}$ is an $\mathcal{O}_{\mathbf{P}}$-sheaf on $\mathbf{P}_n$. In addition we define

$$\mathcal{O}_{\mathbf{M}}(E)[-m] := \mathcal{O}_{\mathbf{M}}(E \otimes (\det S^-)^m)$$
$$\mathcal{O}_{\mathbf{M}}(E)[-m]' := \mathcal{O}_{\mathbf{M}}(E \otimes (\det S^+)^m).$$

Note that these are the negatives of the conformal weights found in Eastwood, Penrose and Wells (1981), but agree with the common usage of the notion of conformal weight.

Let us introduce the abstract index notation for the tensor algebra generated by $S^\pm$, $S_\pm$. We let

$$\mathcal{O}^A := \mathcal{O}_{\mathbf{M}}(S^-), \qquad \mathcal{O}^{A'} := \mathcal{O}_{\mathbf{M}}(S^+),$$
$$\mathcal{O}_A := \mathcal{O}_{\mathbf{M}}(S_-), \qquad \mathcal{O}_{A'} := \mathcal{O}_{\mathbf{M}}(S_+).$$

Then we have

$$\mathcal{O}^{AA'} = \mathcal{O}(S^{AA'}),$$
$$\cong \mathcal{O}(S^- \otimes S^+),$$
$$\cong \mathcal{O}(T(\mathbf{M})),$$

from Proposition 7.1.2, while

$$\mathcal{O}_{AA'} \cong \Omega^1_{\mathbf{M}}$$

and the exterior differentiation

$$\mathcal{O}_{\mathbf{M}} \xrightarrow{d} \Omega^1_{\mathbf{M}}$$

has the form

$$\mathcal{O} \xrightarrow{\nabla_{AA'}} \mathcal{O}_{AA'},$$

or

$$\mathcal{O} \xrightarrow{\nabla_a} \mathcal{O}_a,$$

if lower case indices represent the usual tensor algebra on M, as before. We drop the subscript M on the spinor sheaves $\mathcal{O}_A, \mathcal{O}_{AA'}$, etc., on M as this is the manifold on which they are defined. We shall also write $\mathcal{O}_{AA'}$ for $\nu^*\mathcal{O}_{AA'}$, as a pullback sheaf on F. We shall be clear about which manifold we are working on, and in the case of ambiguity we can write $\mathcal{O}_{\mathbf{M}}(S_{AA'})$ or $_{\mathbf{M}}\mathcal{O}_{AA'}$, to indicate which structure sheaf is indicated. We have a canonical isomorphism

$$\det S_+[-1]' = \det S_+ \otimes \det S^+ \cong \mathbf{M} \times \mathbf{C},$$

or, in the other notation,

$$\mathcal{O}_{[A'B']}[-1]' = \mathcal{O}_{[A'B']} \otimes \mathcal{O}^{[A'B']} = \mathcal{O}.$$

Let $\varepsilon_{A'B'}$ be the section over M of $\mathcal{O}_{[A'B']}[-1]'$ corresponding to the section $[s \equiv 1] \in \Gamma(\mathbf{M}, \mathcal{O})$. Similarly, we define

$$\varepsilon_{AB} \in \Gamma(\mathbf{M}, \mathcal{O}_{[AB]}[1]),$$
$$\varepsilon^{A'B'} \in \Gamma(\mathbf{M}, \mathcal{O}^{[A'B']}[1]'),$$
$$\varepsilon^{AB} \in \Gamma(\mathbf{M}, \mathcal{O}^{[AB]}[1]).$$

These are to be regarded as conformally weighted εs to raise and lower indices. Normally, a spin bundle S has $\det S$ trivial, and $\varepsilon \in \Gamma(M, \det S)$ is a nonvanishing section of this bundle. Here we have $\varepsilon \in \det S[1]$, which is *conformally weighted*, and which has a global nonvanishing section even if $\det S$ itself does not. So we shall manipulate the conformally weighted spinor fields in the same manner as spinor fields on a manifold with an $SL(2, \mathbf{C})$-bundle spin structure (cf. §§2.3 and 2.5). As we have seen, these $GL(2, \mathbf{C})$-bundles on M can restrict to usual spin bundles on S^4 or M (§3.5). Since $\det S_+$ restricted to M or S^4 becomes trivial, then the εs here will restrict to the usual εs as described in Penrose and Rindler (1984) for real space-time.

Let us see how we can raise and lower indices with these εs. For instance

$$\mathcal{O}^{A'} \cong \mathcal{O}^{A'} \otimes \mathcal{O} \cong \mathcal{O}^{A'} \otimes \mathcal{O}_{[A'B']}[1]' \cong \mathcal{O}_{B'}[1],$$
$$f^{A'} \leftrightarrow f^{A'} \otimes 1 \leftrightarrow f^{A'} \otimes \varepsilon_{A'B'} \leftrightarrow f^{A'}\varepsilon_{A'B'}.$$

We now want to look at the de Rham complex on $\mathbf{M}$ in terms of spinors. The de Rham complex on $\mathbf{M}$ has the form

$$\mathcal{O} \xrightarrow{d} \Omega^1 \xrightarrow{d} \Omega^2 \xrightarrow{d} \Omega^3 \xrightarrow{d} \Omega^4.$$

We have already identified $\Omega^1 \cong \mathcal{O}_{AA'}$ above. Let us look at Ω^2. We can write any two-form F_{ab} in terms of its self-dual and anti-self-dual parts (see (4.2.4))

$$F_{ab} = F_{AA'BB'} = \varphi_{A'B'}\varepsilon_{AB} + \psi_{AB}\varepsilon_{A'B'},$$

where $\varphi_{A'B'} = \varphi_{(A'B')}, \psi_{AB} = \psi_{(AB)}$, i.e., these are symmetric spinor forms. In terms of sheaves, letting Ω^2_+ and Ω^2_- denote the self-dual and anti-self-dual two-forms on $\mathbf{M}$,

$$\begin{aligned}
\Omega^2 &\cong \mathcal{O}_{[ab]} \cong \Omega^2_+ \oplus \Omega^2_-, \\
&\cong \mathcal{O}_{(A'B')[AB]} \oplus \mathcal{O}_{(AB)[A'B']}, \\
&\cong \mathcal{O}_{(A'B')}[1] \oplus \mathcal{O}_{(AB)}[1]', \\
&\cong \mathcal{O}(\odot^2 S_+[1]) \oplus \mathcal{O}(\odot^2 S_-[1]'),
\end{aligned}$$

and the exterior derivative splits into its self-dual and anti-self-dual parts:

$$\begin{aligned}
\mathcal{O}_{AA'} &\xrightarrow{d} \mathcal{O}_{(A'B')}[1] \oplus \mathcal{O}_{(AB)}[1]', \\
\varphi_{AA'} &\mapsto \nabla^A_{(B'}\varphi_{A')A} + \nabla^{A'}_{(B}\varphi_{A)A'}
\end{aligned}$$

where $\nabla^A_{B'}$ and $\nabla^{A'}_{B}$ are described below. Recall that $\nabla_{AA'} = \nabla_a$ is the usual exterior derivative, and we see that by the definition of $\varepsilon_{AB'}$, etc., we have $\nabla_{AA'}\varepsilon_{BC} = 0$, and similarly for the other εs (recall from §4.2 that we need to have covariantly constant εs). Now we define

$$\begin{aligned}
\nabla^{B'}_A &:= \nabla_{AA'}\varepsilon^{A'B'}, \\
\nabla^A_{B'} &:= \varepsilon^{AB}\nabla_{BB'},
\end{aligned}$$

just as if these operators were ordinary spinors (noting that the εs commute with ∇ as operators by the discussion above), and we obtain the expressions above for the exterior derivatives of a one-form in spinorial form. We find the three-forms and four-forms in an analogous manner obtaining the full de Rham sequence in spinor form:

$$
\mathcal{O} \xrightarrow{\nabla_{AA'}} \mathcal{O}_{AA'} \begin{array}{c} \xrightarrow{\nabla^A_{B'}} \\ \xrightarrow[\nabla^{A'}_B]{} \end{array} \begin{array}{c} \mathcal{O}_{(A'B')}[1] \\ \oplus \\ \mathcal{O}_{(AB)}[1]' \end{array} \begin{array}{c} \xrightarrow{i\nabla^{A'}_B} \\ \xrightarrow[i\nabla^A_{B'}]{} \end{array} \mathcal{O}_{BB'}[1][1]' \xrightarrow{\nabla^{BB'}} \mathcal{O}[2][2]',
$$

where $d : \Omega^2 \to \Omega^3$ has the form

$$
(\varphi_{A'B'}, \psi_{AB}) \mapsto i\nabla^{A'}_B \varphi_{A'B'} + i\nabla^A_{B'}\psi_{AB},
$$

and $d : \Omega^3 \to \Omega^4$ has the form

$$
\varphi_{BB'} \mapsto \nabla^{BB'}\varphi_{BB'}.
$$

Now we want to relate the relative de Rham complex on $\mathbb{F}$ to pullbacks of sheaves from $\mathbb{P}$ and $\mathbb{M}$. On $\mathbb{P}$ we have the powers of the hyperplane section bundle H, and on $\mathbb{M}$ we have the conformally weighted spinor bundles. On $\mathbb{F}$ we shall write, for instance

$$
\mathcal{O}_A[1]'(1) := \nu^*\mathcal{O}_A[1]' \otimes \mu^*\mathcal{O}_{\mathbb{P}}(1),
$$

suppressing the pullbacks, where the notation makes it clear what the contributions to the sheaf is. We have the following lemma.

Lemma 7.1.3. *The relative de Rham sequence (7.1.2) on $\mathbb{F}$ can be canonically identified with the following pullback sheaves:*
  (a) $\Omega^0_\mu = \mathcal{O}_{\mathbb{F}}$,
  (b) $\Omega^1_\mu \cong \mathcal{O}_A(1)[1]' = \mathcal{O}_{\mathbb{F}}(\nu^*(S_- \otimes \det S_+) \otimes \mu^*(H))$,
  (c) $\Omega^2_\mu \cong \mathcal{O}_{[AB]}(2)[2]' = \mathcal{O}_{\mathbb{F}}(\nu^*(\det S_- \otimes (\det S_+)^2) \otimes \mu^*(H^2))$.

Proof: First we want to construct a bundle homomorphism on $\mathbb{F}$

$$
h : H^{-1} \to S_+,
$$

where $H := \mu^*(H)$, and $S_+ := \nu^*(S_+)$ as before. By definition the fibers of H^{-1} and S_+ over $(L_1, L_2) \in \mathbb{F}$ are precisely L_1 and L_2, respectively (they are both universal bundles). So we define the mapping h by the inclusion $L_1 \hookrightarrow L_2$. Now let $\bigwedge^1_\mu$ be the vector bundle associated with Ω^1_μ, and recall that $S_- \otimes S_+ \cong T(\mathbf{M})$. Let $p \in \mathsf{F}$, and let $v \in \bigwedge^1_\mu$, then $v = [\omega]$, where $\omega \in \bigwedge^1 T^*(\mathsf{F})$, and $[\omega]$ is the equivalence class of this vector modulo vectors in $\mu^* T^*(\mathbb{P})$. Choose a frame $f = (e_1, \ldots, e_5)$ for $T^*(\mathsf{F})$ at p adapted to $Y_p :=$ $\mu^{-1}(\mu(p))$, i.e., adapted to the 'α-plane' Y_p in F, passing through p. That is, e_1 and e_2 are tangent to Y_p at p and e_3, e_4, e_5 vanish when restricted to Y_p at p. Then we can choose ω, the representative for v to be of the form $\omega = c^j e_j$, and $c^j(p) = 0$, $j = 3, 4, 5$. On the other hand, we can choose a frame at p of the form $\tilde{e}_1, \ldots, \tilde{e}_5$, where $\tilde{e}_1, \ldots, \tilde{e}_4 \in S_- \otimes S_+$, and $\tilde{e}_5|_{Y_p}$ vanishes. Then $\omega = d^j \tilde{e}_j$ and $d^5(p) = 0$. The coefficients $d^j(p)$ depend only on v and not on the choice of adapted frame $f = (e_1, \ldots, e_5)$. Thus we have a well-defined mapping

$$\bigwedge^1_\mu \to S_- \otimes S_+.$$

We can then tensor this with h yielding

$$\bigwedge^1_\mu \otimes H^{-1} \to S_- \otimes S_+ \otimes S_+,$$

which we can compose with $S_+ \otimes S_+ \to \det S_+$, yielding a homomorphism

$$\bigwedge^1_\mu \otimes H^{-1} \to S_- \otimes \det S_+.$$

One can show that this mapping is surjective, and that the bundles have the same dimension, and thus we obtain the desired isomorphism (b).

The isomorphism (c) is proved in a similar manner. Here we note that a two-form tangent to an α-plane in $\mathbf{M}$ is self-dual, so proceeding as above we have a mapping

$$\bigwedge^2_\mu \to \bigwedge^2(S_- \otimes S_+),$$

but the image will be in the subbundle $\odot^2 S_+ \otimes \det S_-$ (i.e., covectors of the form $\varphi_{A'B'}\varepsilon_{AB}$, cf. (4.2.3), since the image of $\bigwedge^2_\mu$ will be covectors tangent to the α-planes (lifted to F). Thus we have

$$\bigwedge^2_\mu \to \odot^2 S_+ \otimes \det S_-.$$

Now tensoring with H^{-2} and mapping with the homomorphism h^2 we obtain

$$\bigwedge^2_\mu \otimes H^{-2} \to \odot^2 S_+ \otimes S_+ \otimes S_+ \otimes \det S_-,$$

which maps naturally to $(\det S_+)^2 \otimes \det S_-$. This map is again surjective, and both are line bundles, so we obtain (c). ∎

Next we need to compute the direct images of the sheaves $\Omega^p_\mu(n)$. We first have the following special case where $p = 0$.

Theorem 7.1.4. *The direct images $\nu^q_* \mathcal{O}(n)$ of the sheaves $\mathcal{O}(n)$ on F may be canonically identified as:*

(a) *for $n \geq 0$, $q \geq 1$,*

$$\nu^0_* \mathcal{O}(n) \cong \mathcal{O}^{(A'B'\ldots D')} \cong \mathcal{O}(\odot^n S^+),$$
$$\nu^q_* \mathcal{O}(n) = 0,$$

(b) *for all $q \geq 0$,*

$$\nu^q_* \mathcal{O}(-1) = 0,$$

(c) *for $n \geq 0$, $q \neq 1$,*

$$\nu^1_* \mathcal{O}(n-2) \cong \mathcal{O}_{(A'B'\ldots D')}[1]' \cong \mathcal{O}(\odot^n S_+ \otimes \det S_+),$$
$$\nu^q_* \mathcal{O}(n-2) = 0.$$

Proof: We need to compute $H^q(\nu^{-1}(U), \mathcal{O}(n))$ for open sets U in M, but it suffices to restrict ourselves to arbitrarily small convex sets U in a coordinate chart over which $\nu^{-1}(U)$ can be trivialized, i.e., $\nu^1(U) \cong U \times \mathbf{P}_1$. If we first look at $H^q(\mathbf{P}_1, \mathcal{O}(n))$ we recall that we can use a covering $V_0 \cup V_1$, and compute $H^q(\mathbf{P}_1, \mathcal{O}(n))$ by means of a Laurent series expansion (see Example 3.3.8). By covering $U \times \mathbf{P}_1$ with the same type of covering by two open sets we get the same Laurent series expansion, but where the coefficients of the series depend on the points of the open set U. Then we find that

$$H^q(\nu^1(U), \mathcal{O}(n)) \cong \{\text{holomorphic } f : U \to H^q(\mathbf{P}_1, \mathcal{O}(n))\}. \tag{7.1.9}$$

Using the vanishing theorem (3.3.18), we find that

$$H^q(\mathbf{P}_1, \mathcal{O}(n)) = 0, \quad q \geq 1, \quad n \neq -2, \tag{7.1.10a}$$
$$H^q(\mathbf{P}_1, \mathcal{O}(1)) = 0, \quad q \geq 0, \tag{7.1.10b}$$
$$H^q(\mathbf{P}_1, \mathcal{O}(n-2)) = 0, \quad q \neq 1, \quad n \geq 0. \tag{7.1.10c}$$

These all follow by Laurent series arguments as in Example 3.3.8.

It follows from (7.1.9) that the direct images $\nu_*^q \mathcal{O}(n)$ are sections of vector bundles whose fibers can be identified with

$$H^q(\nu^{-1}(z), \mathcal{O}_{\nu^{-1}(z)}(n)).$$

We could calculate the dimensions of these fibers to see that we get dimensional agreement with the vector bundles on the right hand side of the assertions in the theorem, but this is clearly not sufficient (except for the case of dimension zero!).

Let us consider the simplest case $\nu_*^0 \mathcal{O}(1)$. For any point $z \in \mathbf{M}$, the fiber $\nu^{-1}(z) = \mathbf{P}(S_{+,z})$, where $S_{+,z}$ is the fiber of S_+ over the point z. Consider the hyperplane section sheaf $\mathcal{O}_{\mathbf{P}(S_{+,z})}(1)$ over $\mathbf{P}(S_{+,z})$. The line bundle $H_{\mathbf{P}(S_{+,z})}(1)$ associated with $\mathcal{O}_{\mathbf{P}(S_{+,z})}(1)$ is the dual of the universal bundle over $\mathbf{P}(S_{+,z})$, i.e., each vector in $H_{\mathbf{P}(S_{+,z})}(1)$ is a linear functional on the lines in $S_{+,z}$. The vector space S_z^+ is the dual of $S_{+,z}$; and any vector $v \in S_z^+$ restricted to the lines of $S_{+,z}$ give a holomorphic section of the sheaf $\mathcal{O}_{\mathbf{P}(S_{+,z})}(1)$. Thus there is a canonical homomorphism

$$S_z^+ \to \Gamma(\mathbf{P}(S_{+,z}), \mathcal{O}_{\mathbf{P}(S_{+,z})}(1)),$$

which one can see, by Laurent series arguments again, is an isomorphism. Thus we can identify

$$S_z^+ \cong H^0(\nu^{-1}(z), \mathcal{O}_{\nu^{-1}(z)}(1)),$$

as desired. The general case

$$H^0(\nu^{-1}(z), \mathcal{O}_{\nu^{-1}(z)}(n)) \cong \odot^n S^+,$$

is proved in a similar manner using homogeneous polynomials of degree n instead of linear functionals.

For case (c) we need a special case of Serre duality. Serre duality on any compact complex manifold X asserts that if E is a holomorphic vector bundle on X, then

$$H^p(X, \Omega^q(E)) \cong H^{n-p}(X, \Omega^{n-q}(E^*))$$

(see, e.g., Wells 1980). We need the special case

$$H^1(\mathbf{P}_1, \Omega^1) \cong H^0(\mathbf{P}_1, \mathcal{O}) \cong \mathbf{C}.$$

We can prove Serre duality in this case by using Laurent expansions of both sides and identifying the single nonzero coefficient in the expansions. But we need a proof of this which will work for $\mathcal{O}(-2)$ instead of Ω^1. These sheaves are isomorphic on $\mathbf{P}_1$, but not canonically. Letting $\mathbf{P}_1 = \mathbf{P}(S_+)$, where we let $S_+ = S_{+,z}$ in the following argument, we have the exact sequence of sheaves on $\mathbf{P}_1$:

$$0 \longrightarrow \Omega^1_{\mathbf{P}_1} \longrightarrow \mathcal{O}_{\mathbf{P}_1}(S^+ \otimes H^{-1}) \xrightarrow{{}^t h} \mathcal{O}_{\mathbf{P}_1} \longrightarrow 0, \qquad (7.1.11)$$

where ${}^t h$ is dual to the mapping $h : \mathcal{O}(H^{-1}) \to \mathcal{O}(S_+)$ used in the proof of Lemma 1.7.3. The sequence (7.1.11) follows from (7.1.5) as is easy to see. The long exact sequence on cohomology coming from (7.1.11) yields

$$H^0(\mathbf{P}_1, \mathcal{O}(S^+ \otimes H^{-1})) \to H^0(\mathbf{P}_1, \mathcal{O}) \to H^1(\mathbf{P}_1, \Omega^1)$$
$$\to H^1(\mathbf{P}_1, \mathcal{O}(S^+ \otimes H^{-1})). \qquad (7.1.12)$$

But the first and last terms of (7.1.12) are zero by (7.1.10.b) which gives the isomorphism $H^1(\mathbf{P}_1, \Omega^1) \cong H^0(\mathbf{P}_1, \mathcal{O})$. But $H^0(\mathbf{P}_1, \mathcal{O})$ is canonically isomorphic to $\mathbf{C}$ by evaluation at any point. We need to mimic this proof since we have $\mathcal{O}(-2)$ instead of Ω^1. We have instead of (7.1.11) the exact sequence on $\mathbf{P}_1$,

$$0 \longrightarrow \mathcal{O}(-2) \xrightarrow{{}^t h} \mathcal{O}(S_+)(-1) \xrightarrow{{}^t h} \mathcal{O}(\det S_+) \longrightarrow 0, \qquad (7.1.13)$$

or using the abstract index notation and letting ${}^t h$ be denoted by $\pi_{A'} : \mathcal{O}(1) \to \mathcal{O}_{A'}$, we have the same sequence as (7.1.13)

$$0 \longrightarrow \mathcal{O}(-2) \xrightarrow{\pi_{A'}} \mathcal{O}_{A'}(-1) \xrightarrow{\pi_{B'}} \mathcal{O}_{[A'B']} \longrightarrow 0 \qquad (7.1.14)$$

Using the long exact sequence in cohomology which comes from (7.1.13) or (7.1.14) we obtain the canonical isomorphism

$$H^1(\nu^{-1}(z), \mathcal{O}_{\nu^{-1}(z)}(-2)) \cong S_{[A'B'],z} \cong \wedge^2 S_{+,z}.$$

In fact, if we considered (7.1.14) as a sequence on $\mathbf{F}$, we obtain the long exact sequence of direct images (see Exercise 7.1.7):

$$\begin{array}{ccccccc}
\nu^0_* \mathcal{O}_{A'}(-1) & \to & \nu^0_* \mathcal{O}_{[A'B']} & \to & \nu^1 \mathcal{O}(-2) & \to & \nu^1_* \mathcal{O}_{A'}(-1) \\
\| & & \| & & & & \| \\
0 & & \mathcal{O}_{[A'B']} & & & & 0
\end{array}$$

which also shows that $\nu_*^1 \mathcal{O}(-2) \cong \mathcal{O}_{[A'B']}$. To identify $\nu_*^1 \mathcal{O}(-n-2)$ we replace (7.1.14) with

$$0 \to \mathcal{O}(-n-2) \xrightarrow{\ \pi_{A'} \ldots \pi_{E'}\ } \mathcal{O}_{(A'\ldots D'E')}(-1)$$
$$\xrightarrow{\ \pi_{F'}\ } \mathcal{O}_{(A'\ldots D')[E'F']} \to 0, \quad (7.1.15)$$

which is the same as

$$0 \to \mathcal{O}(-n-2) \xrightarrow{\ ({}^t h)^n\ } \mathcal{O}(\odot^n S_+)(-1),$$
$$\xrightarrow{\ {}^t h\ } \mathcal{O}(\odot^{n-1} S_+ \otimes \det S_+) \to 0.$$

Thus we obtain, by using the long exact sequence on cohomology again,

$$\nu_*^1 \mathcal{O}(-n-2) \cong \mathcal{O}_{(A'\ldots D')[E'F']}$$
$$= \mathcal{O}_{(A'\ldots D')}[1]' \cong \mathcal{O}(\odot^{n-1} S_+)[1]'.$$

∎

We now conclude from Lemma 7.1.3 and Theorem 7.1.4 the following general result.

Theorem 7.1.5. *The direct image sheaves $\nu_*^p \Omega_\mu^q(n)$ are given by:*
(a)

$$\nu_*^0 \Omega_\mu^0(n) \cong \nu_*^0 \mathcal{O}(n) \cong \mathcal{O}^{(A'B'\ldots D')}$$
$$\cong \mathcal{O}(\odot^n S^+), \quad \text{for } n \geq 0,$$

(b)

$$\nu_*^0 \Omega_\mu^1(n) \cong \nu_*^0 \mathcal{O}_A(n+1)[1]' \cong \mathcal{O}_A^{(A'B'\ldots E')}[1]$$
$$\cong \mathcal{O}(\odot^{n+1} S^+ \otimes S_-), \quad \text{for } n \geq -1,$$

(c)

$$\nu_*^0 \Omega_\mu^1(n) \cong \nu_*^0 \mathcal{O}(n+2)[1][2]' \cong \mathcal{O}^{(A'B'\ldots F')}[1][2]'$$
$$\cong \mathcal{O}(\odot^{n+2} S^+)[1][2]', \quad \text{for } n \geq -2,$$

(d)

$$\nu_*^1\Omega_\mu^0(n-2) \cong \nu_*^1\mathcal{O}(-n-2) \cong \mathcal{O}_{(A'B'\dots D')}[1]'$$
$$\cong \mathcal{O}(\odot^n S_+)[1]', \quad for\ n \geq 0,$$

(e)

$$\nu_*^1\Omega_\mu^1(-n-2) \cong \nu_*^1\mathcal{O}_A(-n-1)[-1]' \cong \mathcal{O}_{A(B'\dots D')}[2]'$$
$$\cong \mathcal{O}(S_+ \otimes \odot^{n-1} S_-)[2]', \quad for\ n \geq 1,$$

(f)

$$\nu_*^1\Omega_\mu^2(n-2) \cong \nu_*^1\mathcal{O}(-n)[1][2]' \cong \mathcal{O}_{(C'\dots D')}[1][3]'$$
$$\cong \mathcal{O}(\odot^{n-2} S_+)[1][3]', \quad for\ n \geq 2,$$

(g)   *all other possibilities vanish.*

Now the relative de Rham sequence on $\mathbb{F}$ has the form

$$\Omega_\mu^0(n) \xrightarrow{d_\mu} \Omega_\mu^1(n) \xrightarrow{d_\mu} \Omega_\mu^2(n) \tag{7.1.16}$$

and this induces the direct image sequence (see Exercise 7.1.7)

$$\nu_*^0\Omega_\mu^0(n) \xrightarrow{\tilde{d}_\mu} \nu_*^0\Omega_\mu^1(n) \xrightarrow{\tilde{d}_\mu} \nu_*^0\Omega_\mu^2(n), \tag{7.1.17}$$

which is the same as

$$\mathcal{O}(\odot^n S^+) \xrightarrow{\tilde{d}_\mu} \mathcal{O}(S_- \otimes \odot^{n+1} S^+)[1]' \xrightarrow{\tilde{d}_\mu} \mathcal{O}(\odot^{n+2} S^+)[1][2]'. \tag{7.1.18}$$

We want to identify $\tilde{d}_\mu$ in (7.1.17) in terms of specific spinor differential operators.

Lemma 7.1.6. (a) *If*

$$f = f^{A'B'\dots D'} \in \mathcal{O}^{(A'B'\dots D')} \cong \mathcal{O}(\odot^n S^+),$$

then

$$\tilde{d}_\mu f = \nabla_A^{(E'} f^{A'B'\dots D')},$$

(b) *if*

$$g = g_A^{A'B'\dots E'} \in \mathcal{O}_A^{(A'B'\dots E')}[1]' \cong \mathcal{O}(S_- \otimes \odot^{n+1} S^+)[1]',$$

then

$$\tilde{d}_\mu g = \nabla^{A(F'} g_A^{A'B'...E')}$$

Proof: Let us use local coordinates in $F^I = M^I \times P_1$ of the form $(z^{AA'}, [\pi_{A'}])$, and compute d_μ in terms of these coordinates. If f is a local section of $\mathcal{O}_F$ defined on F^I, then

$$d_\mu f = \pi^{A'} \nabla_{AA'} f(z^{A'}, \pi'_A), \qquad (7.1.19)$$

which defines a local section of $\mathcal{O}_A(1)[1]'$. To see this, we introduce local affine coordinates on $P_0^I = P^I \cap \{\pi_{0'} \neq 0\}$:

$$q^A := \omega^A / \pi_{0'}, \quad A = 0, 1,$$
$$r := \pi_{1'} / \pi_{0'},$$

and on $F_0^I = \mu^{-1}(P_0^I)$:

$$q^A := (iz^{AA'} \pi_{A'}) / \pi_{0'}, \quad A = 0, 1,$$
$$r := \pi_{1'} / \pi_{0'},$$
$$s^A := z^{A1'}, \quad A = 0, 1.$$

The projection $F \to P$ in these coordinates is given by $(q^A, r, s^A) \to (q^A, r)$. Thus in these coordinates we have for a function f

$$d_\mu f = \frac{\partial f}{\partial s^A} ds^A.$$

Now using the chain rule for q^A, r, and s^A, we find that

$$\nabla_A^{0'} := -\frac{\partial}{\partial z_{0'}^A} = \frac{\partial}{\partial s^A} - ir \frac{\partial}{\partial q^A},$$

$$\nabla_A^{1'} := -\frac{\partial}{\partial z_{1'}^A} = i \frac{\partial}{\partial q^A}.$$

The components of $d_\mu f$ are $\frac{\partial f}{\partial s^A}$, $A = 0, 1$. We claim that the components of $\pi_{A'} \nabla_A^{A'} f$, $A = 0, 1$, coincide with $\frac{\partial f}{\partial s^A}$ in these affine coordinates. Namely, $\pi_{A'} \nabla_A^{A'} f$ in these affine coordinates is given by

$$\nabla_A^{0'} f + (\pi_{1'}/\pi_{0'}) \nabla_A^{1'} f = \frac{\partial f}{\partial s^A} - ir \frac{\partial f}{\partial q^A} + ri \frac{\partial f}{\partial q^A}$$

$$= \frac{\partial f}{\partial s^A}.$$

Therefore we can express the sequence (7.1.16) on F^I by

$$\mathcal{O} \xrightarrow{\;\pi_{A'}\nabla^{A'}_A\;} \mathcal{O}_A(1)[1]' \xrightarrow{\;\pi_{A'}\nabla^{AA'}\;} \mathcal{O}(2)[1][2]', \qquad (7.1.20)$$

since d_μ acting on Ω^1_μ can be shown to be represented by $\pi_{A'}\nabla^{AA'}$ in the same manner.

By a change in sign and raising indices, we have also

$$\mathcal{O} \xrightarrow{\;\pi_{A'}\nabla^{AA'}\;} \mathcal{O}^A(1)[1][1]' \xrightarrow{\;\pi_{A'}\nabla^{A'}_A\;} \mathcal{O}(2)[1][2]'. \qquad (7.1.21)$$

The operator $\pi_{A'}$ was introduced earlier in this section (see (7.1.13) and (7.1.14)), and makes sense on F independent of choices of coordinates, and the operator $\nabla^{AA'}$ makes sense on M, but not on F. However one can make sense of the composition

$$\pi_{A'}\nabla^{AA'} \qquad (7.1.22)$$

on F. This is just another name for d_μ, and it has the local coordinate expressions given in (7.1.20) and (7.1.21). Under changes of spinor coordinates for T it will have the same expression. One can also give this expression an invariant meaning by seeing how the connection ∇_a changes with changes in the metric and how this change is annihilated after contracting with the homomorphism $\pi_{A'}$ (see Eastwood, Penrose and Wells 1981 for a discussion of this subtle point). We shall use (7.1.22) as a mnemonic symbol for the operator d_μ in the abstract index context, recalling that it has the same form in local coordinates. Using this convention we can rewrite (7.1.21) as

$$\mathcal{O} \xrightarrow{\;\pi_{A'}\nabla^{AA'}\;} \mathcal{O}^A(1)[1][1]' \xrightarrow{\;\pi_{A'}\nabla^{A'}_A\;} \mathcal{O}(2)[1][2]'. \qquad (7.1.23)$$

We now have a similar description for the higher twists, i.e.,

$$\pi_{E'}\nabla^{E'}_A : \mathcal{O}(n) \to \mathcal{O}_A(n+1)[1]'$$

gives on F^I

$$\pi_{A'}\pi_{B'}\dots\pi_{D'}f^{A'B'\dots D'} \mapsto \pi_{A'}\pi_{B'}\dots\pi_{D'}\pi_{E'}\nabla^{E'}_A f^{A'B'\dots D'},$$

which induces on $\nu_*^0 \mathcal{O}(n) = \mathcal{O}^{(A'B'...D')}$

$$g^{A'...D'} \mapsto \nabla_A^{(E'} g^{A'B'...D')}.$$

Note that $f^{A'B'...D'}$ depends on the coordinates $z^{AA'}$ and $\pi_{A'}$ while $g^{A'...D'}$ defined on $\mathbf{M}^I$ has no dependence on $\pi_{A'}$. The direct image process 'eliminates' the $\pi_{A'}$-parameter, i.e., we have integrated over the fibers parametrized by $\pi_{A'}$ and isomorphic to $\mathbf{P}_1$ (the integration is in the Serre duality, which can be proved explicitly by integration of $(1,1)$-forms over $\mathbf{P}_1$ using Dolbeault's theorem to represent the cohomology classes or by means of contour integrals representing the Laurent series coefficients). Now we have a similar lemma for first direct images sheaves. Consider the sequence

$$\Omega_\mu^0(-n-2) \xrightarrow{d_\mu} \Omega_\mu^1(-n-1) \xrightarrow{d_\mu} \Omega_\mu^2(-n-2),$$

which is the same as, by Lemma 7.1.3,

$$\mathcal{O}(-n-2) \xrightarrow{d_\mu} \mathcal{O}_A(-n-1)[1]' \xrightarrow{d_\mu} \mathcal{O}(-n)[1][2]',$$

and this induces the sequence of direct images sheaves

$$\nu_*^1 \mathcal{O}(-n-2) \xrightarrow{\tilde{d}_\mu} \nu_*^1 \mathcal{O}_A(-n-1)[1]' \xrightarrow{\tilde{d}_\mu} \nu_*^1 \mathcal{O}(-n)[1][2]', \quad (7.1.24)$$

which is the same as, by Theorem 7.1.5,

$$\mathcal{O}_{(A'B'...D')}[1] \xrightarrow{\tilde{d}_\mu} \mathcal{O}_{A(B'...D')}[2]' \xrightarrow{\tilde{d}_\mu} \mathcal{O}_{(C'...D')}[1][3]'. \quad (7.1.25)$$

Lemma 7.1.7. *If* $\varphi = \varphi_{A'B'...D'}$ *is a section of* $\mathcal{O}_{(A'B'...D')}[1]'$ *in* (7.1.25), *then*

$$\tilde{d}_\mu \varphi = \nabla_A^{A'} \varphi_{A'B'...D'},$$

and if $\psi = \psi_{AB'C'...D'}$ *is a section of* $\mathcal{O}_{A(B'...D')}[2]'$, *then*

$$\tilde{d}_\mu \psi = \nabla^{AB'} \psi_{AB'C'...D'}.$$

Proof: Just as in the proof of Lemma 7.1.6, we choose local coordinates for $\mathbf{F}^I$. We then use the exact sequence (7.1.15) and the local expression for $\tilde{d}_\mu$ to obtain the following commutative diagram on F^I:

$$
\begin{array}{ccccc}
0 & \longrightarrow & \mathcal{O}(-n-2) & \xrightarrow{\pi_{A'}\cdots\pi_{E'}} & \mathcal{O}_{(A'B'\cdots D'E')}(-1) \\
& & \downarrow{\scriptstyle \pi_{A'}\nabla_A^{A'}} & & \downarrow{\scriptstyle \nabla_A^{A'}} \\
0 & \longrightarrow & \mathcal{O}_A(-n-1)[1]' & \xrightarrow{\pi_{B'}\cdots\pi_{E'}} & \mathcal{O}_{A(B'\cdots D'E')}(-1)[1]'
\end{array}
$$

$$
\begin{array}{ccccc}
& \xrightarrow{\pi^{E'}\epsilon_{F'G'}} & \mathcal{O}_{(A'B'\cdots D')[F'G']} & \longrightarrow & 0 \\
& & \downarrow{\scriptstyle \nabla_A^{A'}} & & \\
& \xrightarrow{\pi^{E'}\epsilon_{F'G'}} & \mathcal{O}_{(A'B'\cdots D')[F'G']}[1]' & \longrightarrow & 0.
\end{array}
$$

$$(7.1.26)$$

If we represent the diagram (7.1.26) symbolically as

$$
\begin{array}{ccccccccc}
0 & \longrightarrow & \mathcal{A} & \longrightarrow & \mathcal{B} & \longrightarrow & \mathcal{C} & \longrightarrow & 0 \\
& & \downarrow{\scriptstyle d_1} & & \downarrow{\scriptstyle d_2} & & \downarrow{\scriptstyle d_3} & & \\
0 & \longrightarrow & \mathcal{R} & \longrightarrow & \mathcal{S} & \longrightarrow & \mathcal{T} & \longrightarrow & 0
\end{array}
\quad ,
$$

then the direct image exact sequence becomes

$$
\begin{array}{ccccccccc}
\longrightarrow & \nu_*^0(\mathcal{B}) & \longrightarrow & \nu_*^0(\mathcal{C}) & \longrightarrow & \nu_*^1(\mathcal{A}) & \longrightarrow & \nu_*^1(\mathcal{B}) & \longrightarrow \\
& \downarrow{\scriptstyle \nu_*^0(d_2)} & & \downarrow{\scriptstyle \nu_*^0(d_3)} & & \downarrow{\scriptstyle \nu_*^1(d_1)} & & \downarrow{\scriptstyle \nu_*^1(d_2)} & \\
\longrightarrow & \nu_*^0(\mathcal{S}) & \longrightarrow & \nu_*^0(\mathcal{T}) & \longrightarrow & \nu_*^1(\mathcal{R}) & \longrightarrow & \nu_*^1(\mathcal{S}) & \longrightarrow
\end{array}
\quad ,
$$

where $\nu_*^q(d_j)$ are the induced mappings. In (7.1.26) we see that we have $\nu^0(\mathcal{B}) = \nu_*^1(\mathcal{B}) = \nu_*^0(\mathcal{S}) = \nu_*^1(\mathcal{S}) = 0$ by the vanishing part of Theorem 7.1.5. Thus it follows that $\nu_*^0(\mathcal{C}) \cong \nu_*^1(\mathcal{A}), \nu_*^0(\mathcal{T}) \cong \nu_*^1(\mathcal{R})$, with corresponding mappings of induced operators. Moreover $\nu_*^0(\mathcal{C})$ and $\nu_*^0(\mathcal{T})$ can be calculated by Theorem 7.1.5, and we obtain (reversing the order) explicitly:

$$\nu_*^1 \mathcal{O}(-n-2) \quad \cong \quad \nu_*^0 \mathcal{O}_{(A'B'...D')[E'F']}$$

$$\downarrow \nu_*^1(\pi_{A'} \nabla_A^{A'}) \qquad\qquad \downarrow \nu_*^0(\nabla_A^{A'})$$

$$\nu_*^1 \mathcal{O}_A(-n-1)[1]' \quad \cong \quad \nu_*^0 \mathcal{O}_{A(B'...D')[E'F']}[1]'$$

$$= \qquad \mathcal{O}_{(A'B'...D')[E'F']}$$

$$\downarrow \nabla_A^{A'}$$

$$= \qquad \mathcal{O}_{A(B'...D')[E'F']}[1]'$$

$$= \qquad \mathcal{O}_{(A'B'...D')}[1]'$$

$$\downarrow \nabla_A^{A'}$$

$$= \qquad \mathcal{O}_{A(B'...D')}[2]'.$$

Thus we have completed the proof of the lemma.   ∎

Exercises for Section 7.1

7.1.1. Show that $\mu^{-1}(\mathcal{F})$ and $\mathcal{O}_F$ are well-defined $\mathcal{O}_P$ modules and that

$$\mu^* \mathcal{F} := \mu^{-1}(\mathcal{F}) \otimes_{\mathcal{O}_P} \mathcal{O}_F$$

is a well-defined sheaf of $\mathcal{O}_F$-modules on F.

7.1.2. If $X \xrightarrow{f} Y$ is a differentiable mapping, and $E \to Y$ is a differentiable vector bundle, then we define the *pullback vector bundle* $f^*E \to X$ by defining the fibers

$$(f^*E)_x = E_{f(x)}$$

and, by using the transition functions

$$f^* g_{\alpha\beta}$$

on the open covering $\{f^{-1}(U_\alpha)\}$ of X, where $\{g_{\alpha\beta}\}$ is a set of transition functions for the vector bundle E with respect to the covering $\{U_\alpha\}$ of Y. Show that this data defines uniquely a differentiable vector bundle f^*E over X.

7.1.3. Show that, if E is a holomorphic vector bundle on $\mathbf{P}$, then

$$\mu^* \mathcal{O}_{\mathbf{P}}(E) = \mathcal{O}_{\mathbf{F}}(\mu^* E),$$

where $\mu^* E$ is the pullback of the vector bundle E under the mapping μ.

7.1.4. Show that if $H_{\mathbf{P}}^n$ is the hyperplane section bundle on $\mathbf{P}$, then

$$(\mu^* H_{\mathbf{P}}^n)|_{\mu^{-1}(x)} \cong H_{P_1}^n,$$

recalling that $\mu^{-1}(x) \cong \mathbf{P}_1$.

7.1.5. Show that, for $x \in \mathbf{M}$,

$$[\nu_*^1 \mu^* \mathcal{O}(-n-2)]_x \cong \mathcal{O}_{\mathbf{M},x} \otimes_C \odot^n \mathbf{C}^2.$$

7.1.6. Show that (7.1.2.) is exact at Ω_μ^1 and Ω_μ^2.

7.1.7. Let $X \xrightarrow{f} Y$ be a proper continuous mapping, and let

$$0 \to \mathcal{A} \to \mathcal{B} \to \mathcal{C} \to 0$$

be an exact sequence of sheaves over X, then show that

$$0 \to f_*^0 \to f_*^0 \mathcal{B} \to f_*^0 \mathcal{C} \to f_*^1 \mathcal{A} \to f_*^1 \mathcal{B} \to f_*^1 \mathcal{C} \to f_*^2 \mathcal{A} \to \cdots$$

is an exact sequence of sheaves on Y.

7.2   The linear Penrose transform

In this section we want to pull back cohomology from $\mathbf{P}$ to $\mathbf{F}$ and then push it down to $\mathbf{M}$. The composition will then be the Penrose transform mapping cohomology groups on $\mathbf{P}$ to solutions of the massless field equations on $\mathbf{M}$. First we shall study the pullback of cohomology from $\mathbf{P}$ to $\mathbf{F}$. This is essentially a topological problem, as it turns out. Let W' be an open set in $\mathbf{F}$ and let $W = \mu(W')$, which is an open subset of $\mathbf{P}$. We want to determine conditions on W and W' so that data can be transformed from W to W' without loss of information. If $\mathcal{S}$ is any sheaf on W, then the topological pullback sheaf $\mu^{-1}\mathcal{S}$ is a well-defined sheaf on W' (see §7.1). We have an isomorphism

$(\mu^{-1}\mathcal{S})_q \cong \mathcal{S}_{\mu(q)}$, for all $q \in W'$, and this gives rise to a natural mapping

$$\mu^* : \Gamma(W, \mathcal{S}) \to \Gamma(W', \mu^{-1}\mathcal{S})$$

defined by $\mu^* f := f \circ \mu$. We have identified $(\mu^{-1}\mathcal{S})_q$ with $\mathcal{S}_{\mu(q)}$ in this definition. We want to extend this pullback of global sections of $\mathcal{S}$ to pullback of cohomology with coefficients in $\mathcal{S}$. We can use either Čech cohomology or suitable resolutions of $\mathcal{S}$ to effect the pullback. For a Čech covering $\mathfrak{U}$ of W, simply pull back the cover $\mathfrak{U}$ to a covering $\mathfrak{U}'$ of W' and pull back the sections on appropriate intersections to pull back representatives of cocyles of a given degree. We want to use specific resolutions, however, and we shall develop that point of view.

We are interested in pulling back $H^q(W, \mathcal{O}(E))$ for some holomorphic vector bundle E. Dolbeault's theorem (Theorem 3.3.4) allows us to compute cohomology in terms of the fine resolution

$$0 \to \mathcal{O}(E) \to \mathcal{E}^{0,0}(E) \xrightarrow{\bar{\partial}} \mathcal{E}^{0,1}(E) \xrightarrow{\bar{\partial}} \cdots \qquad (7.2.1)$$

on W, i.e.,

$$H^q(W, \mathcal{O}(E)) \cong H^q(\Gamma(W, \mathcal{E}^{0,*}(E))),$$

(see (2.2.29). We want to use this resolution to effect the pullback of cohomology. Suppose we look at this same situation a little more abstractly. Let

$$0 \to \mathcal{S} \to \mathcal{R}^0 \to \mathcal{R}^1 \to \cdots$$

be a resolution of a sheaf $\mathcal{S}$ on W. We denote this as

$$0 \to \mathcal{S} \to \mathcal{R}^\bullet,$$

where we let the dot '$\bullet$' denote the index of the complex of sheaves (to avoid overusing the symbol '$*$' which we shall use for pullback). Now we suppose that we can compute the cohomology of $\mathcal{S}$ in terms of the resolution, i.e.,

$$H^p(W, \mathcal{S}) \cong H^p(\Gamma(W, \mathcal{R}^\bullet)), \qquad (7.2.2)$$

as in the example of the Dolbeault representation of $H^p(W, \mathcal{O}(E))$ given above. That is, as in the above example, $\mathcal{R}^\bullet$ is a complex of flabby, soft, or fine sheaves or more generally acyclic sheaves; the Dolbeault resolution is, for instance fine, and therefore acyclic (see

Theorem 3.3.5 and Example 3.6.5). Note that resolutions of $\mathcal{S}$ with the property (7.2.2) always exist. We can pullback the resolution $\mathcal{R}^\bullet$ obtaining on W'

$$0 \to \mu^{-1}\mathcal{S} \to \mu^{-1}\mathcal{R}^\bullet. \tag{7.2.3}$$

The differential sheaf (7.2.3) will be a resolution of $\mu^{-1}\mathcal{S}$, since it is still exact at the stalk level, but it will not necessarily calculate cohomology. For instance, if $\mathcal{R}^\bullet$ is fine, then $\mu^{-1}\mathcal{R}^\bullet$ certainly will not be, since the sections of $\mu^{-1}\mathcal{R}^\bullet$ will be constant along the fibers, and this would not be preserved by multiplication by a cutoff function along the fiber direction. Even though the resolution (7.2.3) may not compute the cohomology groups $H^p(W, \mu^{-1}\mathcal{S})$, there is a relation given by the spectral sequence of the resolution (7.2.3) (Example 3.6.5). Namely, there is a canonical homomorphism

$$H^p(\Gamma(W', \mu^{-1}\mathcal{R}^\bullet)) \to H^p(W', \mu^{-1}\mathcal{S}), \tag{7.2.4}$$

which is given by Exercises 3.6.1, 3.6.2. This is a generalization of the abstract de Rham theorem. Namely, if (7.2.3) is acyclic, then (7.2.4) is an isomorphism, and this is simply the abstract de Rham theorem (3.6.18). Suppose now, in addition, that

$$\mu^* : \mathcal{R}^\bullet \to \mu^{-1}\mathcal{R}^\bullet \text{ is a homomorphism of complexes}, \tag{7.2.5}$$

that is to say, the diagrams,

$$
\begin{array}{ccc}
\mu^{-1}\mathcal{R}^p & \longrightarrow & \mu^{-1}\mathcal{R}^{p+1} \\
\big\uparrow{\mu^*} & & \big\uparrow{\mu^*} \\
\mathcal{R}^p & \longrightarrow & \mathcal{R}^{p+1}
\end{array}
$$

are commutative. It follows from (7.2.5) that there is an homomorphism of complexes

$$\Gamma(W, \mathcal{R}^\bullet) \xrightarrow{\mu^*} \Gamma(W', \mu^{-1}\mathcal{R}^\bullet),$$

and hence a homomorphism of the associated cohomology

$$H^p(\Gamma(W, \mathcal{R}^\bullet)) \to H^p(\Gamma(W', \mu^{-1}\mathcal{R}^\bullet)). \tag{7.2.6}$$

Now using (7.2.2), (7.2.4), and (7.2.6) we see that we then have an induced canonical mapping

$$H^p(W, \mathcal{S}) \xrightarrow{\mu^*} H^p(W', \mu^{-1}\mathcal{S}). \tag{7.2.7}$$

Now that we have the desired pullback mapping (7.2.7) of cohomology, we want to investigate its behavior. If $\mu : W' \to W$ has connected fibers, then it is clear that

$$H^0(W, \mathcal{S}) = H^0(W', \mu^{-1}\mathcal{S}).$$

We want to give similar but higher order topological conditions on the fibers of μ so that (7.2.7) is an isomorphism for $p \geq 1$. Our major interest for our applications will be the case $p = 1$. Let us say that the mapping $\mu : W' \to W$ is *elementary* if the fibers of this mapping $\mu^{-1}(p) = Y_p \subset W'$ are all connected, and, moreover have vanishing first Betti number, i.e., $H^1(Y_p, \mathbf{C}) = 0$, for all $p \in W$.

Remark: Suppose that $U \subset \mathbf{M}, U' = \nu^{-1}(U), \widehat{U} = \mu \circ \nu^{-1}(U)$, as in Chapter 1, then each fiber Y_p of the mapping $\mu : U' \to \widehat{U}$, for $p \in \widehat{U}$ is biholomorphic to $\tilde{p} \cap U$. Suppose, for instance that U is convex in $\mathbf{M}^I$, then $\tilde{p} \cap \mathbf{M}^I$ is a complex two-plane in $\mathbf{M}^I \cong \mathbf{C}^4$, and $\tilde{p} \cap U$ is therefore convex. Hence $\mu : U' \to \widehat{U}$ for this U will be an elementary mapping. This shows that the condition of being elementary is not too difficult to check in various cases.

We now have the following important lemma. If V is a smooth vector bundle over $W \subset \mathbf{P}$, then let $\mathcal{E}(V)$ be the sheaf of smooth sections of V.

Lemma 7.2.1. *Suppose that $\mu : W' \to W$ is elementary, and that V is a smooth vector bundle over W, then*

$$H^1(W', \mu^{-1}\mathcal{E}(V)) = 0.$$

Proof: We define a fine resolution of $\mu^{-1}\mathcal{E}(V)$ by the sheaves of smooth relative p-forms $\mathcal{E}_\mu^p(V)$ on W' in analogy with (7.1.3):

$$0 \longrightarrow \mu^{-1}\mathcal{E}(V) \longrightarrow \mathcal{E}_\mu^0(V) \xrightarrow{d_\mu} \mathcal{E}_\mu^1(V) \xrightarrow{d_\mu} \mathcal{E}_\mu^2(V) \longrightarrow \cdots.$$
$$(7.2.8)$$

This is a fine resolution, so by the abstract de Rham theorem (3.6.18)

$$H^p(W', \mu^{-1}\mathcal{E}(V)) \cong H^p(\Gamma(W', \mathcal{E}_\mu^\bullet(V)).$$

Thus we need to show that

$$\Gamma(W', \mathcal{E}_\mu^0(V)) \xrightarrow{d_\mu} \Gamma(W', \mathcal{E}_\mu^1(V)) \xrightarrow{d_\mu} \Gamma(W', \mathcal{E}_\mu^2(V)), \quad (7.2.9)$$

is exact. Suppose for each $p \in W$ there is a neighborhood N of p such that the sequence

$$\Gamma(\mu^{-1}(N) \cap W', \mathcal{E}^0_\mu(V)) \xrightarrow{d_\mu} \Gamma(\mu^{-1}(N) \cap W', \mathcal{E}^1_\mu(V))$$

$$\xrightarrow{d_\mu} \Gamma(\mu^{-1}(N) \cap W', \mathcal{E}^2_\mu(V)) \quad (7.2.10)$$

is exact. Let $\mathfrak{U} = \{U_\alpha\}$ be an open covering of W with the property that (7.2.10) is exact for each $N \in \mathfrak{U}$. Then let $\omega \in \Gamma(W', \mathcal{E}^1_\mu(V))$, such that $d_\mu \omega = 0$, and let $f_\alpha \in \Gamma(\mu^{-1}(U_\alpha), \mathcal{E}^1_\mu(V))$ satisfy $d_\mu f_\alpha = \omega$, for all α. Now let $\{\varphi_\alpha\}$ be a partition of unity with respect to the covering $\mathfrak{U}$, then $\mu^* \varphi_\alpha$ is a partition of unity with respect to the covering $\mathfrak{U}' = \{\mu^{-1} U_\alpha\}$ of W'. Define

$$f = \sum_\alpha (\mu^* \varphi_\alpha) f_\alpha.$$

We see that

$$d_\mu f = \sum_\alpha ((\mu^* \varphi_\alpha) d_\mu f_\alpha + d_\mu (\mu^* \varphi_\alpha) f_\alpha)$$

$$= \sum_\alpha (\mu^* \varphi_\alpha \omega + 0)$$

$$= \omega,$$

where $d_\mu(\mu^* \varphi_\alpha) = 0$ since $\mu^* \varphi_\alpha$ is constant along the fibers of μ. Thus (7.2.9) is exact if (7.2.10) is exact for neighborhoods of arbitrary points of **M**.

Now we shall check that (7.2.10) is exact for suitable neighborhoods of a given point p. Choose such a neighborhood N so that $\mu^{-1}(N) \cong N \times \mathbf{P}_2$, i.e., the fibration μ is trivial over N, and also such that the vector bundle V is trivial over N. Now let $N' := \mu^{-1}(N) \cap W'$, then we see that

$$H^1(N', \mu^{-1}\mathcal{E}(V)) \cong H^1(N', \mu^{-1}\mathcal{E})^r,$$

where r is the rank of V. Thus we need to show that (using (7.2.8) in the case where V is the trivial line bundle)

$$\Gamma(N', \mathcal{E}^0_\mu) \xrightarrow{d_\mu} \Gamma(N', \mathcal{E}^1_\mu) \xrightarrow{d_\mu} \Gamma(N', \mathcal{E}^2_\mu) \quad (7.2.11)$$

is exact. Choose $\omega \in \Gamma(N', \mathcal{E}^1_\mu)$ such that $d_\mu \omega = 0$. Then letting $q \in N$, we can consider ω as a smooth family of one-forms on open subsets of $\mathbf{P}_2$ with coefficients depending smoothly on the parameter $q \in N$. We write $\omega(q)$ as the one-form defined on $(\{q\} \times \mathbf{P}_2) \cap W'$. Now define, for $(q, s) \in N \times \mathbf{P}_2$,

$$f(q, s) = \int_{(q, s_0)}^{(q, s)} \omega(q),$$

where the integral is along any path in $(\{q\} \times \mathbf{P}_2) \cap W'$, joining (q, s_0) to (q, s). By the assumption that μ is elementary, we see that the integral is independent of the path, and hence we obtain a solution of the equation

$$d_\mu f = \omega$$

on N', and (7.2.11) is exact, as desired.   ∎

We now have the following fundamental result.

Theorem 7.2.2. *Suppose $W' \subset \mathsf{F}$ is an open set, $W = \mu(W')$, and E is a holomorphic vector bundle on W. If $\mu : W' \to W$ is elementary, then*

$$\mu^* : H^1(W, \mathcal{O}(E)) \to H^1(W', \mu^{-1}\mathcal{O}(E))$$

is a canonical isomorphism.

Proof: The spectral sequence of the resolution

$$0 \to \mu^{-1}\mathcal{O}(E) \to \mu^{-1}\mathcal{E}^{0,\bullet}(E)$$

has the form (see (3.6.15) in Example 3.6.5)

$$E_2^{pq} = H^p(H^q(W', \mu^{-1}\mathcal{E}^{0,\bullet})) \Rightarrow H^r(W', \mu^{-1}\mathcal{O}(E)).$$

Since $H^1(W', \mu^{-1}\mathcal{E}^{0,\bullet}) = 0$, by Lemma 7.2.1, the E_2-term of the spectral sequence has the form, letting $\mathcal{A}^\bullet = \mu^{-1}\mathcal{E}^{0,\bullet}(E)$,

$$\begin{array}{ccc} * & * & * \\ 0 & 0 & 0 \\ H^0(\Gamma(W', \mathcal{A}^\bullet)) & H^1(\Gamma(W', \mathcal{A}^\bullet)) & H^2(\Gamma(W', \mathcal{A}^\bullet)). \end{array}$$

It follows that $E_2^{10} = H^1(\Gamma(W', \mathcal{A}^\bullet))$, $E_\infty^{01} = 0$, and hence that

$$H^1(W', \mu^{-1}\mathcal{O}(E)) \cong H^1(\Gamma(W', \mu^{-1}\mathcal{E}^{0,\bullet})),$$

i.e., that (7.2.4) is an isomorphism. Now $\mu : W' \to W$ has connected fibers implies that

$$\Gamma(W', \mu^{-1}\mathcal{E}^{0,q}(E)) = \Gamma(W, \mathcal{E}^{0,q}(E)).$$

Hence, μ^* is the composition of three isomorphisms

$$H^1(W, \mathcal{O}(E)) \xrightarrow{\cong} H^1(\Gamma(W, \mathcal{E}^{0,\bullet}(E)))$$
$$\xrightarrow{\cong} H^1(\Gamma(W', \mu^{-1}\mathcal{E}^{0,\bullet}(E))) \xrightarrow{\cong} H^1(W', \mu^{-1}\mathcal{O}(E)),$$

and the theorem follows. ∎

Remark: Buchdahl (1985) has a generalization of this theorem for higher degree cohomology.

Now let X be any open subset of $\mathbb{F}$, then there is a spectral sequence associated to the resolution (7.1.3) of the form

$$E_1^{pq} = H^q(X, \Omega^p_\mu(E)) \Rightarrow H^r(X, \mu^{-1}\mathcal{O}(E)). \tag{7.2.12}$$

This follows from (3.6.15) and will be a principal tool in the Penrose transform, as we shall see below. We have now pulled back $H^1(W, \mathcal{O}(E))$ to $H^1(W', \mu^{-1}\mathcal{O}(E))$ (isomorphically, if $\mu : W' \to W$ is elementary). We now want to push down $H^1(W', \mu^{-1}\mathcal{O}(E))$ to M. We shall do this for different special choices of W' and the vector bundle E. Our first result relates to right-handed massless fields on M. Let U be an open subset of M. We define

$$\mathcal{Z}'_n := \ker \nabla^{A'}_A : \mathcal{O}_{(A'B'...D')}[1]' \to \mathcal{O}_{A(B'...D')}[2]',$$

where the differential operator $\nabla^{A'}_A$ maps conformally weighted spinor fields to conformally weighted spinor fields as described in §7.1. We call $\mathcal{Z}'_n$ the *sheaf of holomorphic right-handed massless free fields of helicity $n/2$*. If we consider the sections of $\mathcal{Z}'_n$ on an open subset of $\mathbb{M}^I$ we have the right-handed massless fields described in §4.2, i.e., symmetric spinor fields which are solutions of

$$\nabla^{AA'}\varphi_{A'B'...D'} = 0.$$

Now for any open set $U \subset \mathbb{M}$ we let $U' := \mu^{-1}(U)$, $\widehat{U} := \mu \circ \nu^{-1}(U)$, as before. We then have the following basic result.

Theorem 7.2.3. *Let U be open in $\mathbf{M}$, and suppose $n \geq 1$, then there is a canonical linear transformation*

$$\mathcal{P} : H^1(\widehat{U}, \mathcal{O}(-n-2)) \to \Gamma(U, \mathcal{Z}'_n).$$

If $\mu : U' \to \widehat{U}$ has connected fibers, then $\mathcal{P}$ is injective, and if $M : U' \to \widehat{U}$ is elementary, then $\mathcal{P}$ is an isomorphism.

Proof: We already know from Theorem 7.2.2 that

$$\mu^* : H^1(\widehat{U}, \mathcal{O}(-n-2)) \to H^1(U', \mu^{-1}\mathcal{O}(-n-2))$$

exists, and is injective or an isomorphism according to the topological hypotheses of the theorem. We shall show that there is a natural isomorphism

$$H^1(U', \mu^{-1}\mathcal{O}(-n-2)) \cong \Gamma(U, \mathcal{Z}'_n),$$

and the theorem will be proved. We shall use the spectral sequence (7.2.12) to compute $H^1(U', \mu^{-1}\mathcal{O}(-n-2))$ in terms of the cohomology groups $H^p(U', \Omega^s_\mu(-n-2))$. Then we shall use the Leray spectral sequence of the fibration $\mu : U' \to U$,

$$\widetilde{E}_2^{pq} = H^p(U, \mu^q_*\Omega^s_\mu(-n-2)) \Rightarrow H^r(U', \Omega^s_\mu(-n-2)) \quad (7.2.13)$$

to relate the cohomology groups $H^r(U', \Omega^s_\mu(-n-2))$ to cohomology groups on U. Let us handle the Leray spectral sequence first, since it is somewhat simpler.

For a fixed s we have from Theorem 7.1.5

$$\nu^q_*\Omega^s_\mu(-n-2) = 0, \quad q \neq 1,$$

and $\nu^1_*\Omega^s_\mu(-n-2)$ are particular nontrivial spinor sheaves which are given specifically in Theorem 7.1.5. Thus the spectral sequence $\widetilde{E}_2^{pq}$ is degenerate at the second level, i.e., we have $\widetilde{E}_2^{pq} = 0$, for $q \neq 1$. Hence $\widetilde{E}_\infty^{pq} = \widetilde{E}_2^{pq}$, and it follows from Exercise 3.6.4 that

$$H^r(U', \Omega^s_\mu(-n-2)) \cong H^{r-1}(U, \nu^1_*\Omega^s_\mu(-n-2)), \quad (7.2.14)$$

for $r \geq 1$, and $H^0(U', \Omega^s_\mu(-n-2)) = 0$. This is a version of integration over the fibers of the mapping ν.

Now returning to the spectral sequence (7.2.12) we can write down the E_1-term as follows (letting $m = -n - 2$, for simplicity)

$$\vdots \qquad\qquad \vdots \qquad\qquad \vdots$$

$$H^2(U', \Omega_\mu^0(m)) \xrightarrow{d_\mu} H^2(U', \Omega_\mu^1(m)) \xrightarrow{d_\mu} H^2(U', \Omega_\mu^2(m)) \longrightarrow$$

$$H^1(U', \Omega_\mu^0(m)) \xrightarrow{d_\mu} H^1(U', \Omega_\mu^1(m)) \xrightarrow{d_\mu} H^1(U', \Omega_\mu^2(m)) \longrightarrow$$

$$H^0(U', \Omega_\mu^0(m)) \xrightarrow{d_\mu} H^0(U', \Omega_\mu^1(m)) \xrightarrow{d_\mu} H^0(U', \Omega_\mu^2(m)) \longrightarrow$$

Replacing the terms in this array by the corresponding groups on U under the isomorphism (7.2.13), the mappings by the induced mappings on the direct image sheaves (7.1.17), and we see that the E_1-term looks like

$$\vdots \qquad\qquad \vdots \qquad\qquad \vdots$$

$$H^1(U, \nu_*^1 \Omega_\mu^0(m)) \xrightarrow{\tilde{d}_\mu} H^1(U, \nu_*^1 \Omega_\mu^1(m)) \xrightarrow{\tilde{d}_\mu} H^1(U, \nu_*^1 \Omega_\mu^2(m))$$

$$H^0(U, \nu_*^1 \Omega_\mu^0(m)) \xrightarrow{\tilde{d}_\mu} H^0(U, \nu_*^1 \Omega_\mu^1(m)) \xrightarrow{\tilde{d}_\mu} H^0(U, \nu_*^1 \Omega_\mu^2(m))$$

$$0 \qquad\qquad\qquad 0 \qquad\qquad\qquad 0$$

Now we see that

$$E_2^{01} = \ker\{H^0(U, \nu_*^1 \Omega_\mu^0(m)) \xrightarrow{\tilde{d}_\mu} H^0(U, \nu_*^1 \Omega_\mu^1(m))\},$$

and $E_1^{10} = E_2^{10} = E_\infty^{10} = 0$. Moreover, we see from the above diagram that $E_2^{01} = E_\infty^{01}$. Thus, using Theorem 3.6.2, we see that

$$H^1(U', \mu^{-1}\mathcal{O}(m)) \cong E_2^{01}.$$

Finally we can use Theorem 7.1.5 and Lemma 7.1.7 to see that we obtain

$$H^1(U', \mu^{-1}\mathcal{O}(m)) \cong \ker \Gamma(U, \mathcal{O}_{(A'B'...D')}[1]')$$

$$\xrightarrow{\nabla_A^{A'}} \Gamma(U, \mathcal{O}_{A(B'...D')}[2]') \cong \Gamma(U, \mathcal{Z}_n').$$

∎

If we let $U = \mathbf{M}^\pm$ or $\mathbf{M}^I$, then we see that the topological conditions in Theorem 7.2.3 are satisfied. Thus we have the following corollary.

Corollary 7.2.4. *For $n \geq 1$,*

$$\text{(a)} \quad \mathcal{P} : H^1(\mathbf{P}^\pm, \mathcal{O}(-n-2)) \to \Gamma(\mathbf{M}^\pm, \mathcal{Z}'_n),$$
$$\text{(b)} \quad \mathcal{P} : H^1(\mathbf{P}^I, \mathcal{O}(-n-2)) \to \Gamma(\mathbf{M}^I, \mathcal{Z}'_n),$$

are canonical isomorphisms.

Remark: Theorem 7.2.3 describes every right-handed holomorphic massless field locally as each point of $\mathbf{M}$ has convex neighborhoods which will satisfy the topological hypotheses of the theorem.

We now turn our attention to the solution of the wave equation, the helicity zero case. This is less straightforward then the positive helicity case treated above, as it involves second order differential equations. Recall that the wave equation has the form on $\mathbf{M}^I$

$$\Box\varphi = 0, \tag{7.2.15}$$

where $\Box := \nabla^a\nabla_a = \nabla^{AA'}\nabla_{AA'}$, and φ is a scalar function. This is the helicity zero massless field equation (see §4.2). We want to extend this to act on conformally weighted scalar fields (or functions) on all of $\mathbf{M}$. We note that

$$\mathcal{O} \xrightarrow{\nabla_{AA'}} \mathcal{O}_{AA'},$$

$$\mathcal{O} \xrightarrow{\nabla^{AA'}} \mathcal{O}^{AA'}[1][1]',$$

since $\nabla^{AA'} = \varepsilon^{AB}\varepsilon^{A'B'}\nabla_{BB'}$, and from this we see that $\Box = \nabla^{AA'}\nabla_{AA'}$ is a well-defined mapping

$$\Box : \mathcal{O} \to \mathcal{O}[1][1]'.$$

We can now tensor this with $\mathcal{O}[1]'$ and obtain the mapping $\Box \otimes \text{id}$ which we still denote by $\Box$ and we have

$$\Box : \mathcal{O}[1]' \to \mathcal{O}[1][2]'. \tag{7.2.16}$$

We see that (7.2.16) is a well-defined differential operator on sections of conformally weighted bundles, and the solutions of this equation $\Box\varphi = 0$ on open sets of $\mathbf{M}^I$ can be identified with the solutions of (7.2.15) on that same set. The mapping (7.2.16) is conformally invariant. There are no choices of coordinates in its definition. It depends only on exterior differentiation ($\nabla_{AA'}$) and the global εs, all of which are conformally invariant operators.

We now use (7.2.16) to define an appropriate sheaf of solutions of the global wave equation. Namely, we define

$$\mathcal{Z}_0' := \ker\{\square : \mathcal{O}[1]' \to \mathcal{O}[1][2]'\}$$

and we call $\mathcal{Z}_0'$ the sheaf of *massless fields of helicity zero on* **M**. The sections of $\mathcal{Z}_0'$ on open subsets U of $\mathbf{M}^I$ will then be solutions of the wave equation (7.2.15) on U. We now have the following result.

Theorem 7.2.5. *Let U be open in* **M**, *then there is a canonical linear transformation*

$$\mathcal{P} : H^1(\widehat{U}, \mathcal{O}(-2)) \to \Gamma(U, \mathcal{Z}_0').$$

If $M : U' \to \widehat{U}$ has connected fibers, then $\mathcal{P}$ is injective, and if $\mu : U' \to \widehat{U}$ is elementary, then $\mathcal{P}$ is an isomorphism.

Proof: The proof is identical to the proof of Theorem 7.2.3 except that we have to treat the spectral sequence (7.2.12) for the case $n = 0$ somewhat differently, as it does not degenerate to first order as it did in that case. The E_1-term of (7.2.12) for $n = 0$ has the form

$$
\begin{array}{ccccccc}
\vdots & & \vdots & & \vdots & & \\
H^2(U', \Omega_\mu^0(-2)) & \to & H^2(U', \Omega_\mu^1(-2)) & \to & H^2(U', \Omega_\mu^2(-2)) & \to \\
H^1(U', \Omega_\mu^0(-2)) & \to & H^1(U', \Omega_\mu^1(-2)) & \to & H^1(U', \Omega_\mu^2(-2)) & \to \\
H^0(U', \Omega_\mu^0(-2)) & \to & H^0(U', \Omega_\mu^1(-2)) & \to & H^0(U', \Omega_\mu^2(-2)) & \to
\end{array}
$$
$$(7.2.17)$$

We need to calculate each column of (7.2.17) separately in terms of cohomology on U. For $E_1^{0q} = H^q(U', \Omega_\mu^0(-2))$, we have the only nonvanishing direct image of $\Omega_\mu^0(-2)$ is $\nu_*^1 \Omega_\mu^0(-2) \cong \mathcal{O}[1]'$, from Theorem 7.1.5. Therefore we see from the Leray spectral sequence (7.2.13) that

$$H^q(U', \Omega_\mu^0(-2)) \cong H^{q-1}(U, \mathcal{O}[1]'), \quad \text{for } q \geq 1,$$
$$H^0(U', \Omega_\mu^0(-2)) = 0.$$

For the second column of (7.2.17) Theorem 7.1.5 implies that

$$\nu_*^r \Omega_\mu^1(-2) = 0, \quad r \geq 0,$$

and hence from (7.2.13) we find that

$$H^q(U', \Omega_\mu^1(-2)) = 0, \quad q \geq 0.$$

For the third column, we see, again by Theorem 7.1.5, that the only nonvanishing direct image of $\Omega_\mu^2(-2)$ is given by

$$\nu_*^0 \Omega_\mu^2(-2) \cong H^q(U, \mathcal{O}[1][2]'), \quad q \geq 0.$$

Now there are only two nonzero columns in (7.2.17) and they are separated by a zero column. It follows that d_1 vanishes identically on E_1^{pq}, and hence that $E_2^{pq} = E_1^{pq}$. We define the mapping D to be the mapping induced on E_2^{01} by the spectral sequence mapping d_2, i.e.,

$$D := d_2 : E_2^{01} \rightarrow E_2^{20}.$$

Taking into account our calculations above, we see that the following array represents the E_2-term of the spectral sequence and the mapping D has been singled out from the family of mappings which constitute d_2.

$$
\begin{array}{ccc}
H^1(U, \mathcal{O}[1]') & 0 & H^2(U, \mathcal{O}[1][2]'), \\
H^0(U, \mathcal{O}[1]') \diagdown & 0 \underset{D}{} & H^1(U, \mathcal{O}[1][2]\mathcal{O}), \\
0 & 0 \diagup & H^0(U, \mathcal{O}[1][2]').
\end{array}
\qquad (7.2.18)
$$

Thus we see that

$$D : H^0(U, \mathcal{O}[1]') \rightarrow H^0(U, \mathcal{O}[1][2]').$$

One can see that D is the composition of first order differential operators, and hence is a second order differential operator. It is also conformally invariant as all of the operators in the spectral sequence are conformally invariant. We shall identify it with the wave operator shortly. First we note that

$$E_3^{01} = \ker\{D : H^0(U, \mathcal{O}[1]') \rightarrow H^0(U, \mathcal{O}[1][2]')\}. \qquad (7.2.19)$$

Moreover, we find that $E_3^{01} = E_\infty^{01}$, and $E_1^{10} = E_\infty^{10} = 0$, so by Theorem 3.6.2,

$$H^1(U', \mu^{-1}\mathcal{O}(-2)) \cong E_3^{01}. \qquad (7.2.20)$$

Once we have identified D with a constant multiple of $\square$, we see that (7.2.19) and (7.2.20) will prove the theorem.

We now turn to the verification of this last fact. Consider the exact sequence of sheaves on F

$$0 \to \mathcal{O}(-2) \xrightarrow{\pi_{A'}} \mathcal{O}_{A'}(-1) \xrightarrow{\pi^{A'}} \mathcal{O}[1]' \to 0, \qquad (7.2.21)$$

which is equivalent to (7.1.14). This is a complex of sheaves, and we can map it linearly to a second complex in the following manner:

$$
\begin{array}{ccccc}
\mathcal{O}(-2) & \xrightarrow{\pi_{A'}} & \mathcal{O}_{A'}(-1) & \xrightarrow{\pi^{A'}} & \mathcal{O}[1]' \\
\downarrow & & \downarrow{\scriptstyle \nabla^A_{a'}} & & \downarrow{\scriptstyle \frac{1}{2}\Box} \\
\mathcal{O}(-2) & \xrightarrow{\pi_{A'}\nabla^{A'}_A} & \mathcal{O}_A(-1)[1]' & \xrightarrow{\pi_{A'}\nabla^{AA'}} & \mathcal{O}[1][2]'
\end{array}
\qquad (7.2.22)
$$

where the second sequence is the same as the relative de Rham sequence

$$\Omega^0_\mu(-2) \xrightarrow{\mu} \Omega^1_\mu(-2) \xrightarrow{d_\mu} \Omega^2_\mu(-2)$$

(see Lemma 7.1.3). Each of the horizontal sequences in (7.2.22) is a differential sheaf and as such each has a spectral sequence associated with it. The spectral sequence of the bottom sequence is the one we have been considering ((7.2.12) for $n = 0$). The vertical mapping induces mappings on the spectral sequences, and, in particular, at the E_2 level. Let $\widetilde{E}^{pq}_r$ be the spectral sequence (3.6.15) of the top horizontal complex of sheaves. Noting that $H^0(U', \mathcal{O}[1]') \cong H^0(U, \mathcal{O}[1]')$, and computing $\widetilde{E}^{pq}_r$ in terms of cohomology on U, we can express the mapping between the two spectral sequences at the E_2-level for the relevant terms in the following manner:

$$
\begin{array}{c}
\widetilde{E}_2 \\
\downarrow \\
E_2
\end{array}
\quad
\begin{array}{ccc}
H^0(U, \mathcal{O}[1]') \diagdown & 0 & \\
0 \downarrow{\scriptstyle \beta} & & 0 \xrightarrow{\ \alpha\ } H^0(U, \mathcal{O}[1]') \\
H^0(U, \mathcal{O}[1]') \diagdown & 0 & \downarrow{\scriptstyle \frac{1}{2}\Box} \\
0 & & 0 \xrightarrow[\ D\ ]{} H^0(U, \mathcal{O}[1][2]')
\end{array}
$$

We see that since (7.2.21) is exact, this particular differential sheaf is a resolution of 0, and hence $\widetilde{E}^{01}_\infty = 0$, which implies that $\widetilde{E}^{01}_3 = 0$, and hence that α is an isomorphism. In fact, α must be the identity mapping, since it is a canonical isomorphism. We see that β is also the identity, and it follows from the commutativity of the diagram that $D = \frac{1}{2}\Box$. ∎

We have now represented all massless fields of nonnegative helicity in terms of cohomological data on $\mathbf{P}$. We now want to consider the case of negative helicity. One approach to studying negative helicity is to use the same methodology above to represent left-handed fields in terms of cohomology on dual projective twistor space $\mathbf{P}^* = \mathbf{P}(\mathbf{T}^*)$. However, it is also possible to define the Penrose transform directly on $\mathbf{P}$. Let, for $n \geq 1$,

$$\mathcal{Z}'_{-n} := \ker\{\nabla^{DD'} : \mathcal{O}_{(AB...CD)}[1] \to \mathcal{O}_{(AB...C)}[2][1]'\}, \quad (7.2.23)$$

be the *sheaf of left-handed massless fields on* $\mathbf{M}$ *of helicity* $-n/2$. What we want to describe is a Penrose transform of the form, for $n > 0$, and for U open in $\mathbf{M}$,

$$\mathcal{P} : H^1(\widehat{U}, \mathcal{O}(n-2)) \to \Gamma(U, \mathcal{Z}'_{-n}). \quad (7.2.24)$$

Let us reconsider the positive helicity case for a moment to illustrate the difficulty in defining (7.2.24). We shall contrast the cases of helicity 2 and -2 for simplicity (self-dual and anti-self-dual Maxwell fields, respectively). If

$$\mathcal{P} : H^1(\widehat{U}, \mathcal{O}(-4)) \to \Gamma(U, \mathcal{Z}'_2),$$

is given by Theorem 7.2.3, then let $\omega \in H^1(\widehat{U}, \mathcal{O}(-4))$, and consider the value of $\mathcal{P}(\omega)(x)$, for a specific point x in U. The fiber of the bundle whose sections are massless fields is given by $S_{(A'B')}[1]'$, $_x$ which, by the results in §7.1, is isomorphic to $H^1(\hat{x}, \mathcal{O}_{\hat{x}}(-4))$ (see the proof of Theorem 7.1.4). Using this isomorphism, we see that $\mathcal{P}(\omega)(x)$ is obtained by taking the cohomology class ω and restricting it to the projective line $\hat{x}$ obtaining an element of the vector space $H^1(\hat{x}, \mathcal{O}_{\hat{x}}(-4))$. The Serre duality argument then gives the more normal spinor representation of this same vector, which is the value of the field $\mathcal{P}(\omega)$ at x. Thus, in summary, restricting the cohomology class to $\hat{x}$ gives the value of the field. If ω is represented in terms of differential forms, then this is the usual restriction of a differential form to a submanifold. Now we consider the same situation for the case of negative helicity. Take an element $\omega \in H^1(\widehat{U}, \mathcal{O})$, and consider the restriction $\omega|_{\hat{x}} \in H^1(\hat{x}, \mathcal{O}_{\hat{x}})$. Then we see that since $H^1(\mathbf{P}_1, \mathcal{O}_{\mathbf{P}_1}) = 0$, then the restriction is necessarily *zero*, and so this cannot be the value of the field. So simple restrictions of cohomology groups will not yield anything in this case. The field at x defined by ω turns out to depend on the restriction of ω to *infinitesimal neighborhoods* of the submanifold $\hat{x} \subset \mathbf{P}$. The original integral formulae of Penrose (see §7.3) for this negative helicity case involve not only integration (that is, the Serre duality part), but also differentiation of the integrand. If we think of the cohomology class ω as represented by a differential form say, then we can expand the differential form in a Taylor series with respect to coordinates normal to the submanifold $\hat{x}$ (at least locally this makes sense in terms of usual power series). The fact that $\omega|_{\hat{x}}$ vanishes says the leading term of this series vanishes. It turns out that the value of the field is given by the first nonvanishing coefficient of this expansion. If we have helicity $-n/2$, then the first nonvanishing coefficient turns out to be at the nth order. This has to all be given meaning in terms of the cohomological data. We shall use ideal sheaves to represent the normal bundle and essentially expand cohomology classes in terms of powers of the conormal bundle. Let us describe this briefly here in a general context, and we shall see a specific parametrized version of this in the proof of the next theorem.

Suppose that Y is a complex submanifold of a complex manifold X. Let $\mathcal{I}$ be the ideal sheaf of the submanifold Y, and consider the exact sequence

$$0 \to \mathcal{I} \to \mathcal{O} \to \mathcal{O}/\mathcal{I} \to 0.$$

We shall see below that the quotient sheaf $\mathcal{O}/\mathcal{I}$ can be identified with the sheaf $\mathcal{O}_Y$, the sheaf of holomorphic functions on the submanifold Y. More generally there is an isomorphism which has the form

$$\mathcal{O}|_Y \cong \mathcal{O}/\mathcal{I} \oplus \mathcal{I}/\mathcal{I}^2 \oplus \mathcal{I}^2/\mathcal{I}^3 \oplus \cdots, \qquad (7.2.25)$$

which is given by the Taylor expansion of a germ of f at any point $x \in Y$. Namely, if we choose local coordinates (z, w) for X at a point $x \in Y$ such that

$$Y = \{(z, w) : w_1 = \cdots = w_r = 0)\},$$

then we have for any $f \in \mathcal{O}_x$ the expansion

$$f(x) = f(z, w) = \sum_{i_1 + \cdots + i_r = p} a_{i_1 \ldots i_r}(z) w_1^{i_1} \ldots w_r^{i_r}. \qquad (7.2.26)$$

We see that the monomials $w_1^{i_1} \ldots w_r^{i_r}$ for $i_1 + \cdots + i_r = p$, give a basis for the quotient stalk $\mathcal{I}_x^p / \mathcal{I}_x^{p+1}$, which is a finite-dimensional vector space, for each p. The 'constant term' has the form $a_0(z) \in \mathcal{O}/\mathcal{I}$, and this shows that $\mathcal{O}/\mathcal{I} \cong \mathcal{O}_Y$; these are the intrinsic holomorphic functions on Y. The *normal bundle* $N = N_Y$ of the embedding Y in X is defined by the quotient bundle

$$0 \to T(Y) \to T(X)|_Y \to N_Y \to 0.$$

The dual of the normal bundle is N^*, and one can show (Exercise 7.2.1) that

$$\mathcal{O}(N^*) \cong \mathcal{I}/\mathcal{I}^2. \tag{7.2.27}$$

Moreover, the symmetric powers have the form

$$\mathcal{O}(\odot^p N^*) \cong \mathcal{I}^p/\mathcal{I}^{p+1}.$$

Thus we can express the expansion (7.2.25) in the form

$$\mathcal{O}|_Y \cong \sum_p \mathcal{O}_Y(\odot^p N^*), \tag{7.2.28}$$

where $\mathcal{O}_Y(\odot^0 N^*) := \mathcal{O}_Y$. The expansion (7.2.28) is a formal version of (7.2.26) in coordinates, where the homogeneous powers of w_i in (7.2.26) are replaced by the symmetric powers of the vector bundle N^*. A section of the left hand side of (7.2.28) will correspond to a sequence of sections intrinsic to Y of the bundles on the right hand side. This is the same as the local correspondence

$$f \mapsto \{a_{i_1 \ldots i_r}(z)\},$$

in the expansion (7.2.26).

There are two approaches to understanding (7.2.24). The first is with potentials. If we apply the spectral sequence machine to the relative de Rham complex for the sheaf $\mathcal{O}(n-2)$ we obtain fields on **M** which do not satisfy the massless field equations. However, if we differentiate these fields appropriately (differentiate to nth order) then we find that we generate massless fields of negative helicity, and, at least locally, all such fields can be so represented. So the cohomological machine plus differentiation yields the desired Penrose transform. On the other hand, one can obtain the transform directly without potentials in the following manner (Wells 1979b). We shall show how to calculate the field at a point $x \in U$ by expanding a cohomology class in a power series about the submanifold $\hat{x} \subset \mathbf{P}$. Let $\omega \in H^1(\widehat{U}, \mathcal{O}(n-2))$, then ω defines naturally by restriction an element of $H^1(\hat{x}, \mathcal{O}(n-2))$, which we still call ω. Using the expansion (7.2.28) we can write formally

$$\omega = \omega_0 + \omega_1 + \cdots + \omega_p + \cdots ,$$

where

$$\omega_p \in H^1(\hat{x}, \mathcal{O}_{\hat{x}}(H^{n-2} \otimes \odot^p(N_{\hat{x}}^*))). \qquad (7.2.29)$$

But one can calculate (Exercise 7.2.2) that $N_{\hat{x}} \cong H \oplus H$ (where here H is the hyperplane section bundle of $\hat{x} \cong \mathbf{P}_1$). Recalling that $H^* = H^{-1}$, we find that

$$H^{n-2} \otimes \odot^p(H^{-1} \oplus H^{-1}) = H^{n-2} \otimes H^{-p} \otimes \odot^p(\mathbf{C}^2)$$
$$= H^{n-2-p} \otimes \odot^p(\mathbf{C}^2).$$

But on $\mathbf{P}_1$ we have $\mathcal{O}(-2) \cong \Omega^1$, so we find that, making the substitutions in (7.2.29),

$$\omega_p \in H^1(\hat{x}, \Omega^1_{\hat{x}}(n-p)) \otimes \odot^p(\mathbf{C}^2).$$

We can evaluate the cohomology groups by using (3.3.17) and (3.3.18), and we obtain

$$H^1(\hat{x}, \Omega^1_{\hat{x}}(n-p)) \otimes \odot^p(\mathbf{C}^2) \cong \begin{cases} 0, & p < n, \\ \odot^n(\mathbf{C}^2), & p = n. \end{cases}$$

Thus we define

$$\mathcal{P}(\omega)(x) := \omega_n,$$

then this is a well defined symmetric n-spinor at x, and is the value of the field. As we shall see in the proof of the next theorem, in fact we have

$$\mathcal{P}(\omega)(x) \in [\odot^n S_- \otimes \det S_+]_x = [S_{(AB...D)}[1]']_x.$$

Here we just wanted to illustrate how to get the field at a single point in terms of an expansion. Of course, there's no indication there that this field so defined should satisfy any equations. That's a much more difficult issue which we'll consider later.

To study the negative helicity case we shall need an additional analytical hypothesis which was not necessary in the nonnegative case. We shall summarize some important facts from the theory of functions of several complex variables, which is relevant in this regard. A *Stein manifold* X is a complex manifold with the property that there exists a function $\varphi : X \to \mathbf{R}$, such that $X_c := \{x \in X : \varphi(x) < c\}$ is relatively compact for all $c > 0$ and such that the $(1,1)$form $\partial\bar{\partial}\varphi$ has a positive-definite coefficient matrix at each point of X. This is a generalization of the notion of a convex set in $\mathbf{R}^n$, and any convex set in $\mathbf{C}^n$ is Stein, for instance, although the definition is much broader. See Hörmander (1973), Krantz (1982), and Range (1986), for instance, for a thorough analysis of these important complex manifolds. They are never compact, and they have the following important property, which we shall state as a theorem.

Theorem 7.2.6. *Let X be a Stein manifold, then for any holomorphic vector bundle $\to X$,*

$$H^q(X, \mathcal{O}(E)) = 0, \quad q \geq 1.$$

The proof of this can be found in the references above (see the example in §3.3). It is also true for more general sheaves (coherent sheaves, which are natural generalizations of the locally free sheaves appearing in the theorem). Using Dolbeault's theorem (Theorem 3.3.4), the vanishing of cohomology in the theorem means that we can solve the $\bar{\partial}$-equation

$$\bar{\partial}u = \omega,$$

for u, if $\bar{\partial}\omega = 0$, where u and ω are E-valued $(0,q)$- and $(0, q+1)$-forms on X, respectively. In the book by Hörmander we find that these differential equations are solved by means of the L^2-methods of partial differential equations. The more recent works of Krantz and Range show how to solve these same equations using generalizations of the Cauchy integral formula in several variables which have been developed by a number of mathematicians over the past 15 years. The simplest example of a Stein manifold is the unit ball in $\mathbf{C}^n$ centered at the origin $X = \{x \in \mathbf{C}^n : |x| < 1\}$, where we can take $\varphi(x) := -\log|x|^2$, as is easy to check. However, the *difference* of two such balls centered at the origin (an annulus in several variables) is *not* a Stein manifold, provided that $n > 1$. There are also domains which are topologically equivalent to the ball but which are not Stein (take the ball and press a dimple into it; cf. §3.3).

After all of this digression on Stein manifolds we can now state the following theorem.

Theorem 7.2.7. *Let U be an open Stein submanifold of* $\mathbf{M}$, *then there is a canonical linear transformation, for $n > 0$,*

$$\mathcal{P} : H^1(\widehat{U}, \mathcal{O}(n-2)) \to \mathcal{Z}'_{-n}(U).$$

Moreover, if $\mu : U' \to \widehat{U}$ has connected fibers, then $\mathcal{P}$ is injective, and if $\mu : U' \to \widehat{U}$ is elementary, then $\mathcal{P}$ is an isomorphism.

We shall outline the proof in a moment, but first we give a corollary. Since $\mathbf{M}^I$ and $\mathbf{M}^\pm$ are all convex domains in the coordinate chart $\mathbf{M}^I$, we see that they are necessarily Stein, and they also clearly satisfy the topological hypotheses as we have seen before. Thus we have the following immediate consequence of Theorem 7.2.7.

Corollary 7.2.8. *For $n > 0$,*

$$\text{(a)} \quad \mathcal{P} : H^1(\mathbf{P}^\pm, \mathcal{O}(n-2)) \to \Gamma(\mathbf{M}^\pm, \mathcal{Z}'_{-n}),$$
$$\text{(b)} \quad \mathcal{P} : H^1(\mathbf{P}^I, \mathcal{O}(n-2)) \to \Gamma(\mathbf{M}^I, \mathcal{Z}'_{-n}),$$

are canonical isomorphisms.

Proof: We shall give the essential steps of the proof, but refer to the papers Eastwood, Penrose and Wells (1981), Wells (1979b), Penrose (1977b) and Ward (1977b) for some of the details which we leave out. Let us first describe the mapping $\mathcal{P}$. As we said before, there are two descriptions. We shall give both, and we shall first give the direct approach with power series expansions, and then we shall turn to potentials later.

We want to expand cohomology classes in $H^1(\widehat{U}, n-2)$ about projective lines of the form $\hat{x}$, for $x \in U$. But we want to do this in a uniform manner for all points , so we shall work on the product space $\mathbf{P} \times \mathbf{M}$. The correspondence space $\mathbf{F}$ is naturally embedded in $\mathbf{P} \times \mathbf{M}$ by the mapping $(L_1 \subset L_2) \mapsto (L_1, L_2)$, and this defines $\mathbf{F}$ as a two-codimensional submanifold of $\mathbf{P} \times \mathbf{M}$. Let $\mathcal{I}$ be the ideal sheaf of the submanifold $\mathbf{F}$ in $\mathbf{P} \times \mathbf{M}$ We recall from (7.2.27) that the conormal sheaf of the embedding is given by

$$\mathcal{O}(N^*) \cong \mathcal{I}/\mathcal{I}^2|_{\mathbf{F}}.$$

Let us relate this conormal sheaf to the spinor sheaves on $\mathbf{P} \times \mathbf{M}$. Consider the sheaf $\mathcal{O}^A(1)$ on $\mathbf{P} \times \mathbf{M}$ (i.e., the pullback from $\mathbf{P}$ of $\mathcal{O}_\mathbf{P}(1)$ tensored with the pullback from $\mathbf{M}$ of $\mathcal{O}_\mathbf{M}^A$ under the natural projections to each of the factors in the cartesian product). There is a natural section on $\mathbf{P} \times \mathbf{M}$ of this sheaf which we shall call ω^A, whose vanishing will define $\mathbf{F}$ as a submanifold of $\mathbf{P} \times \mathbf{M}$, i.e.,

$$\mathbf{F} = \{(L_1, L_2) \in \mathbf{P} \times \mathbf{M} : \omega^A(L_1, L_2) = 0\}. \qquad (7.2.30)$$

Let $H \otimes S^A$ be the vector bundle associated to the sheaf $\mathcal{O}^A(1)$. Then at the point $(L_1, L_2) \in \mathbf{P} \times \mathbf{M}$ the fiber of $H \otimes S^A$ is given by $L_1^* \otimes (\mathbf{T}/L_2)$. Choose $l \in L_1$, and let l^* be the dual element of L_1^* (i.e., $l^*(l) = 1$), and let $[l]$ be the image of l under the quotient mapping $\mathbf{T} \to \mathbf{T}/L_2$. Then $l^* \otimes [l] \in L_1^* \otimes (\mathbf{T}/L_2)$, and is independent of the choice of l. This defines the value of the section ω^A at the point (L_1, L_2). It is easy to verify that ω^A satisfies (7.2.30). Thus we have on $\mathbf{P} \times \mathbf{M}$ the mapping

$$
\begin{array}{ccc}
\mathcal{O}_A(-1) & \xrightarrow{\omega^A} & \mathcal{I} \\
\downarrow & & \downarrow \\
f_A & \longmapsto & f_A \omega^A,
\end{array}
\qquad (7.2.31)
$$

is a surjective mapping. In terms of components, if $\omega^A = \{\omega^0, \omega^1\}$, then ω^0 and ω^1 are generators of the ideal $\mathcal{I}$, and any function $f \in \mathcal{I}$ has the form $f = f_1 \omega^0 + f_2 \omega^1$. The mapping (7.2.31) expresses this in an abstract form. The surjectivity is equivalent to saying that the Jacobian matrix of the two components ω^0 and ω^1 have maximal rank on the submanifold $\mathbf{F}$. We also have the exact sequence on $\mathbf{P} \times \mathbf{M}$:

$$0 \to \mathcal{I}^2 \to \mathcal{I} \to {}_\mathbf{F}\mathcal{O}_A(-1) \to 0.$$

$$
\begin{array}{ccc}
\cup\!\!\!\cup & & \cup\!\!\!\cup \\
f_A \omega^A & \mapsto & f_A|_\mathbf{M}
\end{array}
\qquad (7.2.32)
$$

We are writing ${}_X\mathcal{O}$ instead of $\mathcal{O}_X$ to avoid confusion with spinor indices. The sequence (7.2.32) shows that ${}_\mathbf{M}\mathcal{O}_A(-1)$ can be identified with the conormal sheaf $\mathcal{O}(N^*) = \mathcal{I}/\mathcal{I}^2$ (this is a parametrized version of the assertion earlier in this section that $N_{\hat{x}} = H \oplus H$). Now suppose that U is an open subset of $\mathbf{M}$, and we consider the construction applied to U' as a submanifold of $\widehat{U} \times U$. We claim that

$$H^1(U' \times U, \mathcal{I}^n(n-2)) \to H^1(\widehat{U} \times U, \mathcal{O}(n-2)) \qquad (7.2.33)$$

is an isomorphism. We shall verify (7.2.32) in the case $n = 2$ for simplicity. On $\widehat{U} \times U$ we have the short exact sequence

$$0 \to \mathcal{I} \to \mathcal{O} \to \mathcal{O}_{U'} \to 0,$$

which yields the long exact sequence on cohomology

$$\Gamma(\widehat{U} \times U, \mathcal{O}) \to \Gamma(U', \mathcal{O}) \to H^1(\widehat{U} \times U, \mathcal{I})$$
$$\to H^1(\widehat{U} \times U, \mathcal{O}) \to H^1(U', \mathcal{O}), \quad (7.2.34)$$

where we recall that the sheaf $\mathcal{O}_{U'}$ is supported on $U' \subset \widehat{U} \times U$, and we have the isomorphism

$$H^q(\widehat{U} \times U, \mathcal{O}_{U'}) \cong H^q(U', \mathcal{O}_{U'}).$$

The first arrow in (7.2.33) is surjective, namely one can check that

$$\Gamma(U', \mathcal{O}_{U'}) \cong \Gamma(U, \mathcal{O}_U),$$
$$\Gamma(\widehat{U} \times U, \mathcal{O}_{\widehat{U} \times U}) \cong \Gamma(U, \mathcal{O}).$$

This follows since any holomorphic function on $\widehat{U}$ is constant since it is constant on each of the projective lines $\hat{x}$, for $x \in U$, and for any two points P, Q in $\widehat{U}$, there is an $x \in U$ so that $P, Q \in \hat{x}$. Similarly, any holomorphic function on U' is constant on the fibers $\nu^{-1}(x)$, for $s \in U$, and hence corresponds to the pullback of a holomorphic function from U. Now $H^1(U', \mathcal{O}) \cong H^1(U, \mathcal{O})$ by the Leray spectral sequence (7.2.13) and Theorem 7.1.5. Since U is Stein we have that $H^1(U, \mathcal{O}) = 0$ by Theorem 7.2.6. Thus we find that

$$H^1(\widehat{U} \times U, \mathcal{I}) \xrightarrow{\cong} H^1(\widehat{U} \times U, \mathcal{O}). \qquad (7.2.35)$$

Now consider the sequence (7.2.32) and its associated long exact sequence. We find

$$\Gamma(U', \mathcal{O}_A(-1)) \to H^1(\widehat{U} \times U, \mathcal{I}^2)$$
$$\to H^1(\widehat{U} \times U, \mathcal{I}) \to H^1(U', \mathcal{O}_A(-1)).$$

Using the Leray spectral sequence (7.2.13) and Theorem 7.1.5 again, we see that the first and fourth terms vanish, and we conclude that

$$H^1(\widehat{U} \times U, \mathcal{I}^2) \xrightarrow{\cong} H^1(\widehat{U} \times U, \mathcal{I}). \qquad (7.2.36)$$

From (7.2.34) and (7.2.35) we deduce (7.2.32) in the case $n = 2$. The general case is a continuation of this argument using successively higher powers of $\mathcal{I}$. Intuitively (7.2.32) being an isomorphism says that the information contained within an element of $H^1(\widehat{U} \times U, \mathcal{O}(n-2))$ depends only on normal derivatives to U' of order at least n. To 'evaluate the field', in effect, we just take the nth normal derivative by factoring out $\mathcal{I}^{n+1}(n - 2)$. That is dividing out by this power *leaves* all of the lower order information intact. This process also restricts to U' since the sheaf $\mathcal{I}^n(n - 2)/\mathcal{I}^{n+2}(n - 2)$ is supported on U'. More explicitly we have the exact sequence on $\widehat{U} \times U$:

$$0 \to \mathcal{I}^{n+1} \quad \to \quad \mathcal{I}^n \quad \to \quad {}_{U'}\mathcal{O}_{(AB...D)}(-2) \to 0$$
$$\cup \qquad\qquad \cup \qquad\qquad\qquad (7.2.37)$$
$$f_{AB...D}\omega^A\omega^B \dots \omega^D \mapsto f_{AB...D}|_{U'}$$

where $f_{AB...D}$ is the Taylor coefficient of the monomials $\omega^A\omega^B \dots \omega^D$ (which are generators of the nth symmetric power of the conormal bundle). It follows from (7.2.36) that

$$\mathcal{I}^n(n - 2)/\mathcal{I}^{n+1}(n - 2) \cong \mathcal{O}_{(AB...D)}(-2)$$

on U'. We now have the Penrose transform (as originally given in Wells 1979b) of the form

$$H^1(\widehat{U} \times U, \mathcal{O}(n - 2)) \xleftarrow{\cong} H^1(\widehat{U} \times U, \mathcal{I}^n(n - 2))$$
$$\Big\downarrow T$$
$$\uparrow \qquad\qquad\qquad H^1(U', \mathcal{O}_{(AB...D)}(-2)) \qquad (7.2.38)$$
$$\Big\downarrow L$$
$$H^1(\widehat{U}, \mathcal{O}(n - 2)) \xrightarrow{\ \mathcal{P}\ } H^0(U, \mathcal{O}_{(AB...D)}[1])$$

where T is the Taylor coefficient mapping given by (7.2.36), L is the Leray spectral sequence mapping which is an isomorphism, the top isomorphism is given by (7.2.32), the left-hand vertical mapping is the natural pullback mapping, and $\mathcal{P}$, the Penrose transform, is defined as the composition. This is the desired mapping from cohomology on $\mathbf{P}$ to spinor fields on $\mathbf{M}$, but it is not clear from this that the images must satisfy any field equations. One way to see that they do would be lift the spectral sequence of the relative de Rham sequence to this product picture, and deduce the differential equations, etc., as before. We shall not carry this out here. The alternative is to see what one gets from the relative de Rham sequence itself. This leads to potentials.

A *potential for a negative helicity field* $\psi_{AB...D} \in \Gamma(U, \mathcal{Z}'_{-n})$ is a spinor field

$$\varphi^{B'...D'} \in \Gamma(U, \mathcal{O}_A^{(B'...D')}[1]'),$$

such that

$$\nabla^{A(A'}\varphi^{B'...D')} = 0$$
$$\psi_{AB...D} = \nabla_{D'(D} \cdots \nabla_{B'B)}\varphi^{B'...D'}, \qquad (7.2.39)$$

where the symmetrization in the definition of ψ refers only to the unprimed indices. One can check in a straightforward but tedious manner that if $\psi = \psi_{AB...D}$ is given by (7.2.38), then ψ satisfies the negative helicity massless field equations (7.2.23). There is a gauge freedom in these potentials, namely, if $\varphi^{B'...D'}$ is a potential for $\psi_{AB...D}$, then so is

$$\varphi_A^{B'...D'} + \nabla_A^{(B'}\gamma^{C'...D')}, \qquad (7.2.40)$$

for any spinor field $\gamma^{C'...D'} \in \Gamma(U, \mathcal{O}^{(C'...D')})$. It follows from constructions in Penrose (1977b) and Ward (1977b), that locally one can always find such a potential, and that (7.2.40) is the only gauge freedom.

Let us specialize to the case of a left-handed Maxwell field, i.e., a solution of

$$\nabla^{BB'}\psi_{AB} = 0. \qquad (7.2.41)$$

Consider the anti-self-dual two-form

$$F_{ab} = \psi_{AB}\varepsilon_{A'B'},$$

then (7.2.40) become simply (as we have seen before!) $df = 0$, and a potential is any one-form ω such that $d\omega = F$. The gauge freedom is simply that $\omega \mapsto d\omega + d\gamma$, for any holomorphic function γ. Thus the local equivalence of field potentials modulo gauge is simply the de Rham sequence. One can rewrite the de Rham sequence in terms of spinors in a manner that one can see how to show that any locally massless field has the description of potential/gauge. See Eastwood, Penrose and Wells (1981) for more details on this.

Finally, we shall show that if we use the spectral sequence of the relative de Rham complex, then we get the potentials which satisfy (7.2.38). Specifically, there is a Penrose transform of the form

$$\mathcal{P} : H^1(\widehat{U}, \mathcal{O}(n-2)) \to \{\text{potentials for helicity} \qquad (7.2.42)$$
$$-n/2 \text{ massless fields}$$
$$\text{on } U\}/\{\text{gauge freedom}\}$$

The mapping $\mathcal{P}$ is an injection or isomorphism depending on the topological properties of μ as hypothesized, as before.

More precisely, the E_1-term in the spectral sequence (7.2.12) for the case in question, has the following form after using the Leray spectral sequence and Theorem 7.1.5. We also use the fact that $H^q(U, \mathcal{O}^{(C'\ldots D')}) = H^q(U, \mathcal{O}_A^{(B'\ldots D')}[1]') = 0$, for $q \geq 1$, since U is Stein.

$$\vdots \qquad \qquad \vdots \qquad \qquad \vdots$$
$$0 \qquad \qquad 0 \qquad \qquad 0$$
$$0 \qquad \qquad 0 \qquad \qquad 0$$
$$\Gamma(U, \mathcal{O}^{(C'\ldots D')}) \xrightarrow{\nabla_A^{B'}} \Gamma(U, \mathcal{O}^{(B'\ldots D')}[1]') \xrightarrow{\nabla^{AA'}} \Gamma(U, \mathcal{O}^{(A'\ldots D')}[1][2]')$$

(where $(C'\ldots D')$, $(B'\ldots D')$, and $(A'\ldots D')$ have $(n-2)$, $(n-1)$, and (n) indices, respectively). Calculating the E_2 term from this E_1-term, we see that (using the definition in (7.2.38))

$$E_2^{10} = \{\text{potentials}\}/\{\text{gauge}\}.$$

Also $E_\infty^{01} = E_2^{01} = 0$, and $E_1^{10} = E_2^{10}$, and thus it follows from Theorem 3.6.2, as before, that

$$H^1(U', \mu^{-1}\mathcal{O}(n-2)) \cong \{\text{potentials}\}/\{\text{gauge}\}.$$

By calculating the field at a point of U one can check that the two Penrose transforms given by (7.2.37) and (7.2.41) differ by a factor of (-2). This uses the helicity raising and lowering operations of Penrose (1975) and Eastwood (1979). We refer the reader to Eastwood, Penrose and Wells (1981) for this final point, and with this we can conclude that the image of (7.2.37) does satisfy the field equations (7.2.23). ∎

Exercises for Section 7.2

7.2.1. If Y is a complex submanifold of a complex manifold X, and if N_Y is the normal bundle to Y in X, then show that

$$\mathcal{O}_X(N^*) \cong \mathcal{I}_Y/\mathcal{I}_Y^2,$$

where $\mathcal{I}_Y$ is the ideal subsheaf of $\mathcal{O}_X$ of all holomorphic functions on open subsets of X which vanish when restricted to Y.

7.2.2. Let $Y \subset \mathbf{P}_3$ be a complex projective line embedded in $\mathbf{P}_3$, and let N_Y be its normal bundle, then show that

$$N_Y \cong (H \otimes H)|_Y,$$

where H is the hyperplane section bundle on $\mathbf{P}_3$.

7.2.3. Suppose that if

$$\varphi_A^{B'} \in \Gamma(U, \mathcal{O}^{B'}[1]')$$

is a potential for the spinor field

$$\psi_{AB} = \nabla_{B'B}\varphi_A^{B'},$$

(that is, $\varphi_A^{B'}$ satisfies $\nabla^{A(A'}\varphi_A^{B')} = 0$), then show that ψ_{AB} satisfies the self-dual Maxwell's equations

$$\nabla^{AA'}\psi_{AB} = 0.$$

7.3 Integral formulas for massless fields

In the previous section, we saw how massless fields correspond to elements of twistor cohomology groups. Historically, this result arose out of something more down-to-earth, namely contour integral expressions for massless fields (Penrose 1968a, 1969; Penrose and MacCallum 1973). In fact, for the scalar case (the wave equation), an equivalent integral formula was given by Bateman (1904), who in turn was developing work of Whittaker (1903). In this section, we shall work backward, by showing how the Penrose transform of §7.2 gives rise to contour integral formulas.

An element of the sheaf cohomology group $H^1(\widehat{U}, \mathcal{O}(-2s-2))$ can be represented by the (Čech) cocycle $f_{ij}(Z^\alpha)$ defined on overlap regions $W_i \cap W_j$, where $\{W_i\}$ is an open cover of $\widehat{U}$. The corresponding massless field can be expressed as a *branched contour integral* of f_{ij} (see, for example, Penrose 1979a; Penrose and Rindler 1986, pp. 156–160). To keep things simple, let us consider only the special case where $\widehat{U}$ is covered by *two* coordinate charts W and $\underline{W}$, so that there is only one overlap region $W \cap \underline{W}$, and the representative cocycle consists of a single function $f(Z^\alpha)$ defined on this overlap. Then we shall get an ordinary contour integral, rather than a branched one.

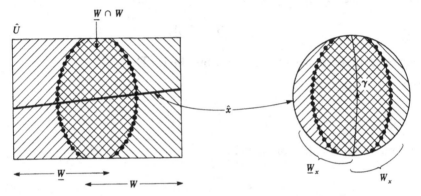

Fig. 7.3.1. Open covering restricted to a projective line.

The picture, therefore, is as in Figure 7.3.1. In this figure, $\widehat{U}$ is a region of twistor space $\mathbb{P}$ derived from a region U in M. The two open sets W and $\underline{W}$ are chosen so that for any line $\hat{x}$ in $\widehat{U}$, the intersections

$$W_x = W \cap \hat{x},$$
$$\underline{W}_x = \underline{W} \cap \hat{x},$$

overlap in an annular region of $\hat{x}$ (i.e., $W_x \cap \underline{W}_x$ is homeomorphic to $S^1 \times \mathbb{R}$). The function $f(Z^\alpha)$ is holomorphic on $W \cap \underline{W}$, and homogeneous of degree $-n-2$ in Z^α. Of course, if h and $\underline{h}$ are functions holomorphic on W and $\underline{W}$, respectively, then $\underline{h} - h$ represents a coboundary, and

$$f' = f + \underline{h} - h \tag{7.3.1}$$

is regarded as being equivalent to f.

We are thinking in terms of the double-fibration picture

and we want to use f to construct a massless field on U. The first step is to pull f back to U', and this is easily done: the pullback of $f(\omega^A, \pi_{A'})$ is

$$g(x^a, \pi_{A'}) := f(ix^{AA'}\pi_{A'}, \pi_{A'}). \tag{7.3.2}$$

Here $(x^a, \pi_{A'})$ are being thought of as coordinates on U'. The function g is holomorphic on the intersection of the two sets $\mu^{-1}(W)$ and $\mu^{-1}(\underline{W})$, which cover U'. And g is homogeneous of degree $-n-2$ in $\pi_{A'}$.

The second, and final, step is to integrate out the π-dependence. For the moment, take $n \geq 0$. Then we set

$$\varphi_{A' \ldots C'}(x^d) = \frac{1}{2\pi i} \oint_\gamma \pi_{A'} \ldots \pi_{C'} g(x^d, \pi_{D'}) \triangle \pi. \tag{7.3.3}$$

The integrand of (7.3.3) contains n factors of $\pi_{A'}$, and $\triangle\pi$ is defined by

$$\triangle \pi := \pi_{E'} d\pi^{E'}. \tag{7.3.4}$$

The canonical holomorphic one-form $\triangle\pi$ on the fibers of ν, is homogeneous of degree two; so the integrand is homogeneous of degree zero in $\pi_{A'}$. This means that, although expressed in terms of the homogeneous coordinates $\pi_{A'}$, the integrand of (7.3.3) is actually defined on the *projective* π-space $\hat{x}$ (see Exercise 7.3.1). Think of x as being fixed, and consider the contour of integration γ as depicted in Figure 7.3.1: it is a circle winding around the annular region $W_x \cap \underline{W}_x$.

By Cauchy's theorem, the value of the integral does not change if we continuously deform γ. In fact, it depends only on the winding number of γ, which we take to be unity (with orientation as depicted in Figure 7.3.1). Also by Cauchy's theorem, the value of the integral does not change if we make the coboundary transformation (7.3.1). So $\varphi_{A' \ldots C'}$ is indeed uniquely defined in terms of the cohomology class $[f]$. Moreover, $\varphi_{A' \ldots C'}$ satisfies the massless free-field equations

$$\nabla^{AA'}\varphi_{A' \ldots C'} = 0, \quad \text{or} \quad \Box\varphi = 0,$$

as we see by differentiating under the integral sign and using

$$\nabla_{AA'} g = i\pi_{A'} \frac{\partial}{\partial \omega^A} f. \tag{7.3.5}$$

For the case of negative helicity (i.e., $n < 0$), we replace (7.3.3) by the integral

$$\varphi_{A...C}(x^d) = \frac{1}{2\pi i} \oint_\gamma \frac{\partial}{\partial \omega^A} \cdots \frac{\partial}{\partial \omega^C} g(x^d, \pi_{D'}) \Delta\pi, \tag{7.3.6}$$

in which $-n$ factors of $\frac{\partial}{\partial \omega^A}$ appear. The remarks we made about the integral (7.3.3) are equally true for (7.3.6), and so this is an explicit version of the Penrose transform for negative helicity massless fields.

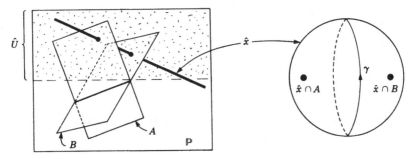

Fig. 7.3.2. Separation of singularities in twistor space.

Let us look at a particular example. Take $f(Z^\alpha)$ to be

$$f(Z^\alpha) = \frac{1}{2Z^0 Z^1}. \tag{7.3.7}$$

So f is singular on the two planes A (where $Z^0 = 0$) and B (where $Z^1 = 0$) (see Figure 7.3.2). We restrict to a region $\widehat{U}$ of twistor space which is disjoint from the line where A and B intersect. So on $\widehat{U}$, the singularities of f are separated. To put this another way, the two open sets $W = \widehat{U} - A$ and $\underline{W} = \widehat{U} - B$ cover $\widehat{U}$, and f is homogeneous of degree -2, so this is the scalar case (i.e., $n = 0$).

Now we simply substitute (7.3.7) into the formulas (7.3.2)–(7.3.4). Writing $\zeta = \pi_{0'}/\pi_{1'}$, we get $\Delta\pi = -d\zeta$, and

$$\varphi(x^{AA'}) = \frac{1}{4\pi i} \oint_\gamma \frac{d\zeta}{(x^{00'}\zeta + x^{01'})(x^{10'}\zeta + x^{11'})}.$$

Here the contour γ separates the two simple poles of the integrand. Evaluating this integral is simply a matter of computing the residue at one of these poles (the orientation of the contour γ is as depicted in Figure 7.3.2), and gives

$$\varphi(x^{\mathsf{a}}) = \frac{1}{x_{\mathsf{a}}x^{\mathsf{a}}}. \tag{7.3.8}$$

This is the elementary solution of the wave equation, based at the origin $x^{\mathsf{a}} = 0$ (note that the line $A \cap B$ in $\mathbf{P}$ corresponds to the origin in $\mathbf{M}$). A slight generalization of equation (7.3.7) gives the elementary solutions based at an arbitrary point y^{α}, that is,

$$\varphi(x^{\mathsf{a}}, y^{\mathsf{a}}) = \frac{1}{(x_{\mathsf{a}} - y_{\mathsf{a}})(x^{\mathsf{a}} - y^{\mathsf{a}})}, \tag{7.3.9}$$

which is the Green's function for the wave operator $\square$ in Minkowski space-time. See Atiyah (1981) for a cohomological discussion of such Green's functions.

More generally still, one has the fields arising from twistor functions of the form

$$f(Z^{\alpha}) = \frac{(C_{\gamma}Z^{\gamma})^{c}(D_{\delta}Z^{\delta})^{d}}{(A_{\alpha}Z^{\alpha})^{a+1}(B_{\beta}Z^{\beta})^{b+1}}, \tag{7.3.10}$$

where a, b, c, d are nonnegative integers, and A_{α}, B_{α}, C_{α}, D_{α} are four fixed dual twistors (elements of the dual twistor space). From the homogeneity of this function, we see that it gives rise to a massless field of helicity $\frac{1}{2}(a + b - c - d)$, which has a pole on the null cone of some point p in $\mathbf{M}$. The line $\hat{p}$ corresponding to this point p is the intersection of the two planes $A_{\alpha}Z^{\alpha} = 0$ and $B_{\alpha}Z^{\alpha} = 0$. A massless field generated in this manner is known as an *elementary state* (Penrose and MacCallum 1973).

Exercises for Section 7.3

7.3.1. Let $\omega = g(\pi_{A'})\pi_{B'}d\pi^{B'}$ be a one-form on nonprojective π-space. Let Υ denote the Euler homogeneity operator $\pi_{A'}\frac{\partial}{\partial\pi_{A'}}$. Show that ω descends to a one-form on the projective π-space $\mathbf{P}_{1}$ if and only if

$$\Upsilon \lrcorner \omega = 0, \qquad \mathcal{L}_{\Upsilon}\omega = 0$$

($\mathcal{L}$ denotes the Lie derivative), and that this in turn is true if and only if g is homogeneous of degree -2.

7.3.2. Calculate the massless field arising from the f that was given in (7.3.10), taking $a = b = c = d = 0$.

7.4 Hyperfunction solutions of the
massless field equations

In §§7.1 and 7.2 we have seen how to use the Penrose transform to generate all of the holomorphic solutions of the massless field equations on suitable open subsets U of M from homogenous holomorphic functions on corresponding subsets of P. In §7.3 we have seen how the Penrose transform can be realized in terms of explicit integral formulas. We now want to consider the solutions of these same equations on real Minkowski space M^4 or on its conformal compactification M. For purposes of this initial discussion, let us consider the wave equation $\Box\varphi = 0$. If φ is a holomorphic solution of this equation on say M^I, then $\varphi|_{M^4}$ will be a real-analytic solution of this equation on M^4. However, $\Box$ is a *hyperbolic* differential equation, it admits solutions which are not smooth (see, e.g., Courant and Hilbert 1965 for a discussion of classical differential equations, including this same wave equation). Thus, we need to admit as candidates for solutions generalized functions which are not necessarily smooth in the usual sense. The standard examples of classes of generalized functions on $\mathbf{R}^n$ used in solving partial differential equations include $L^2_{\text{loc}}(\mathbf{R}^n)$, $\mathcal{D}'(\mathbf{R}^n)$, and $\mathcal{B}(\mathbf{R}^n)$. These are the *locally square-integrable functions* on $\mathbf{R}^n$, the *distributions* on $\mathbf{R}^n$, and the *hyperfunctions* on $\mathbf{R}^n$, respectively. We shall describe each of these spaces briefly and give references for further information. A measurable function f is in $L^2_{\text{loc}}(\mathbf{R}^n)$ if for any $\varphi \in C^\infty_0(\mathbf{R}^n)$, $\int_{\mathbf{R}^n}(\varphi f)^2 dx < \infty$. To define distributions, we define $\mathcal{D}(\mathbf{R}^n)$ to be the vector space of C^∞_0 functions on $\mathbf{R}^n$ with the following topology: a sequence φ_j converges to a limit φ_0 in $\mathcal{D}(\mathbf{R}^n)$ and only if $\operatorname{supp}\varphi_j \subset K$, for some fixed K compact in $\mathbf{R}^n$, and

$$D^\alpha_{\varphi_j} \to D^\alpha_{\varphi_0},$$

uniformly on K for all multiindices α. A *distribution* T is an element of the dual space $\mathcal{D}^*(\mathbf{R}^n)$ which is continuous, i.e., $T(\varphi_j) \to T(\varphi_0)$ if $\varphi_j \to \varphi_0$ in $\mathcal{D}(\mathbf{R}^n)$. The set of such continuous linear functionals on $\mathcal{D}(\mathbf{R}^n)$ is denoted by $\mathcal{D}'(\mathbf{R}^n)$. The last class of generalized functions is the hyperfunctions, and they are more complicated to describe. Intuitively, hyperfunctions can be described in terms of linear functionals on real-analytic functions with a suitable (complete) topology, or as generalized boundary values of holomorphic functions defined on appropriate open subsets of the complement of $\mathbf{R}^n$ in its complexification $\mathbf{C}^n$. This latter point of view, due to Sato, is the most convenient one for our purposes. We shall mention the analytic linear functional point of view later in our discussion. To facilitate the discussion of boundary behavior of holomorphic functions on $\mathbf{R}^n$, we shall need the notion of relative cohomology, a natural generalization of the usual sheaf cohomology employed till now. If $\mathcal{F}$ is a sheaf of abelian groups on a space X, then let $0 \to \mathcal{F} \to \mathcal{C}^\bullet$ be its canonical resolution (see (3.6.11)). Then the ordinary sheaf cohomology can be defined by (as we saw before in §3.6)

$$H^q(X, \mathcal{F}) = H^q(\Gamma(X, \mathcal{C}^\bullet)).$$

Now, if we define for any closed subset $S \subset X$, and any sheaf $\mathcal{S}$ on X,

$$\Gamma_S(X, \mathcal{S}) := \{\gamma \in \Gamma(X, \mathcal{S}) : \operatorname{supp} \gamma \subset S\},$$

then we have the *sections of $\mathcal{S}$ over X with support in S*. Thus, we have for the resolution $0 \to \mathcal{F} \to \mathcal{C}^\bullet(\mathcal{F})$ the complex of sections with support in S, and we define the cohomology of X with coefficients in $\mathcal{F}$ and with support in S to be

$$H_S^q(X, \mathcal{F}) := H^q(\Gamma_S(X, \mathcal{C}^\bullet(\mathcal{F}))).$$

Moreover, we have the exact sequence of complexes

$$0 \to \Gamma_S(X, \mathcal{C}^\bullet(\mathcal{F})) \to \Gamma(X, \mathcal{C}^\bullet(\mathcal{F})) \to \Gamma(X - S, \mathcal{C}^\bullet(\mathcal{F})) \to 0,$$

since each $\mathcal{C}^q(\mathcal{F})$ is a flabby sheaf (see Exercise 7.4.1), and this induces a long exact sequence on cohomology (cf. the proof of Theorem 3.3.3c)

$$
\begin{aligned}
0 \to H_S^0(X, \mathcal{F}) &\to H^0(X, \mathcal{F}) \to H^0(X - S, \mathcal{F}) \\
&\to H_S^1(X, \mathcal{F}) \to H^1(X, \mathcal{F}) \to H^1(X - S, \mathcal{F}) \to \cdots.
\end{aligned} \tag{7.4.1}
$$

Let us look at this in a special case. Let $X = \mathbf{C}$, and $S = \mathbf{R}$, the real-axis embedded in the complex plane in the usual manner, and let $\mathcal{F} = \mathcal{O}$, the sheaf of holomorphic functions on **C**. Then we have from (7.4.1)

$$0 \to H_{\mathbf{R}}^0(\mathbf{C}, \mathcal{O}) \to H^0(\mathbf{C}, \mathcal{O}) \to H^0(\mathbf{C} - \mathbf{R}, \mathcal{O})$$
$$\to H_{\mathbf{R}}^1(\mathbf{C}, \mathcal{O}) \to H^1(\mathbf{C}, \mathcal{O}) \to \cdots .$$

Now $H_{\mathbf{R}}^0(\mathbf{C}, \mathcal{O})$ is the set of holomorphic functions on **C** supported on **R**, hence is zero. For different reasons $H^1(\mathbf{C}, \mathcal{O}) = 0$ (see Example 3.3.8), thus we have

$$H_{\mathbf{R}}^1(\mathbf{C}, \mathcal{O}) \cong H^0(\mathbf{C} - \mathbf{R}, \mathcal{O})/H^0(\mathbf{C}, \mathcal{O}).$$

Let us write $\mathbf{C} - \mathbf{R} = \mathbf{C}^+ \cup \mathbf{C}^-$, the union of the upper and lower half-planes. Then we see that

$$H_{\mathbf{R}}^1(\mathbf{C}, \mathcal{O}) = [H^0(\mathbf{C}^+, \mathcal{O}) \oplus H^0(\mathbf{C}^-, \mathcal{O})]/H^0(\mathbf{C}, \mathcal{O}). \qquad (7.4.2)$$

Thus the relative cohomology group on the left side of (7.4.2), which has *a priori* only the abstract definition, has a concrete representation in terms of holomorphic functions on $\mathbf{C}^{\pm}$ and **C**. We define

$$\mathcal{B}(\mathbf{R}) := H_{\mathbf{R}}^1(\mathbf{C}, \mathcal{O}) \qquad (7.4.3)$$

and call $\mathcal{B}(\mathbf{R})$ the *hyperfunctions* on **R**. We see that $\mathcal{B}(\mathbf{R})$ consists of holomorphic functions in the upper and lower half plane modulo entire functions. If we let $H^0(\overline{\mathbf{C}^+}, \mathcal{O})$ and $H^0(\overline{\mathbf{C}^-}, \mathcal{O})$ denote the holomorphic functions on the closed half spaces $\overline{\mathbf{C}^+}$ and $\overline{\mathbf{C}^-}$, respectively, then we can define

$$\widetilde{\mathcal{A}}(\mathbf{R}) := [H^0(\overline{\mathbf{C}^+}, \mathcal{O}) \oplus H^0(\overline{\mathbf{C}^-})]/H^0(\mathbf{C}, \mathcal{O})$$

which is a subvector space of $\mathcal{B}(\mathbf{R})$. Then we can define, for $[f^+, f^-] \in \widetilde{\mathcal{A}}(\mathbf{R})$,

$$f(x) := \lim_{y \to 0}[f^+(x + iy) - f^-(x - iy)], \qquad (7.4.4)$$

which will be a well-defined real-analytic function on **R**. Thus, f defined by (7.4.4) is the difference of the boundary values on **R** of the pair of functions f^+ and f^-. It is in this manner that ordinary functions can be considered as hyperfunctions as defined by (7.4.3) using the decomposition (7.4.4). If $[f^+, f^-]$ defines a hyperfunction $\tilde{f}$, then we can take the boundary value (7.4.4) in larger classes, $L^2(\mathbf{R})$, $L^p(\mathbf{R})$, $\mathcal{D}'(\mathbf{R})$, etc. The largest class is, by definition, the formal pair (f^+, f^-), where there are no restrictions on the growth of f^+ and f^- near the real axis. If $(f^+, f^-) \in \tilde{\mathcal{A}}(\mathbf{R})$, then we see that $f(x)$ defined by (7.4.4) satisfies

$$\frac{d}{dx} f(x) = \lim_{y \to 0} \left(\frac{\partial f^+}{\partial z}(x + iy) - \frac{\partial f^-}{\partial z}(x - iy) \right).$$

So the derivative of the function $f(x)$ with domain $\mathbf{R}$ can be defined in terms of the holomorphic derivative of the functions (f^+, f^-) representing f. This gives rise to the definition of a derivative of an arbitrary hyperfunction. We define

$$\frac{d}{dx} : \mathcal{B}(\mathbf{R}) \to \mathcal{B}(\mathbf{R}),$$

by

$$\frac{d}{dx} \tilde{f} = \frac{d}{dx}(f^+, f^-) = \left(\frac{\partial f^+}{\partial z}, \frac{\partial f^-}{\partial z} \right),$$

and if f^+, f^- have nice boundary values, then the derivatives will be the usual derivatives.

This is completely analogous to the theory of differentiation of distributions given by

$$\left\langle \frac{dT}{dx}, \varphi \right\rangle = - \left\langle T, \frac{d\varphi}{dx} \right\rangle$$

for $T \in \mathcal{D}'(\mathbf{R})$, $\varphi \in \mathcal{D}(\mathbf{R})$, which agrees with the usual differentiation of functions if T is a smooth function, by integrating by parts. Let us give an example. If we let δ be the Dirac measure on $\mathbf{R}$ centered at $0 \in \mathbf{R}$ (Dirac delta function), then $\delta \in \mathcal{D}'(\mathbf{R})$, and is defined by $\langle \delta, \varphi \rangle = \varphi(0)$, for $\varphi \in \mathcal{D}(\mathbf{R})$. The derivatives of δ are defined by

$$\langle \delta^{(j)}, \varphi \rangle = (-1)^j \varphi^{(j)}(0).$$

Now consider the hyperfunction

$$\tilde{f} := (f^+, f^-)$$

$$f^\pm(z) = \frac{1}{2\pi z}, \quad z \in \mathbf{C}^\pm.$$

We see that at all points $x \neq 0$ on $\mathbf{R}$,

$$f(x) = [bf^+ - bf^-](x) = 0,$$

so $\tilde{f}$ has support at $x = 0$. But as a function, $(bf^+ - bf^-)(0)$ is not defined. Now we can show that the data $f^\pm = \frac{1}{2\pi i z}$ defines a continuous linear functional on $\mathcal{A}(\mathbf{R})$ (which is dense in $\mathcal{C}(\mathbf{R})$ with respect to uniform convergence on compact subsets). Namely, if $\varphi \in \mathcal{A}(\mathbf{R})$, then let $\tilde{\varphi}$ be a holomorphic extension of φ to a neighborhood of $\mathbf{R} \subset \mathbf{C}$, and define

$$\langle \tilde{f}, \varphi \rangle = \int_\gamma \tilde{f}(z)\tilde{\varphi}(z)dz,$$

where γ is a contour which has winding number $+1$, and which encircles $0 \in \mathbf{C}$, and is contained in the domain of definition of $\tilde{\varphi}$. We see that

$$\langle \tilde{f}, \varphi \rangle = \varphi(0),$$

and this is independent of the contour, and this shows that the hyperfunction $\tilde{f}$ agrees with the δ measure on this dense subset of functions. This example is carried out in detail to show that the theory of hyperfunctions is quite concrete and closely tied to the classical theory of residues. An example of a hyperfunction with support at the origin which is *not* a distribution is given by $\tilde{f} = (f^+, f^-)$, where

$$f^\pm = e^{1/z}, \quad z \in \mathbf{C}^\pm.$$

(Exercise 7.4.2). This corresponds to an infinite series of derivatives of the δ-measure, while any distribution with point support is a finite sum of derivatives of the δ-measure. It is in this sense that hyperfunctions are more general than distributions. They allow local singularities of infinite order, which distributions do not.

More generally, we define

$$\mathcal{B}(\mathbf{R}^n) := H^n_{\mathbf{R}^n}(\mathbf{C}^n, \Omega^n),$$

(noting that $\Omega^n \cong \mathcal{O}$ on $\mathbf{C}^n$, but we shall use this same definition on a curved manifold). One can also represent $\mathcal{B}(\mathbf{R}^n)$ as boundary values of holomorphic functions, but not from an upper and lower half plane. One needs a finite set of cones (or wedges) which emanate from $\mathbf{R}^n$ in $\mathbf{C}^n$, e.g.,

$$\operatorname{Im} z_1 > 0, \ldots, \operatorname{Im} z_n > 0,$$

$$\operatorname{Im} z_n > 0, \ldots, \operatorname{Im} z_{n-1} > 0, \operatorname{Im} z_n < 0, \text{ etc.}$$

If we cover $\mathbf{C}^n - \mathbf{R}^n$ with a suitable Čech covering of such open cones, we can define an algebraic sum of limits of boundary values of holomorphic functions on $\mathbf{R}^n$. The generalization of (7.4.2) is given by

$$H^{n-1}(\mathbf{C}^n, \Omega^n) \to H^{n-1}(\mathbf{C}^n - \mathbf{R}^n, \Omega^n)$$
$$\to H^n_{\mathbf{R}^n}(\mathbf{C}^n, \Omega^n) \to H^n(\mathbf{C}^n, \Omega^n)$$

But for $n > 1, H^{n-1}(\mathbf{C}^n, \Omega^n) = H^n(\mathbf{C}^n, \Omega^n) = 0$, so

$$H^{n-1}(\mathbf{C}^n - \mathbf{R}^n, \Omega^n) = H^n_{\mathbf{R}^n}(\mathbf{C}^n, \Omega^n) \qquad (7.4.5)$$

generalizes (7.4.2), and if we use a suitable Čech cover for the left hand side we can take boundary values as in (7.4.4) to obtain ordinary functions where the boundary values exist (see the references cited below). In the context of twistors, we shall find representations like (7.4.5) for the hyperfunction differential forms we consider. The fundamental fact about these classes of generalized functions is that there is a sequence of strict inclusions

$$\mathcal{A}(\mathbf{R}^n) \subset C^\infty(\mathbf{R}^n) \subset L^2_{\text{loc}}(\mathbf{R}^n) \subset \mathcal{D}'(\mathbf{R}^n) \subset \mathcal{B}(\mathbf{R}^n). \qquad (7.4.6)$$

We have indicated above how these inclusions came about in the case of $\mathbf{R}$, but the general case requires a fair amount of work and we refer to Hörmander (1985), Treves (1975) or Schwartz (1959) for the theory of distributions, to de Rham (1960) for the theory of differential forms with distribution coefficients (currents), and to Schapira (1970), Komatsu (1973), Guillemin et al. (1979), or Hörmander (1985) for the theory of hyperfunctions.

Let Y be a real-analytic manifold of real dimension n, then there always exists a complexification of Y, Y^c, which is a complex manifold of complex dimension n, with $Y \subset Y^c$, and $T(Y^c)|_Y \cong T(Y) \otimes \mathbf{C}$, i.e., we can choose local, holomorphic coordinates $(z_1, \ldots, z_n)$ for Y^c at a point $p \in Y$ so that Y has the form $[\text{Im } z_1 = \cdots = \text{Im } z_n = 0]$ near p. Such complexifications always exist and are not unique. Using such a complexification, we can define hyperfunctions on the manifold Y by setting

$$\mathcal{B}(Y) := H^n_Y(Y^c, \Omega^n).$$

It is a fact that $\mathcal{B}(Y)$ does not depend on which complexification is chosen. It follows from their definition that hyperfunctions on Y

form a module over $\mathcal{A}_Y$, the sheaf of real-analytic functions on Y. However, just as for distributions, there is no well-defined notion of multiplication of hyperfunctions by themselves. If $\mathcal{A}^p$ is the sheaf of real-analytic p-forms on Y, then we can write

$$\mathcal{B}^p := \mathcal{B} \otimes_A \mathcal{A}^p,$$

the tensor product of the real-analytic p-forms and the hyperfunctions. These are the hyperfunction-valued p-forms.

Now we return to twistor geometry. We recall that the set $P = \mathbf{P}^0 = \{L_1 \subset \mathbf{T} : \Phi(L_1) = 0\}$ corresponds to $M = \mathbf{M}^0 = \{L_2 \in \mathbf{T} : \Phi(L_2) = 0\}$, under the Klein correspondence. For suitable $U \subset \mathbf{M}$, the basic Penrose transform has the form that first cohomology on $\tilde{U}$ with the sheaf of solutions of the Cauchy–Riemann equations of homogeneity $(-2n - 2)$ on $\tilde{U}$ maps to holomorphic solutions of the massless field equations of helicity $n/2$. We shall generalize to a mapping of data from P to M by considering the hyperfunction solutions of the *tangential* Cauchy–Riemann equations on P. On an open set in $\mathbf{C}^n$, the hyperfunctions solutions of the Cauchy–Riemann equations will be holomorphic, but on submanifolds this need not be the case with the tangential Cauchy–Riemann equations, as we shall see shortly. Let us now look at the notion of tangential Cauchy–Riemann equations more carefully.

Consider a real hypersurface Y in a complex manifold X of complex dimension n. Suppose $Y = \{x \in X : \rho(x) = 0\}$, and $d\rho \neq 0$ on Y, where $\rho \in C^\infty(X)$. Consider $\bigwedge^{0,q} T^*(X)|_Y$. We want to define the vector bundle of $(0, q)$-covectors which are tangential to Y.

Consider a real hypersurface Y embedded in a complex manifold X of complex dimension n, and let $\iota : Y \to X$ denote the embedding. Consider the complex of complex-valued real-analytic forms on $U \subset Y$:

$$\cdots \to \mathcal{A}_Y^q(U) \xrightarrow{d} \mathcal{A}_Y^{q+1}(U) \xrightarrow{d} \cdots.$$

We let

$$\mathcal{A}_Y^{0,q}(U) := \iota^*(\mathcal{A}_X^{0,q}(\tilde{U})),$$

where $\tilde{U}$ is an open set in X such that $\tilde{U} \cap Y = U$, and where ι^* is the pullback mapping to the submanifold Y. We see that there is a natural induced mapping

$$\bar{\partial}_Y : \mathcal{A}_Y^{0,q}(U) \to \mathcal{A}_Y^{0,q+1}(U),$$

so that the diagram

$$
\begin{array}{ccc}
\mathcal{A}_Y^{0,q}(U) & \xrightarrow{\ \bar{\partial}_Y\ } & \mathcal{A}_Y^{0,q+1}(U) \\[2pt]
\iota^* \big\uparrow & & \big\uparrow \iota^* \\[2pt]
\mathcal{A}_X^{0,q}(\widetilde{U}) & \xrightarrow{\ \bar{\partial}\ } & \mathcal{A}_X^{0,q+1}(U)
\end{array}
$$

commutes. This is the induced Dolbeault complex on $Y \subset X$. It is a fact, which we won't verify here, that this complex of sheaves is *intrinsic* to the geometry of Y (see Polking and Wells 1978 for details). This definition is well-suited for computation, however. To calculate $\bar{\partial}_Y$ acting on a given $(0,q)$-form on Y, just take any extension to the ambient space, calculate $\bar{\partial}$ of that extension, and then pull back to Y again. If $\mathcal{B}$ is the sheaf of hyperfunctions on Y, then $\mathcal{B}$ is a module over the sheaf of rings $\mathcal{A}$, and hence we can define the complex of $\mathcal{B}$-valued $(0,q)$-forms on Y, letting

$$
\mathcal{B}_Y^{0,q} := \mathcal{B}_Y \otimes_{\mathcal{A}_Y} \mathcal{A}_Y^{0,q}
$$

$$
\cdots \longrightarrow \mathcal{B}_Y^{0,q-1} \xrightarrow{\ \bar{\partial}_Y\ } \mathcal{B}_Y^{0,q} \xrightarrow{\ \bar{\partial}_Y\ } \mathcal{B}_Y^{0,q+1} \longrightarrow \cdots . \qquad (7.4.7)
$$

We can tensor (7.4.7) with the restriction of any holomorphic vector bundle on X to Y, since $\bar{\partial}_Y$ will annihilate the transition functions of E on Y. We can then define the *hyperfunction Dolbeault groups* on Y as

$$
{}'H_\mathcal{B}^{0,q}(Y, E) := H^q(\Gamma(Y, \mathcal{B}_Y^{0,q}(E))).
$$

The " '' " on ${}'H_\mathcal{B}^{0,q}(Y, E)$ indicates the Dolbeault groups are *intrinsic* to the submanifold Y. Similarly, we can define the real-analytic Dolbeault group ${}'H_\mathcal{A}^{0,q}(Y, E)$ in the same manner. These cohomology groups depend only on the tangential Cauchy–Riemann structure of Y, and does not involve any other differential equations.

We now look at real Minkowski space and we let $\mathcal{Z}_{n,\mathcal{A}}'$ and $\mathcal{Z}_{n,\mathcal{B}}'$ denote the sheaf of real-analytic and hyperfunction solutions of the massless field equations of helicity $n/2$ on real compactified Minkowski space M, for $n \geq 0$, respectively, i.e.,

$$
\mathcal{Z}_{n,\mathcal{B}}' = \{ \varphi_{A'...D'} \in \mathcal{B}_{A'...D'} : \nabla^{AA'} \varphi_{A'...D'} = 0 \}
$$

and similarly for the sheaf $\mathcal{A}$ (we are dropping the notation for the conformal weights, but they are implicitly there). Our object is to show that all of the hyperfunction solutions of massless field equations on real Minkowski space can be generated by means of a Penrose transform from the intrinsic hyperfunction tangential Dolbeault groups $'H_{\mathcal{B}}^{0,1}(P, H^{-n-2})$ with coefficients in the appropriate power of the hyperplane section bundle. Moreover, there is an injection of the intrinsic real-analytic tangential Dolbeault group $'H_{\mathcal{A}}^{0,1}(P, H^{-n-2})$ into $'H_{\mathcal{B}}^{0,1}(P, H^{-n-2})$ and the restriction of $\mathcal{P}$ to this subspace yields precisely the real-analytic solutions of the same field equation, i.e., there is a diagram of the following sort,

$$
\begin{array}{ccc}
'H_{\mathcal{B}}^{0,1}(P, H^{-n-2}) & \overset{\mathcal{P}}{\underset{\cong}{\longrightarrow}} & \mathcal{Z}'_{n,\mathcal{B}}(M)[1][1]' \\
\uparrow & & \uparrow \\
'H_{\mathcal{A}}^{0,1}(P, H^{-n-2}) & \overset{\mathcal{P}}{\underset{\cong}{\longrightarrow}} & \mathcal{Z}'_{n,\mathcal{A}}(M)[1][1]'.
\end{array}
\tag{7.4.8}
$$

In this section we shall outline the proofs of these results, and indicate extensions of this to representation of hyperfunction solutions on M^4 in terms of the hyperfunction data on P^I, which is somewhat more difficult. Finally, we shall obtain as a corollary the fact that on M or M^4 any solution φ of the massless field equations of nonnegative helicity can be written as the sum of solutions of positive and negative frequency $\varphi = \varphi^+ + \varphi^-$, in the sense of Streater and Wightman (1978). It is not at all evident that in general $'H_{\mathcal{A}}^{0,q}(Y, E)$ is injected into $'H_{\mathcal{B}}^{0,q}(Y, E)$, although this will be true in a specific twistor context. Our first task is to relate the intrinsic Dolbeault groups on P in (7.4.8) to cohomology groups in the ambient space. If S is a closed subset of a complex manifold X, then we define the (extrinsic) Dolbeault groups on S by

$$
H^{0,q}(S, E) := \varinjlim_{U \supset S} H^{0,q}(U, E),
$$

for E a holomorphic vector bundle on X, and where U is an arbitrary subset of X containing S. We have the following lemma which is an analog of (7.4.2) in one variable.

Lemma 7.4.1. *Suppose that*

$$
H^1(\mathbf{P}, \mathcal{O}_{\mathbf{P}}(E)) = H^2(\mathbf{P}, \mathcal{O}_{\mathbf{P}}(E)) = 0,
$$

then

(a) $\quad 'H_{\mathcal{A}}^{0,1}(P,E) \cong H^{0,1}(\overline{\mathbf{P}^+},E) \oplus H^{0,1}(\overline{\mathbf{P}^-},E),$

$\qquad\qquad \cong H^1(\overline{\mathbf{P}^+},\mathcal{O}_{\mathbf{P}}(E)) \oplus H^1(\overline{\mathbf{P}^-},\mathcal{O}_{\mathbf{P}}(E)),$

(b) $\quad 'H_{\mathcal{B}}^{0,1}(P,E) \cong H^{0,1}(\mathbf{P}^+,E) \oplus H^{0,1}(\mathbf{P}^-,E),$

$\qquad\qquad \cong H^1(\mathbf{P}^+,\mathcal{O}_{\mathbf{P}}(E)) \oplus H^1(\mathbf{P}^-,\mathcal{O}_{\mathbf{P}}(E)).$

Proof: We use the results in Polking and Wells (1978), namely Theorems 5.6 and 6.6 of that paper, extended to vector bundle coefficients (see also Hill and MacKichan 1977). We obtain

$$\cdots \longrightarrow H^{0,1}(\mathbf{P},E) \longrightarrow H^{0,1}(\overline{\mathbf{P}^+},E) \oplus H^{0,1}(\overline{\mathbf{P}^-},E)$$
$$\xrightarrow{b_{\mathcal{A}}} 'H_{\mathcal{A}}^{0,1}(P,E) \longrightarrow H^{0,2}(\mathbf{P},E) \longrightarrow \cdots,$$

$$\cdots \longrightarrow H^{0,1}(\mathbf{P},E) \longrightarrow H^{0,1}(\mathbf{P}^+,E) \oplus H^{0,1}(\mathbf{P}^-,E)$$
$$\xrightarrow{b_{\mathcal{B}}} 'H_{\mathcal{B}}^{0,1}(P,E) \longrightarrow H^{0,2}(\mathbf{P},E) \longrightarrow \cdots.$$

Using the hypotheses, and Dolbeault's theorem (Theorem 3.3.4) we obtain the desired conclusion. The mappings $b_{\mathcal{A}}$ and $b_{\mathcal{B}}$ represent the 'jumps' in the boundary values of the cohomology classes on $\mathbf{P}^{\pm}$, completely analogous to the jumps in the boundary values of f^+, f^- in (7.4.2) given by (7.4.4). ∎

Corollary 7.4.2. *If $E = H^n, n \in \mathbf{Z}$, then*

(a) $\quad 'H_{\mathcal{A}}^{0,1}(P,H^n) \cong H^1(\overline{\mathbf{P}^+},\mathcal{O}_{\mathbf{P}}(n)) \oplus H^1(\overline{\mathbf{P}^-},\mathcal{O}_{\mathbf{P}}(n)),$

(b) $\quad 'H_{\mathcal{B}}^{0,1}(P,H^n) \cong H^1(\mathbf{P}^+,\mathcal{O}_{\mathbf{P}}(n)) \oplus H^1(\mathbf{P}^-,\mathcal{O}_{\mathbf{P}}(n)).$

Proof: It follows from Example 3.3.9 that

$$H^j(\mathbf{P},\mathcal{O}_{\mathbf{P}}(n)) = 0, \quad \text{for } j = 1,2$$

and for all integers n. Thus the hypotheses of Theorem 7.4.1 are satisfied. ∎

We now have the following lemma, which shows that the real-analytic data on P injects into the hyperfunction data on P.

Lemma 7.4.3. *The natural homomorphism*

$$'H_{\mathcal{A}}^{0,1}(P, H^m) \to 'H_{\mathbf{B}}^{0,1}(P, H^m) \tag{7.4.9}$$

is an injection for all integers m.

Proof: We see that the Penrose transform $\mathcal{P}$ yields the commutative diagram

$$
\begin{array}{ccc}
H^1(\mathbf{P}^\pm, \mathcal{O}(-2n-2)) & \xrightarrow[\cong]{\mathcal{P}} & \mathcal{Z}_n'(\mathbf{M}^\pm) \\
\uparrow & & \uparrow \\
H^1(\overline{\mathbf{P}^\pm}, \mathcal{O}(-2n-2)) & \xrightarrow[\cong]{\mathcal{P}} & \mathcal{Z}_n'(\overline{\mathbf{M}^\pm}).
\end{array}
$$

Since the right-hand vertical arrow is injective, then the left-hand vertical arrow is also injective. Using Lemma 7.4.1, we see that the mapping in (7.4.9) is injective. ∎

We can define the boundary value mapping

$$b : \mathcal{Z}_n'(\overline{\mathbf{M}^+}) \oplus \mathcal{Z}_n'(\overline{\mathbf{M}^-}) \to \mathcal{Z}_{n,\mathcal{A}}'(M^4)$$

by setting

$$b(f^+, f^-) = \lim_{y \to 0} [f^+(x+iy) - f^-(x-iy)],$$

for $x \in M^4$, $y >> 0$, where we have represented $\mathbf{M}^\pm$ as in Proposition 1.6.2. We have the composition

$$'H_{\mathcal{A}}^{0,1}(P, H^{-2n-2}) \xrightarrow{\cong} \begin{array}{c} H^1(\overline{\mathbf{P}^+}, \mathcal{O}(-2n-2)) \\ \oplus \\ H^1(\overline{\mathbf{P}^-}, \mathcal{O}(-2n-2)) \end{array}$$

$$\xrightarrow{\mathcal{P}} \begin{array}{c} \mathcal{Z}_n'(\overline{\mathbf{M}^+}) \\ \oplus \\ \mathcal{Z}_n'(\overline{\mathbf{M}^-}) \end{array} \xrightarrow{b} \mathcal{Z}_{n,\mathcal{A}}'(M^4)$$

This composition is then a Penrose transform of the form

$$\mathcal{P} : 'H_{\mathcal{A}}^{0,1}(P, H^{-2n-2}) \to \mathcal{Z}_{n,\mathcal{A}}'(M^4). \tag{7.4.10}$$

The hyperfunction Penrose transform is an extension of (7.4.10). We formulate this as a theorem. It is illustrated in Figure 7.4.1.

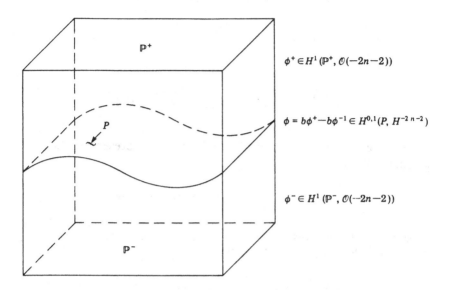

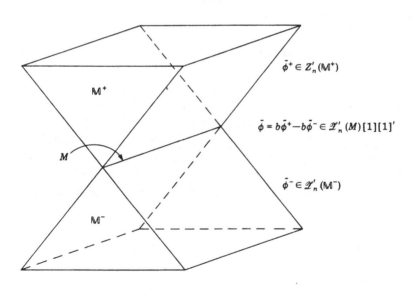

Fig. 7.4.1. The hyperfunction Penrose transform.

Theorem 7.4.4. *For $n \geq 0$, there exist Penrose transforms $\mathcal{P}_A$ and $\mathcal{P}_B$ which are isomorphisms and a commutative diagram*

$$
\begin{array}{ccc}
'H_A^{0,1}(P, H^{-2n-2}) & \xrightarrow{\mathcal{P}_A} & \mathcal{Z}'_{n,A}(M)[1][1]' \\
\downarrow & & \downarrow \\
'H_B^{0,1}(P, H^{-2n-2}) & \xrightarrow{\mathcal{P}_B} & \mathcal{Z}'_{n,B}(M)[1][1]'.
\end{array}
$$

This theorem was proved in Wells (1981) by representing the hyperfunction data in the form

$$'H^{0,1}(P, H^{-2n-2}) = H_P^2(\mathbf{P}, \mathcal{O}(-2n-2)).$$

This is not difficult to see from the representation given in Corollary 7.4.2. One then computes the pullback

$$\mu^* : H_P^2(\mathbf{P}, \mathcal{O}(-2n-2)) \to H_F^5(\mathbf{F}, \mu^{-1}\mathcal{O}(-2n-2)),$$

and then uses a generalized Leray direct image theorem to get a mapping of the form

$$\nu_* : H_F^5(\mathbf{F}, \mathcal{F}') \to H_M^4(\mathbf{M}, \mathcal{F}'')$$

for suitable sheaves $\mathcal{F}'$ and $\mathcal{F}''$. Essentially the machinery of Eastwood, Penrose and Wells (1981) is generalized to the level of relative cohomology. This does not work for the case of relative cohomology on the noncompact set P^I, i.e., for the cohomology groups $H_{P^I}^2(\mathbf{P}^I, \mathcal{O}(-2n-2))$, which is the hyperfunction data on P^I expressed in the form of relative cohomology. In Bailey, Ehrenpreis and Wells (1982) the following more general result is proven using different techniques.

Theorem 7.4.5. *There is a canonical linear isomorphism*

$$\mathcal{P} : H_{P^I}^2(\mathbf{P}^I, \mathcal{O}_{\mathbf{P}}(-2n-2)) \to \mathcal{Z}'_{n,B}(M^I).$$

This is in turn a consequence of the following splitting theorem.

Theorem 7.4.6. *Given $u \in \mathcal{Z}'_{n,\mathcal{B}}(M^I)$ there exists $u^\pm \in \mathcal{Z}_n(\mathsf{M}^\pm)$ with $u = bu^+ - bu^-$.*

Remark: In the physics terminology we would say that any hyperfunction massless field is the sum of positive and negative frequency massless fields (cf. Streater and Wightman 1978).

Proof: We shall only outline the proof, as it depends on some delicate points in the theory of microlocal analysis of Sato (the complete details are contained in Bailey, Ehrenpreis and Wells 1982). We shall consider only the case $n = 0$, as that has all the essential difficulties. Consider any hyperfunction solution u of $\Box u = 0$ on $M^4(= M^I)$, then from the general theory of characteristics of partial differential equations in the sense of Sato's microfunctions, we can represent u as the boundary values of holomorphic solutions in truncated cones which are contained in $\mathsf{M}^\pm$. More precisely, consider the open cones in M^I defined by

$$T^\pm := \{x^a - iy^a : 2(y^0)^2 - (y^1)^2 - (y^2)^2 - (y^3)^2 > 0, \pm y^0 > 0\}.$$

These open cones are open subsets of $\mathsf{M}^\pm$. Let $T^\pm$ represent a generic truncation of the cones $T^\pm$ as pictured in Figure 7.4.2.

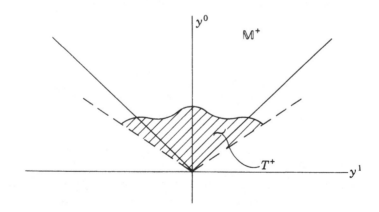

Fig. 7.4.2. Truncation of the cones for hyperfunctions represented as boundary values of holomorphic functions.

The general theory of Sato (using the fundamental results in Sato, Kashiwara and Kawai 1973; for details see Bailey, Ehrenpreis and Wells 1982) shows that because the characteristics of $\Box$ are the null cones (the boundary of $M^{\pm}$), there are holomorphic functions $w^{\pm}$ defined in a suitable truncation of $T^{\pm}$, which we denote by $T_0^{\pm}$, so that $u = bw^+ - bw^-$. In general a hyperfunction on $\mathbf{R}^4$ is the boundary value of holomorphic functions in cones which approach $\mathbf{R}^4$ from *all* directions, but since $\Box u = 0$, it follows from the general theory that the holomorphic functions can be restricted to lie in the truncated cones specified above (in other words, the holomorphic functions in the complementary cones can be taken to be zero). Now we apply $\Box$ to $w^{\pm}$, obtaining $b\Box w^+ - b\Box w^- = 0$. So if we define $f := b\Box w^+$, it follows from the 'edge of the wedge theorem' (see Komatsu 1973), that f is holomorphic in an open neighborhood of M^4. In Figure 7.4.3 the edge of the wedge theorem is pictured for $M^{\pm}$.

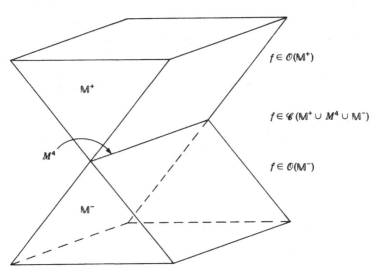

Fig. 7.4.3. Edge of the wedge theorem.

It asserts that: any holomorphic function in $M^{\pm}$ which is continuous across $\partial M^+ \cap \partial M^- = M^4$ is holomorphic across M^4. This is a generalized Hartogs–Riemann Removable Singularity type theorem. We have used a localized version of this above. Now, by using the integral formula for the solution of the equation $\Box u = f$ on M^4 (see, e.g., Courant and Hilbert II 1965), we can solve the equation $\Box u = f$, for the f given above, for real-analytic u on M^4. Such a u must be the restriction of a holomorphic function defined near M^4 in M^I, which we also denote by u. We now set $u^{\pm} := w^{\pm} - u$, and we see that $u^{\pm}$ is holomorphic in $T_0^{\pm}$ (a possibly smaller truncation), and $u^{\pm}$ satisfies $\Box u^{\pm} = 0$, where defined. Now we can shrink $T_0^{\pm}$ to new truncations $T_1^{\pm}$ which will satisfy the criterion of Theorem 7.2.3, namely that the intersection of α-planes with $T_1^{\pm}$ are one-connected. We now consider $\widehat{T}_1^{\pm}$ in **P**, and we have the Penrose isomorphism from Theorem 7.2.3.

$$\mathcal{P} : H^1(\widehat{T}_1^{\pm}, \mathcal{O}(-2)) \xrightarrow{\cong} \mathcal{Z}_0(T_1^{\pm}),$$

and we define

$$\hat{v} := \mathcal{P}^{-1}(u^{\pm}) \in H^1(\widehat{T}_1^{\pm}, \mathcal{O}(-2)).$$

We now see that if $T := T_1^+ \cup M^4 \cup T_1^-$, then $\widehat{T}$ is a neighborhood of P^I in $\mathbf{P}^I$. We let $\widehat{T}^{\pm} := \mathbf{P}^{\pm} \cap \widehat{T}$, and then we have the commutative diagram

$$
\begin{array}{ccc}
& & H^1(\mathbf{P}^+, \mathcal{O}(-n-2)) \\
H^1(\mathbf{P}^I, \mathcal{O}(-n-2)) & \xrightarrow{\eta} & \oplus \\
& & H^1(\mathbf{P}^-, \mathcal{O}(-n-2)) \\
& & \quad\downarrow \epsilon \\
& & H^1(\widehat{T}^+, \mathcal{O}(-n-2)) \\
H^1(\widehat{T}, \mathcal{O}(-n-2)) & \xrightarrow{\kappa} & \oplus \\
& & H^1(\widehat{T}^-, \mathcal{O}(-n-2))
\end{array}
$$

$$
\begin{array}{ccccc}
\xrightarrow{\alpha} & H^2_{P^I}(\mathbf{P}^I, \mathcal{O}(-n-2)) & \to & H^2(\mathbf{P}^I, \mathcal{O}(-n-2)) \\
& \quad\downarrow \beta & & \quad\downarrow \\
\xrightarrow{\gamma} & H^2_{P^I}(\widehat{T}, \mathcal{O}(-n-2)) & \xrightarrow{\zeta} & H^2(\widehat{T}, \mathcal{O}(-n-2)),
\end{array}
$$

where the rows are the relative cohomology exact sequence for the pairs $(\mathbf{P}^I, \mathbf{P}^I - P^I)$ and $(\widehat{T}, \widehat{T} - P^I)$, respectively (where we note that $\mathbf{P}^I - P^I = \mathbf{P}^+ \cup \mathbf{P}^-$, etc.). All of the vertical mappings are restrictions and β is an isomorphism (the 'excision isomorphism', see Komatsu 1973). Essentially the hyperfunction data on P^I does not depend on which ambient complex manifold is used, in this case **P** or $\widehat{T}$, and this is a general fact. Since $\mathbf{P}^I$ can be covered by two Stein manifolds it follows that $H^2(\mathbf{P}^{\pm}, \mathcal{O}_{\mathbf{P}}(-2)) = 0$ (see Exercise

7.4.4). The mapping η is injective as can be seen by considering the Minkowski space interpretation via the Penrose transform. If we select an arbitrary inverse α^{-1} for α we observe that

$$\varepsilon\alpha^{-1}\beta^{-1}\gamma(\hat{v}^+ + \hat{v}^-) - (\hat{v}^+ + \hat{v}^-) \in \text{im } \kappa.$$

Thus in Minkowski space, if we set

$$u^\pm := \mathcal{P}\alpha^{-1}\beta^{-1}\gamma\hat{v}^\pm,$$

we obtain

$$bu^+ - bu^- = b\mathcal{P}(\hat{v}^+) - b\mathcal{P}(\hat{v}^-)$$
$$= u.$$

■

Corollary 7.4.7. *There is a canonical isomorphism*

$$\mathcal{P}_\mathcal{B} : H^2_{P^I}(\mathbf{P}^I, \mathcal{O}_\mathbf{P}(-2n-2)) \to \mathcal{Z}'_{n,\mathcal{B}}(M^4).$$

Proof: This follows directly from the proof of the above theorem. ■

It is not too difficult to show from Theorem 7.4.6 that Theorems 7.4.4 and 7.4.5 hold, but we shall omit the details here (see Bailey, Ehrenpreis and Wells 1982).

Exercises for Section 7.4

7.4.1. Show that $C^q(\mathcal{F})$ is a flabby sheaf for each $q \geq 1$.

7.4.2. Verify that the hyperfunction on $\mathbf{R}$ defined by

$$f^\pm = e^{1/z}, \quad \mathbf{C}^\pm,$$

is not a distribution. Express this hyperfunction as an infinite series of derivatives of the δ-measure.

7.4.3. Verify Theorem 7.4.4 by using the representation

$$'H^{0,1}(P, H^{-2n-2}) = H^2_P(\mathbf{P}, \mathcal{O}(-2n-2)).$$

7.4.4. Show that $\mathbf{P}^I$ can be covered by two Stein manifolds, and that from this it follows that

$$H^2(\mathbf{P}^\pm, \mathcal{O}_\mathbf{P}(-2)) = 0.$$

8

Self-dual gauge fields

This chapter deals with the twistor construction of solutions of the anti-self-dual Yang–Mills equations $*F_{ab} = -F_{ab}$ on Euclidean four-space E^4. Of course, the construction applies equally well to the self-duality equations $*F_{ab} = F_{ab}$, simply by switching orientation on E^4, or (equivalently) by using dual twistors instead of twistors. And it also applies to the Lorentzian signature versions $*F_{ab} = \pm i F_{ab}$, by analytic continuation. But, as we saw in Chapter 5, the most important applications of (anti)-self-dual gauge fields (namely instantons and magnetic poles) have involved the Euclidean version.

The basic theorem relates anti-self-dual gauge fields to holomorphic vector bundles over projective twistor space. This correspondence is described in detail in §8.1. It enables one to obtain, in principle at least, all solutions of the semi-linear system of first order partial differential equations $*F_{ab} = -F_{ab}$.

At the end of this section is a discussion of massless fields defined on an anti-self-dual gauge field background, combining the description of such gauge fields in terms of vector bundles over twistor space with the cohomological description of massless fields given in Chapter 7.

One way of constructing holomorphic vector bundles is to extend vector bundles of lower rank. For example, extending a line bundle by another line bundle gives a vector bundle of rank two. In general, not all vector bundles are obtainable by this process. But large classes of them are, including in particular the bundles corresponding to $SU(2)$-instantons and to $SU(2)$ magnetic poles. Indeed, it was in the context of instantons that the idea of extensions was first applied to anti-self-dual gauge fields (Atiyah and Ward 1977; Corrigan et al. 1978b): it gives rise to a sequence of 'ansätze' which convert solutions of linear equations (namely certain massless field equations) into solutions of the anti-self-dual Yang–Mills equations. These ansätze (a more general version of those appearing in Atiyah and Ward 1977, and Corrigan et al. 1978b) are described in §8.2.

Then in §8.3 we come to the specific problem of instantons, or, more precisely, anti-self-dual gauge fields on S^4 (cf. §5.2). One can construct the appropriate holomorphic vector bundles over projective twistor space in two different ways: either as extensions, or by the 'monad' method of Horrocks (1964). The latter has become well-known, in the context of instantons, as the Atiyah–Drinfeld–Hitchin–Manin (ADHM) construction (Atiyah et al. 1978). We shall give the essential features, and some of the details, of both methods of obtaining instanton solutions.

These two constructions have analogues which apply to the problem of magnetic poles, and §8.4 deals with this subject.

In these applications (namely instantons and magnetic poles), we shall generally restrict our attention to the gauge group SU(2), this being the 'simplest' nonabelian compact group. Much of the theory extends to more general gauge groups, but things become more complicated, and in many cases have not yet been fully worked out.

8.1 Correspondence between self-dual gauge fields and holomorphic bundles

This section describes the correspondence between solutions of $*F_{ab} = -F_{ab}$ and holomorphic vector bundles over twistor space. There are two alternative descriptions of this correspondence: either in terms of the 'double fibration' picture $\mathbf{M} \leftarrow \mathbf{F} \rightarrow \mathbf{P}$, or in terms of the 'single fibration' picture $\mathbf{P} \rightarrow S^4$ (cf. Chapter 1). We shall mainly use the former description, and give an outline of what things look like in terms of the latter.

Thus, to begin, let us consider the 'complexified' problem of gauge fields on $\mathbf{M}^I \cong \mathbf{C}^4$ (complexified Euclidean space), with gauge group $GL(n, \mathbf{C})$, satisfying the equations $*F_{ab} = -F_{ab}$. Previously, we only defined gauge theory on M^4 and E^4, but its analytic extension to $\mathbf{M}^I$ is evident. The gauge potential Φ_a is now a holomorphic function on $\mathbf{M}^I$ and all of the equations extend by complex linearity (note that we are writing Φ_a for the gauge potential instead of the often used A_a, which we also used in Chapter 5; we therefore have less of a conflict with the spinorial substitution $a \mapsto AA'$ which will play a major role in this chapter). Recall that projective twistors (that is, points of $\mathbf{P} \cong \mathbf{P}_3(\mathbf{C})$) correspond to totally-null self-dual complex two-planes in $\mathbf{M}$ (α-planes). The following proposition is the key to the twistor description of anti-self-dual gauge fields. In it, let U be an open set in $\mathbf{M}$ which is *elementary*, i.e., such that for every self-dual plane $\widetilde{Z}$ that intersects U, the intersection $\widetilde{Z} \cap U$ is connected and simply-connected; and let Φ_a denote a gauge potential, D_a the corresponding derivative operator (connection) acting on sections of a bundle V over U, and F_{ab} the corresponding gauge field (curvature). The gauge group is irrelevant, but for the sake of being specific, let it be $GL(n, \mathbf{C})$.

Proposition 8.1.1. *The curvature F_{ab} is anti-self-dual if and only if for every self-dual plane $\widetilde{Z}$ that intersects U, the restriction of D_a to $U \cap \widetilde{Z}$ is integrable.*

Proof: The basic idea is that self-dual two-forms are orthogonal to anti-self-dual two-forms, so the restriction of an anti-self-dual curvature to a self-dual two-plane vanishes, and therefore the restricted connection is integrable, or flat. By $D_a|_{U \cap \widetilde{Z}}$ being integrable we mean that the condition

$$v^a D_a \psi = 0$$

for all vector fields v^a tangent to $\widetilde{Z}$ (ψ being a section of the vector bundle V over U), uniquely determines ψ on $U \cap \widetilde{Z}$ in terms of its value at any given point of $U \cap \widetilde{Z}$. Since $U \cap \widetilde{Z}$ is connected and simply-connected, this is equivalent to the condition that the restriction of the curvature form F_{ab} to $U \cap \widetilde{Z}$ should vanish, i.e., that

$$v^a w^b F_{ab} = 0 \tag{8.1.1}$$

for all vectors v^a and w^b tangent to $\widetilde{Z}$.

Now if $\widetilde{Z}$ corresponds to the point in twistor space whose homogeneous coordinates are $(\omega^A, \pi_{A'})$, then any vector v^a tangent to $\widetilde{Z}$ has the form $v^a = \lambda^A \pi^{A'}$, for some spinor λ^A. So equation (8.1.1) is equivalent to

$$\varphi_{A'B'} \pi^{A'} \pi^{B'} = 0,$$

where $\varphi_{A'B'} \varepsilon_{AB}$ is the self-dual part of F_{ab} (cf. §4.2). This is the condition for integrability on $\widetilde{Z}$ (the spinor $\pi_{A'}$ being associated with that particular $\widetilde{Z}$). Therefore integrability for *all* $\widetilde{Z}$ is equivalent to $\varphi_{A'B'} = 0$, namely the vanishing of the self-dual part of F_{ab}. ∎

This simple geometric characterization of anti-self-duality is at the heart of the following basic theorem. In it, we use the same notation as in Proposition 8.1.1; in addition, let $\widehat{U}$ denote that open set in $\mathbf{P}$ whose points correspond to self-dual planes that intersect U, and let $\hat{x}$ denote the line in $\mathbf{P}$ corresponding to $x \in \mathbf{M}^I$.

Theorem 8.1.2. *Let U be an elementary open set in* **M**. *There is a natural one-to-one correspondence between*

   (a) *anti-self-dual* $\mathrm{GL}(n, \mathbf{C})$-*gauge fields on* U; *and*
   (b) *holomorphic rank-n vector bundles E over $\widehat{U}$, such that E restricted to $\hat{x}$ is trivial for all $x \in U$.*

Proof: We shall describe how to go from (a) to (b) and then how to go from (b) to (a). It is worth remarking first, though, that the objects in (a) and (b) should be regarded as being specified 'modulo' the usual equivalence relations. In other words, 'gauge fields' means 'gauge fields modulo gauge transformations', and 'bundles' means 'bundles modulo equivalence'.

Suppose, then, that we are given an anti-self-dual $\mathrm{GL}(n, \mathbf{C})$ vector bundle V defined on $U \subseteq \mathbf{M}^I$. To construct a vector bundle E over $\widehat{U}$, we must assign a copy of the vector space $\mathbf{C}^n$ to each point Z of $\widehat{U}$, this $\mathbf{C}^n$ being the fiber E_Z over Z. This assignment is given by the definition

$$E_Z := \{\psi : D\psi = 0 \text{ on } \widetilde{Z}\}.$$

Here ψ is a section of the vector bundle V restricted to $\widetilde{Z} \subset U$, and the definition requires it to be covariantly constant on $\widetilde{Z}$, i.e., to satisfy the propagation equation

$$v^a D_a \psi = 0 \tag{8.1.2}$$

for all v^a tangent to $\widetilde{Z}$. In view of Proposition 8.1.1, equation (8.1.2) is integrable: given ψ at one point on $\widetilde{Z}$, the equation provides a consistent rule for propagating ψ all over $\widetilde{Z}$. To put this another way, the space E_Z of solutions to (8.1.2) is isomorphic to $\mathbf{C}^n$. Since the whole procedure is holomorphic, we have constructed a holomorphic vector bundle E over $\widehat{U} \subset \mathbf{P}$.

Furthermore, the bundle E defined in this way possesses the desired property of being trivial when restricted to a line $\hat{x}$. To see this, let x be a point in **M**. Then a vector $\psi \in V_x$ at x determines a parallel field ψ on each of the self-dual planes through x, and hence determines a point in E_Z for every point Z on the line $\hat{x}$ in $\mathbf{P}$. In other words, each $\psi \in V_x \cong \mathbf{C}^n$ determines a holomorphic section of $E|_{\hat{x}}$. In particular, choosing n linearly independent ψs in V_x gives us n linearly independent sections of $E|_{\hat{x}}$. So $E|_{\hat{x}}$ is trivial.

This completes the proof of the first part of the theorem. We shall now give a more explicit construction for E, by 'patching'. Although this is not an essential part of the proof, it is perhaps useful to see how the gauge potential Φ_a determines transition functions for the holomorphic vector bundle E defined on $\widehat{U}$. If $\{W_0, W_1, \dots\}$ is an open covering of $\widehat{U}$, for which $E|_{W_j}$ is trivial, then there are transition functions F_{ij} which are matrix-valued functions of the form

$$F_{ij} : W_i \cap W_j \to \mathrm{GL}(n, \mathbf{C})$$

so that sections of E are given by vector-valued functions ψ_i defined on W_i and satisfying

$$\psi_i = F_{ij}\psi_j,$$

as we saw in (2.1.4–6) and Proposition 2.1.3. In general, transition functions must satisfy a compatibility condition $F_{ij}F_{jk}F_{ki} = \mathrm{id}$, as in (2.1.6), so we are not free to choose arbitrary F_{ij} to be transition functions. However, in the case of a covering by *two* open sets on each of which the bundle is trivial, the compatibility condition becomes void. This happens in certain special cases as we see below. We then let $F = F_{01}$ for the two open sets $W_0 \cup W_1 = \widehat{U}$, and refer to F as the *transition matrix*.

For the sake of simplicity, let us take $U = \mathsf{M}^I$ and $\widehat{U} = \mathsf{P}^I$. In other words, we assume that Φ_a is holomorphic on the whole of M^I. Let $Z^\alpha = (\omega^A, \pi_{A'})$ be homogeneous coordinates for P as usual (cf. §2.3). Let P and Q be points in $\widehat{U}$ with coordinates $P^\alpha = (0, 0, 1, 0)$ and $Q^\alpha = (0, 0, 0, 1)$, respectively. As usual, $\widetilde{P}$ and $\widetilde{Q}$ denote the corresponding two-planes in U. If $\widetilde{Z}$ is a third two-plane in U, given by the equation $\omega^A = ix^{AA'}\pi_{A'}$, does $\widetilde{Z}$ intersect $\widetilde{P}$ or $\widetilde{Q}$? To find the intersection of $\widetilde{Z}$ and $\widetilde{P}$ means solving the simultaneous equations

$$\omega^A = ix^{AA'}\pi_{A'},$$
$$0 = ix^{A0'},$$

for x; and these equations have a unique solution (corresponding to a point p_Z in U) if $\pi_{1'} \neq 0$. Similarly, if $\pi_{0'} \neq 0$, then $\widetilde{Z}$ and $\widetilde{Q}$ intersect in a single point q_Z. The geometry of all this is depicted in Figure 8.1.1.

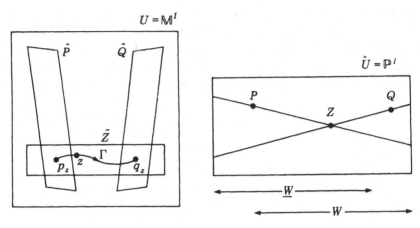

Fig. 8.1.1. Geometry of the Klein correspondence.

Define two open subsets W and $\underline{W}$ of $\widehat{U}$ by

$$W = \{(\omega^A, \pi_{A'}) : \pi_{1'} \neq 0\}$$
$$\underline{W} = \{(\omega^A, \pi_{A'}) : \pi_{0'} \neq 0\}.$$

These two coordinate charts cover $\widehat{U} \cong \mathbf{P}^I$, since the line I is given by $\pi_{A'} = 0$. In order to avoid a profusion of subscripts, we are using the somewhat simpler notation W, $\underline{W}$ instead of W_0, W_1. Note that if Z is any point in W, then the intersection p_Z of $\widetilde{Z}$ and $\widetilde{P}$ exists. As a consequence, the vector bundle E, restricted to W, is trivial. The reason for this is that a covariantly constant field ψ on $\widetilde{Z}$ is determined by giving its value ψ_p at the point p_Z, so $E|_W$ is holomorphically isomorphic to $W \times \mathbf{C}^n$, where the components of ψ_p serve as coordinates on $\mathbf{C}^n$ (think of ψ as a column n-vector). Similarly, $E|_{\underline{W}}$ is a product $\underline{W} \times \mathbf{C}^n$, where the components of ψ_q serve as coordinates on $\mathbf{C}^n$. On the overlap region $W \cap \underline{W}$, these two sets of coordinates ψ_p and ψ_q are related by the integrated version of the propagation rule (8.1.2), namely

$$\psi_q = F(Z^\alpha)\psi_p,$$

where $F(Z^\alpha)$ is given (in physicists' parlance) by the path-ordered exponential integral

$$F(Z) = \mathcal{P} \exp\left(-\int\limits_\Gamma \Phi_a dx^a\right),$$

Γ being a path in the two-plane $\widetilde{Z}$ from p_Z to q_Z (see Figure 8.1.1). The choice of path does not affect F, precisely because the connection is integrable over $\widetilde{Z}$. Thus we see that the transition matrix F which determines E, is given by integrating the gauge potential Φ_a, or (more accurately) by solving a set of linear ordinary differential equations in which the components of Φ_a appear as coefficients.

Let us move on now to the second part of the theorem, namely showing how to go from (b) to (a). Let E be a holomorphic rank-n vector bundle over $\widehat{U}$, such that $E|_{\hat{x}}$ is trivial for all $x \in U$. We have to construct a gauge potential Φ_a on U. Once again, we shall first give a geometric construction, and then a more explicit version using transition matrices. An abstract version, analogous to the linear Penrose transform, is given in the Remarks following this proof.

Since $E|_{\hat{x}}$ is trivial and $\hat{x}$ is a Riemann sphere, the space $\Gamma(\hat{x}, E|_{\hat{x}})$ of holomorphic sections of $E|_{\hat{x}}$ is isomorphic to $\mathbf{C}^n$, by Liouville's theorem. This assigns a copy of $\mathbf{C}^n$ to every point x; in other words, we have a rank-n vector bundle $\widetilde{E}$ over U. We now want to define a connection D_a on this bundle, which means knowing how to parallel-propagate vectors ψ in $\widetilde{E}$ along curves in U. This in turn is equivalent to knowing how to propagate ψ in *null* directions, since the null vectors span the whole tangent space of $\mathbf{M}^I$. Any null vector v^a in $\mathbf{M}^I$ is tangent to a unique self-dual two-plane $\widetilde{Z}$, and the various points on $\widetilde{Z}$ correspond to the lines in $\widehat{U}$ passing through Z (see Figure 8.1.2).

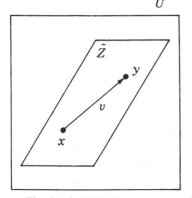
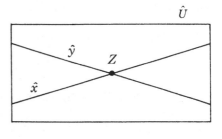

Fig. 8.1.2. Null line corresponding to intersection of two projective lines.

If $\hat{x}$ and $\hat{y}$ are two lines intersecting at Z, then clearly there is a natural identification between the sections of $E|_{\hat{x}}$ and the sections of $E|_{\hat{y}}$: namely, identify the sections at the point Z. So one knows how to propagate a vector ψ from x to y, and hence one knows the connection on $\widetilde{E}$. By construction, this connection is integrable (flat) on all self-dual two-planes, and therefore is a solution of the anti-self-dual Yang–Mills equations.

Actually, the above argument is incomplete, since we should check that the propagation rule really *does* define a connection, i.e., that it satisfies the various axioms (such as linearity) that a connection ought to possess. We could prove this abstractly (see Exercise 8.1.4), but instead we shall give an explicit description of how to extract the gauge potential Φ_a from the transition matrix $F(Z^\alpha)$ for the bundle E.

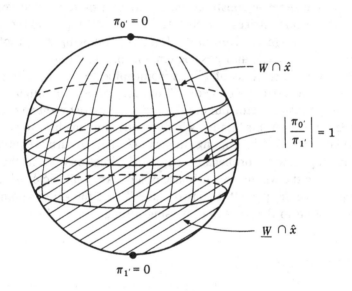

Fig. 8.1.3. Patching in the projective line $\hat{x}$.

Suppose that the twistor space $\widehat{U}$ is covered by two coordinate charts W and $\underline{W}$, with the property that for every line $\hat{x}$ in $\widehat{U}$ (given by $\omega^A = ix^{AA'}\pi_{A'}$) the intersection $W \cap \underline{W} \cap \hat{x}$ contains the circle $|\pi_{0'}/\pi_{1'}| = 1$. (See Figure 8.1.3.) Suppose the point $\pi_{0'} = 0$ lies in $W \cap \hat{x}$, while the point $\pi_{1'} = 0$ lies in $\underline{W} \cap \hat{x}$. The cover of $\mathbb{P}^I$ introduced in the first part of the proof satisfies these requirements; indeed, in that case, we had $W \cap \hat{x} = \hat{x} - \{\pi_{1'} = 0\}$ and $\underline{W} \cap \hat{x} = \hat{x} - \{\pi_{0'} = 0\}$, so that the overlap $W \cap \underline{W} \cap \hat{x}$ is the complement of the south and north poles on $\hat{x}$. More generally, for the local problem (where we are interested in gauge fields on some neighborhood U of a point $x \in \mathbf{M}^I$), it does not represent a serious loss of generality to assume that the corresponding twistor space $\widehat{U}$ admits a cover as in Figure 8.1.3.

The vector bundle E is specified by giving a holomorphic $n \times n$ transition matrix $F(Z^\alpha)$ on the intersection $W \cap \underline{W}$. The transition relation is

$$\underline{\xi} = F(Z)\xi,$$

where ξ and $\underline{\xi}$ are column n-vectors whose components serve as co-ordinates on the fibers of E above W and $\underline{W}$, respectively. The first step is to restrict E to a line $\hat{x}$ in $\widehat{U}$. This is achieved by substituting $\omega^A = ix^{AA'}\pi_{A'}$ into $F(Z^\alpha)$, thereby obtaining the transition matrix

$$G(x, \pi_{A'}) = F(ix^{AA'}\pi_{A'}, \pi_{A'}) \tag{8.1.3}$$

for the bundle $E|_{\hat{x}}$ over $\hat{x}$. We must now find the holomorphic sections of $E|_{\hat{x}}$, and this can be done as follows. Find nonsingular $n \times n$ matrices $H(x, \pi_{A'})$ and $\underline{H}(x, \pi_{A'})$, with H holomorphic for all $\pi_{A'} \in W \cap \hat{x}$ and $\underline{H}$ holomorphic for all $\pi_{A'} \in \underline{W} \cap \hat{x}$, such that the *splitting formula*

$$G = \underline{H}H^{-1} \tag{8.1.4}$$

is valid on $W \cap \underline{W} \cap \hat{x}$. Since $E|_{\hat{x}}$ is trivial, such matrices H and $\underline{H}$ must exist. Each section of $E|_{\hat{x}}$ is then given by

$$\xi = H\psi,$$
$$\underline{\xi} = \underline{H}\psi,$$

where ψ is a constant n-vector; that is, $\psi \in \mathbf{C}^n$.

The connection is now obtained by differentiating along a null direction $v^a = \lambda^A \pi^{A'}$ (cf. Figure 8.1.2). This gives

$$
\begin{aligned}
0 &= v^a D_a \psi && (\psi \text{ parallel on } \widetilde{Z}) \\
&= v^a \nabla_a \psi + v^a \Phi_a \psi \\
&= (v^a \nabla_a H^{-1})\xi + v^a \Phi_a \psi && (\psi = H^{-1}\xi) \\
&= -H^{-1}(v^a \nabla_a H)\psi + v^a \Phi_a \psi \\
&= \lambda^A \pi^{A'}(-H^{-1}\nabla_{AA'}H + \Phi_{AA'})\psi,
\end{aligned}
$$

which holds for all ψ and for all λ^A. So we deduce that Φ_a is given by

$$\pi^{A'}\Phi_{AA'} = H^{-1}\pi^{A'}\nabla_{AA'}H. \tag{8.1.5}$$

For this to make sense (that is, for the connection $\Phi_a(x)$ to be well-defined), we must prove that the right-hand side of (8.1.5) is linear in $\pi_{A'}$. Since H is a function of $\pi_{A'}$, this is not immediately obvious. To prove it, operate on the splitting formula (8.1.4) with $\pi^{A'}\nabla_{AA'}$ (this operator annihilates G, as can be seen from its definition (8.1.3) in terms of F):

$$
\begin{aligned}
0 &= \pi^{A'}\nabla_{AA'}G \\
&= \pi^{A'}\nabla_{AA'}(\underline{H}H^{-1}) \\
&= (\pi^{A'}\nabla_{AA'}\underline{H})H^{-1} - \underline{H}H^{-1}(\pi^{A'}\nabla_{AA'}H)H^{-1},
\end{aligned}
$$

so that

$$
H^{-1}\pi^{A'}\nabla_{AA'}H = \underline{H}^{-1}\pi^{A'}\nabla_{AA'}\underline{H}. \tag{8.1.6}
$$

Now the left-hand side of (8.1.6) is holomorphic for $\pi_{A'} \in W \cap \hat{x}$, while the right-hand side is holomorphic for $\pi_{A'} \in \underline{W} \cap \hat{x}$. Thus both sides are holomorphic on the whole Riemann sphere $\hat{x}$, and in addition are homogeneous of degree one in $\pi_{A'}$. So (essentially by Liouville's theorem) both sides must be linear in $\pi_{A'}$, which was what we wanted to prove.

A couple of remarks are in order here. First, there is some freedom in the choice of the matrices $\underline{H}$ and H, namely $H \mapsto H\Lambda$, $\underline{H} \mapsto \underline{H}\Lambda$, where Λ is a nonsingular $n \times n$ matrix of functions of x only (not of $\pi_{A'}$). Clearly $\underline{H}H^{-1}$ is invariant under this change. The formula (8.1.5) reveals that the corresponding transformation on Φ_a is given by

$$
\Phi_a \mapsto \Lambda^{-1}\Phi_a\Lambda + \Lambda^{-1}\nabla_a\Lambda,
$$

which is exactly a $\mathrm{GL}(n, \mathbf{C})$-gauge transformation.

Secondly, we can see directly that the gauge field is anti-self-dual by operating on (8.1.5) with $\pi^{B'}\nabla^A_{B'}$; for this leads to

$$
\nabla^A{}_{(B'}\Phi_{A')A} + \Phi^A{}_{(B'}\Phi_{A')A} = 0, \tag{8.1.7}
$$

which is a spinor version of the anti-self-duality equation $*F_{ab} = -F_{ab}$; indeed, the left-hand side of (8.1.7) is precisely $\varphi_{A'B'}$, where $\varphi_{A'B'}\varepsilon_{AB}$ is the self-dual part of F_{ab}. Thus if we take any twistor matrix $F(Z^\alpha)$, split it into $\underline{H}$ and H, and then apply the formula (8.1.5), we obtain an anti-self-dual gauge field; and *all* anti-self-dual gauge fields are obtainable in this way. In the next three sections we shall see some explicit examples of this procedure.

We have now described a map (a) $\rightarrow$ (b) from gauge fields to vector bundles, and a map (b) $\rightarrow$ (a) from vector bundles to gauge fields. It remains to be checked that these maps are inverses of each other; that is, that each of the compositions (a) $\rightarrow$ (b) $\rightarrow$ (a) and (b) $\rightarrow$ (a) $\rightarrow$ (b) is the identity. This is an immediate consequence of the geometric definitions of the maps. We shall not write out the detailed argument here, but shall end this lengthy proof by using our 'explicit formulas' to show that (a) $\rightarrow$ (b) $\rightarrow$ (a) is the identity.

Suppose we begin with Φ_a and construct $F(Z^\alpha)$ according to

$$F(Z) = \mathcal{P} \exp \left(- \int_\Gamma \Phi_a dx^a \right).$$

Let x be a point on the curve Γ (see Figure 8.1.1). Then matrices H and $\underline{H}$ which split F are

$$H = \mathcal{P} \exp \left(\int_{pz}^{x} \Phi_a dx^a \right),$$

$$\underline{H} = \mathcal{P} \exp \left(\int_{qz}^{x} \Phi_a dx^a \right),$$

obtained by partitioning the curve Γ into two segments. And the gauge potential $\tilde{\Phi}_a$ obtained from H is given by

$$\pi^{A'} \tilde{\Phi}_{AA'} = H^{-1} \pi^{A'} \nabla_{AA'} H$$
$$= \pi^{A'} \Phi_{AA'},$$

whence $\tilde{\Phi}_a = \Phi_a$, as expected.   ∎

Remarks:

(1) It is worth emphasizing that in the M-picture one has a vector bundle with connection, while in the P-picture, one has only a vector bundle (and *no* connection). All information about the connection D on $V \rightarrow U$ is coded into the holomorphic structure of the bundle $E \rightarrow \hat{U}$.

(2) The correspondence described by the theorem is conformally invariant, in the sense that the maps (a) $\rightarrow$ (b) and (b) $\rightarrow$ (a) commute with the action of the complexified conformal group on M and the

action of $PGL(4, \mathbf{C})$ on $\mathbf{P}$. It is also gauge-invariant, in an analogous sense.

(3) The open sets W and $\underline{W}$, are both biholomorphic to $\mathbf{C}^3$, as one can easily check. All holomorphic vector bundles on a convex open set in $\mathbf{C}^n$ are trivial, by a theorem of Grauert (Grauert 1960; see, e.g., Gunning and Rossi 1965), but the explicit constructions in this proof show immediately that $E|_W$ and $E|_{\underline{W}}$ are trivial, and Grauert's theorem is not needed.

(4) The transformation $E \to (\widetilde{E}, D)$ given by (b) $\to$ (a) in Theorem 8.1.2 is a special case of the *Penrose transform*, transforming data on a twistor space to solutions of field equations on Minkowski space.

(5) Consider the double fibration

$$
\begin{array}{ccc}
 & U' & \\
{}^{\mu}\swarrow & & \searrow^{\nu} \\
\widehat{U} & & U
\end{array}
$$

and the relative de Rham complex

$$\Omega^0_\mu(E) \xrightarrow{d_\mu} \Omega^1_\mu(E) \xrightarrow{d_\mu} \Omega^2_\mu(E)$$

and the induced sequence on direct image sheaves

$$\nu^0_*\Omega^0_\mu(E) \xrightarrow{\tilde{d}_\mu} \nu^0_*\Omega^1_\mu(E) \xrightarrow{\tilde{d}_\mu} \nu^0_*\Omega^2_\mu(E),$$

as in §§7.1–7.2. Then

$$\mathcal{O}_\mathbf{M}(\widetilde{E})) = \nu^0_*\Omega^0_\mu(E),$$
$$D = \tilde{d}_\mu$$

are the induced vector bundle $\widetilde{E}$ (in the form of its associated locally free sheaf), and the connection D. In other words, D is the first order operator induced from the canonical relative de Rham operator d_μ in the relative de Rham sequence on U' (see Exercise 8.1.4).

Example 8.1.3: The case $n = 1$. Here the gauge group, namely $GL(1, \mathbf{C})$, is abelian, so we are dealing with (complexified) Maxwell theory. Thus Theorem 8.1.2 says that anti-self-dual Maxwell fields correspond to certain holomorphic line bundles over twistor space.

This is something that we 'already knew' before proving Theorem 8.1.2, as the following argument demonstrates.

We recall from §7.1 that anti-self-dual Maxwell fields on U correspond to elements of the sheaf cohomology group $H^1(\widehat{U}, \mathcal{O})$. This cohomology group is related to holomorphic line bundles on $\widehat{U}$ (see Chapter 3): namely, the short exact sequence of sheaves

$$0 \to \mathbf{Z} \to \mathcal{O} \xrightarrow{\exp} \mathcal{O}^* \to 0$$

leads to a long exact sequence of cohomology groups, part of which is

$$\to H^1(\widehat{U}, \mathbf{Z}) \to H^1(\widehat{U}, \mathcal{O}) \xrightarrow{\exp} H^1(\widehat{U}, \mathcal{O}^*) \xrightarrow{c_1} H^2(\widehat{U}, \mathbf{Z}) \to .$$

Now $H^1(\widehat{U}, \mathcal{O}^*)$ is exactly the space of holomorphic line bundles over $\widehat{U}$; and if the region U in $\mathbf{M}$ is 'sufficiently nice' (for example, if it is convex), then $H^1(\widehat{U}, \mathbf{Z}) = 0$ and $H^2(\widehat{U}, \mathbf{Z}) \cong \mathbf{Z}$ (cf. §3.1). The image $c_1(E) \in H^2(\widehat{U}, \mathbf{Z})$ of a line bundle $E \in H^1(\widehat{U}, \mathcal{O}^*)$ is the Chern class of E, and $c_1(E) = 0$ if and only if $E|_{\hat{x}}$ is trivial for every $x \in U$. (This is because the second homology of $\widehat{U}$ is generated by the lines $\hat{x}$ in $\widehat{U}$; and the line bundles over each Riemann sphere $\hat{x}$ are completely characterized by their Chern class, so $E|_{\hat{x}}$ is trivial if and only if $c_1(E|_{\hat{x}}) = 0$.) Thus the set of anti-self-dual Maxwell fields on U is

$$\{\varphi_{AB} \text{ on } U : \nabla^{AA'}\varphi_{AB} = 0\}$$
$$\cong H^1(\widehat{U}, \mathcal{O})$$
$$\cong \{E \in H^1(\widehat{U}, \mathcal{O}^*) : c_1(E) = 0\}$$
$$\cong \{\text{holomorphic line bundles } E \text{ on } \widehat{U},$$
$$\text{with } E|_{\hat{x}} \text{ trivial for all } x \in U\}.$$

This is precisely the $n = 1$ version of Theorem 8.1.2, which we can therefore view as a nonlinear generalization of the Penrose transform for massless fields of helicity one.

Let us now see how the splitting construction works in this abelian case. As usual, assume $\widehat{U}$ to be covered with two coordinate charts W and $\underline{W}$ of the type described in Theorem 8.1.2. The line bundle E is specified by a nonsingular 1×1 matrix $F(Z^\alpha)$ on $W \cap \underline{W}$, that is, by a nowhere-vanishing holomorphic function on $W \cap \underline{W}$. The condition

$c_1(E) = 0$ implies that there exists a holomorphic function $f(Z^\alpha)$ on $W \cap \underline{W}$ such that $F = \exp(f)$. This follows from the exactness of the sequence

$$0 \to H^1(\widehat{U}, \mathcal{O}) \xrightarrow{\text{exp}} H^1(\widehat{U}, \mathcal{O}^*) \xrightarrow{c_1} H^2(\widehat{U}, \mathbf{Z});$$

the function f is a cocycle representing a class in $H^1(\widehat{U}, \mathcal{O})$.

In order to split $G(x, \pi_{A'}) = F(ix^{AA'}\pi_{A'}, \pi_{A'})$ multiplicatively as $\underline{H}H^{-1}$, we may split $g(x, \pi_{A'}) = f(ix^{AA'}\pi_{A'}, \pi_{A'})$ additively as $\underline{h} - h$, where $h(x, \pi_{A'})$ and $\underline{h}(x, \pi_{A'})$ are, respectively, holomorphic for $\pi_{A'} \in W_x = W \cap \hat{x}$ and for $\pi_{A'} \in \underline{W}_x = \underline{W} \cap \hat{x}$. Then $H = \exp(h)$ and $\underline{H} = \exp(\underline{h})$. Now the splitting $g = \underline{h} - h$ is a 'Taylor–Laurent' splitting, g being defined on an annular region, namely a neighborhood of the circle $|\pi_{0'}/\pi_{1'}| = 1$ on the Riemann sphere $\hat{x}$ (cf. Figure 8.1.3). So we can obtain h and $\underline{h}$ via Cauchy's integral formula, as follows.

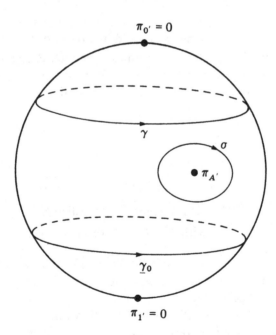

Fig. 8.1.4. Using Cauchy's integral formula.

Let γ and $\underline{\gamma}$ denote contours as depicted in Figure 8.1.4, and choosing a frame for the spinors on M, we set

$$h(x, \pi_{A'}) = \frac{1}{2\pi i} \int_\gamma (\xi_{0'} \xi^{B'} \pi_{B'})^{-1} g(x, \xi_{A'}) \pi_{0'} \Delta \xi,$$

$$\underline{h}(x, \pi_{A'}) = \frac{1}{2\pi i} \int_{\underline{\gamma}} (\xi_{0'} \xi^{B'} \pi_{B'})^{-1} g(x, \xi_{A'}) \pi_{0'} \Delta \xi,$$

where $\Delta \xi$ is the one-form $\xi_{A'} d\xi^{A'}$. The factors $\xi_{0'}^{-1}$ and $\pi_{0'}$ are included in order to ensure that the integrand be homogeneous of degree zero in both $\xi_{A'}$ and $\pi_{A'}$, and choosing them represents a choice of gauge (recall that splitting is not unique, and that the choice of a particular splitting corresponds to a choice of gauge). It is clear that h and $\underline{h}$ are holomorphic on W_x and $\underline{W}_x$, respectively, and that

$$\underline{h} - h = \int_{\underline{\gamma}} - \int_\gamma = \int_\sigma = g$$

by Cauchy's integral formula.

According to (8.1.5), the gauge potential Φ_a is determined by

$$\pi^{A'} \Phi_{AA'} = H^{-1} \pi^{A'} \nabla_{AA'} H$$
$$= \pi^{A'} \nabla_{AA'} h$$
$$= \frac{-i}{2\pi i} \int_\gamma \xi_{0'}^{-1} \partial_A g(x, \xi_{A'}) \pi_{0'} \Delta \xi \qquad (8.1.8)$$

where ∂_A represents the operator $\partial/\partial \omega^A$, i.e.,

$$\partial_A g(x, \xi_{A'}) = \left[\frac{\partial}{\partial \omega^A} f(\omega^B, \xi_{B'}) \right]_{\omega^B = ix^{BB'} \pi_{B'}}.$$

We see that the right-hand side of (8.1.8) is indeed linear in $\pi_{A'}$, as expected. After 'lifting off' $\pi_{A'}$ and then replacing $\xi_{A'}$ by $\pi_{A'}$ we obtain

$$\Phi_{AA'} = \frac{1}{2\pi i} \int i \pi_{0'} o_{A'} \partial_A g(x, \pi_{B'}) \Delta \pi,$$

where $o_{A'}$ is the constant spinor $(1, 0)$, so that $\pi_{0'} = \pi_{A'} o^{A'}$. This, then, is the integral formula for Φ_a, in a particular gauge.

Let us now compute the corresponding gauge field F_{ab}, which will be anti-self-dual: $F_{AA'BB'} = \varphi_{AB}\varepsilon_{A'B'}$, where

$$\varphi_{AB} = \nabla_{A'(A}\Phi_{B)}{}^{A'}$$

$$= -\frac{1}{2\pi i}\int \partial_A\partial_B f(ix^{CC'}\pi_{C'},\pi_{C'})\Delta\pi$$

by differentiating (8.1.9). In other words, the $n = 1$ version of the (b) $\rightarrow$ (a) transform of Theorem 8.1.2 is the same as the Penrose transform of §7.2.

We see, therefore, that the whole procedure linearizes in this case (not surprisingly, since the gauge group is abelian). In general, however, things are not quite so easy, as we shall see later in this chapter.

We return now to the general (nonabelian) case, and discuss the question of how to reduce the gauge group from $\mathrm{GL}(n,\mathbf{C})$ to a subgroup of $\mathrm{GL}(n,\mathbf{C})$. That is, we want versions of Theorem 8.1.2 which involve anti-self-dual gauge fields with gauge groups such as $\mathrm{SL}(n,\mathbf{C})$ or $\mathrm{SU}(n)$. Such versions exist for *all* subgroups of $\mathrm{GL}(n,\mathbf{C})$, but we shall only discuss the two special cases $\mathrm{SL}(n,\mathbf{C})$ and $\mathrm{SU}(n)$. First, we reduce from $\mathrm{GL}(n,\mathbf{C})$ to $\mathrm{SL}(n,\mathbf{C})$ as follows.

Theorem 8.1.4. *There is a natural one-to-one correspondence between:*
  (a) *anti-self-dual* $\mathrm{SL}(n,\mathbf{C})$-*gauge fields on* U; *and*
  (b) *holomorphic rank-n vector bundles E over $\widehat{U}$, such that $E|_{\hat{x}}$ is trivial for all $x \in U$, and $\det E$ is trivial.*

Remark: The determinant of a vector bundle E is a line bundle whose transition functions are $\det F_{ij}$, where F_{ij} are the transition matrices for E. The condition that $\det E$ be trivial is equivalent to demanding that each fiber of E may be equipped with a volume element (that is, a nonvanishing n-form) which varies holomorphically from fiber to fiber (cf. Example 2.3.1). This n-form is, in effect, a nowhere-vanishing holomorphic section of $\det E$. If $F = F(Z^\alpha)$ is a transition matrix for E, then $\det F$ is a transition function for $\det E$; so $\det E$ is trivial if and only if coordinates on the bundle can be chosen in such a way that the transition matrix $F(Z^\alpha)$ is unimodular (has determinant 1) for all Z^α.

Proof: We simply make a small addition to Theorem 8.1.2. Having an $\mathrm{SL}(n,\mathbf{C})$-gauge field on U means having an $\mathrm{SL}(n,\mathbf{C})$-connection on the n-vector bundle V over U, and this in turn means that each fiber of V can be equipped with a nonvanishing n-form, such that the resulting n-form field is covariantly constant over U. The space E_Z of covariantly constant sections of $V|_{\widetilde{Z}}$ inherits this n-form. Thus each fiber E_Z of E has a nonvanishing n-form, and this form varies holomorphically with Z. So $\det E$ is trivial.

Conversely, suppose that E admits a holomorphic nonvanishing n-form. If $\hat{x}$ is a line in $\widehat{U}$, then $V_x = \Gamma(\hat{x}, E|_{\hat{x}})$ inherits this n-form. By the construction of the connection D_a on V, this n-form is covariantly constant in all null directions, hence in all directions; so D_a is an $\mathrm{SL}(n, \mathbf{C})$-connection.

Let us describe the situation in terms of the splitting construction for the gauge field. Suppose that F is a transition matrix for E, and that $\det F = 1$. Then one can choose the matrices H and $\underline{H}$ which split F to satisfy $\det H = 1 = \det \underline{H}$. From the definition of the gauge potential Φ_a in terms of H, it then follows that $\operatorname{tr} \Phi_a = 0$, which means that Φ_a is an $\mathrm{SL}(n, \mathbf{C})$-gauge potential. ∎

We want next to consider the reduction of the gauge group from $\mathrm{GL}(n, \mathbf{C})$ to the 'real form' $\mathrm{U}(n)$. Incorporating the previous theorem will then reduce the gauge group to $\mathrm{SU}(n)$. In order to achieve a 'real' reduction such as this, we have to restrict the gauge field to a suitable 'real' subset of M, as opposed to the (four-complex-dimensional) open set U we have used up till now. (This is essentially because a holomorphic function on a complex manifold cannot be real-valued without being constant.) So we shall restrict to the Euclidean space E^4, or to its compactification S^4, as this is where most applications lie.

The other possibility is to restrict to a copy of $\mathbf{R}^4$ equipped with a metric of signature $+ + - -$; but the subject of self-dual fields on $(+ + - -)$-space has not, hitherto, been of much interest, and we shall not pursue it here. Note that an $\mathbf{R}^4$ with a metric of Lorentzian signature $+ - - -$ is no good: there are no nontrivial $\mathrm{SU}(n)$ anti-self-dual gauge fields on Minkowski space-time (cf. §5.1).

In order to facilitate our discussion, it is convenient at this point to bring in the 'single-fibration' picture $\mathsf{P} \to S^4$. Recall from §1.6 that the main details of this are as follows. Let $\sigma : Z^\alpha \mapsto \sigma(Z)^\alpha$ denote the antiholomorphic involution on P defined by

$$(\sigma(Z)^0, \sigma(Z)^1, \sigma(Z)^2, \sigma(Z)^3) = (\overline{Z^1}, -\overline{Z^0}, \overline{Z^3}, -\overline{Z^2}).$$

For any point Z in $\mathbb{P}$, the line $\hat{p}$ joining Z to $\sigma(Z)$ is a *real* line; that is, $\hat{p}$ is invariant under σ. All real lines arise in this way, and no two real lines intersect, so $\mathbb{P}$ is fibered by its real lines, and the quotient space (the space of all real lines) is S^4. The line $\hat{I}$ given by $Z^2 = 0 = Z^3$ is real, and corresponds to the 'point at infinity' on S^4; removing $\hat{I}$ gives a fibring of $\mathbb{P} - \hat{I} = \mathbb{P}^I$ over E^4. The real lines in $\mathbb{P}^I$ are given by $\omega^A = ix^{AA'}\pi_{A'}$, where $x^{AA'}$ is defined as

$$\begin{pmatrix} x^{00'} & x^{01'} \\ x^{10'} & x^{11'} \end{pmatrix} = \frac{1}{\sqrt{2}} \begin{pmatrix} -ix^0 + x^3 & x^1 - ix^2 \\ x^1 + ix^2 & -ix^0 - x^3 \end{pmatrix}$$

in terms of the four real coordinates (x^0, x^1, x^2, x^3) on E^4. Note that the reality condition on $x^{AA'}$ is

$$x^{11'} = -\overline{x^{00'}},$$
$$x^{10'} = \overline{x^{01'}}.$$

Given any open set U in S^4, let $\widehat{U}$ denote the corresponding open set of $\mathbb{P}$. That is, $\widehat{U}$ is the inverse image of U under the projection map from $\mathbb{P}$ to S^4. We can also think of $\widehat{U}$ as the bundle of almost complex structures on U (cf. §9.1).

Now given a complex n-vector bundle V over U, with connection D, we can pull both V and D back to $\widehat{U}$, thereby obtaining a vector bundle E over $\widehat{U}$, with connection $\widehat{D}$. At this stage, E is only a differentiable bundle, since V is only differentiable (not holomorphic, since, in particular, U is not a complex manifold). We want to endow E with a holomorphic structure. As a start towards this aim, there is a natural almost complex structure on E, namely the following.

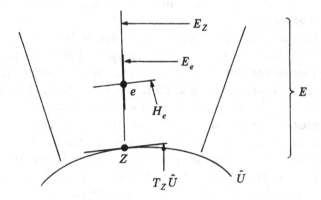

Fig. 8.1.5. The almost-complex structure on the vector bundle E.

Let e be a point of E, and let $T_e(E)$ denote its real tangent space (so $T_e(E)$ has real dimension $6 + 2n$). The connection $\widehat{D}$ on E gives us a decomposition of $T_e(E)$ into horizontal and vertical subspaces H_e and E_e, of dimension 6 and $2n$, respectively:

$$T_e(E) = H_e \oplus E_e$$

(cf. §2.4). Since each fiber E_Z is a copy of $\mathbf{C}^n$, E_e is endowed with a natural complex structure (by definition, E_e is tangent to, and isomorphic with, the fiber E_Z; see Figure 8.1.5). Moreover, H_e is naturally isomorphic to $T_Z(\widehat{U})$, which also comes equipped with a complex structure, since $\widehat{U}$ is a complex manifold. Thus $T_e(E)$ has a natural complex structure, so E has a natural *almost* complex structure. For E to be a holomorphic vector bundle, this almost complex structure must be *integrable*. (This is the Newlander–Nirenberg theorem of complex manifold theory; see, e.g., Hörmander 1973 for a proof of this theorem and a formulation of the notion of integrability of almost complex structures. See also Exercise 2.2.7.) The condition for integrability turns out to be precisely that the original connection D should be anti-self-dual. This is demonstrated in Atiyah, Hitchin and Singer (1978, §5) and leads to an alternative proof of Theorem 8.1.2, one adapted to the 'fibration' correspondence $\mathbf{P} \to S^4$. Theorem 8.1.2 was given in a 'local' form, for open subsets of $\mathbf{M}^I$, but there is no difficulty in extending it to obtain global versions; in particular, we have, in view of the above discussion, the following reformulation of Theorem 8.1.2.

Theorem 8.1.5. *There is a natural one-to-one correspondence between:*

(a) *anti-self-dual* $\mathrm{GL}(n, \mathbf{C})$*-gauge fields on S^4;*
and

(b) *holomorphic rank-n vector bundles E over $\mathbf{P} \cong \mathbf{P}_3(\mathbf{C})$, such that $E|_{\hat{x}}$ is trivial for all $x \in S^4$.*

The theorem in this form is clearly relevant to instantons, which 'live' on S^4 (cf. §5.2). But the point of introducing it here is to facilitate the reduction of the gauge group to a real form, so let us return to this problem.

Suppose that we have an anti-self-dual $\mathrm{GL}(n, \mathbf{C})$-gauge field over an open set $U \subset S^4$, that is, an anti-self-dual connection on the rank-n vector bundle V over U. To reduce the gauge group to $\mathrm{U}(n)$, we equip each fiber of V with a positive-definite Hermitian form, requiring that this form be preserved by the connection. An equivalent way of saying this is: the connection should preserve an antilinear isomorphism $\tau_0 : V \to V^*$ from V to its dual V^*, such that $\langle \psi, \tau_0 \varphi \rangle$ is a positive-definite Hermitian form on V. Here $\langle\,,\,\rangle$ denotes the natural pairing between V and V^*.

Now we use τ_0 to lift the involution σ on $\widehat{U}$ to a map $\tau : E \to E^*$, where E is the vector bundle over $\widehat{U}$ corresponding to the given anti-self-dual gauge field. In other words, $\tau : E \to E^*$ is an antilinear isomorphism from E to its dual E^*, such that the following diagram commutes:

$$
\begin{array}{ccc}
E & \stackrel{\tau}{\longrightarrow} & E^* \\
\downarrow & & \downarrow \\
\widehat{U} & \stackrel{\sigma}{\longrightarrow} & \widehat{U}
\end{array}
\qquad (8.1.10)
$$

So τ maps E_Z (the fiber of E over Z) to $E^*_{\sigma(Z)}$ (the fiber of E^* over $\sigma(Z)$). Conversely, having such a structure τ on Z guarantees that the corresponding gauge field on U admits an Hermitian structure, that is, it is a $U(n)$-gauge field. Let us give a more precise statement of all this, without proof (which may be found in Atiyah, Hitchin and Singer 1978, Theorem 5.2; see also Atiyah 1979, Theorem 2.9).

Let E be a holomorphic vector bundle over $\widehat{U} \subset \mathbb{P}$. An antilinear isomorphism $\tau : E \to E^*$ which is such that the above diagram (8.1.10) commutes, and such that

$$
\langle \xi, \tau \eta \rangle = \overline{\langle \eta, \tau \xi \rangle} \quad \text{for all } \xi \in E_Z, \eta \in E_{\sigma(Z)},
$$

is called a *real form* on E. Assume that $E|_{\hat{x}}$ is trivial for all $x \in U$; then τ induces a nondegenerate Hermitian form τ_0 on the space $V_x = \Gamma(\hat{x}, E|_{\hat{x}})$ of holomorphic sections of $E|_{\hat{x}}$, for each $x \in U$. If τ_0 is positive-definite, then we say that the real form τ is *positive*. We can now state the theorem towards which we have been aiming.

Theorem 8.1.6. *There is a natural one-to-one correspondence between:*
   (a) *anti-self-dual* $SU(n)$*-gauge fields on U;*
       and
   (b) *holomorphic rank-n vector bundles E over $\widehat{U}$, such that*
       (i) $E|_{\hat{x}}$ *is trivial, for all $x \in U$*
       (ii) $\det E$ *is trivial*
       (iii) E *admits a positive real form.*

Remarks:

(1) Here, as in Theorem 8.1.5, U denotes an elementary open set in S^4 (possibly all of S^4), and $\widehat{U}$ denotes the corresponding open set in $\mathbf{P} \cong \mathbf{P}_3(\mathbf{C})$.

(2) As usual, it is to be understood that the $\mathrm{SU}(n)$ gauge-equivalence is factored out of the set (a), and that the set (b) is factored out by the vector bundle isomorphisms which preserve all the structure that E is required to possess.

(3) Choosing a particular real form on E would not really represent any additional information. In the generic (irreducible) situation, τ, if it exists, is unique up to scale (see Remark (2) on p. 442 of Atiyah, Hitchin and Singer 1978).

(4) Reduction to other real forms of $\mathrm{GL}(n, \mathbf{C})$ can be effected in a similar sort of way. For a description of how to deal with the symplectic groups $\mathrm{Sp}(n)$, see Atiyah (1979), p. 50; the treatment of orthogonal groups $\mathrm{SO}(n)$ is not very different.

Let us study the reality condition ((iii) in Theorem 8.1.6) in terms of the transition matrix construction. Let W and $\underline{W}$ be two coordinate charts covering a region $\widehat{U}$ of twistor space, in such a way that $\sigma : W \rightarrow \underline{W}$ is a diffeomorphism. (See, for example, Figure 8.1.3: $\sigma|_{\hat{x}}$ is the antipodal map on $\hat{x}$, and maps the northern hemisphere W_x to the southern hemisphere $\underline{W}_x$.) Suppose the vector bundle E over $\widehat{U}$ is specified by the transition relation $\underline{\xi} = F(Z)\xi$. The dual E^* of E is described by the transition relation $\underline{\mu} = \mu F(Z)^{-1}$ for *row* n-vectors μ and $\underline{\mu}$. The natural pairing between vectors and dual vectors is defined by

$$(\xi, \mu) \mapsto \mu\xi \in \mathbf{C},$$

which is clearly preserved by the transition relation: $\underline{\mu}\underline{\xi} = \mu\xi$.

We define a map $\tau : E \rightarrow E^*$ as follows. If $Z \in W$, then

$$\tau : (Z, \xi) \mapsto (\sigma(Z), \xi^*)$$

while if $Z \in \underline{W}$, then

$$\tau : (Z, \underline{\xi}) \mapsto (\sigma(Z), \underline{\xi}^*).$$

Here $*$ denotes complex conjugate transpose, so if ξ is a column vector, then ξ^* is a row vector whose components are the complex conjugates of ξ. In order for the above definition of τ to make sense, we need a consistency condition: if $Z \in W \cap \underline{W}$, then $\underline{\xi} = F(Z)\xi$ should imply $\xi^* = \underline{\xi}^* F(\sigma(Z))^{-1}$. This requires that

$$F(Z)^* = F(\sigma(Z)). \qquad (8.1.11)$$

It is clear that τ is a real form on E. It may, however, not be positive. We shall return to this question later; first, let us see what condition Φ_a satisfies as a consequence of the condition (8.1.11) on F.

Given a holomorphic matrix-valued function $K(Z^\alpha)$, let $K^\dagger(Z^\alpha)$ denote the holomorphic matrix function defined by

$$K^\dagger(Z) = K(\sigma(Z))^*.$$

Therefore, the condition (8.1.11) can be rewritten as $F^\dagger = F$. Given a matrix $K(x^a, \pi_{B'})$ (using the coordinates described above), define

$$K^\dagger(x^a, \pi_{B'}) = K(x^a, \sigma(\pi)_{B'})^*,$$

where $\sigma(\pi)_{0'} = \overline{\pi_{1'}}$ and $\sigma(\pi)_{1'} = -\overline{\pi_{0'}}$. Then

$$G(x^a, \pi_{B'}) = F(ix^{AA'}\pi_{A'}, \pi_{B'})$$

satisfies $G^\dagger = G$ if F satisfies $F^\dagger = F$. So when we split G into $\underline{H}H^{-1}$, we obtain

$$\underline{H}H^{-1} = G = G^\dagger = (\underline{H}H^{-1})^\dagger = H^{\dagger -1}\underline{H}^\dagger,$$

whence

$$H^\dagger \underline{H} = \underline{H}^\dagger H. \qquad (8.1.12)$$

Now apply the usual 'globality' argument: the left-hand side of (8.1.12) is holomorphic for $\pi_{A'} \in \underline{W}_x$ and the right-hand side for $\pi_{A'} \in W_x$, so both sides are independent of $\pi_{A'}$. Thus

$$H^\dagger \underline{H} = \underline{H}^\dagger H = \Xi(x^a),$$

and clearly Ξ is Hermitian: $\Xi^* = \Xi$. Under a gauge transformation $H \mapsto H\Lambda$, $\underline{H} \mapsto \underline{H}\Lambda$, Ξ transforms as

$$\Xi \mapsto \Lambda^*\Xi\Lambda;$$

by choosing Λ appropriately, we can therefore diagonalize Ξ so that its diagonal entries are ± 1 (there cannot be any zero eigenvalues, since H and $\underline{H}$ are both nonsingular and therefore so is Ξ). From now on, suppose that such a gauge has been chosen.

From the reality constraints on $x^{AA'}$ we get $\overline{\nabla_{00'}} = -\nabla_{11'}$, $\overline{\nabla_{01'}} = \nabla_{10'}$ and hence

$$\overline{\pi^{A'}\nabla_{0A'}} = \sigma(\pi)^{A'}\nabla_{1A'}.$$

Let us abbreviate $H(x^{\mathfrak{a}}, \pi_{A'})$ to $H(\pi)$, etc. Then we have

$$\begin{aligned}
\pi^{A'}(\Phi^*)_{0A'} &= \{\sigma(\pi)^{A'}\Phi_{1A'}\}^* \\
&= \{\underline{H}(\sigma(\pi))^{-1}\sigma(\pi)^{A'}\nabla_{1A'}\underline{H}(\sigma(\pi))\}^* \\
&= \{\pi^{A'}\nabla_{0A'}\underline{H}^\dagger(\pi)\}\underline{H}^\dagger(\pi)^{-1} \\
&= -\Xi H(\pi)^{-1}\pi^{A'}\nabla_{0A'}H(\pi)\Xi^{-1} \quad \text{since } \underline{H}^\dagger = \Xi H^{-1} \\
&= -\Xi\pi^{A'}\Phi_{0A'}\Xi^{-1}.
\end{aligned}$$

A similar relation holds for $\Phi_{1A'}$, so we conclude that

$$\Phi_a^* = -\Xi\Phi_a\Xi^{-1}.$$

If Ξ is diagonal with p $(+1)$s and q (-1)s, then this says that Φ_a is a U(p,q)-connection.

We see, therefore, that only if $p = 0$ or $q = 0$ is τ positive (or negative, which amounts to the same thing). This is automatic in the case $n = 2$, $\det E$ trivial: for then $\det F = 1 \Rightarrow \det \Xi = 1$ and Ξ is $\pm$ the identity 2×2 matrix. Thus Φ_a is an SU(2)-connection. But if one wants to obtain an SU(n) field with $n > 2$, then the F one begins with should satisfy, in addition to $F^\dagger = F$, the requirement that Ξ be positive- (or negative-) definite.

Note, incidentally, that the condition $F^\dagger = -F$ also leads to U(p,q) fields. For example, if $n = 2$, $\det F = 1$, and $F^\dagger = -F$, then $\Phi_a^* = -\Xi\Phi_a\Xi^{-1}$ where Ξ is an *anti*-Hermitian unimodular matrix; so the eigenvalues of Ξ are $+1$, -1 and Φ_a is an SU(1,1)-connection.

Finally, at the end of this section, we want to indicate how the results of Chapter 7 (on massless free fields) may be extended to massless fields on a given anti-self-dual gauge-field background. (What this means will be explained below.) Instead of repeating all the analysis of Chapter 7 with suitable generalizations, we shall merely outline the ideas; for complete proofs, see Hitchin (1980), and also Rawnsley (1979); Eastwood, Penrose and Wells (1981); and Wells (1982a).

Recall from Chapter 7 that an element of $H^1(\widehat{U}, \mathcal{O}(2s - 2))$ determines a massless free field of helicity $-s$ on U. Let Φ_a be an anti-self-dual gauge potential on U, corresponding to a holomorphic vector bundle E over $\widehat{U}$. To get massless fields coupled to this gauge field, we replace the sheaf $\mathcal{O}(2s - 2) := \mathcal{O}_{\mathbf{P}}(H^{2s-2})$ by

$$\mathcal{O}(E, 2s - 2) := \mathcal{O}_{\mathbf{P}}(E \otimes H^{2s-2}).$$

So we have to study the cohomology group $H^1(\widehat{U}, \mathcal{O}(E, 2s - 2))$.

Suppose that $\widehat{U}$ is covered by several coordinate charts W_i, and that E is determined by the $n \times n$ transition matrices F_{ij}. (This generalizes the situation where there are only two coordinate charts: we are demonstrating here that such a generalization is easily made.) When restricted to a line $\hat{x}$, F_{ij} splits into $H_i H_j^{-1}$, with $H_i(x, \pi_{A'})$ holomorphic for $x \in U$ and $\pi_{A'} \in W_i \cap \hat{x}$. Let f_{ij} be a cocycle representing an element of $H^1(\widehat{U}, \mathcal{O}(E, 2s - 2))$. This means that $f_{ij} = f_{ij}(Z^\alpha)$ is a column n-vector of holomorphic functions on $W_i \cap W_j$, homogeneous of degree $2s - 2$, and satisfying

$$f_{ij} = -F_{ij} f_{ji}$$
$$f_{ij} + F_{ij} f_{jk} + F_{ik} f_{ki} = 0.$$

(In these equations and those that follow, the Einstein summation convention does *not* operate on the indices i, j, k.) So the functions $g_{ij}(x, \pi_{A'})$ defined by

$$g_{ij} = H_i^{-1} f_{ij}$$

satisfy

$$g_{ij} = -g_{ji}$$
$$g_{ij} + g_{jk} + g_{ki} = 0,$$

which says that for fixed x, $g_{ij}(x, \pi_{A'})$ determines n elements of

$$H^1(\hat{x}, \mathcal{O}(2s - 2)).$$

If $s \le 0$, then $\pi_{A'} \ldots \pi_{C'} g_{ij}$ (with $1 - 2s$ factors of the spinor π) is homogeneous of degree -1. Therefore, since $H^1(\mathbf{P}_1, \mathcal{O}(-1))$ vanishes, there exist splitting functions $h_{iA' \ldots C'}(x, \pi)$, holomorphic for $\pi_{A'} \in W_i \cap \hat{x}$, such that

$$\pi_{A'} \ldots \pi_{C'} g_{ij}(x, \pi_{D'}) = h_{iA' \ldots C'}(x, \pi_{D'}) - h_{jA' \ldots C'}(x, \pi_{D'}).$$

Multiplying this equation by $\pi^{A'}$ gives

$$\pi^{A'} h_{iA'B'\ldots D'} = \pi^{A'} h_{jA'B'\ldots D'};$$

whence, by Liouville's theorem (i.e., the fact that $H^0(\mathbf{P}_1, \mathcal{O}) \cong \mathbf{C}$, see Example 3.3.8) we get a spinor field $\varphi_{B'\ldots D'}$ on U, defined by

$$\varphi_{B'\ldots D'}(x) = \pi^{A'} h_{iA'B'\ldots D'}(x, \pi_{E'}).$$

This field $\varphi_{A'\ldots C'}$ has $-2s$ indices and is totally symmetric. To see what equation it satisfies, first note that

$$\pi^{A'} \nabla_{AA'} g_{ij} = -H_i^{-1}(\pi^{A'} \nabla_{AA'} H_i) H_i^{-1} f_{ij}$$
$$= -\pi^{A'} \Phi_{AA'} g_{ij},$$

so that

$$\pi^{A'} D_{AA'} g_{ij} = 0. \tag{8.1.13}$$

If $s \leq -1$, this implies that

$$D^{AA'} h_{iA'B'\ldots D'} = D^{AA'} h_{jA'B'\ldots D'}.$$

Since $H^0(\mathbf{P}_1, \mathcal{O}(-1)) = 0$, we therefore have $D^{AA'} h_{iA'B'\ldots D'} = 0$ and so

$$D^{AA'} \varphi_{A'B'\ldots D'} = 0. \tag{8.1.14}$$

Also, equation (8.1.13) implies that $D_a D^a g_{ij} = 0$, and so in the $s = 0$ case we find, by a similar argument, that

$$D_a D^a \varphi = 0. \tag{8.1.15}$$

We see, therefore, that for $s \leq 0$, one gets a massless field of helicity $|s|$ coupled to the background anti-self-dual gauge field, in the sense that the gauge field is unaltered, but the 'ordinary' derivative ∇_a is replaced by the covariant derivative D_a when acting on $\varphi_{A'\ldots C'}$. This is not (in general) a 'complete' coupling as defined by physicists, since the gauge field remains source-free. Compare with Example 5.1.1 (with $m = 0$), where there is a source term $J_a = -[\varphi, D_a\varphi]$ for the gauge field.

We have exhibited a map from $H^1(\widehat{U}, \mathcal{O}(E, 2s - 2))$ to the space of solutions of (8.1.15) (if $s = 0$), or of (8.1.14) (if $s \leq -1$), on U. This map is, in fact, an isomorphism. For a proof, see Hitchin (1980), Theorem 3.1; Wells (1982a).

What about the $s > 0$ cases? Let us investigate these in the same sort of way. Since $H^1(\mathbf{P}_1, \mathcal{O}(2s - 2))$ vanishes for $s \geq \frac{1}{2}$, we can split g_{ij}:

$$g_{ij} = h_i - h_j.$$

Operating on this with $\pi^{A'} D_{AA'}$ yields a field $\psi_{AA'...C'}(x)$, with $2s-1$ primed indices, defined by

$$\psi_{AA'...C'}(x)\pi^{A'} ... \pi^{C'} = \pi^{A'} D_{AA'} h_i(x, \pi_{D'}). \qquad (8.1.16)$$

If $s = \frac{1}{2}$, then the resulting field ψ_A satisfies the coupled equation

$$D^{AA'}\psi_A = 0. \qquad (8.1.17)$$

(See Exercise 8.1.1.) But when $s \geq 1$, things get slightly more complicated. This is to be expected, because of the algebraic constraints that arise when one 'naively' tries to couple massless fields of helicity $-s \leq -1$ to an anti-self-dual gauge field (see end of §4.3). Our field $\psi_{AA'...C'}$ represents a massless field whose coupling does not merely consist of the replacement $\nabla_a \mapsto D_a$, and thereby avoids the constraints.

First, applying $\pi^{D'} D_{AD'}$ to equation (8.1.16), and using anti-self-duality, gives

$$D_{A(A'}\psi^A{}_{B'...D')} = 0. \qquad (8.1.18)$$

Secondly, the field $\psi_{AA'...C'}$ is not uniquely defined, since the h_i are not unique: they admit the freedom

$$h_i(x, \pi_{D'}) \mapsto h_i(x, \pi_{D'}) + \pi^{A'} ... \pi^{C'} \Lambda_{A'...C'}(x)$$

where $\Lambda_{A'...C'}$ is an arbitrary field on U, with $2s - 2$ indices. The corresponding change in $\psi_{AA'...C'}$ is

$$\psi_{AA'...C'} \mapsto \psi_{AA'...C'} + D_{A(A'}\Lambda_{B'...C')}. \qquad (8.1.19)$$

So the theorem (of which we have here demonstrated a part) is that $H^1(U, \mathcal{O}(E, -1))$ is isomorphic to the space of solutions of (8.1.17) on U; and that $H^1(\widehat{U}, \mathcal{O}(E, 2s - 2))$ for $s \geq 1$ corresponds to the space of solutions of (8.1.18) on U, modulo the equivalence transformations (8.1.19). (This latter space is therefore the first cohomology group of a certain complex on U.) See Hitchin (1980), Theorem 4.1 for more details of this, and a complete proof.

To see how the above description of the $s \geq 1$ case is related to the usual description of massless fields, let us take E to be the trivial line-bundle and define $\varphi_{A...C}$ (with $2s$ indices) by differentiating $\psi_{AA'...C'}$ $2s - 1$ times:

$$\varphi_{AB...D} = \nabla_{(B}{}^{B'} ... \nabla_D{}^{D'} \psi_{A)B'...D'}. \tag{8.1.20}$$

Then this $\varphi_{A...C}$ satisfies the massless free-field equations (see Exercise 8.1.2).

This section has dealt with the correspondence between self-dual gauge fields on four-space, and holomorphic bundles over $\mathbb{P}$ which, when restricted to lines, are trivial. There are many ways of generalizing the correspondence. Two avenues of generalization are as follows.

1. One can study bundles over $\mathbb{P}$ which, when restricted to lines, are not trivial, but rather isomorphic to some given vector bundle over $\mathbb{P}_1$. These correspond to the solutions of certain nonlinear equations on $\mathbf{M}$, analogous to the self-dual gauge-field equations. See Sparling (1978) and Leiterer (1983).

2. One can generalize to higher dimensions, and study holomorphic bundles over higher-dimensional twistor spaces. These correspond to solutions of certain gauge-field equations on higher-dimensional space-times, which (like the self-duality equations in dimension four) take the form of linear conditions on the gauge field F_{ab}. See Ward (1984) and Corrigan et al. (1985).

Exercises for Section 8.1

8.1.1. Verify equation 8.1.17, using $D_a D^a g_{ij} = 0$ and the anti-self-duality condition $D_{(A'}{}^A D_{B')A} = 0$.

8.1.2. In the case where the gauge field is zero, check that $\varphi_{A...C}$ defined by (8.1.20) is a massless free field, and that it is invariant under the transformation (8.1.19).

8.1.3. Complete the proof of the fact that $H^1(\widehat{U}, \mathcal{O}(E, 2s - 2))$ corresponds to massless fields coupled to an anti-self-dual gauge field. Use the techniques of Hughston and Ward (1979), §§2.6 and 2.8, rather than the methods of Hitchin (1980).

8.1.4. Verify that the Penrose transform of a locally free sheaf $\mathcal{E} \to \widehat{U}$ which is trivial on each submanifold $\hat{x} \subset \widehat{U}$, for $x \in U$, has the form

$$\mathcal{P}(\mathcal{E}) = (\nu_0^* \mu^*(\mathcal{E}), \nu_0^* d_\mu),$$

where

$$\mu^*(\mathcal{E}) \xrightarrow{d_\mu} \mu^*(\mathcal{E}) \otimes_{\mathcal{O}_F} \Omega_\mu^1$$

is the natural relative exterior derivative (cf. Remark (5) following Theorem 8.1.2).

8.2 Ansätze for SU(2)-fields

The previous section described the correspondence between anti-self-dual gauge fields and holomorphic vector bundles over twistor space. Can one use this correspondence to construct solutions of the anti-self-duality equations? The part of the construction which is most difficult to implement is the splitting (8.1.4). In the rank $n = 1$ case, we saw (Example 8.1.3) that the splitting could be carried out by means of a Cauchy-type integral. The problem in this case is linear, as one would expect. But for rank $n \geq 2$, the problem becomes non-linear, and there is no 'splitting formula' in general. This section is devoted to the study of certain special $n = 2$ cases where the splitting *can* be carried out, and where the problem is linearized. These special cases correspond to holomorphic rank-two bundles which are extensions of one line bundle by another. The space-time translation of this idea gives a sequence $\mathcal{A}_1, \mathcal{A}_2, \mathcal{A}_3, \ldots$ of 'ansätze', each of which converts solutions of a linear equation into solutions of the (nonlinear) anti-self-duality equations. So each ansatz generates a subclass of the set of all $\mathrm{SL}(2, \mathbf{C})$ anti-self-dual fields. The remarkable thing is that, as we shall see in the two succeeding sections, every $\mathrm{SU}(2)$-instanton and magnetic pole solution belongs to one (or more) of these special subclasses, and so can be obtained from an ansatz.

After describing what the ansätze are, we shall discuss the problem of how to choose the linear fields so that the anti-self-dual solutions they generate are real, i.e., $\mathfrak{su}(2)$-valued rather than $\mathfrak{sl}(2, \mathbf{C})$-valued.

Let us begin, then, with the idea of extensions. Suppose that the rank-two vector bundle E over a region $\widehat{U}$ of twistor space is the *extension* of a line bundle L_1 by another line bundle L_2. This means that there is an exact sequence of vector bundles of the form

$$0 \to L_1 \to E \to L_2 \to 0.$$

Suppose that $\widehat{U}$ is covered by two coordinate charts W and $\underline{W}$, as in §8.1. Then E is determined by a transition matrix $F(Z^\alpha)$, on the overlap region $W \cap \underline{W}$, of the form

$$F = \begin{pmatrix} \Xi_1 & \Gamma \\ 0 & \Xi_2 \end{pmatrix},$$

where Ξ_1 and Ξ_2 are, respectively, the transition functions for L_1 and L_2, and where Γ should be regarded as defining an element of the sheaf cohomology group $H^1(\widehat{U}, \mathcal{O}(L_1 \otimes L_2^{-1}))$.

Let us take $\det F = 1$, i.e., $\Xi_1 = \Xi_2^{-1}$, so that the gauge group is $\mathrm{SL}(2, \mathbf{C})$. The line bundle L_1 over $\widehat{U}$ is classified topologically by its Chern number, which is obtained by evaluating the Chern class $c_1(L_1)$ on any projective line $\hat{p}$ in $\widehat{U}$. Let k denote the *negative* of the Chern number of L_1. This means that $\Xi_1(Z^\alpha)$ has the form

$$\Xi_1(Z^\alpha) = \zeta^k \exp f(Z^\alpha)$$

where $\zeta = \pi_{0'}/\pi_{1'}$ and where f is a holomorphic function on $W \cap \underline{W}$. Define

$$
\begin{aligned}
g(x, \pi_{A'}) &= f(ix^{AA'}\pi_{A'}, \pi_{A'}), \\
\Omega(x, \pi_{A'}) &= \Gamma(ix^{AA'}\pi_{A'}, \pi_{A'}).
\end{aligned}
\tag{8.2.1}
$$

Our task may then be stated as follows: we must find functions a, b, c, d, $\underline{a}$, $\underline{b}$, $\underline{c}$, $\underline{d}$, of x and $\pi_{A'}$ (the first four holomorphic in W_x, i.e., for $|\zeta| \le 1$, and the last four holomorphic for $|\zeta| \ge 1$), such that

$$
\underbrace{\begin{pmatrix} \zeta^k e^g & \Omega \\ 0 & \zeta^{-k}e^{-g} \end{pmatrix}}_{G} = \underbrace{\begin{pmatrix} \underline{a} & \underline{b} \\ \underline{c} & \underline{d} \end{pmatrix}}_{\underline{H}} \underbrace{\begin{pmatrix} d & -b \\ -c & a \end{pmatrix}}_{H^{-1}}
\tag{8.2.2}
$$

and such that

$$ad - bc = \underline{ad} - \underline{bc} = 1. \tag{8.2.3}$$

First, we deduce from (8.2.2) and (8.2.3) the equations

$$
\begin{aligned}
c &= \underline{c}e^g\zeta^k, \\
d &= \underline{d}e^g\zeta^k.
\end{aligned}
\tag{8.2.4}
$$

To solve these, split g into $\underline{h} - h$ (as in Example 8.1.3), rewrite (8.2.4) as

$$
\begin{aligned}
ce^h &= \underline{c}e^{\underline{h}}\zeta^k, \\
de^h &= \underline{d}e^{\underline{h}}\zeta^k,
\end{aligned}
$$

and then apply the usual Liouville-type argument. This tells us that if $k < 0$, we must have $c = d = 0$ which is impossible: so the $k < 0$ cases are never splittable. From now on, assume that $k \geq 0$. The solution of (8.2.4) is

$$c = e^{-h} \sum_{j=0}^{k} c_j \zeta^j,$$

$$d = e^{-h} \sum_{j=0}^{k} d_j \zeta^j,$$

$$\underline{c} = e^{-\underline{h}} \sum_{j=0}^{k} c_j \zeta^{j-k},$$

$$\underline{d} = e^{-\underline{h}} \sum_{j=0}^{k} d_j \zeta^{j-k},$$

where $c_0, c_1, \ldots, c_k, d_0, \ldots, d_k$ are arbitrary functions of x.

Next, we deduce from (8.2.2) and (8.2.3) an equation for a and $\underline{a}$:

$$e^{-\underline{h}}\underline{a} - \zeta^k e^{-h} a = e^{-\underline{h}-h} \Omega \sum_{j=0}^{k} c_j \zeta^j. \qquad (8.2.5)$$

To solve this, expand $e^{-\underline{h}-h}\Omega$ in a Taylor–Laurent series

$$e^{-\underline{h}-h}\Omega = \sum_{r=-\infty}^{\infty} \Delta_r \zeta^{-r}, \qquad (8.2.6)$$

where the Δ_r are functions of x, and define

$$\theta_r = \sum_{j=0}^{k} c_j \Delta_{j-r};$$

then the unique solution of (8.2.5) is

$$a = -e^h \zeta^{-k} \sum_{r=1}^{\infty} \theta_r \zeta^r,$$

$$\underline{a} = e^{\underline{h}} \sum_{r=-\infty}^{0} \theta_r \zeta^r.$$

(Actually, this solution is only unique if $k > 0$; the $k = 0$ case is special and we shall return to it later. For the time being, assume $k > 0$.) A similar argument gives b and $\underline{b}$:

$$b = -e^h \zeta^{-k} \sum_{r=1}^{\infty} \varphi_r \zeta^r,$$

$$\underline{b} = e^{\underline{h}} \sum_{r=-\infty}^{0} \varphi_r \zeta^r,$$

(8.2.7)

where

$$\varphi_r = \sum_{j=0}^{k} d_j \Delta_{j-r}.$$

Clearly $\underline{a}$ and $\underline{b}$ are holomorphic for $|\zeta| \geq 1$; in order that a and b be holomorphic for $|\zeta| \leq 1$, however, we need

$$\theta_r = \varphi_r = 0 \quad \text{for } 1 \leq r < k. \tag{8.2.8}$$

(These conditions only come into play if $k > 1$.) Also, we need the unimodularity condition $ad - bc = 1$, which is equivalent to

$$c_0 \varphi_k - d_0 \theta_k = 1. \tag{8.2.9}$$

We may make any choice of $c_0, \ldots, d_k$ compatible with conditions (8.2.8) and (8.2.9), and will thereby get the matrices H and $\underline{H}$ which split G. The freedom in the choice of $c_0, \ldots, d_k$ corresponds precisely to the expected gauge freedom. (There is also some freedom in the choice of h and $\underline{h}$, but this can absorbed into $c_0, \ldots, d_k$.) We still have to investigate whether the equations (8.2.8) and (8.2.9) admit any solutions at all. As we shall now see, there is a condition on the Δ_r which guarantees that they do.

Let N denote the $(k + 1) \times (k + 1)$ matrix

$$N = \begin{pmatrix} 1 & 0 & \cdots & 0 \\ \Delta_{-k} & \Delta_{-k+1} & \cdots & \Delta_0 \\ \vdots & \vdots & \ddots & \vdots \\ \Delta_{-1} & \Delta_0 & \cdots & \Delta_{k-1} \end{pmatrix}$$

Suppose $\det N \neq 0$, so that the rows $v_0, v_1, \ldots, v_k$ of N are linearly independent over $\mathbf{C}$. Think of the v_r as vectors in $\mathbf{C}^{k+1}$, and also

of $c = (c_0, \ldots, c_k)$ and $d = (d_0, \ldots, d_k)$ as such vectors. Equation
(8.2.8) says that c and d are orthogonal to each of $v_2, \ldots, v_k$ (with respect to the complexified Euclidean metric on $\mathbf{C}^{k+1}$). In other words,
c and d lie in the complex two-plane P orthogonal to $v_2, \ldots, v_k$ in
$\mathbf{C}^{k+1}$. Additionally, $\det N \neq 0$ also implies that the projections of
v_0 and v_1 into P are linearly independent. It follows that equation
(8.2.9), which can be written

$$(v_0 \cdot c)(v_1 \cdot d) - (v_0 \cdot d)(v_1 \cdot c) = 1 \qquad (8.2.10)$$

(the dot denoting the complex-Euclidean inner product), has solutions for c and d in P.

Conversely, if $\det N = 0$, then we see that v_0 is a linear combina-
tion of $v_1, \ldots, v_k$; in this case, if c and d are orthogonal to $v_2, \ldots, v_k$,
it is impossible to satisfy (8.2.10), since the left-hand side of (8.2.10)
vanishes identically.

We see, therefore, that the condition $\det N \neq 0$ is necessary and
sufficient for the splitting matrices H and $\underline{H}$ to exist. If $k = 1$,
this condition is just $\Delta_0 \neq 0$; while if $k = 2$, the condition is
$\Delta_{-1}\Delta_1 - \Delta_0^2 \neq 0$; and so forth. Remember that we are working
'locally' in complexified space-time, on a region U on $\mathbf{M}$. The functions f and Γ defining E are assumed to be chosen so that g and
Ω (given by equation (8.2.1)) are holomorphic for all $x \in U$ and
for $\pi_{A'}$ in some neighborhood of $|\pi_{0'}/\pi_{1'}| = 1$. It follows that the
$\Delta_r(x)$ are holomorphic on U. The functions $c_0, \ldots, d_k$ can be chosen to be holomorphic on U as well (provided that $\det N(x) \neq 0$ for
all $x \in U$). Consequently, H and $\underline{H}$, and hence the gauge potential
Φ_a, will be holomorphic on U.

It is not necessary for the Δ_r to be smooth in order for Φ_a to be
smooth, as we shall see in §8.3. But it *is* necessary for $\det N$ to be
nonzero. If, for a specific point x, we have $\det N(x) = 0$, then the
gauge field will have a singularity at x. For if Φ_a were smooth at x,
$E|_{\hat{x}}$ would be trivial, hence the splitting matrices H and $\underline{H}$ would
exist, and hence $\det N(x)$ would be nonzero. If $\det N(x)$ is zero,
then the line $\hat{x}$ corresponding to x is called a 'jumping line' of E,
since on this line the bundle E jumps from being trivial to being
nontrivial (Okonek et al. 1980). This is illustrated by the following
example.

Example 8.2.1: Take $k = 1$, $f = 0$ and $\Gamma = \omega^0 \omega^1/(\pi_{0'}\pi_{1'})$. So

$h = \underline{h} = 0$, and

$$
\begin{aligned}
\Delta_0 &= -(x^{00'}x^{11'} + x^{01'}x^{10'}) \\
\Delta_1 &= -x^{01'}x^{11'}, \\
\Delta_{-1} &= -x^{00'}x^{10'}, \\
\Delta_r &= 0, \quad |r| \geq 2.
\end{aligned}
$$

If $\Delta_0(x) = 0$, then the transition matrix for $E|_{\hat{x}}$ is

$$
G(x, \pi_{A'}) = \begin{pmatrix} \zeta & \Delta_1\zeta^{-1} + \Delta_{-1}\zeta \\ 0 & \zeta^{-1} \end{pmatrix}.
$$

And since

$$
\begin{pmatrix} 1 & -\Delta_1 \\ 0 & 1 \end{pmatrix} \begin{pmatrix} \zeta & \Delta_1\zeta^{-1} + \Delta_{-1}\zeta \\ 0 & \zeta^{-1} \end{pmatrix} \begin{pmatrix} 1 & -\Delta_{-1} \\ 0 & 1 \end{pmatrix} = \begin{pmatrix} \zeta & 0 \\ 0 & \zeta^{-1} \end{pmatrix},
$$

we see that $E|_{\hat{x}}$ is not trivial, but is equivalent to $H^{-1} \oplus H$, where H is the hyperplane section bundle (with Chern number $+1$) over $\hat{x}$. If, on the other hand, $\Delta_0(x) \neq 0$, then $E|_{\hat{x}}$ is trivial, since a splitting then exists. As we shall see below, the expression for the gauge potential Φ_a involves $\nabla_a \log \Delta_0$, so Φ_a is singular when (and only when) $\Delta_0 = 0$.

There is a specific choice of gauge which is often very useful, called the R-gauge, after a paper by C. N. Yang (Yang 1977). It consists of taking $d_0 = c_k = 0$, $c_0 = d_k$. The calculation of Φ_a in this gauge leads to the following result.

Theorem 8.2.2. *Let B_a be a potential for an anti-self-dual $\mathrm{GL}(1, \mathbb{C})$-gauge field (i.e., a complex Maxwell field), and let $\{\Delta_r\}_{r=1-k}^{k-1}$ be a set of fields satisfying*

$$
\begin{aligned}
&\text{if } k > 1: \quad (\nabla_{A0'} + 2B_{A0'})\Delta_r = (\nabla_{A1'} + 2B_{A1'})\Delta_{r+1} \\
&\qquad\qquad \text{for } 1 - k \leq r \leq k - 2; \qquad\qquad\qquad (8.2.11) \\
&\text{if } k = 1: \quad (\nabla_a + 2B_a)(\nabla^a + 2B^a)\Delta_0 = 0.
\end{aligned}
$$

Suppose that B_a and Δ_r are holomorphic on the region U of complexified space-time. Let M be the $k \times k$ matrix

$$
M = \begin{pmatrix} \Delta_{1-k} & \cdots & \Delta_0 \\ \vdots & \ddots & \vdots \\ \Delta_0 & \cdots & \Delta_{k-1} \end{pmatrix},
$$

i.e., $M_{rs} = \Delta_{r+s-k-1}$ for $1 \leq r, s \leq k$. Let E, F and G be the corner elements of its inverse: $E = (M^{-1})_{11}$, $F = (M^{-1})_{1k} = (M^{-1})_{k1}$ and $G = (M^{-1})_{kk}$ (assume that $\det M \neq 0$ on U). Finally, define a gauge potential Φ_a by

$$\Phi_{A0'} = \frac{1}{2F} \begin{pmatrix} \eth_{A0'} F & 0 \\ -2\eth_{A1'} G & -\eth_{A0'} F \end{pmatrix},$$

$$\Phi_{A1'} = \frac{1}{2F} \begin{pmatrix} -\eth_{A1'} F & -2\eth_{A0'} E \\ 0 & \eth_{A1'} F \end{pmatrix},$$

where $\eth_a = \nabla_a - 2B_a$. Then Φ_a is the potential for an anti-self-dual $SL(2, \mathbf{C})$-gauge field.

Remarks:

(1) The Maxwell field B_a is that which corresponds to the line bundle whose transition function is e^f (cf. Example 8.1.3). It is defined by

$$\pi^{A'} B_{AA'} = \pi^{A'} \nabla_{AA'} h = \pi^{A'} \nabla_{AA'} \underline{h}$$

in terms of the h and $\underline{h}$ that split f.

(2) The Δ_r defined by (8.2.6) automatically satisfy (8.2.11). The Δ_r for $|r| \leq k - 1$ should be thought of as the components of a massless field of helicity $k-1$ coupled to the Maxwell field $2B_a$; this, after all, is what an element of the sheaf cohomology group

$$H^1(\widehat{U}, \mathcal{O}(L_1 \otimes L_2^{-1})) = H^1(\widehat{U}, \mathcal{O}(L_1^2))$$

corresponds to (see §7.1). For example, in the $k = 2$ case, the spinor field $\varphi_{A'B'}$ defined by

$$\varphi_{0'0'} = \Delta_1, \qquad \varphi_{0'1'} = \Delta_0, \qquad \varphi_{1'1'} = \Delta_{-1}$$

satisfies $(\nabla^{AA'} + 2B^{AA'})\varphi_{A'B'} = 0$.

(3) Clearly $\det M = \det N$, so the condition $\det N \neq 0$ encountered earlier guarantees that M^{-1} exists. But note that, *a priori*, Φ_a may have singularities where the function $F = (M^{-1})_{1k}$ vanishes. If so, these singularities can be removed by a gauge transformation, since we know from our earlier discussion that there must be *some* gauge in which Φ_a is smooth.

(4) Putting $f = 0$ (and therefore $B_a = 0$) gives the ansätze originally presented in Atiyah and Ward (1977) and Corrigan et al. (1978b). The Δ_r are now the components of a massless *free* field

of helicity $k - 1$. Let us look at the $k = 1$ case in more detail. Here $E = F = G = \Delta_0^{-1}$, and if we work in terms of Euclidean coordinates x^a, then a straightforward calculation gives

$$\Phi_a = i\tilde{\sigma}_{ab} \nabla^b \log \Delta_0,$$

where the 2×2 matrices $\tilde{\sigma}_{ab} = -\tilde{\sigma}_{ba}$ are defined by

$$\tilde{\sigma}_{01} = \quad \tilde{\sigma}_{23} = -\tfrac{1}{2}\sigma_1 := \tfrac{1}{2} \begin{pmatrix} 0 & -1 \\ -1 & 0 \end{pmatrix},$$

$$\tilde{\sigma}_{02} = -\tilde{\sigma}_{13} = -\tfrac{1}{2}\sigma_2 := \tfrac{1}{2} \begin{pmatrix} 0 & i \\ -i & 0 \end{pmatrix},$$

$$\tilde{\sigma}_{03} = \quad \tilde{\sigma}_{12} = \quad \tfrac{1}{2}\sigma_3 := \tfrac{1}{2} \begin{pmatrix} 1 & 0 \\ 0 & -1 \end{pmatrix},$$

in terms of the three Pauli matrices $\sigma_1, \sigma_2, \sigma_3$. Note that $\tilde{\sigma}_{ab}$ is self-dual: $*\tilde{\sigma}_{ab} = \tilde{\sigma}_{ab}$. Thus our ansatz $\mathcal{A}_1$ with $B_a = 0$ is simply the orientation-reversed version of (5.2.1).

An alternative, and perhaps neater, description of this ansatz is the following one. It involves identifying the vector bundle V over U (on which $D_a = \nabla_a + \Phi_a$ is a connection) with the primed spin-bundle S_+ over U. Explicitly, we regard a spinor field $\alpha_{A'}$ as corresponding to a section α of V, according to

$$\alpha = \begin{pmatrix} \alpha_{0'} \\ -\alpha_{1'} \end{pmatrix}.$$

It is then a simple matter to check that the connection D_a is given, in terms of Δ_0, by

$$D_{AA'}\alpha_{B'} = \nabla_{AA'}\alpha_{B'} - \alpha_{A'} \nabla_{AB'} \log \Delta_0 + \tfrac{1}{2}\alpha_{B'} \nabla_{AA'} \log \Delta_0;$$

and to check directly that this connection is anti-self-dual, namely that

$$D^A{}_{(C'} D_{A')A} \alpha_{B'} = 0.$$

N. M. J. Woodhouse (1983, 1985) has generalized this type of description, so that it applies to the higher ansätze $\mathcal{A}_k$, $k \geq 2$. It applies to the most general case where $B_a \neq 0$, but for the sake of simplicity we shall give it in the case when $B_a = 0$. Let $\varphi_{A'...D'}$ be a massless free field of helicity $k - 1$ on a region U of complexified space-time (so it has $2k - 2$ indices). Let W be the trivial vector bundle of rank $k + 1$ over U. Denote the sections of W by $\alpha^{A'\ldots C'}$, a totally symmetric spinor field with k indices.

Now define a connection D_a on W by

$$D_{AA'}\alpha^{B'\ldots D'} = \nabla_{AA'}\alpha^{B'\ldots D'} + \varepsilon_{A'}{}^{(B'}\xi_A{}^{C'\ldots D')},$$

where ∇_a is (as usual) the flat connection and where $\xi_A{}^{C'\ldots D'}$, a spinor symmetric in its $k - 1$ primed indices, is given by

$$(\nabla_{AA'}\varphi_{B'\ldots C'D'E'\ldots F'})\alpha^{D'\ldots F'} = \varphi_{B'\ldots C'A'E'\ldots F'}\xi_A{}^{E'\ldots F'}.$$

In order for $\xi^{C'\ldots D'}$ to be well-defined, it is necessary for $\varphi_{A'\ldots C'}$ to be algebraically 'sufficiently general'. This condition is the nonvanishing of a scalar function S, a polynomial of degree k in φ. For $k = 2$, we have

$$S = \varphi_{A'B'}\varphi^{A'B'},$$

while for $k = 3$, S is given by

$$S = \varphi_{A'B'C'D'}\varphi^{C'D'E'F'}\varphi_{E'F'}{}^{A'B'}.$$

As long as $S(x) \neq 0$ for all $x \in U$, the connection D_a is well-defined. In fact, S is nothing other than the function $\det M = \det N$ that we encountered before, with the Δ_r being thought of as the components of $\varphi_{A'\ldots C'}$.

For any section $\alpha^{A'\ldots C'}$ of W, we have (by a straightforward calculation)

$$D^A{}_{(A'}D_{B')A}\alpha^{C'\ldots E'} = 0;$$

in other words, D_a is an anti-self-dual connection on W. To get an anti-self-dual connection on a rank-two bundle V, we choose V to be an appropriate subbundle of W, and restrict D_a to V. The sections of V consist, by definition, of those spinor fields $\alpha^{A'\ldots C'}$ which satisfy

$$\varphi_{A'\ldots C'D'\ldots F'}\alpha^{D'\ldots F'} = 0. \tag{8.2.12}$$

This amounts to $k - 1$ equations on the $k + 1$ components of $\alpha^{A'\ldots C'}$, and $S \neq 0$ guarantees that these equations are independent of one another, so V is indeed of rank two. Furthermore, if (8.2.12) holds, then so does

$$\varphi_{B'\ldots C'D'\ldots F'}D_{AA'}\alpha^{D'\ldots F'} = 0,$$

which means that the connection D_a restricts to the subbundle V. This restricted connection is still anti-self-dual. Also, V admits a covariantly constant two-form ω, defined by

$$\omega(\alpha, \hat{\alpha}) = \varphi_{B'\dots C'D'\dots F'} \alpha^{A'B'\dots C'} \hat{\alpha}_{A'}{}^{D'\dots F'},$$

where $\alpha^{A'\dots C'}$ and $\hat{\alpha}^{A'\dots C'}$ are sections of V. So beginning with a massless free field $\varphi_{A'\dots C'}$, we have constructed an anti-self-dual $SL(2, \mathbf{C})$-gauge field, depending nonlinearly on $\varphi_{A'\dots C'}$.

We shall now sketch the derivation of Theorem 8.2.2. Let the matrix M and the functions E, F, and G be defined as in the statement of the theorem. The unique solution of the conditions on $c_0, \dots, d_k$ and the R-gauge conditions $d_0 = c_k = 0$, $c_0 = d_k$ is

$$d_r = F^{-\frac{1}{2}} (M^{-1})_{1r},$$

$$c_{r-1} = F^{-\frac{1}{2}} (M^{-1})_{kr}$$

for $r = 1, 2, \dots, k$. Denote by H_0 the matrix H evaluated at $\zeta = 0$, and by $\underline{H}_\infty$ the matrix $\underline{H}$ evaluated at $\zeta = \infty$. Then a straightforward calculation gives

$$\Phi_{A0'} = H_0^{-1} \nabla_{A0'} H_0$$
$$= \begin{pmatrix} \frac{1}{2} F^{-1} \eth_{A0'} F & 0 \\ P & -\frac{1}{2} F^{-1} \eth_{A0'} F \end{pmatrix},$$

where $\eth_a = \nabla_a - 2B_a$ and where

$$P = c_0 \nabla_{A0'} \theta_k - \theta_k \nabla_{A0'} c_0 + 2c_0 \theta_k B_{A0'};$$

and

$$\Phi_{A1'} = \underline{H}_\infty^{-1} \nabla_{A1'} \underline{H}_\infty$$
$$= \begin{pmatrix} -\frac{1}{2} F^{-1} \eth_{A1'} F & Q \\ 0 & \frac{1}{2} F^{-1} \eth_{A1'} F \end{pmatrix},$$

where

$$Q = c_0 \nabla_{A1'} \varphi_0 - \varphi_0 \nabla_{A1'} c_0 + 2c_0 \varphi_0 B_{A1'}.$$

The functions θ_k and φ_0 are given, respectively, by

$$\theta_k = F^{-\frac{1}{2}} \sum_{r=1}^{k} (M^{-1})_{rk} \Delta_{r-k-1},$$

$$\varphi_0 = F^{-\frac{1}{2}} \sum_{r=1}^{k} (M^{-1})_{r1} \Delta_r.$$

It remains only to prove that

$$P = -F^{-1}\eth_{A1'}G$$
$$Q = -F^{-1}\eth_{A0'}E.$$

To do so involves using equation (8.2.11), and some messy matrix manipulations. The technique is described in Appendix B of Corrigan et al. (1978b).   ∎

We still have to deal with the case $k = 0$. This is easily done: the transition matrix can be split as

$$\begin{pmatrix} e^g & \Omega \\ 0 & e^{-g} \end{pmatrix} = \begin{pmatrix} e^{\underline{h}} & \underline{b} \\ 0 & e^{-\underline{h}} \end{pmatrix} \begin{pmatrix} e^{-h} & -b \\ 0 & e^h \end{pmatrix} \qquad (8.2.13)$$

where b and $\underline{b}$ satisfy

$$e^{-\underline{h}}\underline{b} - e^{-h}b = \Omega e^{-\underline{h}-h}$$

and can therefore be expressed in terms of the Δ_r (the Taylor–Laurent coefficients of $\Omega e^{-\underline{h}-h}$). It is not really necessary to go any further, since we see immediately from (8.2.13) that the gauge potential Φ_a will be upper-triangular, and consequently cannot give any $SU(2)$ fields (except diagonal, and therefore abelian, ones). Our main interest is in $SU(2)$ solutions, and we shall now show that the ansätze $\mathcal{A}_k$ for $k \geq 1$ do indeed generate many such solutions.

We saw in §8.1 that if the transition matrix F satisfies the reality condition $F^\dagger = F$ (as well as det $F = 1$), then it generates an $SU(2)$-gauge field. But if F is upper triangular, as it is in the case of the ansätze, then it cannot satisfy $F^\dagger = F$ without being diagonal. We require instead that F be *equivalent* to a matrix $\widetilde{F}$ satisfying $\widetilde{F}^\dagger = \widetilde{F}$, i.e., that

$$\widetilde{F} = \underline{K}^{-1}FK$$

for some unimodular matrices K and $\underline{K}$ holomorphic on W and $\underline{W}$, respectively. Of course, F and $\widetilde{F}$ determine the same bundle E (up to equivalence) and hence the same gauge field.

So given F, we have to look for K and $\underline{K}$, such that $\underline{K}^{-1}FK$ is real. Without loss of generality, we may take $\underline{K} = 1$: for if $\widetilde{F}$ is real, then so is

$$\underline{K}\widetilde{F}K^\dagger = FK\underline{K}^\dagger;$$

note that $\underline{K}^\dagger$ is holomorphic on W.

Example 8.2.3: Let k be an odd integer and suppose that f is imaginary and Γ is real (i.e., $f^\dagger = -f$ and $\Gamma^\dagger = \Gamma$). Then the gauge field generated by the corresponding ansatz is SU(2)-valued, for

$$\widetilde{F} = \begin{pmatrix} \zeta^k e^f & \Gamma \\ 0 & \zeta^{-k} e^{-f} \end{pmatrix} \begin{pmatrix} 0 & -1 \\ 1 & 0 \end{pmatrix} = \begin{pmatrix} \Gamma & -\zeta^k e^f \\ \zeta^{-k} e^{-f} & 0 \end{pmatrix}$$

satisfies $\widetilde{F}^\dagger = \widetilde{F}$. (Note that $\dagger$ maps ζ to $-\zeta^{-1}$.) Compare with Remark (4) above: there, $k = 1$ and $f = 0$; and if $\Gamma^\dagger = \Gamma$, then Δ_0 is real-valued and Φ_a is manifestly SU(2)-valued. (It just so happens there that the R-gauge of Theorem 8.2.2 is a 'real' gauge, but this is a fluke; the condition $\widetilde{F}^\dagger = \widetilde{F}$ only guarantees that there is *some* gauge in which Φ_a is real, and it may not be the gauge of Theorem 8.2.2.)

Example 8.2.4: Let k be any positive integer, and suppose that f is real and Γ is given by

$$\Gamma = Q^{-1}\{e^f + (-1)^k e^{-f}\},$$

where $Q = (\pi_{0'}\pi_{1'})^{-k}P$, and $P(Z^\alpha)$ is a homogeneous polynomial of degree $2k$ satisfying $P^\dagger = (-1)^k P$. Then again we get a real gauge field, for

$$\widetilde{F} = \begin{pmatrix} \zeta^k e^f & \Gamma \\ 0 & \zeta^{-k} e^{-f} \end{pmatrix} \begin{pmatrix} 0 & -1 \\ 1 & \zeta^k Q \end{pmatrix} = \begin{pmatrix} \Gamma & (-\zeta)^k e^{-f} \\ \zeta^{-k} e^{-f} & Q e^{-f} \end{pmatrix}$$

satisfies $\widetilde{F}^\dagger = \widetilde{F}$. Note that $\zeta^k Q$ is holomorphic on W, as required. We shall see in §8.4 that the SU(2) magnetic pole solutions are obtained from this class of transition matrices.

From these two examples it is clear that many anti-self-dual SU(2)-gauge fields can be obtained from the ansätze $\mathcal{A}_1, \mathcal{A}_2, \ldots$. It should be emphasized, however, that these examples do not exhaust all the possibilities for getting real gauge fields from the ansätze.

In this section, we have dealt only with rank two, this being the simplest nonabelian case. One can extend these ideas to bundles of higher rank, but inevitably things become more complicated. (See Ward 1981d, 1982; Athorne 1982, 1983; Burzlaff 1982 for some higher-rank examples.)

8.3   The twistor construction of instantons

This section is devoted to describing the construction of anti-self-dual Yang–Mills instantons. Recall from §5.2 that these are anti-self-dual gauge fields on S^4, classified topologically by a positive integer c_2. For the sake of simplicity, we shall restrict our attention to the gauge group SU(2); however, the ideas extend to other gauge groups, and some remarks on this will be given towards the end of the section. We know (from the Atiyah–Hitchin–Singer theorem quoted in §5.2) that the moduli space (i.e., the parameter space) of anti-self-dual SU(2)-instantons of charge $-c_2$ is a real manifold of dimension $8c_2 - 3$. Our task is, as far as feasible, to describe this manifold and construct the instanton solutions. To do so, we use Theorem 8.1.6, with $U = S^4$ and $n = 2$: the instantons correspond to certain holomorphic rank-two vector bundles (which henceforth we call 'instanton bundles') over $\mathbb{P} \cong \mathbf{P}_3(\mathbf{C})$.

The first thing to note is that the results of Serre (1956) tell us that holomorphic bundles over $\mathbf{P}_3(\mathbf{C})$ are necessarily *algebraic* (that is to say, one can choose coordinates so that the transition matrices which determine them are rational functions of the homogeneous coordinates Z^α). So we are in the realm of algebraic geometry. Algebraic geometers have developed two methods of constructing bundles over $\mathbf{P}_3$: the method of curves, and the method of monads. The first of these is related to the ansätze of §8.2, while the second leads to the Atiyah–Drinfeld–Hitchin–Manin (Atiyah et al. 1978) construction. We shall discuss each, beginning with the method of curves.

The technique of associating vector bundles with curves was first used by Serre, and was applied to the instanton problem in Atiyah and Ward (1977). In what follows, we shall state several results without proof; for proofs and more details, the reader is referred to the papers of Hartshorne (1978a, b).

Suppose E is an instanton bundle on $\mathbb{P}$, with second Chern number $c_2 > 0$. Note that c_2 is the same as the second Chern number of the bundle $\tilde{E}$ over S^4 on which the instanton connection lives, since E (as a topological bundle) is the pullback to $\mathbb{P}$ of $\tilde{E}$ over S^4. (See the discussion leading to Theorem 8.1.5 in §8.1.) The other topological invariant of E is $c_1(E)$, which is zero for instanton bundles, because of the condition that E restricted to real lines be trivial. The bundle E has no global sections (except zero), i.e., $H^0(\mathbb{P}, \mathcal{O}(E)) = 0$. This can be proved either in the $\mathbb{P}$-picture (once one has the necessary machinery), or by converting to the space-time picture, as demonstrated in Exercise 8.3.1.

So an instanton bundle E has no nontrivial global sections. But if we twist E by sufficiently many powers of the hyperplane-section line-bundle H, then eventually there *will* be a section. In other words: given E, there exists a positive integer n such that the twisted bundle $E(n) = E \otimes H^n$ admits a nontrivial global section.

Theorem 8.3.1. *Let E be an instanton bundle with second Chern class $c_2 > 0$. If n is an integer bigger than $(3c_2 + 1)^{\frac{1}{2}} - 2$, then $E(n)$ admits a nontrivial global section.*

For a proof see Hartshorne (1978b, §8). The following table sets out the minimum value of n for c_2 up to 12.

c_2	1	2	3	4	5	6	7	8	9	10	11	12
n	1	1	2	2	3	3	3	4	4	4	4	5

If s is a section of $E(n)$, let Y denote the zero-set of s (i.e., the set of points in $\mathbb{P}$ where s vanishes). Roughly speaking, Y is a curve, since s consists of two functions, and their simultaneous vanishing defines a set of codimension two in $\mathbb{P}$. But the word 'curve' has to be interpreted in a sufficiently liberal sense, as meaning a subvariety of $\mathbb{P}$ which is possibly reducible (having several components), possibly with nilpotent elements in its structure sheaf, and which is locally a complete intersection (having the correct number of defining equations). If n is sufficiently large (possibly larger that the minimum value specified in Theorem 8.3.1), and s is sufficiently general, then the curve Y will be nonsingular and connected, i.e., as nice as one could wish for. (See Hartshorne 1978b, §1 for more details on all this.) But it is often more convenient to allow s to have, for example, several disjoint components, in order to keep the value of n as low as possible.

What all this is leading up to is the following: the curve Y (if it is connected) *determines* E, so that given Y, one can 'reconstruct' E. More generally, if Y has several disjoint components $Y_1, \ldots, Y_r$, then we must specify, in addition, r nonzero constants $\lambda_1, \ldots, \lambda_r$ (up to an overall multiplicative factor) in order to determine E. (In general, the same E will arise from many different sets of data of this type: the correspondence is not one-to-one. This is, in part, because the section s of $E(n)$ may not be unique.) The curves Y that arise in this construction satisfy various restrictions. In particular, if the ith component curve Y_i has degree d_i and genus g_i, then the relations

$$\sum_{i=1}^{r} d_i = c_2 + n^2,$$

$$g_i = (n-2)d_i + 1 \tag{8.3.1}$$

are satisfied. Presently, we shall give examples of curves satisfying these constraints. But first, let us see how to construct E from the 'curve-data'.

Since $E(n)$ has a section s which vanishes only along the curve Y, this section s is nowhere-vanishing on the complement $Y' = \mathbb{P} - Y$ of Y, and so it generates a trivial line bundle on Y'. Thus $E(n)|_{Y'}$ admits a trivial line bundle I as subbundle, and is therefore an extension of I by H^{2n}. Such extensions correspond to elements Γ of $H^1(Y', \mathcal{O}(-2n))$, and once we are *given* Γ, we know from §8.2 how to compute the corresponding gauge potential A_a. The data $\{n, Y, \lambda_1, \ldots, \lambda_r\}$ determines Γ uniquely. A proof of this is given, for example, in Hartshorne (1978b, Theorem 1.1). We shall content ourselves here with studying some examples.

Example 8.3.2: Take $n = 1$. Then (bearing in mind that the degree of a curve is a positive integer and its genus a nonnegative integer) the only solution of (8.3.1) is $d_i = 1$, $g_i = 0$, $c_2 = r - 1$. A curve of degree one and genus zero is a straight line, so Y is a disjoint union of r straight lines Y_i in $\mathbb{P}$, with $r \geq 2$ (in order that c_2 be positive). If $r = 1$, so that Y is a single straight line, then E turns out to be trivial. This is a special case of the following result: if Y is a complete intersection (i.e., the intersection of two hypersurfaces in $\mathbb{P}$), then E is a direct sum of line bundles (Hartshorne 1978b, Corollary 1.2).

Given any line Y_i, there is an 'elementary solution'

$$\Gamma_i \in H^1(\mathbb{P} - Y_i, \mathcal{O}(-2))$$

based on that line (see §7.3). And $\Gamma = \sum_{i=1}^{r} \lambda_i \Gamma_i$ is the element of $H^1(Y', \mathcal{O}(-2))$ which determines the vector bundle E as an extension. In order for E to be 'real', we require each of the lines Y_i to be real (i.e., preserved by the antiholomorphic involution σ on $\mathbb{P}$) and also their weights λ_i to be real numbers.

So the anti-self-dual instanton potential corresponding to E is given by the ansatz $A_a = i\tilde{\sigma}_{ab}\nabla^b \log \Delta_0$ (see Remark (4) after Theorem 8.2.2), where Δ_0 is (cf. equation (5.2.2) of §5.2)

$$\Delta_0(x) = \sum_{i=1}^{r} \frac{\lambda_i}{(x - x_i)^2},$$

the x_i being distinct points on S^4 corresponding to the real lines Y_i. From the table at the end of §5.2 we see that for $c_2 \geq 3$, not all instantons are obtained by this construction (i.e., from (5.2.1) and (5.2.2)). But for $c_2 = 1$ or 2, all anti-self-dual instantons *do* arise in this way. In the $c_2 = 1$ case, this was proved using differential-geometric techniques (Atiyah, Hitchin and Singer 1978). It was proved using algebraic geometry for both $c_2 = 1$ and $c_2 = 2$ (Hartshorne 1978b, 8.4.1 and 9.6). See the two Theorems 8.3.3 and 8.3.4 stated below.

For all possible values of the weights λ_i, we get a bundle E over $\mathbb{P}$. But we have to make sure that the condition

$$E|_{\hat{x}} \text{ is trivial,} \quad \text{for all } x \in S^4$$

holds, and this is *not* true for all values of the λ_i. There are always some lines $\hat{x}$ in $\mathbb{P}$ on which E is not trivial, and we have to ensure that none of these lines is real. Lines $\hat{x}$ on which E is not trivial are called *jumping lines*. For instanton bundles E (not just in the $n = 1$ case we are examining at the moment, but in general), the set of jumping lines of E corresponds to a divisor D in complexified compactified space-time $M \cong G_{2,4}(\mathbb{C})$, i.e., a subvariety of codimension one in M; and the degree of D is equal to $c_2(E)$ (see Barth 1977). In the present example, this variety is the set of all points x in M where $\Delta_0(x)$ vanishes (in the language of the previous section, and in the general n case, the variety in question is the set of points where the function $S = \det M = \det N$ vanishes). At these points, the gauge potential A_a is singular. We have to make sure that the singularity locus D does not intersect S^4. In the present example, this is simply the condition that the λ_i all be *positive*. Clearly this ensures that $\Delta_0(x)$ is positive for all real points x (even at $x = \infty$, if we take into account the conformal scaling of Δ_0).

To sum up: this example involves choosing r distinct points on S^4, and r positive numbers λ_i (up to an overall scale), and yields an instanton of charge $-c_2 = 1 - r$.

Theorem 8.3.3 (Hartshorne 1978a, Theorem 4.2). *Example 8.3.2, with $r = 2$, gives all anti-self-dual instantons with $c_2 = 1$. The moduli space $M(1)$ of such instantons is the interior of S^4 in $\mathbf{R}^5$ (i.e., the five-dimensional open ball).*

Theorem 8.3.4 (Hartshorne 1978a, Theorem 4.4). *Example 8.3.2, with $r = 3$, gives all anti-self-dual instantons with $c_2 = 2$. The moduli space $M(2)$ of such instantons is a connected, but not simply-connected, real 13-dimensional manifold.*

Example 8.3.5: If we let $n = 2$, then the constraint equations (8.3.1) imply that $g_i = 1$, $\sum d_i = c_2 + 4$. Thus, the Y_i are elliptic curves (see Fulton 1969, Hartshorne 1977). In particular, Y might be a single elliptic curve of degree $d = c_2 + 4$. The instantons obtained in this way form a family of dimension

$$8c_2 - 3 \quad \text{if } c_2 \leq 4;$$
$$4c_2 + 16 \quad \text{if } c_2 \geq 5.$$

Note that $4c_2 + 16 < 8c_2 - 3$ for $c_2 \geq 5$, so this elliptic curve construction does not yield all instantons with $c_2 \geq 5$. If $c_2 = 1$ or 2, then we *do* get all instantons (in other words, this is an alternative, and more complicated, way of constructing the same bundles as in Example 8.3.2 with $c_2 = 1$ or 2). It is conceivable that this elliptic curve construction may give all instantons with $c_2 = 3$ or 4, since the number of parameters is correct. But whether it does or not is unknown (the problem is a complicated one requiring analysis of several degenerate special cases).

Another reason why this $n = 2$ example is more difficult than the $n = 1$ case of Example 8.3.2, is that the reality condition is obscure. In the $n = 1$ case we applied Example 8.2.3, where reality is straightforward, but this only works for *odd* n. In the $n = 2$ case, it is not simply a matter of requiring that $\Gamma \in H^1(Y', \mathcal{O}(-4))$ be real, or equivalently that the corresponding self-dual Maxwell field $\varphi_{A'B'}$ be real. The conditions on Γ or on $\varphi_{A'B'}$, in order to ensure reality, are more complicated. Also, the singularity set of $\varphi_{A'B'}$ in M (i.e., the set of jumping lines) is more complicated. The details of all this have not yet been worked out.

If $n \geq 3$, the task becomes even more difficult, since the genus of the curve Y grows rapidly. Not much is known about these cases.

Let us move on now to the second method of constructing vector bundles over $\mathbb{P}$, namely the method of monads. This technique is due to Horrocks (1964). A 'monad' is a sequence of vector bundles

$$F \xrightarrow{A} G \xrightarrow{B} H$$

with linear maps A and B between them, such that A is injective, B is surjective, and BA vanishes. The idea is to construct a vector bundle E as a quotient

$$E = \ker B / \operatorname{Im} A,$$

the point being that the bundles F, G and H are taken to be 'simple' bundles that one already knows about. In fact, in the case of instanton bundles, they are particularly simple; this follows from the work of Barth and Hulek (1978); see also Barth (1982). Drinfeld and Manin, and (independently) Atiyah and Hitchin, applied this work to obtain a construction for instantons; see Drinfeld and Manin (1978a, b, 1979), Atiyah et al. (1978), Atiyah (1979), Madore, Richard and Stora (1979).

In this special (instanton) case, the monad is self-dual, in the following sense. The bundle G is equipped with a symplectic form, which enables one to identify G with its dual; and H, B are respectively the duals of F, A. So the data consists only of F, G (with a symplectic form) and A. For instantons of second Chern class k, G is the trivial bundle of rank $2k + 2$ over $\mathbb{P}$, while F consists of the direct sum of k copies of the standard line-bundle $\mathcal{O}(-1)$. Thus, only A need be specified. With respect to given bases in F and G, A is clearly a $(2k + 2) \times k$ matrix of linear forms in the homogeneous coordinates Z^α on $\mathbb{P}$. The construction one ends up with may be described as follows.

Let V and W be complex vector spaces of respective dimensions $2k + 2$ and k. Let $(\, , \,)$ be a symplectic form (nondegenerate skew bilinear form) on V. Let

$$A(Z) : W \to V$$

be a linear map, depending linearly on Z^α. So $A(Z) = A_\alpha Z^\alpha$, where each of A_0, A_1, A_2, A_3 is a constant linear map from W to V. For any subspace U of V, let U^0 denote its annihilator, i.e.,

$$U^0 = \{v \in V : (u, v) = 0 \text{ for all } u \in U\}.$$

We impose the following condition on the map $A(Z)$:

> For all $Z^\alpha \neq 0$, $\quad U_Z := A(Z)W$ has dimension k,
> and is isotropic (i.e., $U_Z \subset U_Z^0$).
$$(8.3.2)$$

For each Z^α, let E_Z be the quotient vector space $E_Z := U_Z^0/U_Z$. Then E_Z depends algebraically on Z^α; in fact, it clearly depends only on the proportionality class of Z^α, i.e., $E_{\lambda Z} = E_Z$ for any nonzero complex number λ. Since U_Z^0 has dimension $k+2$ (being given by k equations in $2k+2$ unknowns) and the subspace U_Z has dimension k, it follows that E_Z has dimension two. Thus we have constructed a rank-two vector bundle E over the complex projective space $\mathbf{P}$.

The symplectic form $(\,,\,)$ on V is inherited by E_Z; in other words, $(\,,\,)$ gives rise to a well-defined and nondegenerate two-form on E_Z. So by Theorem 8.1.4, the structure group of E reduces to $\mathrm{SL}(2,\mathbf{C})$.

The restriction of E to the line in $\mathbf{P}$ which joins the two points X and Y is trivial provided $U_X^0 \cap U_Y = 0$, which is equivalent to $U_X \cap U_Y^0 = 0$ (see Atiyah 1979, pp. 60–61). We have to make sure that all the real lines have this property. For this, we need some extra structure, which will enable us to impose a reality condition on $A(Z)$.

Fix antilinear maps

$$\sigma : W \to W \quad \text{with } \sigma^2 = 1,$$
$$\sigma : V \to V \quad \text{with } \sigma^2 = -1.$$

Therefore, on W, we can think of σ as complex conjugation, while on V, we can think of σ as being like the real structure σ on twistor space. We require that σ on V be compatible with the symplectic form, i.e., that

$$(\sigma v_1, \sigma v_2) = \overline{(v_1, v_2)},$$

the overbar denoting complex conjugation. Because of this, V acquires an Hermitian form $\langle\,,\,\rangle$ defined by

$$\langle v_1, v_2 \rangle := -(v_1, \sigma v_2),$$
$$(8.3.3)$$

and we require this to be *positive-definite*. Our reality condition on $A(Z)$ can now be stated: it is

$$\sigma A(Z)w = A(\sigma Z)\sigma w$$
$$(8.3.4)$$

for all $w \in W$. From (8.3.4) we deduce that $U_{\sigma Z} = \sigma U_Z$, while from (8.3.3) it follows that the orthogonal complement $U^\perp$ of any subspace U of V is $(\sigma U)^0$. Hence $U_Z^0 = (U_{\sigma Z})^\perp$, and so by the positive-definiteness of the Hermitian form, $U_Z^0 \cap U_{\sigma Z} = 0$. This means that E, restricted to any real line, is trivial (the real lines being those that join a point Z to σZ).

It remains only to demonstrate that E admits a positive real form τ (see the remarks leading up to Theorem 8.1.6). This is easy:

$$\tau : E_Z \rightarrow E_{\sigma Z}^*$$

is given by

$$\tau(\xi)\eta = \langle \eta, \xi \rangle,$$

where $\xi \in E_Z$ and $\eta \in E_{\sigma Z}$. It is simple to see that τ is well-defined, and that it is indeed a positive real form on E. So we conclude from Theorem 8.1.6 that we have constructed an SU(2)-instanton bundle, depending on the positive integer k, and on the linear map $A(Z)$ satisfying (8.3.2) and (8.3.4).

Clearly we should regard two maps $A(Z)$ and $A'(Z)$ as equivalent if they are linearly related, i.e.,

$$A'(Z) = MA(Z)N,$$

where M, N act on V, W and preserve their structure:

$$M \in \mathrm{Sp}(k+1), \quad N \in \mathrm{GL}(k, \mathbf{R}),$$

where $\mathrm{Sp}(k+1)$ is the subgroup of $\mathrm{GL}(2k, \mathbf{C})$ preserving the symplectic form $(\,,\,)$ and the real form σ on V, while $\mathrm{GL}(k, \mathbf{R})$ preserves the real form σ on W. Certainly A and A' lead to equivalent vector bundles. If we now count parameters, while taking account of this equivalence, we discover that for each k, the construction yields an $(8k - 3)$-parameter family of instantons.

This suggests that the number k is the same as the second Chern class $c_2(E)$, and that the construction yields *all* SU(2)-instantons. This is indeed the case. The proof of these statements may be found, for instance, in Chapter 7 of Atiyah (1979). It follows from a theorem of Barth (Barth and Hulek 1978), together with a vanishing theorem (Rawnsley 1979, Wells 1982a). Drinfeld and Manin (1979) have reformulated Barth's work, and have shown that, given the vector bundle E, each of the vector spaces W and V may be defined as a cohomology group $H^1(\mathbb{P}, E \otimes Q)$ where Q is some 'standard' bundle over $\mathbb{P}$. Such a cohomology group may also, of course, be thought of as the space of solutions of an appropriate linear field equation on S^4 (describing some massless field coupled to the anti-self-dual gauge field). So one can also construct the spaces W and V directly in the S^4-picture, as spaces of solutions of linear equations on S^4 (Witten 1977, Osborn 1982). This provides an alternative proof of completeness (i.e., that the monad construction yields all instanton bundles).

How does one find $(2k + 2) \times k$ matrices $A(Z)$ satisfying (8.3.2) and (8.3.4)? Following Christ (1979), this problem can be restated as follows. Choose bases on V and W such that

(i) σ on W is complex conjugation;
(ii) the symplectic form on V is represented by the skew $(2k + 2) \times (2k + 2)$ matrix

$$\Omega = \begin{pmatrix} 0 & 1 & & & & & \\ -1 & 0 & & & & & \\ & & 0 & 1 & & & \\ & & -1 & 0 & & & \\ & & & & \ddots & & \\ & & & & & 0 & 1 \\ & & & & & -1 & 0 \end{pmatrix},$$

i.e., $(v, u) = v^t \Omega u$, where v and u are represented as column $(2k + 2)$-vectors and v^t denotes the transpose of v;
(iii) σ on V is given by $\sigma v = \Omega \bar{v}$, where $\bar{v}$ is the complex conjugate of v.

Note that by (8.3.3), the Hermitian form on V is $\langle v, u \rangle = v^t \bar{u}$, which is indeed positive-definite.

The reality condition (8.3.4) can now be stated as

$$\Omega \overline{A(Z)} = A(\sigma Z). \tag{8.3.5}$$

Define matrices $B^{A'}$ and C_A by

$$A(Z) = B^{A'} \pi_{A'} - C_A \omega^A;$$

then the reality condition (8.3.5) becomes

$$\begin{aligned} \Omega \overline{B^{0'}} &= -B^{1'}, \\ \Omega \overline{B^{1'}} &= B^{0'}, \\ \Omega \overline{C_0} &= -C_1, \\ \Omega \overline{C_1} &= C_0. \end{aligned} \tag{8.3.6}$$

The trick now is to introduce quaternions. Think of quaternions as 2×2 complex matrices, generated by

$$1 = \begin{pmatrix} 1 & 0 \\ 0 & 1 \end{pmatrix},$$

$$\mathbf{i} = -i\sigma^1 = \begin{pmatrix} 0 & -i \\ -i & 0 \end{pmatrix},$$

$$\mathbf{j} = -i\sigma^2 = \begin{pmatrix} 0 & -1 \\ 1 & 0 \end{pmatrix},$$

$$\mathbf{k} = -i\sigma^3 = \begin{pmatrix} -i & 0 \\ 0 & i \end{pmatrix}.$$

Note that $\mathbf{i}^2 = -1$, $\mathbf{ij} = \mathbf{k}$, etc. A point x^a in E^4 is taken to correspond to a quaternion x according to

$$\sqrt{2}x = x^0 + i\vec{x} \cdot \vec{\sigma} = x^0 1 - x^1 \mathbf{i} - x^2 \mathbf{j} - x^3 \mathbf{k}. \tag{8.3.7}$$

Note that by our usual conventions, the 2×2 matrix x is equal to $ix^{AA'}$, where the index A labels rows, and A′ labels columns. Let $B^{A'}_{p,q}$ denote the spinor appearing in the pth row and the qth column of the matrix $B^{A'}$, and consider, for fixed m, n, the 2×2 matrix $B^{A'}_{2m+B'-1,n}$ (where B′ labels rows and A′ labels columns). The reality condition (8.3.6) says precisely that this 2×2 matrix is a quaternion. So we can think of $B^{A'}$ as a $(k+1) \times k$ matrix of quaternions. Do the same thing for C_A, and put $\omega^A = ix^{AA'}\pi_{A'}$. The end-result is that $A(Z)$ is represented by the $(k+1) \times k$ matrix of quaternions

$$M(x) = B - Cx$$

(with Cx denoting quaternionic multiplication). The reality condition (8.3.4) having been taken care of, we only have to worry about the condition (8.3.2). This is simple to formulate in terms of M. Let M^* denote the conjugate transpose of M (the conjugate of a quaternion $h = h_0 1 + h_1 \mathbf{i} + h_2 \mathbf{j} + h_3 \mathbf{k}$ being $\overline{h} = h_0 1 - h_1 \mathbf{i} - h_2 \mathbf{j} - h_3 \mathbf{k}$), and consider the $k \times k$ quaternionic matrix M^*M. Condition (8.3.2) is equivalent to

For each x, $M(x)^*M(x)$ should be real and nonsingular. (8.3.8)

By real we mean that each of the k^2 quaternionic entries h in M^*M should be real-valued (i.e., $\overline{h} = h$).

As an example, consider

$$M(x) = \begin{pmatrix} \lambda_1 & \lambda_2 & \cdots & \lambda_k \\ x_1 - x & 0 & \cdots & 0 \\ 0 & x_2 - x & & 0 \\ \vdots & & \ddots & \\ 0 & 0 & & x_k - x \end{pmatrix}, \qquad (8.3.9)$$

where the λ_i are nonzero real numbers, and the x_i are distinct quaternions. This $M(x)$ satisfies (8.3.8), and in fact corresponds to the 't Hooft solutions ((5.2.1), (5.2.2)), as we shall see below.

The anti-self-dual gauge potential on E^4 can be obtained in a fairly direct way from $M(x)$. We want to construct a bundle F over S^4 or E^4, and a connection on F; this is done as follows. If $x \in S^4$ corresponds to the line $\hat{x} \in \mathbb{P}$ joining Z to σZ, then F_x is the two-dimensional subspace of V which is orthogonal to both U_Z and $U_{\sigma Z}$ (see Atiyah 1979 for more details on this and on what follows). In other words, F is obtained as a subbundle of the trivial rank-$(2k+2)$ bundle $S^4 \times V$. And the instanton connection on F is simply the orthogonal projection to F of the flat connection ∇_a on this trivial bundle.

Let $v(x)$ be a quaternionic $(k+1)$-column-vector satisfying

$$\begin{aligned} v^* v &= 1, \\ v^* M &= 0. \end{aligned} \qquad (8.3.10)$$

Then $P = vv^*$, which is a quaternionic $(k+1) \times (k+1)$ matrix depending on x, is the orthogonal projection from V to F_x. Note that $P^2 = P$. The covariant derivative D_a on F is defined by projecting ∇_a via P. In other words, if f is a section of F, then vf takes values in V, and so

$$\begin{aligned} D_a(vf) &= P\nabla_a(vf) \\ &= vv^*\nabla_a(vf) \\ &= v\{\nabla_a f + (v^*\nabla_a v)f\}. \end{aligned}$$

Thus the gauge potential $A_a = D_a - \nabla_a$ is

$$A_a = v^*\nabla_a v. \qquad (8.3.11)$$

This is a quaternion-valued one-form. In fact, it takes values in the imaginary quaternions, since

$$A_a^* = (\nabla_a v^*)v$$
$$= -v^* \nabla_a v$$
$$= -A_a,$$

by virtue of (8.3.10). We have identified the imaginary quaternions (generated by $\mathbf{i}, \mathbf{j}, \mathbf{k}$) with the Lie algebra of $SU(2)$ (generated by the Pauli matrices σ_i), so A_a is an $SU(2)$-gauge potential. This construction of A_a via $M(x)$ and $v(x)$ is called the ADHM construction.

By way of example, for the matrix M given by (8.3.9), a solution v of the conditions (8.3.10) is given by

$$v^* = \varphi^{-\frac{1}{2}} \left[1 \quad \frac{\lambda_1}{x - x_1} \cdots \frac{\lambda_k}{x - x_k} \right],$$

$$\varphi = 1 + \sum_{i=1}^{k} \lambda_i^2 \|x_i - x\|^{-2}.$$

Computing A_a according to (8.3.11) leads to the 't Hooft solutions (5.2.1).

The ADHM construction has many remarkable features, some of which are discussed by Corrigan et al. (1978a), Christ et al. (1978), and Corrigan and Goddard (1984). One such feature (for which, see also Atiyah 1979), is that there is a simple expression for the Green's function of a massless scalar field coupled to the gauge field (in the fundamental representation of the gauge group). The Green's function $G(x, y)$ by definition satisfies

$$D^2 G(x, y) = -\delta(x - y)$$

with the boundary condition $G(x, y) \to 0$ as $\|x\| \to \infty$. In terms of $v(x)$, it is simply given by

$$G(x, y) = \frac{v(x)^* v(y)}{4\pi^2 \|x - y\|^2}.$$

All of the above was for the gauge group $SU(2)$. Another nice feature of the ADHM construction is that it generalizes very easily to other gauge groups, such as $Sp(n), SU(n)$ and $SO(n)$; see, for example, Atiyah (1979). And the construction has been used to obtain many new instanton solutions, for SU(2) as well as for other groups: see Christ et al. (1978), Meyers and de Roo (1978), and McCarthy (1981a, b, 1983).

Donaldson (1984a) has shown that the ADHM construction enables one to describe Yang–Mills instantons in terms of holomorphic bundles over $\mathbf{P}_2$ (rather than over twistor space $\mathbf{P}_3$). For applications of this description, see Atiyah (1984) and Hurtubise (1986).

Exercise for Section 8.3

8.3.1. Let U be an open set in S^4, with $\widehat{U}$ the corresponding open subset of $\mathbf{P}$. Show that the elements of $H^0(\widehat{U}, \mathcal{O}(E))$ correspond to covariantly constant sections ζ of the bundle $\widetilde{E}$ over U. That is, $\zeta = \zeta(x)$ is a section of $\widetilde{E}$, satisfying $D_a \zeta = 0$. Using the SU(2) structure, show that if ζ is such a field, then its complex conjugate ζ^* is another, so that $\{\zeta, \zeta^*\}$ forms a linearly independent set of covariantly constant sections of V. Hence if $H^0(\widehat{U}, \mathcal{O}(E))$ is nonzero, then the corresponding anti-self-dual connection on U is flat.

8.4 Magnetic poles: solutions

In this section, we shall see how twistor theory has been applied to the problem of finding magnetic poles: solutions of the Bogomolny equations (5.3.4). As usual, we shall concentrate on the case where the gauge group is SU(2), and merely refer to the literature for generalizations to larger groups.

First, however, let us mention the Maxwell case, and Dirac magnetic poles. Strictly speaking, these have a magnetic field B only, and no electric field E; whereas an anti-self-dual Maxwell field (for which a twistor description exists) has $E = iB$ (see §4.1). So we should try to find a twistor description of 'complex, anti-self-dual' Dirac poles with magnetic charge $2\pi n$, and therefore electric charge $2\pi in$. Such a description does indeed exist. In terms of line bundles over (part of) twistor space, it was first given by Penrose and Sparling (1979). More recently, Bailey (1985) has discussed the problem in terms of relative cohomology groups. The interested reader may refer to these papers for further details.

We turn, now, to the nonabelian case. To recap: this involves finding an SU(2)-gauge potential A_i on $\mathbf{R}^3$, and a scalar field φ in the adjoint representation, with

$$B_i := \tfrac{1}{2}\varepsilon_{ijk}F^{jk} = -D_i\varphi, \qquad (8.4.1)$$

$$|\varphi| = 1 - n/r + O(r^{-2}) \qquad (8.4.2)$$

as $r \to \infty$ in $\mathbf{R}^3$, where n is a positive integer. By the argument following equation (5.3.5), such a configuration has finite energy $E = 8\pi n$.

The key observation which enables one to apply a twistor construction is that (8.4.1) can be obtained from the anti-self-dual gauge-field equations by 'dimensional reduction'. If we consider a gauge potential A_a in Euclidean four-space E^4, and impose the condition that A_a be independent of x^0 (and so a function only of x^i, $i = 1, 2, 3$), then the anti-self-duality equations $*F_{ab} = -F_{ab}$ are equivalent to

$$B_i := \tfrac{1}{2}\varepsilon_{ijk}F^{jk} = D_iA_0.$$

These are the same as (8.4.1) if we identify A_0 with the negative of the Higgs field φ. This identification is consistent with gauge invariance, in the sense that if we restrict to gauge transformations $g(x^a)$ which are independent of x^0, then A_0 transforms homogeneously:

$$A_0 \mapsto g^{-1}A_0g;$$

in other words, just as φ does (being in the adjoint representation).

It is worth remarking that it makes little difference whether we begin with the self-dual or anti-self-dual equations; whether we identify A_0 with φ or with $-\varphi$; and whether we end up with $B = D\varphi$ or $B = -D\varphi$. It all just amounts to a choice of orientation, or to the choice of 'monopoles' as opposed to 'antimonopoles'.

So solutions to (8.4.1) and (8.4.2) correspond to certain holomorphic vector bundles, of rank two, over (part of) twistor space. The problem is to characterize and construct these bundles. It was observed by Manton (1978) that the known $n = 1$ solution (the Prasad–Sommerfield monopole) could be obtained from the ansatz $\mathcal{A}_1$ of §8.2 (in fact, from the Corrigan–Fairlie–'t Hooft–Wilczek ansatz: (5.2.1) and §8.2 Remark (4)). Knowing this, it is easy to see explicitly what the bundle is that corresponds to the Prasad–Sommerfield solution

(Example 8.4.1 below). Manton also pointed out that no *other* solutions could be obtained from the CF'tHW ansatz. But it turns out that the 'higher' ansätze $\mathcal{A}_n$ for $n \geq 2$ produce new solutions.

First, $\mathcal{A}_2$ was used to produce an axially-symmetric solution of charge $n = 2$ (Ward 1981a). This depends on five parameters, and so it is not the most general solution of charge two (which depends on $4 \times 2 - 1 = 7$ parameters). Prasad (1981) and Prasad and Rossi (1981) extended this by showing that for each $n \geq 3$, one can find a 'candidate' axially-symmetric solution, of charge n, depending as before on five parameters. See Example 8.4.2. The qualification 'candidate' is used because they found fields (A_i, φ) satisfying (8.4.1) and (8.4.2), but did not prove that these fields are smooth on $\mathbf{R}^3$, as is required. In terms of the notation of §8.2, they had functions Δ_r which were manifestly smooth, but had yet to show that the determinant $\det M = \det N$ was nowhere-vanishing. The next development was the observation that $\mathcal{A}_2$ produces a solution of charge two depending on the full seven parameters (Ward 1981b, c) (see Example 8.4.3). This in turn was extended to charge $n \geq 3$, to obtain a 'candidate' solution depending on the full $4n - 1$ parameters (Corrigan and Goddard 1981).

Our examples are special cases of Example 8.2.4, and we shall use the notation of that example (writing n for k). We have to specify two twistor functions f and Q, which are 'real' in the sense that $f^\dagger = f$, $Q^\dagger = Q$. We want the gauge field to be independent of x^0, and this is achieved by requiring that f and Q depend on ω^A only via the combination

$$\gamma = i\omega^1/\pi_{1'} - i\omega^0/\pi_{0'}. \qquad (8.4.3)$$

For substituting $\omega^A = ix^{AA'}\pi_{A'}$ into this gives a function $\gamma(x^a, \zeta)$ which is independent of x^0. Here $\zeta = \pi_{0'}/\pi_{1'}$. Note that $\gamma^\dagger = \gamma$. In the examples that follow, $f(\gamma, \zeta)$ and

$$\Gamma(\gamma, \zeta) = Q^{-1}\{e^f + (-1)^n e^{-f}\}$$

are holomorphic for $|\zeta|$ near 1, and for *all* γ. As a result, the functions Δ_r are smooth (real-analytic) on $\mathbf{R}^3$, and one has only to check that $\det M$ is nowhere-zero on $\mathbf{R}^3$.

Example 8.4.1: Letting $n = 1$ and $f = Q = \gamma/\sqrt{2}$ gives the Prasad–Sommerfield solution. Note here, for example, that

$$\Gamma = 2f^{-1}\sinh f$$

is an entire function of γ. Three parameters (arising from translations in $\mathbf{R}^3$) can be introduced into this solution. In terms of the notation of §8.2, we have

$$g = -\tfrac{1}{2}(x^1 + ix^2)\zeta + x^3 + \tfrac{1}{2}(x^1 - ix^2)\zeta^{-1},$$
$$\Delta_0 = r^{-1}\sinh r, \quad (r^2 = x_i x^i), \tag{8.4.4}$$
$$\varphi = ir^{-2}(r\coth r - 1)(x^1\sigma_1 + x^2\sigma_2 - x^3\sigma_3).$$

Note that $\det M = \Delta_0$ is nowhere-zero, and that

$$|\varphi| = \coth r - r^{-1}$$

is smooth on $\mathbf{R}^3$, vanishes only at $r = 0$, and has the required asymptotic behavior $|\varphi| \cong 1 - r^{-1}$ as $r \to \infty$.

Example 8.4.2: Let $n \geq 2$, $f = \gamma/\sqrt{2}$ and

$$Q = \prod_{k=1}^{n}\{f - \tfrac{1}{2}i\pi(n + 1 - 2k)\};$$

then we obtain axisymmetric solutions of charge n. Five parameters can be introduced into the solution by using rigid motions of $\mathbf{R}^3$. If $n = 2$, then one can see explicitly that $\det M$ is nowhere-zero (Ward 1981a, c). More sophisticated methods can be used to show that the same thing is true (and hence that one gets a smooth solution on $\mathbf{R}^3$) for all $n > 2$ (Hitchin 1983, Theorem 8.2).

Example 8.4.3: Let $n = 2$, $f = \tfrac{1}{2}\pi\gamma/\sqrt{2\delta}$ and $Q = f^2 + \gamma$, where

$$\delta = \tfrac{1}{2}pq(\zeta - \zeta^{-1}) + q.$$

Here p and q are real parameters, with q determined in terms of p by the elliptic integral

$$\sqrt{q} = \tfrac{1}{4}\int_0^{2\pi}(1 + ip\sin\theta)^{-\frac{1}{2}}d\theta.$$

We choose that branch of $\sqrt{\delta}$ which has positive real part, and then f and Q have the required analyticity properties. If $p = 0$, this example reduces to the previous one (with $n = 2$). If $p \neq 0$, it is not axisymmetric. Rigid motions of $\mathbf{R}^3$, together with p, give a family of solutions depending on seven parameters. By a continuity argument, one knows that det M is nowhere-zero, provided p is sufficiently close to zero. For large p, one has to use other methods to establish that the solution is nonsingular (see the discussion later in this section). For $n > 2$, the general $(4n - 1)$-parameter case is far more complicated than this: see Corrigan and Goddard (1981).

A solution of charge n is thought of as representing n magnetic poles, located at the points where the Higgs field φ vanishes. In the axisymmetric cases of Examples 8.4.1 and 8.4.2, φ vanishes at one point only: the n poles are superimposed on one another. But in Example 8.4.3, for $p \neq 0$, φ vanishes at two distinct points: the solution represents two separated poles. The existence theorem of Taubes (1981) (see §5.3) is built on the idea of having n 'widely separated' poles.

Because the ansatz $\mathcal{A}_n$ gives a 'candidate' solution with the correct number of parameters, it is plausible that *every* n-pole solution is obtained in this way. That this is indeed the case was proved by Hitchin (1982a). He used a 'minitwistor' description in which the redundant x^0-dependence is factored out from the beginning. The vector field $\partial/\partial x^0$ on E^4 corresponds to the nowhere-vanishing holomorphic vector field

$$V = \pi_{0'} \frac{\partial}{\partial \omega^0} + \pi_{1'} \frac{\partial}{\partial \omega^1}$$

on $\mathbf{P}^I$. Note that $V(\gamma) = 0$, where γ is defined by (8.4.3). The quotient of $\mathbf{P}^I$ by V is a two-dimensional complex manifold $T\mathbf{P}_1$ called minitwistor space. In fact, $T\mathbf{P}_1$ is the total space of the holomorphic tangent bundle of the Riemann sphere $\mathbf{P}_1$, hence the notation.

Hitchin's approach involves going directly from $\mathbf{R}^3$ to $T\mathbf{P}_1$, rather than via E^4 and $\mathbf{P}^I$:

$$
\begin{array}{ccc}
E^4 & \longleftrightarrow & \mathbf{P}^I \\
{\scriptstyle \frac{\partial}{\partial x^0}} \downarrow & & \downarrow V \\
\mathbf{R}^3 & \longleftrightarrow & T\mathbf{P}_1
\end{array}
$$

Namely, $T\mathbf{P}_1$ is thought of as being the space of oriented straight lines in $\mathbf{R}^3$. Any such straight line is uniquely specified by a unit vector $\overrightarrow{u}$ parallel to it, and a perpendicular vector $\overrightarrow{v}$ giving its displacement from the origin O (see Figure 8.4.1). The set of all pairs $(\overrightarrow{u}, \overrightarrow{v})$ of three-vectors satisfying $\overrightarrow{u} \cdot \overrightarrow{u} = 1$, $\overrightarrow{u} \cdot \overrightarrow{v} = 0$ can clearly be identified with the tangent bundle of the unit sphere S^2. If $\mathbf{R}^3$ is equipped with an orientation, then this space of directed lines can be given a complex structure in a natural way, and this is precisely the complex structure that $T\mathbf{P}_1$ has by virtue of being the holomorphic tangent bundle of $\mathbf{P}_1 \cong S^2$. See Hitchin (1982a) for details.

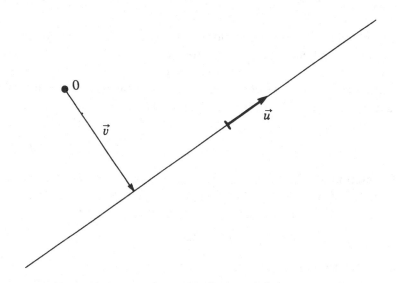

Fig. 8.4.1. Parametrizing oriented straight lines in $\mathbf{R}^3$.

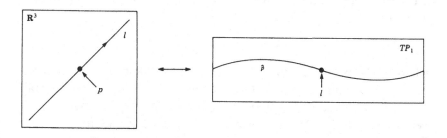

Fig. 8.4.2. The correspondence between a point p and a holomorphic section $\hat{p}$ of $T\mathbf{P}_1$.

A point p in $\mathbf{R}^3$ is determined by the set of all directed lines l through it, and this set is a holomorphic section $\hat{p}$ of $T\mathbf{P}_1$ (i.e., a holomorphic vector field on $\mathbf{P}_1$; see Figure 8.4.2). The set of *all* holomorphic sections of $T\mathbf{P}_1$ depends on three *complex* parameters, and corresponds to the complexification $\mathbf{C}^3$ of $\mathbf{R}^3$. There is a natural 'real structure' on $T\mathbf{P}_1$ (an antiholomorphic involution σ), which enables one to pick out the 'real' sections corresponding to points of $\mathbf{R}^3$.

We can describe the above more explicitly, using local coordinates (ζ, γ) on $T\mathbf{P}_1$. Let ζ be the usual stereographic coordinate on $\mathbf{P}_1$, and γ the coordinate corresponding to the vector $\gamma\zeta\frac{\partial}{\partial\zeta}$ at $\zeta \in \mathbf{P}_1$. (These coordinates ζ and γ are consistent with those introduced earlier, cf. equation (8.4.3).) The correspondence between $T\mathbf{P}_1$ and $\mathbf{R}^3$ (with coordinates (x^1, x^2, x^3)) is expressed in the basic equation

$$\tfrac{1}{\sqrt{2}}\gamma = -\tfrac{1}{2}(x^1 + ix^2)\zeta + x^3 + \tfrac{1}{2}(x^1 - ix^2)\zeta^{-1} \qquad (8.4.5)$$

(cf. equation 8.4.4). If we fix γ and ζ, then this is the equation of a line in $\mathbf{R}^3$; if on the other hand we fix x^i, then (8.4.5) defines a real section of $T\mathbf{P}_1$. The antiholomorphic involution on $T\mathbf{P}_1$ is

$$\sigma : (\zeta, \gamma) \mapsto (-\bar{\zeta}^{-1}, \bar{\gamma}),$$

and a 'real section' means one preserved by σ.

As an aside, we remark that the twistor solution procedure for the Laplace equation on E^4 (or for the wave equation on M^4) can be reduced to one for the Laplace equation on $\mathbf{R}^3$ (see Exercise 8.4.1).

The corresponding contour integral formula is as follows: given a function $f(\gamma, \zeta)$ holomorphic on a suitable region of $T\mathbf{P}_1$, define $\varphi(x^i)$ by

$$\varphi(x) = \oint f(\gamma, \zeta)d\zeta,$$

where γ denotes the expression on the right-hand side of (8.4.5). Then φ is a solution of the Laplace equation on $\mathbf{R}^3$. This general solution of the Laplace equation is (in essence) the famous one of Whittaker (1903). Another, even older, application of the $T\mathbf{P}_1 \leftrightarrow \mathbf{R}^3$ correspondence is the solution by Weierstrass of the minimal surface equations in $\mathbf{R}^3$ (see Hitchin 1982a, Appendix).

We return now to the Bogomolny equations $B_i = -D_i\varphi$. As remarked earlier, a solution of these equations corresponds to a holomorphic vector bundle E over $\mathbf{P}^I$. Because of the x^0-invariance, this bundle actually comes from a holomorphic vector bundle $\widetilde{E}$ over the minitwistor space $T\mathbf{P}_1$ (i.e., E is the pullback of $\widetilde{E}$ along the canonical surjection $\mathbf{P}^I \to T\mathbf{P}_1$). The bundle $\widetilde{E}$ can be constructed directly from the data on $\mathbf{R}^3$, as follows.

Given an $SU(2)$-connection D_j and an adjoint scalar field φ on $\mathbf{R}^3$, Hitchin defines $\widetilde{E}$ by saying that the fiber $\widetilde{E}_l$ of $\widetilde{E}$ above $l \in T\mathbf{P}_1$ is the solution space of the ordinary differential equation

$$(u^j D_j - i\varphi)s = 0. \tag{8.4.6}$$

Here u^j is the unit tangent vector pointing (in the positive direction) along the oriented line l, and s is a two-component function on l. So the solution space of (8.4.6) is two-dimensional, as required. It follows from the Bogomolny equations that $\widetilde{E}$ has a natural holomorphic structure: see Hitchin (1982a), Theorem (4.2). He shows also, as one might expect, that from $\widetilde{E}$ one can 'reconstruct' A_j and φ.

This construction is, of course, closely related to that of §8.1. There, the bundle E was defined in terms of the solution of linear equations on totally-null two-planes; the anti-self-duality equations were needed to ensure integrability, while the holomorphic structure came automatically. Here, by contrast, there is no integrability problem, since l is one-dimensional; but the Bogomolny equations are needed to guarantee a holomorphic structure.

For each line l, the space of solutions which decay exponentially at $+\infty$ is one-dimensional. (Since the line is oriented, one can distinguish $+\infty$ from $-\infty$.) This one-dimensional subspace of $\widetilde{E}_l$ varies holomorphically with l, and therefore defines a holomorphic line-bundle which is a subbundle of $\widetilde{E}$. But this, as we saw in §8.2, means that the gauge field is generated by one of the ansätze $\mathcal{A}_n$. It turns out, in fact, that all magnetic pole solutions of charge n can be obtained from the ansatz $\mathcal{A}_n$.

The set of those lines l for which (8.4.6) has a solution decaying to zero at both ends, forms a compact holomorphic (in fact algebraic) curve S in $T\mathbf{P}_1$, called the *spectral curve*. It has genus $g = (n-1)^2$. The important point is that S determines $\widetilde{E}$, and hence (A_j, φ).

The curve S is given by the vanishing of a polynomial Q in γ, ζ and ζ^{-1}, of degree n. This is the same Q which appears in the ansätze, cf. Examples 8.4.1–3. Given that Q has to be real (i.e., preserved by the involution σ), and that one may ignore an overall scale factor, such polynomials depend on $n^2 + 2n$ real parameters. However, the curves S that arise in the magnetic pole problem are constrained by $g = (n-1)^2$ conditions, and so the parameter space has dimension

$$n^2 + 2n - (n-1)^2 = 4n - 1,$$

as we know it should. In general, the solutions involve abelian functions defined on a complex g-dimensional torus, the Jacobian of S.

If $n = 1$, then S is a rational curve, in fact a section of $T\mathbf{P}_1$ corresponding to a point in $\mathbf{R}^3$ (cf. Example 8.4.1). For the axisymmetric cases of Example 8.4.2, S is reducible, and is the union of n sections of $T\mathbf{P}_1$. Finally, note that in the generic $n = 2$ case of Example 8.4.3, one encounters an elliptic integral: the genus g is $(2 - 1)^2 = 1$ in this case. This last case is discussed in detail by Hurtubise (1983).

Many of the properties of multi-pole configurations have been investigated by O'Raifeartaigh et al. (1981, 1982, 1982a, b, 1983).

It is worth remarking at this point that Murray (1983, 1984) has extended Hitchin's results to arbitrary (connected, compact, simple) gauge groups. And some examples of SU(3) solutions have been investigated in detail in Ward (1981d, 1982), Athorne (1982, 1983).

The above approach to constructing magnetic pole solutions, using spectral curves and the ansätze $\mathcal{A}_n$, is quite effective. Its drawback is that it is difficult to decide whether or not a given solution is everywhere smooth. An alternative approach, due to Nahm (1981, 1982a, b), guarantees smoothness from the outset. This method is the analogue of the ADHM construction for instantons discussed in the second half of §8.3. We recall that the ADHM construction involved taking a certain quaternionic matrix $M(x)$ depending linearly on x, finding a quaternionic vector $v(x)$ satisfying $v^*v = 1, v^*M = 0$, and then defining a gauge potential by $A_a = v^*\nabla_a v$. In Nahm's adaptation, M and v become infinite-dimensional. More specifically, v takes values in $\mathbf{C}^n \otimes \mathbf{C}^2 \otimes \mathcal{L}^2(0, 2)$; thinking of $\mathbf{C}^2$ as being the quaternions, we can write $v = v(x^i, z)$, where v is a quaternion-valued n-vector field on $\mathbf{R}^3 \times (0, 2)$, and is square-integrable on $(0, 2)$ for each x^i. The condition $v^*v = 1$ is replaced by

$$\int_0^2 v^*v\, dz = 1. \tag{8.4.7}$$

And M becomes a differential operator, the algebraic equation $v^*M = 0$ being replaced by

$$\left(i\frac{d}{dz} + x^* + iT^*\right)v = 0, \tag{8.4.8}$$

where x is the quaternion corresponding to x^a (see equation (8.3.7)), and $T = T(z)$ is an $n \times n$ matrix with values in the imaginary quaternions. So we can write

$$T(z) = T_1(z)\mathbf{i} + T_2(z)\mathbf{j} + T_3(z)\mathbf{k},$$

where **i**, **j**, **k** span the imaginary quaternions (cf. equation (8.3.7)). The $T_i(z)$ are $n \times n$ matrices of complex-valued functions of $z \in (0, 2)$.

The idea is that given a suitable $T_i(z)$, one finds a solution v of (8.4.7) and (8.4.8), and then defines (A_j, φ) by

$$A_j = \int_0^2 v^* \nabla_j v \, dz,$$

$$\varphi = i \int_0^2 z v^* v \, dz. \tag{8.4.9}$$

This is the ADHMN construction: Nahm's adaptation of the ADHM construction. In order for (A_j, φ) to be a solution of the Bogomolny equations and to satisfy the necessary regularity and boundary conditions, one has to impose conditions on T_i, namely:

C1. $T_i(z)$ is analytic on $(0, 2)$ with simple poles at $z = 0$ and $z = 2$;

C2. $T_i^* = -T_i$ (T^* being the conjugate transpose of T);

C3. $T_i(z) = -\overline{T}_i(2 - z)$ ($\overline{T}$ being the complex conjugate of T);

C4. at each pole, the residues of (T_1, T_2, T_3) define an irreducible representation of SU(2);

C5. $T_i(z)$ satisfies the differential equation

$$\frac{dT_i}{dz} = \tfrac{1}{2}\varepsilon_{ijk}[T_j, T_k]. \tag{8.4.10}$$

It is proved in Hitchin (1983) that the Nahm construction provides an exact equivalence between magnetic pole solutions of charge n, and $n \times n$ matrices T_i satisfying C1–5. It is this equivalence which was referred to earlier, when we mentioned that 'more sophisticated methods' could be used to show that the Examples 8.4.2 are all smooth.

Some examples of the Nahm construction have been worked out in detail: see Brown, Panagopoulos and Prasad (1982); Panagopoulos (1983); Rouhani (1984). The construction generalizes to other gauge groups, and some examples of its application to higher groups may be found in Bowman (1983); Bowman et al. (1984); Horvath and Rouhani (1984).

The Nahm equations (8.4.10) are of considerable interest in themselves. First, (8.4.10) can be obtained by reducing the self-duality equations $*F_{ab} = F_{ab}$ from four dimensions down to one. For if we specify that A_a depend only on x^0 (and not on x^1, x^2, x^3), then $*F_{ab} = F_{ab}$ becomes

$$D_0 A_i = \tfrac{1}{2}\varepsilon_{ijk}[A_j, A_k]. \qquad (8.4.11)$$

Now we can choose a gauge such that $A_0 = 0$: then (8.4.11) is the same as (8.4.10), with $x^0 = z$ and $A_i = T_i$. It is a curious fact that Nahm's construction gives a link between the self-duality equations in three dimensions (the Bogomolny equations) and the self-duality equations in one dimension (the Nahm equations). This 'reciprocity' is discussed in greater depth by Corrigan and Goddard (1984).

The second interesting point is that the Nahm equations can be solved by considering a linear flow on an abelian variety, the Jacobian of an algebraic curve S (Hitchin 1983, Atiyah 1982). This curve S is none other than the spectral curve which appeared earlier in this section. The idea of linear flow on a torus is, of course, familiar in the context of completely-integrable Hamiltonian systems (cf. Adler and van Moerbeke 1980).

Donaldson (1984b) showed that solutions of the Nahm conditions C1–5 correspond to rational maps $f : \mathbf{P}_1(\mathbf{C}) \rightarrow \mathbf{P}_1(\mathbf{C})$ of degree n. This gives a very convenient description of the moduli space M_n (the space of solutions of charge n). The connection between the rational maps and the (A_i, φ) configuration in $\mathbf{R}^3$ can also be seen more directly (Hurtubise 1985).

Exercise for Section 8.4

8.4.1. Show that the space of complex-valued harmonic functions on $\mathbf{R}^3$ is isomorphic to $H^1(T\mathbf{P}_1, \mathcal{O}(-2))$.

9

Twistors for self-dual space-time

In previous chapters, we have discussed twistors for flat (or conformally flat) space-time. The question arises: can one define twistors for curved space-time? In general, there is no completely satisfactory way of doing so. The reason why difficulties occur can be illustrated by studying the scattering of twistors through an impulsive gravitational wave (Penrose and MacCallum 1973): the complex structure of twistor space gets 'shifted', so that there is no unique complex structure. Another way of seeing this involves hypersurface twistors (Penrose 1975, Penrose and Ward 1980, Penrose 1983). Here one associates a 'curved' twistor space with each hypersurface in space-time. But the complex structure of this twistor space depends, in general, on the choice of the hypersurface, and shifts as one moves from one hypersurface to another.

An approach which bypasses these problems is to use 'ambi-twistors', which correspond to the complex null geodesics in a complex space-time. The ambitwistor space *does* have a natural complex structure: it is a five-dimensional complex manifold (LeBrun 1983). But some of the power of twistor theory is lost. We shall return to this subject in Chapter 10. In the present chapter, we shall deal with *anti-self-dual space-times* (i.e., those in which the Weyl tensor is anti-self-dual, see §6.2). With each such space-time $\mathcal{M}$, there is associated a curved twistor space $\mathcal{P}$ which contains, in its complex structure, all the information about the conformal structure of $\mathcal{M}$. Furthermore, some additional data on the twistor space enables one to describe the metric of $\mathcal{M}$ (not just its conformal structure), in such a way that this metric is necessarily Einstein ($R_{ab} = \frac{1}{4}Rg_{ab}$) or vacuum ($R_{ab} = 0$). Finally, endowing the twistor space with a 'real structure' enables one to pick out a positive-definite 'real slice' of $\mathcal{M}$. All these results are discussed in §9.1, while in §9.2 we show how they may be used to generate anti-self-dual Einstein or vacuum metrics such as those mentioned in §6.2.

9.1   Correspondence between self-dual space-times and curved twistor spaces

This section describes how the correspondence between flat space-time **M** and flat twistor space **P** can be generalized to a correspondence between curved anti-self-dual space-times and curved twistor spaces. This is sometimes called the 'nonlinear graviton' construction, after the title of Penrose's (1976) paper. We shall also see how the twistor description of massless free fields (Chapter 7) generalizes to massless fields on an anti-self-dual space-time background.

There are two ways of describing the 'curved' correspondence: a holomorphic way which generalizes the idea of self-dual complex planes in M^I, and a positive-definite way which generalizes the fibration $P \rightarrow S^4$. We shall concentrate on the first, and then give a brief outline of the second.

Let $(\mathcal{M}, g_{ab})$ be a complex space-time (see §6.2). We want to pick out in $\mathcal{M}$ a family of holomorphic two-surfaces $\{S\}$ which generalize the three-dimensional family of totally-null self-dual two-planes (α-planes) in flat space-time **M**. Let S be a totally-null two-surface in $\mathcal{M}$, that is to say, S is a two-dimensional complex submanifold of $\mathcal{M}$ with the property that if p is any point on S, and v^a and w^a are any two tangent vectors at p, then $v^a w_a = 0$. The vectors tangent to S at p must have the form $\lambda^A \pi^{A'}$, with either $\pi^{A'}$ fixed (in which case S is self-dual) or λ^A fixed (in which case S is anti-self-dual). So we can say that S is a *self-dual surface* if there exists a primed spinor field $\pi_{A'}$ on S, such that each vector field tangent to S has the form $\lambda^A \pi^{A'}$, for some spinor field λ^A on S.

The existence of self-dual surfaces in $\mathcal{M}$ imposes a condition on its curvature. To see what this is, let $v^a = \lambda^A \pi^{A'}$ and $u^a = \mu^A \pi^{A'}$ be two vector fields tangent to S. Frobenius' theorem requires that the Lie bracket of v^a and u^a be a linear combination of v^a and u^a, namely:

$$v^a \nabla_a u^b - u^a \nabla_a v^b = \varphi v^b + \psi u^b, \qquad (9.1.1)$$

for some scalar functions φ and ψ. Substituting the spinor expressions for v^a and u^a into (9.1.1), one sees that this condition is equivalent to

$$\pi^{A'} \nabla_{AA'} \pi_{B'} = \xi_A \pi_{B'}, \qquad (9.1.2)$$

for some spinor field ξ_A. Operating on (9.1.2) with $\pi^{B'} \pi^{C'} \nabla^A_{C'}$ and using the spinor Ricci identity (6.1.5), then gives

$$\widetilde{\Psi}_{A'B'C'D'} \pi^{A'} \pi^{B'} \pi^{C'} \pi^{D'} = 0. \qquad (9.1.3)$$

This is the integrability condition for S.

In flat space $\mathbf{M}^I$, there are a three-complex-parameter family of self-dual surfaces (one for each point of $\mathbf{P}^I$). Let us require that our curved space-time $\mathcal{M}$ should also admit a three-complex-parameter family of self-dual surfaces. In other words, for every point p in $\mathcal{M}$, and for every spinor $\pi_{A'}$ at p, there should exist a self-dual surface S through p, with tangent vectors of the form $\lambda^A \pi^{A'}$ at p. In view of (9.1.3), this clearly implies that $\tilde{\Psi}_{A'B'C'D'} = 0$, i.e., that the conformal curvature of $\mathcal{M}$ be anti-self-dual.

Note that the spinor field $\pi_{A'}$ can be rescaled (i.e., $\pi_{A'} \mapsto \lambda\pi_{A'}$) so that it is covariantly constant over S, satisfying

$$\pi^{A'}\nabla_{AA'}\pi_{B'} = 0. \tag{9.1.4}$$

The integrability condition for (9.1.4) is obtained by operating on it with $\pi^{C'}\nabla^A_{C'}$, giving $\tilde{\Psi}_{A'B'C'D'}\pi^{A'}\pi^{B'}\pi^{C'} = 0$, which, of course, is satisfied if $\tilde{\Psi}_{A'B'C'D'} = 0$.

The necessary condition $\tilde{\Psi}_{A'B'C'D'} = 0$ is also sufficient: if $(\mathcal{M}, g_{ab})$ *is* anti-self-dual, then it admits a three-complex-parameter family of self-dual surfaces. To see this, one may argue as follows. Given any point p in $\mathcal{M}$ and a spinor $\xi_{A'}$ at p, one can find a spinor field $\pi_{A'}$ on $\mathcal{M}$, satisfying (9.1.4), and equal to $\xi_{A'}$ at p. This field $\pi_{A'}$ then defines a (holomorphic) two-dimensional distribution on $\mathcal{M}$, namely the distribution spanned by the vector fields of the form $\lambda^A \pi^{A'}$. And this distribution is integrable by virtue of (9.1.4). So, in particular, there exists a self-dual surface through p, with tangent vectors of the form $\lambda^A \xi^{A'}$ at p. Considering the dimensions involved shows that the space $\mathcal{P}$ of all self-dual surfaces in $\mathcal{M}$ is therefore three-complex-dimensional. We call $\mathcal{P}$ the *twistor space* of $\mathcal{M}$.

In order to ensure that $\mathcal{P}$ is a nice Hausdorff manifold, we need to impose a convexity assumption on $\mathcal{M}$. For the sake of simplicity, let us assume that $\mathcal{M}$ is such that its twistor space $\mathcal{P}$ has the same topology as $\mathbf{P}^I$ and call such an $\mathcal{M}$ *civilized*. Every point in an anti-self-dual complex space-time possesses a civilized neighborhood: any 'small, convex' neighborhood will do. Note that flat space-time $\mathbf{M}^I$ is civilized, by definition.

The twistor space $\mathcal{P}$ depends only on the *conformal* structure of $\mathcal{M}$; in other words, $(\mathcal{M}, g_{ab})$ and $(\mathcal{M}, \Omega^2 g_{ab})$, where Ω is a nonvanishing holomorphic function on $\mathcal{M}$, have the same self-dual surfaces, and hence the same twistor space. One can see this from the fact that

(9.1.4) is conformally invariant, with $\pi_{A'}$ having conformal weight zero (see (6.1.13)).

The twistor space contains a four-parameter family of holomorphic rational curves, one for each point of $\mathcal{M}$ (by a *rational* curve in a complex manifold we mean a one-dimensional embedded complex submanifold which is biholomorphic to the complex projective line $\mathbf{P}_1$). Namely, given a point p in $\mathcal{M}$, the self-dual surfaces passing though p are parametrized by $\mathbf{P}(S_{A'})$, and since $S_{A'} \cong \mathbf{C}^2$, we have a rational curve of surfaces passing though p, which we denote (in parallel to the flat case) by $\hat{p}$. An important property of these curves is that each has normal bundle isomorphic to $H \oplus H$, where H is the hyperplane-section line bundle on $\mathbf{P}_1$ (this is explained in more detail below). We can now state the first basic theorem.

Theorem 9.1.1. *There is a natural one-to-one correspondence between:*

(a) *civilized anti-self-dual spaces* $(\mathcal{M}, g_{ab})$*, up to conformal rescaling of* g_{ab}*;*
    and
(b) *three-dimensional complex manifolds* $\mathcal{P}$ *(homeomorphic to* $\mathbf{P}^I$ *), containing a four-parameter family of holomorphic rational curves, each with normal bundle isomorphic to* $H \oplus H$*.*

Proof: We have already demonstrated most of (a) $\rightarrow$ (b), but still need to check that the curves have the right normal bundle. We recall the definition of the normal bundle of a complex submanifold Y of a complex manifold X given in §7.2 as $N_Y = (T(X)|_Y)/T(Y)$.

We shall investigate tangent vectors in $\mathcal{P}$ by using their 'local twistor' description in $\mathcal{M}$. Let Z be a point in $\mathcal{P}$, and Q a tangent vector at Z. Think of Z as being joined to a neighboring point Y by the displacement vector εQ, where ε is a small parameter. As usual, let $\widetilde{Z}$ and $\widetilde{Y}$ denote the corresponding self-dual surfaces in $\mathcal{M}$ (see Figure 9.1.1). Let v^a be a vector field on $\widetilde{Z}$, such that εv^a is a connecting vector field joining $\widetilde{Z}$ to the neighboring surface $\widetilde{Y}$. Thus the Lie derivative of v^a along $\widetilde{Z}$, is tangent to $\widetilde{Z}$; i.e., if $\pi_{A'}$ is the spinor field associated with $\widetilde{Z}$ and λ^A is any field on $\widetilde{Z}$, then

$$\lambda^B \pi^{B'} \nabla_{BB'} v^{AA'} - v^{BB'} \nabla_{BB'} (\lambda^A \pi^{A'}) = \xi^A \pi^{A'},$$

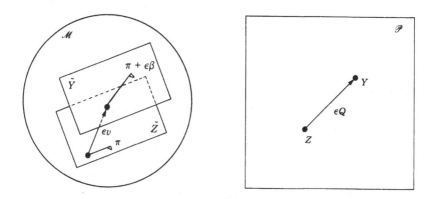

Fig. 9.1.1. Self-dual surfaces corresponding to points.

for some field ξ^A. Therefore,

$$\pi_{A'} \lambda^B \pi^{B'} \nabla_{BB'} v^{AA'} = \lambda^A \beta^{A'} \pi_{A'}, \qquad (9.1.5)$$

where

$$\beta_{A'} = v^b \nabla_b \pi_{A'}. \qquad (9.1.6)$$

Note that $\varepsilon\beta_{A'}$ represents the change in $\pi_{A'}$ between $\widetilde{Z}$ and $\widetilde{Y}$. The connecting vector v^a is only defined up to the addition of vectors tangent to $\widetilde{Z}$, i.e., of the form $\mu^A \pi^{A'}$, so its 'invariant part' is

$$\alpha^A = iv^{AA'}\pi_{A'}. \qquad (9.1.7)$$

The factor of i in (9.1.7) is conventional, and is inserted because it ensures compatibility with our standard flat-space conventions. The vector Q in twistor space corresponds to the pair $(\alpha^A, \beta_{A'})$ of spinor fields on $\widetilde{Z}$. These satisfy propagation equations over $\widetilde{Z}$, which determine them everywhere on $\widetilde{Z}$ if they are specified at one point. Namely, (9.1.4), (9.1.5), and (9.1.7) give

$$\lambda^B \pi^{B'} \nabla_{BB'} \alpha^A = -i\lambda^A \pi^{B'} \beta_{B'}, \qquad (9.1.8)$$

while from (9.1.6) we obtain

$$\lambda^B \pi^{B'} \nabla_{BB'} \beta_{A'} = \lambda^B \pi^{B'} v^{CC'} [\nabla_{BB'}, \nabla_{CC'}] \pi_{A'}$$
$$= -i\lambda^B \alpha^C \Phi_{BCA'B'} \pi^{B'} + i\Lambda \lambda_C \alpha^C \pi_{A'}, \qquad (9.1.9)$$

using (6.1.5) and (6.1.6). These are the equations of 'local twistor transport' over $\widetilde{Z}$, with $(\alpha^A, \beta_{A'})$ being the local twistor representing Q (Penrose and MacCallum 1973).

Notice, however, that this representation is slightly redundant, since we are in *projective* twistor space. The pair $(\alpha^A, \beta_{A'}) = (0, \pi_{A'})$ corresponds to the *zero* vector $Q = 0$ (since for this pair, we have $\widetilde{Y} = \widetilde{Z}$, and the spinor $\pi_{A'}$ is unchanged up to a constant factor). So we must factor out by pairs of the form $(0, k\pi_{A'})$, with k constant. Notice also that α^A and $\beta_{A'}$ are homogeneous of degree one in $\pi_{A'}$, in the sense that if we multiply $\pi_{A'}$ by a constant k, then α^A becomes $k\alpha^A$, and $\beta_{A'}$ becomes $k\beta_{A'}$.

Now let $\hat{p}$ be a curve in $\mathcal{P}$ corresponding to a point p in $\mathcal{M}$. Let $Q = Q(\pi_{B'})$ be a vector field on $\hat{p}$ (not necessarily tangent to $\hat{p}$), represented by fields $\alpha^A(x, \pi_{B'})$ and $\beta_{A'}(x, \pi_{B'})$ on the various self-dual surfaces which pass through p. Here, of course, $\pi_{B'}$ are being used as homogeneous coordinates for $\hat{p} \cong \mathbf{P}_1$, and to parametrize the self-dual surfaces through p. Evaluate these fields at the point p, to obtain a pair of spinors $(\Omega^A(\pi_{B'}), \Pi_{A'}(\pi_{B'}))$; this pair represents the tangent vector $Q(\pi_{B'})$. More precisely, a local holomorphic section of $T(\mathcal{P})|_{\hat{p}}$ consists of a pair $(\Omega^A, \Pi_{A'})$ of functions of $\pi_{B'}$, homogeneous of degree one, modulo pairs of the form $(0, f(\pi_{B'})\pi_{A'})$, where $f(\pi_{B'})$ is homogeneous of degree zero. On the other hand, a vector field tangent to $\hat{p}$ is characterized by α^A vanishing at p, i.e., $\Omega^A = 0$. So a local holomorphic section of $T(\hat{p})$ consists of a pair $(0, \Pi_{A'})$, modulo $(0, f\pi_{A'})$ as before. The normal bundle of $\hat{p}$, namely

$$N_{\hat{p}} = T(\mathcal{P})|_{\hat{p}}/T(\hat{p}),$$

is obtained by taking the quotient of these two . So we see that local sections of $N_{\hat{p}}$ correspond to spinors $\Omega^A(\pi_{B'})$, homogeneous of degree one in $\pi_{B'}$. In other words, $N_{\hat{p}} \cong H \oplus H$, as was claimed; each of $\Omega^0(\pi_{B'})$ and $\Omega^1(\pi_{B'})$ is a section of H (or, equivalently, of the sheaf $\mathcal{O}(1)$).

We have thus completed the proof of half the theorem, demonstrating how to get from (a) to (b). To prove the remaining half is not difficult. Suppose we are given the twistor space $\mathcal{P}$, with its four-complex-parameter family of holomorphic curves. Then define $\mathcal{M}$ to be the parameter space for this family. Note that by a theorem of Kodaira (1962) the family is complete and maximal, since

$$H^1(\hat{p}, \mathcal{O}(N_{\hat{p}})) = 0, \qquad H^0(\hat{p}, \mathcal{O}(N_{\hat{p}})) \cong \mathbf{C}^4,$$

where $\mathcal{O}(N_{\hat{p}}) \cong \mathcal{O}(1) \oplus \mathcal{O}(1)$ is the sheaf of germs of holomorphic sections of $N_{\hat{p}}$ (see Example 3.3.9). Furthermore, there is a natural isomorphism between $H^0(\hat{p}, \mathcal{O}(N_{\hat{p}}))$ and $T_p(\mathcal{M})$, which can be used to define a conformal metric on $\mathcal{M}$. Namely, a vector $V \in T_p(\mathcal{M})$ is defined to be null if the corresponding section $\widehat{V}$ of $N_{\hat{p}}$ has a zero. In terms of our previous notation, $\widehat{V}$ is represented by a pair $\Omega^A(\pi_{B'})$ of linear functions of $\pi_{B'}$; so for $\widehat{V}$ to have a zero, we need Ω^0 and Ω^1 to vanish at the same value of $\pi_{B'}$. This is a *quadratic* condition on the coefficients of Ω^A (in fact, if $\Omega^A(\pi_{B'}) = \Omega^{AA'}\pi_{A'}$, then the condition is $\Omega^{AA'}\Omega_{AA'} = 0$). So the conformal structure is well-defined, i.e., is a genuine 'complex-Riemannian' conformal structure which defines g_{ab} up to a conformal factor.

A 'noninfinitesimal' version of this definition is that two points p and q in $\mathcal{M}$ are defined to be null-separated (i.e., there is a null geodesic joining them) if, in $\mathcal{P}$, the corresponding curves $\hat{p}$ and $\hat{q}$ intersect (cf. Penrose 1976).

Finally, note that each point Z in $\mathcal{P}$ defines a two-surface $\widetilde{Z}$ in $\mathcal{M}$, namely the set of all points p in $\mathcal{M}$ whose corresponding curves $\hat{p}$ pass through Z. Clearly this surface is totally-null (any two points on it are null-separated), since the corresponding curves in $\mathcal{P}$ all intersect at Z. With an appropriate choice of orientation on $\mathcal{M}$, we therefore have a three-parameter family of self-dual surfaces in $\mathcal{M}$. In view of the discussion at the beginning of this section, we conclude that $(\mathcal{M}, g_{ab})$ is anti-self-dual. This completes the proof of the theorem. ∎

We see from this theorem that the 'curved twistor space' $\mathcal{P}$ determines an anti-self-dual conformal structure on $\mathcal{M}$. The remarkable thing is that $\mathcal{P}$, together with some simple additional structure (namely a pair of differential forms), determines a metric g_{ab} on $\mathcal{M}$ (not just a conformal metric), and that this metric is automatically a solution of the Einstein condition $R_{ab} = \frac{1}{4}Rg_{ab}$. As a special case we obtain anti-self-dual vacuum solutions. These two differential forms τ and ρ are defined as follows.

First we shall describe τ, which is a one-form. If $(\alpha^A, \beta_{A'})$ represents a vector Q at Z in $\mathcal{P}$, then we set

$$\tau(Q) := \pi_{A'}\beta^{A'}, \qquad (9.1.10)$$

where, as usual, $\pi_{A'}$ is the spinor field associated with $\widetilde{Z}$, and satisfying (9.1.4). For this definition to make sense, the quantity (9.1.10) must be constant over $\widetilde{Z}$; from equation (9.1.9) we see that

$$\lambda^B \pi^{B'} \nabla_{BB'}(\pi_{A'}\beta^{A'}) = i\lambda^A \alpha^B \pi^{A'} \pi^{B'} \Phi_{ABA'B'},$$

which vanishes provided that $(\mathcal{M}, g_{ab})$ is an Einstein space.

Actually, τ is not a one-form on $\mathcal{P}$ in the usual sense, since it is homogeneous of degree two in $\pi_{A'}$. If it were really a form on $\mathcal{P}$, it would have had to be homogeneous of degree zero. As it is, τ is a section of the bundle $T^*(\mathcal{P}) \otimes L^{-2}$, where L is the 'universal' line-bundle over $\mathcal{P}$ (the fiber L_Z of L above Z is the one-dimensional vector space of all spinor fields $\pi_{A'}$ which 'point along' $\widetilde{Z}$ and are constant over $\widetilde{Z}$—any two such fields being proportional). We say that a 'p-form homogeneous of degree n' is a section of $\Omega^p(\mathcal{P}) \otimes L^{-n}$.

The other form on $\mathcal{P}$ is a three-form homogeneous of degree four. Its existence (unlike that of τ) does not require any condition on the Ricci tensor of $\mathcal{M}$. Let $Q_j = (Q_1, Q_2, Q_3)$ be three tangent vectors at Z in $\mathcal{P}$ represented by $(\alpha_j{}^A, \beta_{jA'})$. Then $\rho(Q_1, Q_2, Q_3)$ is defined by setting

$$\rho_{ijk} := \tfrac{1}{2}(\pi_{A'}\beta_i^{A'})(\alpha_{jA}\alpha_k^A), \qquad (9.1.11)$$

and skew-symmetrizing it over i, j, and k. In other words,

$$\rho(Q_1, Q_2, Q_3) = \tfrac{1}{6}(\rho_{123}+\rho_{312}+\rho_{231} \\ -\rho_{132}-\rho_{213}-\rho_{321}). \qquad (9.1.12)$$

This quantity is constant over $\widetilde{Z}$, and so ρ is well-defined (see Exercise 9.1.1).

Before stating the theorem which makes use of these forms, let us note a crucial property of the form τ: that it is 'transverse' to the holomorphic curves of the form $\hat{p}$. Namely, if Q is a nonzero vector tangent to $\hat{p}$, then $\tau(Q) \neq 0$. For such a Q is represented by the local twistor $(0, \beta_{A'})$, modulo local twistors of the form $(0, f\pi_{A'})$; thus $Q \neq 0$ means that $\beta_{A'}$ is *not* proportional to $\pi_{A'}$, and so $\pi_{A'}\beta^{A'} \neq 0$. Hence $\tau(Q) \neq 0$.

Theorem 9.1.2. *There is a natural one-to-one correspondence between*

   (a) *civilized anti-self-dual Einstein spaces* $(\mathcal{M}, g_{ab})$, *with (constant) scalar curvature* 24Λ;
     and

   (b) *three-dimensional complex manifolds* $\mathcal{P}$, *homeomorphic to* $\mathbb{P}^I$, *possessing the following structure:*

(i) *a four-parameter family of holomorphic rational curves, each with normal bundle $H \oplus H$;*

(ii) *a holomorphic line-bundle L such that $L|_{\hat{p}} \cong H^{-1}$ on each curve $\hat{p}$;*

(iii) *a holomorphic one-form τ homogeneous of degree two, such that $\tau(Q) \neq 0$, for any nonzero Q tangent to a curve of the family* (i);

(iv) *a holomorphic three-form ρ homogeneous of degree four, such that $\tau \wedge d\tau = 2\Lambda\rho$.*

Proof: This theorem was first proved for the special case of $\Lambda = 0$ (Penrose 1976, Curtis et al. 1979). The proof below follows that of Ward (1980).

We have already shown how to get from (a) to (b), except for establishing the relation

$$\tau \wedge d\tau = 2\Lambda\rho. \tag{9.1.13}$$

First, we need an expression for $d\tau$. Let Q_1 and Q_2 be two vectors at Z in $\mathcal{P}$, represented by the connecting vector fields $v_1{}^a$ and $v_2{}^a$ on $\tilde{Z}$, or alternatively by the local twistors $(\alpha_1{}^A, \beta_{1A'})$ and $(\alpha_2{}^A, \beta_{2A'})$ on $\tilde{Z}$. Recall that τ is defined by $\tau(Q) = \pi_{A'}\beta^{A'} = \pi_{A'}v^b\nabla_b\pi^{A'}$, so

$$
\begin{aligned}
d\tau(Q_1, Q_2) &= \tfrac{1}{2}v_1{}^b\nabla_b(\pi_{A'}v_2{}^c\nabla_c\pi^{A'}) - \tfrac{1}{2}v_2{}^b\nabla_b(\pi_{A'}v_1{}^c\nabla_c\pi^{A'}) \\
&= (v_1{}^b\nabla_b\pi_{A'})(v_2{}^c\nabla_c\pi^{A'}) + \tfrac{1}{2}v_1{}^b v_2{}^c\pi_{A'}[\nabla_b, \nabla_c]\pi^{A'} \\
&= \beta_{1A'}\beta_2{}^{A'} + \Lambda\alpha_{1A}\alpha_2{}^A. \tag{9.1.14}
\end{aligned}
$$

We have used the fact that the Lie bracket of $v_1{}^a$ and $v_2{}^a$ must have the form $\mu^A\pi^{A'}$, for some μ^A; and also used the spinor Ricci identities (6.1.5), (6.1.6) with $\tilde{\Psi}_{A'B'C'D'} = 0$, $\Phi_{ABA'B'} = 0$.

It is now a simple matter to compute $\tau \wedge d\tau$:

$$
\begin{aligned}
\tau \wedge d\tau(Q_1, Q_2, Q_3) &= \tfrac{1}{6}\sum_{\sigma}(-1)^{|\sigma|}(\pi_{A'}\beta^{A'}_{\sigma(1)})(\beta_{\sigma(2)B'}\beta^{B'}_{\sigma(3)} + \Lambda\alpha_{\sigma(2)B}\alpha^B_{\sigma(3)}) \\
&= 2\Lambda\rho(Q_1, Q_2, Q_3)
\end{aligned}
$$

by (9.1.11) and (9.1.12), where σ denotes permutations of $\{1, 2, 3\}$. So the passage from (a) to (b) is complete.

The structure list in (b) actually contains some redundancy. The line-bundle L in part (ii) of (b) satisfies $L^4 = K$, where K is the canonical line bundle of $\mathcal{P}$. And L can be *defined* to be $K^{1/4}$, this fourth root being guaranteed to exist as a consequence of part (i) (see Hitchin 1982b). Also, the form ρ, being a three-form of degree four, is a section of $K \otimes L^{-4} \cong I$ (where I denotes the trivial line bundle) and is therefore unique up to a constant. The same goes for $\tau \wedge d\tau$. All that part (iv) does is to identify the constant of proportionality between $\tau \wedge d\tau$ and ρ.

Let us now deal with the converse problem: how to get from (b) to (a), i.e., how to construct anti-self-dual Einstein metrics. We already know, from Theorem 9.1.1, that $\mathcal{P}$ determines $\mathcal{M}$ and its conformal structure. The extra information in τ and ρ will determine g_{ab}. In fact, τ gives us a symplectic form ε_+ on primed spin-space, and then ρ gives us a symplectic form ε_- on unprimed spin-space; the metric is finally obtained as the tensor product of ε_- and ε_+ (cf. the standard expression $g_{ab} = \varepsilon_{AB}\varepsilon_{A'B'}$).

The first step is to identify the spin-bundles over $\mathcal{M}$. Let V be the subbundle of the holomorphic tangent bundle of $\mathcal{P}$ consisting of those vectors which are annihilated by τ. So V has rank two. And if $\hat{p}$ is a holomorphic curve, then (because of the requirement in (iii) that τ be transverse to such curves), the restriction $V|_{\hat{p}}$ of V to $\hat{p}$ is naturally isomorphic to the normal bundle $N_{\hat{p}}$ of $\hat{p}$. We define the spin-spaces S_p^- and S_p^+ at p by

$$S_p^- := \Gamma(\hat{p}, V|_{\hat{p}} \otimes H^{-1}), \qquad (9.1.15)$$
$$S_p^+ := \Gamma(\hat{p}, H). \qquad (9.1.16)$$

Recall that $V|_{\hat{p}} \cong N_{\hat{p}} \cong H \oplus H$, so $V|_{\hat{p}} \otimes H^{-1} \cong I \oplus I$ is trivial and S_p^- is two-dimensional. Clearly S_p^+ is two-dimensional as well. Furthermore, $S_p^- \otimes S_p^+$ is naturally isomorphic to $\Gamma(\hat{p}, V|_{\hat{p}})$ and hence to $T_p(\mathcal{M})$. And, moreover, a vector in $T_p(\mathcal{M})$ is null if and only if it is the tensor product of a primed spinor (element of S_p^+) and an unprimed spinor (element of S_p^-). In other words, this definition of the spin-spaces S_p^- and S_p^+ (and hence of the spin-bundles on $\mathcal{M}$), is consistent with the conformal structure that we already have on $\mathcal{M}$.

Next let us define a two-form ε_+ on S_p^+. Let ξ and μ be two elements of S_p^+. So, by (9.1.16), ξ and μ are functions on $\hat{p}$, homogeneous of degree one. Thus $\xi d\mu - \mu d\xi$ is a one-form of degree two, which, since $\hat{p}$ is one-dimensional, must be proportional to $\tau|_{\hat{p}}$. We define $\varepsilon_+(\xi, \mu)$ by

$$\xi d\mu - \mu d\xi = \varepsilon_+(\xi, \mu)\tau|_{\hat{p}}. \qquad (9.1.17)$$

The transversality assumption in (iii) ensures that $\varepsilon_+(\xi, \mu)$ is well-defined; since it is homogeneous of degree zero on $\hat{p}$, it must be constant.

Finally, we define the two-form ε_- on S_p^-. Let ξ and μ be two elements of S_p^-: so ξ and μ are vector fields on $\hat{p}$, homogeneous of degree -1. Let ν be any other vector field on $\hat{p}$, with $\tau(\nu) \neq 0$. Define $\varepsilon_-(\xi, \mu)$ by

$$\varepsilon_-(\xi, \mu)\tau(\nu) = 2\rho(\nu, \xi, \mu). \tag{9.1.18}$$

The quantity $\varepsilon_-(\xi, \mu)$ is independent of τ, since both $\tau(\nu)$ and $\rho(\nu, \xi, \mu)$ are invariant under the transformation $\nu \mapsto \nu + a\xi + b\mu$, with a and b being scalars, and both are linear in ν. Furthermore, $\varepsilon_-(\xi, \mu)$ is homogeneous of degree zero, and so is a constant.

A simple example of this construction of ε_+ and ε_- may be found in Exercise 9.1.4.

We now have a metric g_{ab}, defined to be the product of ε_- and ε_+, and compatible with the conformal structure. We have to show that it is an anti-self-dual Einstein metric, with scalar curvature $R = 24\Lambda$. We shall simply give an outline of the proof here: details may be found in Ward (1980).

. The first step is to define a connection ∇_a on $\mathcal{M}$. This is done by demonstrating, in the $\mathcal{P}$-picture, how to parallel-propagate spinors in null directions. The next step is to show that ∇_a is torsion-free, and that it preserves ε_- and ε_+; it follows that ∇_a is the unique (Levi–Cevita) connection determined by g_{ab}. Then one shows that ∇_a, acting on the unprimed spin-bundle restricted to an arbitrary self-dual surface, is flat. From this it follows, by the Ricci identity (6.1.7), that the Einstein condition $\Phi_{ABA'B'} = 0$ holds. We already know, from Theorem 9.1.1, that g_{ab} has anti-self-dual conformal curvature. The final step is to verify that the scalar curvature is 24Λ. This is done by looking at ∇_a acting on the primed spin-bundle restricted to a self-dual surface, and using (6.1.5).

The final step in the proof of the theorem is to verify that the composition (a) $\rightarrow$ (b) $\rightarrow$ (a) is the identity.   ∎

Remarks:

(1) The bundle $V \otimes L$ is a vector bundle of rank two over $\mathcal{P}$, and is trivial on any holomorphic curve $\hat{p}$. Therefore it corresponds to an anti-self-dual gauge field on $\mathcal{M}$. (The discussion of Chapter 8 only

involved *flat* space-time, but extends naturally to anti-self-dual gauge fields on anti-self-dual space-times.) The gauge field determined by $V \otimes L$ is, in fact, the unprimed spin-bundle S^- on $\mathcal{M}$, with its natural connection ∇_a.

(2) If $\Lambda = 0$, then $d\tau \wedge \tau = 0$, which means that V is an integrable distribution on $\mathcal{P}$. In this case, $\mathcal{P}$ is fibered holomorphically over $\mathbf{P}_1$, with each fiber being a (two-dimensional) integral surface of V (Penrose 1976). Each of these surfaces is equipped with a two-form ε_- of degree two, defined as in (9.1.18). In the flat case, this fibration is $[\omega^A, \pi_{A'}] \mapsto [\pi_{A'}]$, so each fiber is isomorphic to $\mathbf{C}^2$ parametrized by ω^A (the ω^A-space), and the $\mathbf{P}_1$ is the projective $\pi_{A'}$-space. The symplectic form ε_- is $\varepsilon_{AB} d\omega^A \wedge d\omega^B$.

(3) We shall see in §9.2 that the conditions (i)–(iv) which the twistor space is required to satisfy, are often quite easily arranged. For example, a theorem of Kodaira (1963) guarantees that if we take a neighborhood of a line in flat twistor space and deform its complex structure, then condition (i) (the existence of a suitable family of holomorphic curves) will, for sufficiently mild deformations, be satisfied automatically. See the appendix of Curtis et al. (1979).

(4) The important point to note is that the data in the $\mathcal{P}$-picture is essentially 'free' holomorphic data: in particular, it is not subject to any differential equation. The nonlinear partial differential equations of the space-time picture (such as $R_{ab} = \frac{1}{4} R g_{ab}$) are 'transformed away'; going from $\mathcal{P}$ to $\mathcal{M}$ is the Penrose transform in this context.

(5) We could also view the correspondence in terms of the double-fibration picture

$$\mathcal{P} \leftarrow \mathcal{F} \rightarrow \mathcal{M},$$

where $\mathcal{F}$ is the projective primed spin-bundle on $\mathcal{M}$. If $\mathcal{M}$ is *any* complex space-time, then $\mathcal{F}$ possesses a natural holomorphic two-dimensional distribution: at a point $(p, \pi_{A'})$ of $\mathcal{F}$, this distribution is spanned by the two vectors $\{\pi^{A'} \nabla_{0A'}, \pi^{A'} \nabla_{1A'}\}$ in $T_p(\mathcal{M})$ lifted horizontally to $T_{(p,\pi)}(\mathcal{F})$. The distribution is integrable if and only if $\tilde{\Psi}_{A'B'C'D'} = 0$ (in which case its integral surfaces correspond, of course, to the self-dual surfaces in $\mathcal{M}$).

(6) Theorems 9.1.1 and 9.1.2 are 'local' in the sense that they refer to 'civilized' regions of complex space-time. However, this requirement was just imposed for the sake of convenience, and can be relaxed.

Let us next turn to the question of how to deal with four-real-dimensional *positive-definite* anti-self-dual spaces. One can do this by simply adding an antiholomorphic involution σ on $\mathcal{P}$ to what we had before (cf. §§1.6, 8.1). One has to require that τ and ρ be real (i.e., be preserved by σ), and that there exists a four-real-parameter family of *real* holomorphic rational curves in $\mathcal{P}$. In this positive-definite case, there is a more direct route for constructing the associated twistor space $\mathcal{P}$, one used by Atiyah, Hitchin and Singer (1978). What follows is a description of this.

Let $(\mathcal{M}, g_{ab})$ be a real four-dimensional (positive-definite) Riemannian space. We assume it to be oriented and to possess a spin structure. Let $\pi_{A'}$ be a primed spinor at some point p in $\mathcal{M}$, and take $\pi_{A'}$ to have unit length:

$$\pi^{A'}\sigma(\pi)_{A'} = 1, \tag{9.1.19}$$

where $\sigma(\pi)_{A'}$ is defined, as in §8.1, by $\sigma(\pi)_{0'} = \overline{\pi_{1'}}$, $\sigma(\pi)_{1'} = -\overline{\pi_{0'}}$. The spinor $\pi_{A'}$ defines a complex structure on $T_p(\mathcal{M})$, namely

$$v^a \mapsto J_b{}^a v^b := iv^{AB'}[\pi_{B'}\sigma(\pi)^{A'} + \sigma(\pi)_{B'}\pi^{A'}]. \tag{9.1.20}$$

One can check (see Exercise 9.1.6) that $J_b{}^a$ is a complex structure compatible with the metric, namely that:

 (a) $J_b{}^a$ is a real tensor (i.e., it preserves the reality condition $v^{11'} = -\overline{v^{00'}}$, $v^{10'} = \overline{v^{01'}}$ on v^a),
 (b) $J^2 = -I$ (i.e., $J_b{}^c J_c{}^a = -\delta_b{}^a$),
 (c) J is compatible with the metric (i.e., $g_{ab}J_c{}^a J_d{}^b = g_{cd}$).

Clearly a change of phase $\pi_{A'} \mapsto e^{i\theta}\pi_{A'}$ with θ real, does not affect $J_a{}^b$. So we see that a *projective* spinor at p determines a complex structure of $T_p(\mathcal{M})$. Let $\mathcal{F}$ denote the projective primed spin-bundle of $\mathcal{M}$ (thus the fiber of $\mathcal{F}$ above $p \in \mathcal{M}$ is a complex projective line, consisting of spinors $\pi_{A'}$ at p modulo the proportionality relation $\pi_{A'} \mapsto \lambda\pi_{A'}$, $\lambda \in \mathbf{C}^*$). Then $\mathcal{F}$ has a natural almost complex structure, defined as follows.

Let ξ be a point in $\mathcal{F}$, corresponding to a point $p \in \mathcal{M}$ and a projective spinor $[\pi_{A'}]$ at p. The tangent space $T_\xi(\mathcal{F})$ is decomposed into horizontal and vertical subspaces:

$$T_\xi(\mathcal{F}) = H_\xi \oplus V_\xi,$$

since the spin-bundle S^+ (and hence also $\mathcal{F}$) comes equipped with a connection. Now V_ξ has a natural complex structure (since, as remarked above, the fiber is a complex projective line). Moreover, H_ξ

is naturally isomorphic to $T_p(\mathcal{M})$ which has a complex structure determined by $[\pi_{A'}]$. Therefore $T_\xi(\mathcal{F})$ has a natural complex structure, as claimed.

So far, there has been no condition on the curvature of $\mathcal{M}$. If however, we want the almost-complex structure on $\mathcal{F}$ to be integrable, then a condition is needed, namely that $\tilde\Psi_{A'B'C'D'} = 0$ (Atiyah, Hitchin and Singer 1978, Theorem 4.1). This condition is also sufficient: if the conformal structure of $\mathcal{M}$ is anti-self-dual, then $\mathcal{F}$ becomes a three-dimensional complex manifold which we may denote by $\mathcal{P}$, since it is the twistor space of $\mathcal{M}$. We could also have obtained this $\mathcal{P}$ by complexifying $\mathcal{M}$ and then defining $\mathcal{P}$ to be the space of self-dual surfaces in this complexification. The $\mathcal{F}$-description emphasizes that in the case of a positive-definite anti-self-dual space $\mathcal{M}$, the twistor space is fibered over $\mathcal{M}$, with each fiber $\hat{p}$ being a copy of $\mathbf{P}_1$ (see Figure 9.1.2). Note that $\hat{p}$ is holomorphically embedded in $\mathcal{P}$ (it is a holomorphic curve), but the fibration $\mathcal{P} \to \mathcal{M}$ is *not* holomorphic, merely smooth.

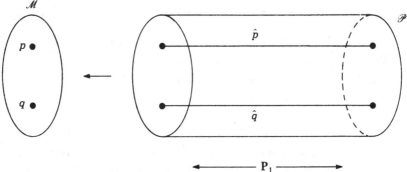

Fig. 9.1.2. The fibration of $\mathcal{P}$ over $\mathcal{M}$.

Finally in this section, we shall indicate how the results of Chapter 7 may be extended to cover massless fields on a given anti-self-dual space-time background. Namely, such fields (of helicity $-s$) correspond to elements of $H^1(\mathcal{P}, \mathcal{O}(2s-2))$, where $\mathcal{O}(2s-2)$ is the sheaf of holomorphic sections of L^{2-2s}. What follows is simply a rough argument, similar to the one given at the end of §8.1; a complete proof (using different techniques) may be found in, say, Hitchin (1980). The reader is also referred to the discussion by Atiyah (1981) of a related matter: the twistor version of the Green's function for the conformally invariant Laplace operator on a self-dual space.

For the sake of simplicity, we restrict to the vacuum case; but the results generalize to spaces with anti-self-dual conformal curvature and no restriction on R_{ab} (see Exercise 9.1.8). Let us use the double-fibration picture of Remark (5) above, and think of a function on $\mathcal{P}$ as a function $f = f(x, \pi_{A'})$ on $\mathcal{F}$ satisfying

$$\pi^{A'} \nabla_{AA'} f(x, \pi_{B'}) = 0. \tag{9.1.21}$$

That is to say, f is constant over each integral surface of the distribution mentioned in Remark (5) (cf. the discussion of this differential operator in §7.1). An important point to note here is that the anti-self-dual vacuum condition enables us to choose a spin-frame for the primed spinors which is covariantly constant, so that the symbol $\pi_{A'}$ may be thought of as denoting *either* a spinor *or* the components of that spinor with respect to a constant spin-frame. The operator $\nabla_{AA'}$, when acting on a function of x and $\pi_{B'}$, means differentiating with respect to x, leaving $\pi_{B'}$ constant, which is then unambiguous. But if we were to drop the vacuum condition, then this abuse of notation would no longer be justified, as covariantly constant basis-spinors would not then necessarily exist (cf. (6.1.5) and (6.1.6)). We would then have to modify (9.1.21).

Let us indicate, first, how massless fields of helicity $s > 0$, that is, solutions of

$$\nabla^{AA'} \varphi_{A'B'...D'} = 0, \tag{9.1.22}$$

on $\mathcal{M}$, where $\varphi_{A'B'...D'}$ has $2s$ indices, correspond to elements of $H^1(\mathcal{P}, \mathcal{O}(-2s - 2))$. Suppose the twistor space is covered by open sets $\mathcal{U}_i$, and that $f_{ij}(x, \pi) := f_{ij}(x, \pi_{A'})$ is a cocycle representing an element of this cohomology group. Let $\mathcal{V}_i$ denote the open set in $\mathcal{F}$ corresponding to $\mathcal{U}_i$ (in other words, the inverse image of $\mathcal{U}_i$ under the projection map from $\mathcal{F}$ to $\mathcal{P}$). Think of f_{ij} as being defined on $\mathcal{V}_i \cap \mathcal{V}_j$, satisfying (9.1.21), and homogeneous of degree $-2s - 2$ in $\pi_{A'}$.

Now $\pi_{A'} \ldots \pi_{C'} f_{ij}(x, \pi)$, with $1 + 2s$ factors of π, is homogeneous of degree -1, and so can be 'split', since $H^1(\mathcal{F}, \mathcal{O}(-1))$ vanishes:

$$\pi_{A'} \ldots \pi_{C'} f_{ij}(x, \pi) = h_{iA'...C'}(x, \pi) - h_{jA'...C'}(x, \pi), \tag{9.1.23}$$

where $h_{iA'...C'}(x, \pi)$ is holomorphic on $\mathcal{V}_i$. We emphasize that $h_{iA'...C'}$ does *not* satisfy (9.1.21), i.e., it is an object on $\mathcal{F}$ which does not descend to $\mathcal{P}$. Multiplying (9.1.23) by $\pi^{A'}$ and using the familiar Louiville-theorem argument, we see that

$$\varphi_{B'...D'} = \pi^{A'} h_{iA'B'...D'}(x, \pi) \tag{9.1.24}$$

is a function only of x (and not of π). Note that $\varphi_{B'\ldots D'}$ has $2s$ indices, and is totally symmetric. Operating on (9.1.23) with $\nabla^{AA'}$ and using (9.1.21) gives

$$\nabla^{AA'} h_{iA'B'\ldots D'} = 0,$$

and hence the massless field equation (9.1.22). For $s = 0$, a similar argument gives $\Box\varphi = 0$.

For negative s, the situation is more complicated, as one would expect, since the algebraic constraints (6.1.10) come into play when $s < -1$. For fields with primed indices, i.e., equation (9.1.22), there is of course no problem because $\tilde{\Psi}_{A'B'C'D'} = 0$ in an anti-self-dual space. We leave this case as an exercise for the reader (see Exercise 9.1.7).

Notes for Section 9.1

(1) There is a twistor procedure for solving Einstein's vacuum equations for space-times possessing two commuting Killing vectors (Ward 1983). However, this is not really a 'twistor description' of such space-times, because it involves the following trick. The vacuum equations with two symmetries can be reduced to an equation which can also be obtained as a reduction of the $SU(2)$ self-dual gauge-field equations in flat space. This fact was first pointed out by L. Witten (1979). It follows that these vacuum solutions can be described in terms of rank-two holomorphic vector bundles over flat twistor space. And one can generate solutions by, for example, using the ansätze described in §8.2.

(2) We have seen that anti-self-dual space-times can be described in terms of curved twistor spaces. Of course, *self*-dual space-times can similarly be described in terms of curved *dual* twistor spaces. But Penrose has also searched for a way of encoding the information of a self-dual space-time in twistor space (i.e., not dual twistor space). This construction he calls 'the googly graviton' (the term 'googly' being one used in the game of cricket). The ultimate hope is to describe general space-times (i.e., neither self-dual nor anti-self-dual) in terms of a twistor space construction. As yet, this program is in its preliminary stages. For details, see Penrose (1979), Law (1985), and numerous articles in the *Twistor Newsletter*.

(3) Hitchin (1981) considered the following question: if $\mathcal{M}$ is a compact oriented anti-self-dual Riemannian four-manifold, does its twistor space $\mathcal{P}$ admit a Kähler structure? He showed that it does so only if $\mathcal{M}$ is conformally equivalent to S^4 or $\mathbf{P}_2$ (with their standard metrics). So $\mathbf{P}_3$ and the flag manifold $\mathbf{F}_{12}(\mathbf{C}^3)$ are the only compact Kähler twistor spaces (this flag manifold being the twistor space of $\mathbf{P}_2$: see §9.2). Other twistor spaces give examples of non-Kähler manifolds; for example, the twistor space of a flat four-dimensional torus $\mathbf{R}^4/\Phi$ is a non-Kähler manifold described by Blanchard (1956).

(4) There are several different ways of generalizing the correspondence described in this section to 'space-times' of dimension greater than four. the most natural generalization associates twistor spaces with quaternionic-Kähler or hyper-Kähler spaces (Salamon 1982). A completely different approach was studied by Bryant (1985).

Exercises for Section 9.1

9.1.1. Show that the three-form ρ defined by (9.2.11) and (9.2.12) satisfies

$$\lambda^B \pi^{B'} \nabla_{BB'} \rho(Q_1, Q_2, Q_3) = 0.$$

(Hint: Use equations (9.1.8) and (9.1.9). Note that an expression such as $\beta_1^{A'} \beta_2^{B'} \beta_3^{C'}$, when skewed over $\{1, 2, 3\}$, gives zero, because of the two-dimensionality of spin-space.)

9.1.2. As in the previous exercise, check that the expression (9.1.14) for $d\tau$ is indeed constant over the self-dual surface $\widetilde{Z}$.

9.1.3. Let $(\mathcal{M}, g_{ab})$ be conformally flat, with metric g_{ab} given by $g_{ab} = \Omega^2 \eta_{ab}$, where η_{ab} is the standard Minkowski metric, and $\Omega = (1 - \frac{1}{2}\Lambda \eta_{ab} x^a x^b)^{-1}$. Show that this space-time satisfies the Einstein condition $R_{ab} = 6\Lambda g_{ab}$. Its twistor space is, of course, the flat twistor space $\mathbf{P}^I$. Show that the forms τ and ρ induced by g_{ab} on $\mathbf{P}^I$ are

$$\tau = \pi_{A'} d\pi^{A'} + \Lambda \omega_A d\omega^A,$$
$$\rho = \frac{1}{6}\varepsilon_{\alpha\beta\gamma\delta} Z^\alpha dZ^\beta \wedge dZ^\gamma \wedge dZ^\delta,$$

in terms of the standard coordinates on $\mathbf{P}^I$.

9.1.4. Take flat twistor space $\mathbb{P}^I$, with τ and ρ given by the expressions in the previous exercise. Show that ε_+ is given, according to the definition of (9.1.17), by

$$\varepsilon_+(\xi, \mu) = (1 - \tfrac{1}{2}\Lambda\eta_{ab}x^a x^b)^{-1}\varepsilon_{A'B'}\xi^{A'}\mu^{B'},$$

where $\xi = \xi^{A'}\pi_{A'}$ and $\mu = \mu^{A'}\pi_{A'}$. In other words, ε_+ is equal to the 'flat' symplectic form multiplied by the conformal factor $\Omega = (1 - \tfrac{1}{2}\Lambda\eta_{ab}x^a x^b)^{-1}$.

An unprimed spinor ξ at x^a has the form

$$\xi = \xi^A \frac{\partial}{\partial \omega^A} - i\Lambda x_{AA'}\xi^A \frac{\partial}{\partial \pi_{A'}}.$$

Check that $\tau(\xi) = 0$, as required. Show that ε_- is given, from (9.1.18), by

$$\varepsilon_-(\xi, \mu) = \Omega\varepsilon_{AB}\xi^A\mu^B.$$

So we see that $\varepsilon_- \otimes \varepsilon_+$ is equal to the Minkowski metric multiplied by the conformal factor Ω^2.

9.1.5. Show that the composition (a) $\to$ (b) $\to$ (a) of steps in Theorem 9.1.2, is the identity.

9.1.6. Prove the statements about the tensor $J_a{}^b$ made in the proposition following (9.1.20).

9.1.7. Let $\mathcal{M}$ be an anti-self-dual vacuum space, and $\mathcal{P}$ the corresponding twistor space. Show that elements of $H^1(\mathcal{P}, \mathcal{O}(2s - 2))$, with $s \geq 1$, correspond to solutions of

$$\nabla_{A(A'}\psi^A_{B'...D')} = 0,$$

where $\psi_{AA'...C'}$ has $2s - 1$ primed indices, modulo the equivalence transformations

$$\psi_{AA'B'...D'} \mapsto \psi_{AA'B'...D'} + \nabla_{A(A'}\Lambda_{B'...D')},$$

where $\Lambda_{B'...D'}$ is an arbitrary field on $\mathcal{M}$, symmetric in its $2s - 2$ indices. If $s = \tfrac{1}{2}$, show that one gets the neutrino equation $\nabla^{AA'}\psi_A = 0$ (hint: see the analogous discussion at the end of §8.1).

9.1.8. Let $(\mathcal{M}, g_{ab})$ be a space with anti-self-dual conformal curvature, and $\mathcal{P}$ its twistor space. Find the field equations in $\mathcal{M}$ whose solutions correspond to elements of $H^1(\mathcal{P}, \mathcal{O}(2s-2))$. Bear in mind the discussion following (9.1.21) concerning nonvacuum space.

9.2   Constructing self-dual space-times

In the previous section, we saw that anti-self-dual Einstein (or vacuum) space-times correspond to curved twistor spaces. In principle, this provides a way of constructing such metrics. The steps are: first, set up a suitable twistor space $\mathcal{P}$; secondly, find the holomorphic curves in $\mathcal{P}$, which then become the points of $\mathcal{M}$; and finally, derive the metric on $\mathcal{M}$. It is often fairly easy to carry out the first step, but one tends to get stuck at the second step, namely that of constructing the holomorphic curves. In this section we shall study some classes of examples in which this hurdle can be overcome.

Example 9.2.1: Let us first consider the linearized version of the construction, in which $\mathcal{P}$ is an 'infinitesimal' deformation of $\mathbb{P}^I$. For the sake of simplicity we shall deal with the vacuum case: thus Remark (2) in the previous section applies.

The two 'hemispheres'

$$U = \{[\omega^A, \pi_{A'}] : |\pi_{0'}/\pi_{1'}| \leq 1\}$$
$$\underline{U} = \{[\underline{\omega}^A, \underline{\pi}_{A'}] : |\underline{\pi}_{0'}/\underline{\pi}_{1'}| \geq 1\},$$

cover $\mathbb{P}^I$. That is to say, identifying U and $\underline{U}$ on the equator $|\pi_{0'}/\pi_{1'}| = 1$ by means of the transition functions

$$\underline{\omega}^A = \omega^A, \qquad \underline{\pi}_{A'} = \pi_{A'},$$

gives $\mathbb{P}^I$. (Actually, we should take each of U and $\underline{U}$ to be slightly bigger, and open, so that $U \cap \underline{U}$ is a neighborhood of the equator.) We deform $\mathbb{P}^I$ by modifying the transition functions. We want to preserve the fibration $\mathbb{P}^I \to \mathbf{P}_1$, where $\mathbf{P}_1$ is the projective $\pi_{A'}$ space. The space $\mathbf{P}_1$ is *rigid*: it cannot be deformed (Morrow and Kodaira 1971). Thus we retain

$$\underline{\pi}_{A'} = \pi_{A'}, \tag{9.2.1}$$

while replacing $\underline{\omega}^A = \omega^A$ with

$$\underline{\omega}^A = \omega^A + \lambda f^A(\omega^B, \pi_{B'}), \tag{9.2.2}$$

where f^A is holomorphic on $U \cap \underline{U}$ and homogeneous of degree one in $(\omega^B, \pi_{B'})$, while λ is a small parameter. In order to preserve the symplectic form $\varepsilon_- = d\omega_A \wedge d\omega^A$ on each ω^A-fiber, the relation (9.2.2) must be a canonical transformation. If we ignore terms of order λ^2, then this amounts to

$$\partial_A f^A = 0,$$

where $\partial_A = \frac{\partial}{\partial \omega^A}$. Thus f^A must have the form

$$f^A = \varepsilon^{AB} \partial_B f, \qquad (9.2.3)$$

where $f(\omega^A, \pi_{A'})$ is homogeneous of degree two. So the (infinitesimal) deformation, and the curved twistor space $\mathcal{P}$ it gives rise to, are determined by the twistor function f. More accurately, we should think of f as describing an element of $H^1(\mathbb{P}^I, \mathcal{O}(2))$, and infinitesimal deformations are also described by a first cohomology group (see Morrow and Kodaira 1971 for a discussion of this).

The next step is to construct the holomorphic curves in $\mathcal{P}$. These are given by expressions

$$\omega^A = h^A(x^{BB'}, \pi_{B'})$$
$$\underline{\omega}^A = \underline{h}^A(x^{BB'}, \pi_{B'}) \qquad (9.2.4)$$

satisfying the transition relation

$$\underline{h}^A(x, \pi) = h^A(x, \pi) + \lambda f^A(h^B(x, \pi), \pi_{B'}). \qquad (9.2.5)$$

Here $h^A(x, \pi)$ is holomorphic for $|\pi_{0'}/\pi_{1'}| \leq 1$, and $\underline{h}^A(x, \pi)$ is holomorphic for $|\pi_{0'}/\pi_{1'}| \geq 1$. The $x^{BB'}$ are four parameters which label the holomorphic curves, and therefore serve as coordinates on $\mathcal{M}$. In the present case, we can get h^A and $\underline{h}^A$ by splitting f^A:

$$f^A(ix^{BB'}\pi_{B'}, \pi_{B'}) = \underline{k}^A(x^{BB'}, \pi_{B'}) - k^A(x^{BB'}, \pi_{B'}), \qquad (9.2.6)$$

and then putting

$$h^A(x, \pi) = ix^{AA'}\pi_{A'} + \lambda k^A(x, \pi),$$
$$\underline{h}^A(x, \pi) = ix^{AA'}\pi_{A'} + \lambda \underline{k}^A(x, \pi). \qquad (9.2.7)$$

The point here is that (9.2.6) and (9.2.7) are equivalent to (9.2.5), *provided* that we ignore terms of order λ^2. Thus in the infinitesimal case, we can find the holomorphic curves.

Next, we compute the conformal metric on $\mathcal{M}$, in terms of the local coordinates $x^{AA'}$. There will exist quantities $\Sigma_{BB'}{}^{AA'}(x)$, which we have to compute, such that null vectors at x have the form

$$v^{AA'} = \xi^B \alpha^{B'}(\varepsilon_B{}^A \varepsilon_{B'}{}^{A'} + \lambda \Sigma_{BB'}{}^{AA'}(x)), \qquad (9.2.8)$$

for some ξ^B and $\alpha^{B'}$. The gradient of h^A in the direction $v^{AA'}$ is, by (9.2.7) and (9.2.8),

$$v^{AA'}\nabla_{AA'}h^B = i\xi^B \alpha^{B'}\pi_{B'} + \lambda \xi^A \alpha^{A'}(\nabla_{AA'}k^B + i\Sigma_{AA'}{}^{BB'}\pi_{B'}). \qquad (9.2.9)$$

The operator $\nabla_{AA'}$ here is the 'flat' one, i.e., partial differentiation with respect to the variable $x^{AA'}$. Since $v^{AA'}$ is meant to be null, (9.2.9) should vanish for some value of $\pi_{A'}$, which (without loss of generality) we can take to be $\pi_{A'} = \alpha_{A'}$. Now by (9.2.6), $\pi^{A'}\nabla_{AA'}k^B$ is a global quantity, homogeneous of degree two in $\pi_{A'}$, so it must be quadratic in $\pi_{A'}$, i.e., it must have the form

$$\pi^{A'}\nabla_{AA'}k^B(x,\pi) = -i\Sigma_{AA'}{}^{BB'}\pi^{A'}\pi_{B'}, \qquad (9.2.10)$$

for some $\Sigma_{AA'}{}^{BB'}(x)$. So we take $\Sigma_{AA'}{}^{BB'}$ to be *defined* by (9.2.10), for then (9.2.9) vanishes at $\pi_{A'} = \alpha_{A'}$. Having $\Sigma_{AA'}{}^{BB'}$ tells us what the null vectors in $\mathcal{M}$ are, and hence what the conformal metric is.

Before proceeding further, note that the splitting (9.2.6) is not unique: the freedom in k^A is clearly given by $k^A \mapsto k^A + t^{AA'}\pi_{A'}$, where $t^{AA'}$ is a function of x only. This freedom can be exploited to ensure that $\nabla_{AA'}k^A = 0$, and hence that $\Sigma_{AA'}{}^{AB'} = 0$. For by (9.2.3), we have $\nabla_{AA'}f^A = 0$, which by (9.2.6) means that

$$\nabla_{AA'}k^A = p^{B'}_{A'}\pi_{B'},$$

with $p^{B'}_{A'}$ being a function of x only. Solving the equation

$$\nabla_{AA'}t^{AB'} = -p^{B'}_{A'}$$

for $t^{AB'}$, tells us how to change k^A in order that $\nabla_{AA'}k^A$ should vanish. Note also that we have to take $\Sigma_{AA'}{}^{BA'} = 0$, since if $\Sigma_{AA'}{}^{BB'}$ had any trace part in its primed indices, this would not be determined by (9.2.10). So $\Sigma_{AA'BB'}$ is symmetric (or, equivalently, $\Sigma_{AA'}{}^{BB'}$ is trace-free) in both its primed and its unprimed indices.

As was said above, $\Sigma_{AA'BB'}$ gives us the conformal metric on $\mathcal{M}$, and from (9.1.8) this is given by the line-element expression

$$ds^2 = (\varepsilon_{AB}\varepsilon_{A'B'} + 2\lambda\Sigma_{AA'BB'})dx^{AA'}dx^{BB'}. \qquad (9.2.11)$$

In other words, (9.2.11) is the metric on $\mathcal{M}$, up to a possible conformal factor. In fact, no conformal rescaling is necessary: (9.2.11) *is* the metric (see Exercise 9.2.1). If we had not chosen a gauge in which $\nabla_{AA'}k^A$ vanished, then a correction *would* have been necessary.

The anti-self-dual Weyl spinor Ψ_{ABCD} obtained from this metric, thought of as a spinor field on flat space-time, is a solution of the massless free-field equations

$$\nabla^{AA'}\Psi_{ABCD} = 0;$$

this follows from the Bianchi identity. On the other hand, the twistor function f introduced in (9.2.3) determines a massless free field φ_{ABCD} on flat space-time. In fact, Ψ_{ABCD} and $\lambda\varphi_{ABCD}$ are the same, up to a constant numerical factor (see Exercise 9.2.2).

Example 9.2.2: These examples are taken from Ward (1978) and Curtis, Lerner and Miller (1978). They are classes of vacuum solutions in which the curvature is *not* infinitesimal. As before, the idea is to take two regions U and $\underline{U}$ of flat twistor space, and glue them together by

$$\underline{\pi}_{A'} = \pi_{A'}, \qquad \underline{\omega}^A = \omega^A + f^A(\omega^B, \pi_{B'}) \qquad (9.2.12)$$

to obtain $\mathcal{P}$. The following special choice for the function f^A is such that one can construct the holomorphic curves in $\mathcal{P}$, and hence the metric on $\mathcal{M}$, explicitly.

Let $p^{AA'\ldots C'}$ be a constant spinor with one unprimed and n primed indices, and put

$$p^A = p^{AA'\ldots C'}\pi_{A'}\ldots\pi_{C'}. \qquad (9.2.13)$$

Let $g(\gamma, \pi_{A'})$ be a holomorphic function of three variables, with the homogeneity property

$$g(\lambda^{n+1}\gamma, \lambda\pi_{A'}) = \lambda^{1-n}g(\gamma, \pi_{A'}), \qquad (9.2.14)$$

and take the transition function f^A to be

$$f^A(\omega^B, \pi_{B'}) = p^A g(p_B\omega^B, \pi_{B'}). \qquad (9.2.15)$$

Then f^A is homogeneous of degree one, as it should be. And (9.2.12), with this choice of f^A, preserves the symplectic form $\varepsilon_- = d\omega_A \wedge d\omega^A$ on the ω^A-space.

One significance of the choice (9.2.15) is that the function $p_A \omega^A$ is a globally-defined holomorphic function on $\mathcal{P}$, homogeneous of degree $n+1$. Indeed, by (9.2.12) and (9.2.15) we have $\underline{p}_A \underline{\omega}^A = p_A \omega^A$. There are also other global functions on $\mathcal{P}$, namely functions lifted from $\mathbf{P}_1$ via the fibration $\mathcal{P} \to \mathbf{P}_1$ (i.e., polynomials in $\pi_{A'}$). But, in general, for a twistor space fibered over $\mathbf{P}_1$, these lifted polynomials would be the *only* global functions on $\mathcal{P}$. The twistor spaces obtained by (9.2.15) are special in admitting an extra one. This extra function corresponds, in the anti-self-dual space-time $\mathcal{M}$, to a nonconstant solution of the twistor equation (or Killing spinor equation)

$$\nabla^{A(A'} \xi^{B'...D')} = 0, \tag{9.2.16}$$

where $\xi^{B'...D'}$ is a symmetric $(n+1)$-index spinor field on $\mathcal{M}$. We note that polynomials in $\pi_{A'}$ correspond to *constant* solutions of (9.2.16).

To construct the holomorphic curves in $\mathcal{P}$, one can proceed as follows. First split the function g:

$$g(ip_A x^{AA'} \pi_{A'}, \pi_{A'}) = \underline{k}(x, \pi_{A'}) - k(x, \pi_{A'}), \tag{9.2.17}$$

where, as usual, k is holomorphic for $|\pi_{0'}/\pi_{1'}| \leq 1$, and so forth. The functions in (9.2.17) are homogeneous of degree $(1-n)$ in $\pi_{A'}$, and $H^1(\mathbf{P}_1, \mathcal{O}(p))$ vanishes for $p \geq -1$, so the splitting (9.2.17) is possible if and only if $1 - n \geq -1$. From now on, therefore, let us assume that $0 \leq n \leq 2$. The holomorphic curves are then given by

$$\begin{aligned} \omega^A &= h^A(x, \pi_{A'}) := i x^{AA'} \pi_{A'} + p^A k(x, \pi_{A'}), \\ \underline{\omega}^A &= \underline{h}^A(x, \pi_{A'}) := i x^{AA'} \pi_{A'} + p^A \underline{k}(x, \pi_{A'}). \end{aligned} \tag{9.2.18}$$

These satisfy the necessary transition relation

$$\underline{h}^A(x, \pi_{A'}) = h^A(x, \pi_{A'}) + f^A(h^B(x, \pi_{A'}), \pi_{B'}),$$

the point being that for our special choice of f^A, we have

$$f^A(h^B(x, \pi_{A'}), \pi_{B'}) = f^A(i x^{BB'} \pi_{B'}, \pi_{B'}).$$

Now that one has the holomorphic curves, it is fairly easy to compute the metric on $\mathcal{M}$. The details may be found in Ward (1978).

For each value of n (i.e., $n = 0$, 1, or 2), one obtains a class of vacuum solutions.

If $n = 0$, the metric is that of an 'anti-self-dual pp-wave' and has the form

$$ds^2 = du\,dv + dx\,dy + f(v, x)dv^2,$$

where $f(v, x)$ is an arbitrary function.

If $n = 1$, we get the metric (6.2.4); the functions V and ω_i appearing in that expression, and satisfying (6.2.5), are generated from the twistor function g. More accurately, of course, one should think of g as a cocycle representing an element of $H^1(\mathbb{P}^I, \mathcal{O}(1 - n))$. In the $n = 1$ case, therefore, g determines an anti-self-dual Maxwell field φ_{AB}, and (V, ω_i) constitutes a potential for φ_{AB}. The metrics (6.2.4) all admit a Killing vector $\frac{\partial}{\partial \tau}$, which is, in fact, the divergence of the Killing spinor field $\xi^{A'B'}$ mentioned in (9.2.16) (see Tod and Ward 1979).

Finally, if $n = 2$, then the metrics one gets can be described as follows. Put

$$\varphi_{A'B'} = \frac{1}{2\pi i} \oint \pi_{A'} \pi_{B'} g_\gamma (i_P {}_C x^{CC'} \pi_{C'}, \pi_{C'}) \pi_{D'} d\pi^{D'},$$

where $g_\gamma = \frac{\partial g}{\partial \gamma}$; and write

$$\theta = 1 - \varphi_{0'1'} \qquad \varphi = \varphi_{0'0'},$$
$$\widetilde{\varphi} = \varphi_{1'1'}, \qquad \Lambda = \theta^2 - \varphi\widetilde{\varphi}.$$

Then in the coordinate system $y^{\mathrm{a}} = (x^{00'}, x^{11'}, x^{01'}, x^{10'})$, the metric is given by

$$g^{\mathrm{ab}} = -\Lambda^{-1} \begin{pmatrix} 2\varphi & -\theta & 0 & \widetilde{\varphi} \\ -\theta & 2\widetilde{\varphi} & \varphi & 0 \\ 0 & \varphi & 0 & \theta \\ \widetilde{\varphi} & 0 & \theta & 0 \end{pmatrix}.$$

This is not very illuminating, and these metrics are possibly not very interesting in themselves, since they do not possess a positive-definite 'slice'. But the example illustrates the way in which an essentially arbitrary function g can generate solutions of the anti-self-dual Einstein equations.

Example 9.2.3: Let $\mathcal{P}$ be the flag manifold $\mathbf{F}_{12}(\mathbf{C}^3)$. In other words, each point Z in $\mathcal{P}$ is a pair (p, l), where p is a point in $\mathbf{P}_2$, and l is a line in $\mathbf{P}_2$ containing p. This twistor space admits a four-parameter family of rational holomorphic curves with the correct normal bundle: these curves are given as follows. Let L be a line in $\mathbf{P}_2$ and P a point not on L; these determine a curve in $\mathcal{P}$, consisting of all pairs (p, l), where l passes through P, and where l and L intersect at p. Locally, P and L are each specified by two complex coordinates, so the space $\mathcal{M}$ of curves described above is indeed four-complex-dimensional.

Once more we shall leave the computation of the metric on $\mathcal{M}$ as an exercise (see Ward 1980). It turns out to be the complexification of the Fubini–Study metric on $\mathbf{P}_2$ (see §6.2). In fact, the twistor space admits a real structure, and so one gets a *positive-definite* anti-self-dual Einstein metric, which is precisely the Fubini–Study metric on $\mathbf{P}_2$.

From the discussion in the previous section, we see that the twistor space $\mathbf{F}_{12}(\mathbf{C}^3)$ is fibered over the corresponding Riemannian space $\mathbf{P}_2$, with each fiber being a holomorphic curve. It is important to note that this is *not* the natural holomorphic fibration of the flag manifold $\mathbf{F}_{12}$ over $\mathbf{P}_2 = \mathbf{F}_1(\mathbf{C}^3)$.

This example, unlike the previous ones, is Einstein rather than vacuum. The forms τ and ρ on $\mathcal{P}$ may be specified as follows. Let $p^j = (p^1, p^2, p^2)$ be homogeneous coordinates for $\mathbf{P}_2$, and $l_j = (l_1, l_2, l_3)$ homogeneous coordinates for its dual $\mathbf{P}_2^*$. Then (p^j, l_j) serve as 'bi-homogeneous' coordinates on $\mathbf{F}_{12}$. In terms of these we can take

$$\tau = \tfrac{1}{2} l_j dp^j - \tfrac{1}{2} p^j dl_j,$$
$$\rho = -\tfrac{1}{2} \tau \wedge d\tau.$$

Remarks:

(1) Recall from §6.2 that an asymptotically locally Euclidean (ALE) space is one whose infinity is like that of E^4, factored by a discrete subgroup Γ of $SO(4)$. Hitchin (1979) has shown how to construct the appropriate twistor spaces, for the case when Γ is cyclic, and hence to construct these particular ALE metrics. They turn out to be given by (6.2.4) and (6.2.6). In a sense, therefore, they are special cases of Example 9.2.2 with $n = 1$. In unpublished work, Hitchin has made some progress towards extending these results to *other* subgroups Γ, and this has been carried further by the work of P. Kronheimer. As one might expect, the difficulty in general is to construct the holomorphic curves in the twistor space.

(2) There is an analogue, one dimension lower down, of the construction we have been discussing in this chapter. This is a correspondence between three-dimensional Einstein–Weyl spaces and two-dimensional 'minitwistor' spaces (Hitchin 1982a). Jones and Tod (1985) have shown how this may be understood in terms of 'dimensional reduction' from the standard construction one dimension higher up.

Exercises for Section 9.2

9.2.1. Verify, using the definition in Theorem 9.1.2, that (9.2.11) is the correct expression for the linearized metric in Example 9.2.1.

9.2.2. Show that the two fields Ψ_{ABCD} and φ_{ABCD} defined at the end of Example 9.2.1 are equal (up to some constant factor).

9.2.3. Generalize Example 9.2.1 to the linearized Einstein (i.e., not vacuum) case.

9.2.4. Compute the metrics in Examples 9.2.2 and 9.2.3.

10

The Penrose transform for
general gauge fields

In the previous three chapters we have seen how the Penrose transform maps essentially free holomorphic data on a twistor space to solutions of self-dual field equations in various contexts. There were two aspects to this program. The first was the formal transform (or correspondence) between holomorphic data and solutions of field equations. The second was *using* the specific transform to generate some (and sometimes all) of the solutions of the equations in a reasonably explicit manner. Some of these solutions, in particular for the nonlinear problems, had been unknown before using the techniques of the Penrose transform.

In the study of non-self-dual problems the results are much weaker. There is a systematic generalization of the Penrose transform which translates holomorphic data on more general twistor spaces to solutions of quite general field equations. These are primarily of the Yang–Mills–Higgs–Dirac type, and not necessarily with any type of self-duality condition. There is as yet no satisfactory transform from the general Einstein equations, although there are some partial results.

In this chapter we want to outline the principal results in this direction. The initial impetus came from the work of Isenberg, Yasskin and Green (1978) and Witten (1978), which represented solutions of the full Yang–Mills equations in terms of holomorphic vector bundles on formal neighborhoods of ambitwistor space. This was generalized by Henkin and Manin (1980) in which they represented solutions of coupled Yang–Mills–Higgs–Dirac systems in terms of algebraically coupled cohomology groups and vector bundles. The nonlinearity of the original differential equations was transformed to a nonlinear cohomology and vector bundle problem, again on formal neighborhoods of ambitwistor space. Henkin (1982) modified this to have simply *vector bundles* on the formal neighborhoods representing the solutions of the equations. The cohomology group information was encoded into the transition functions of Henkin's vector bundles.

The formal neighborhoods of ambitwistor space have structure sheaves with nilpotent elements. The theory of supermanifolds, initiated by Leites, Berezin, Kostant, and others (see §10.2) had a parallel development to the twistor geometry described in this book. The fundamental idea of Witten (1978) from which he derived his formal neighborhood ideas, was to combine the ideas of twistor geometry and supermanifolds into one unified picture. This was carried out by Manin in great detail in his book (1988). What one obtains is a description of the full Yang–Mills–Higgs–Dirac fields on Minkowski space (*or* superfields on super Minkowski space) in terms of holomorphic vector bundles on a 'super twistor manifold' of a suitable type. In §10.1 we review the formal neighborhood development, and in §10.2 we give a brief introduction to supermanifolds, with a summary of the basics results of Manin and others on the representation of super Yang–Mills fields.

We emphasize that these methods in the non-self-dual case have yielded very little in the way of *solutions* of the field equations, but they do give a geometric *derivation* of the field equations themselves in terms of elementary geometric concepts, a result, which, in itself, has some genuine intrinsic value.

10.1 The Penrose transformation on formal neighborhoods of the space of null lines

Let L denote the set of null lines in complexified Minkowski space M. Each null line $l \in$ L is a rational curve ($\cong$ P$_1$) holomorphically embedded in M, and is a projective line in P$_5$ under the Plücker embedding (see §§1.3 and 1.4). Theorem 1.4.1 gives us a parametrization of L as F$_{13}$ (which we also called ambitwistor space A), a five-dimensional complex manifold. There are higher-dimensional analogues of twistor geometry in which the space of null lines and ambitwistor space do not coincide, and, for our purposes, the space of null lines is the important and fundamental concept. We have a double fibration of the form

$$
\begin{array}{ccc}
 & \mathsf{G} & \\
{}^{\sigma}\swarrow & & \searrow^{\rho} \\
\mathsf{L} & & \mathsf{M}
\end{array}
\qquad (10.1.1)
$$

(see (1.2.13)), where G $=$ F$_{123}$. Now using the fact that L $=$ F$_{13}$, we see that we have an embedding

$$
\mathsf{L} \to \mathsf{P} \times \mathsf{P}^* = (\mathsf{F}_1 \times \mathsf{F}_3) \qquad (10.1.2)
$$

given by

$$[L_1 \subset L_3] \xrightarrow{\cong} (L_1, L_3) \in \mathbf{P} \times \mathbf{P}^*.$$

The image of L in $\mathbf{P} \times \mathbf{P}^*$ is a quadric surface. If we let $[Z^\alpha]$ be homogeneous coordinates for $\mathbf{P}$ and $[W_\beta]$ be homogeneous coordinates for $\mathbf{P}^*$, and let $Z \cdot W = Z^\alpha W_\beta$ be the natural duality pairing of T and T^*, we see that

$$\mathsf{L} = \{([Z^\alpha], [W_\beta]) \in \mathbf{P} \times \mathbf{P}^* : Z \cdot W = 0\}.$$

In other words $Z \cdot W = 0$ if and only if the line L_1 represented by Z^α in T is contained in the hyperplane L_3 represented by $W_\beta \in \mathsf{T}^*$.

We want to discuss formal neighborhoods of a submanifold in a general context first. Let Y be a complex submanifold of a complex manifold X, and let $\mathcal{I} = \mathcal{I}_Y$ be the ideal sheaf of all holomorphic functions in the sheaf $\mathcal{O}_X$ which vanish on Y. Then we recall the normal bundle Taylor series expansion in this setting given by (7.2.28)

$$\mathcal{O}_X|_Y \cong \sum_p \mathcal{O}_Y(\odot^p N_Y^*) \cong \sum_p \mathcal{I}^p/\mathcal{I}^{p+1}. \tag{10.1.3}$$

Suppose we define

$$\mathcal{O}_Y^{(k)} := \mathcal{O}_X/\mathcal{I}_Y^{k+1},$$

and we call $(Y, \mathcal{O}^{(k)})$ the kth *formal neighborhood of Y in X*, then we see that $\mathcal{O}_Y^{(k)}$ has a finite expansion similar to (10.1.3) of the form

$$\mathcal{O}_Y^{(k)} \cong \sum_{p=0}^{k} \mathcal{I}^p/\mathcal{I}^{p+1},$$

$$\cong \sum_{p=0}^{k} \mathcal{O}_Y(\odot^p N_Y). \tag{10.1.4}$$

Namely (10.1.4) is simply (10.1.3) modulo $\mathcal{I}_Y^{k+1}$ on both sides of the equation. Thus the kth formal neighborhood contains formal Taylor expansions in the normal bundle direction up to order k. There is a natural inclusion (as ringed spaces) of

$$(Y, \mathcal{O}_Y^{(k)}) \to (U, \mathcal{O}_X),$$

where U is any topological neighborhood of Y contained in X (see Exercise 10.1.1).

A fundamental question in complex analysis is whether one can extend an analytic object from a submanifold Y to a neighborhood of Y in X, or, for that matter, to all of X. One way of attacking this problem is by formal power series normal to the submanifold, and to try to extend the given analytic object one order at a time, i.e., extending it to the first formal neighborhood, then to the second, etc. If at any point no formal extension is possible, then it is clear than an extension to a full topological neighborhood is ruled out. There is a cohomological theory of such formal extensions, developed by Griffiths (1965, 1966), among others. The basic idea is that there are natural cohomological obstructions to extension which are calculable in principle. These are analogous to characteristic classes being obstructions to the triviality of a vector bundle. We shall summarize the extension theory that we need. We shall be interested in extending from Y to the first, second, and third formal neighborhoods for the case of $Y = \mathsf{L}$, $X = \mathsf{P} \times \mathsf{P}^*$ and the quadratic embedding (10.1.2) above. The analytic objects we shall want to extend will be vector bundles on Y.

A holomorphic vector bundle E defined over $Y^{(k)}$ is defined by transition functions

$$g_{\alpha\beta} \in \mathcal{O}^{(k)}(U_\alpha \cap U_\beta) \otimes \mathrm{GL}(r, \mathbf{C}),$$

where r is the rank of the bundle and such that

$$g_{\alpha\beta} g_{\beta\alpha} = I,$$
$$g_{\alpha\beta} g_{\beta\gamma} g_{\gamma\alpha} = I.$$

In other words, $g_{\alpha\beta}$ is represented by functions $\tilde{g}_{\alpha\beta}$ holomorphic on $\tilde{U}_{\alpha\beta} = \tilde{U}_\alpha \cap \tilde{U}_\beta$, and $\{\tilde{U}_\alpha\}$ is an open covering of the closed set Y in X, and

$$\tilde{g}_{\alpha\beta} \tilde{g}_{\beta\alpha} = I + f_{\alpha\beta},$$
$$\tilde{g}_{\alpha\beta} \tilde{g}_{\beta\alpha} \tilde{g}_{\gamma\alpha} = I + f_{\alpha\beta\gamma},$$

where $f_{\alpha\beta}$ and $f_{\alpha\beta\gamma}$ vanish to order $(k+1)$ on Y, i.e., $f_{\alpha\beta}$ and $f_{\alpha\beta\gamma}$ define elements of $\mathcal{I}^{k+1}$ at each point where they are defined. We have the following fundamental extension theorem due to Griffiths (1965).

Theorem 10.1.1. (a) Let $E^{(k)} \to Y^{(k)}$ be a holomorphic vector bundle over the kth infinitesimal neighborhood $Y^{(k)}$ of Y in X. There exists a cohomology class

$$\eta(E^{(k)}) \in H^2(Y, \mathcal{O}(\text{End}(E) \otimes \odot^{k+1} N_Y^*)),$$

with the property that there exists a holomorphic extension $E^{(k+1)}$ of $E^{(k)}$ from $Y^{(k)}$ to $Y^{(k+1)}$ if and only if $\eta(E^{(k)}) = 0$.

(b) If $\eta(E^{(k)}) = 0$, then the cohomology group

$$H^1(Y, \mathcal{O}(\text{End}(E) \otimes \odot^{(k+1)} N_Y^*))$$

acts transitively on the set of all extensions.

See Griffiths (1965) for the proof of this. It is also carried out in detail for the first few orders in Pool (1987) in the special case of $Y = \mathbf{L} \subset \mathbf{P} \times \mathbf{P}^*$. We shall indicate the nature of the class $\eta(E^{(k)})$, in the special case that Y is a hypersurface in X. Let ρ_α be defining functions for Y in U_α, then the ideal sheaf $\mathcal{I}$ is generated in U_α by ρ_α, i.e., any section of $\mathcal{I}$ over U_α is of the form $f\rho_\alpha$, where f is holomorphic in U_α. Let $g_{\alpha\beta}^{(1)}$ be any extension of $g_{\alpha\beta}$ to $U_{\alpha\beta}^{(1)}$, and we see that

$$g_{\alpha\beta}^{(1)} g_{\beta\gamma}^{(1)} g_{\gamma\alpha}^{(1)} = I + \mathcal{I}, \qquad (10.1.5)$$

that is to say,

$$g_{\alpha\beta}^{(1)} g_{\beta\gamma}^{(1)} g_{\gamma\alpha}^{(1)} |_{Y^{(0)}=Y} = I.$$

We want to find such an extension so that

$$g_{\alpha\beta}^{(1)} g_{\beta\gamma}^{(1)} g_{\gamma\alpha}^{(1)} = I + \mathcal{I}^2. \qquad (10.1.6)$$

Define $\eta_{\alpha\beta\gamma}^{(1)}$ by

$$g_{\alpha\beta}^{(1)} g_{\beta\gamma}^{(1)} g_{\gamma\alpha}^{(1)} |_{Y^{(1)}} - I = \eta_{\alpha\beta\gamma} \rho_\alpha,$$

then $\eta_{\alpha\beta\gamma}$ is a well-defined two-cochain with coefficients in

$$\mathcal{O}(\text{End}(E) \otimes N_Y^*).$$

One can check that this cochain is closed and defines an element $\eta(E) \in H^2(Y, \mathcal{O}(\text{End}(E) \otimes N_Y^*))$. It is a fact, which we shall not verify here, that $\eta(E)$ is independent of the initial choice of extension. Similarly, we can define

$$\eta(E^{(k)}) \in H^2(Y, \mathcal{O}(\text{End}(E) \otimes \odot^{k+1} N_Y^*)).$$

These are the fundamental obstructions which occur in Theorem 10.1.1. In principle, one must calculate

$$H^2(Y, \mathcal{O}(\mathrm{End}(E) \otimes \odot^{k+1} N_Y^*)).$$

If this were always to vanish (which is the case for certain classes of bundles on certain spaces Y, see Griffiths 1966), then there are no obstructions to extension of a vector bundle formally to all infinitesimal neighborhoods. Or, what is the same thing, we can find the formal power series expansion in the normal bundle direction for the desired extended vector bundle. The question of whether this expansion is convergent is then a separate issue. In the case of line space $L \subset P \times P^*$, the situation is more delicate, as we shall see.

Now we need to turn to the question of the Penrose transform for the case of the double fibration

$$
\begin{array}{ccc}
 & G & \\
\sigma\swarrow & & \searrow\rho \\
L & & M.
\end{array}
$$

First we shall consider any such double fibration. Namely, suppose we have any double fibration correspondence with Minkowski space of the form

$$
\begin{array}{ccc}
 & S_2 & \\
f_1\swarrow & & \searrow f_2 \\
S_1 & & M,
\end{array}
$$

where f_1 and f_2 are surjective maximal rank mappings. If U is open in M, and $\widehat{U} = f_1 \circ f_2^{-1}(U)$ and $\hat{x} = f_1 \circ f_2^{-1}(x)$, for $x \in U$, then we shall say that a vector bundle $E \to \widehat{U}$ is a *Yang–Mills bundle* if $E|_{\hat{x}}$ is trivial for all $x \in U$. The nomenclature is taken from the fact that in numerous situations such bundles when transformed to M satisfy the Yang–Mills equations. In particular, we have from Theorem 8.1.2 that for the case

$$
\begin{array}{ccc}
 & F & \\
\mu\swarrow & & \searrow\nu \\
P & & M
\end{array}
$$

we have holomorphic Yang–Mills bundles on $\widehat{U} \subset P$ are in one-to-one correspondence with holomorphic vector bundles with connection $(\widetilde{E}, A)$ on an open set U in M which are anti-self-dual. We write $\mathcal{P}(E) = (\widetilde{E}, A)$. Similarly for the double fibration

$$
\begin{array}{ccc}
 & F^* & \\
\swarrow & & \searrow \\
P^* & & M,
\end{array}
$$

we have holomorphic Yang–Mills bundles on $\widehat{U} \subset \mathbf{P}^*$ are in one-to-one correspondence with self-dual holomorphic vector bundles with connection $(\widetilde{E}, A)$ defined on an open set U contained in $\mathbf{M}$. This is proved completely analogously to Theorem 8.1.2.

Looking at the anti-self-dual case in more detail, we recall from Chapter 8 that the locally free sheaf $\mathcal{O}_{\mathbf{M}}(\widetilde{E})$ could be described by

$$\mathcal{O}_{\mathbf{M}}(\widetilde{E}) = \nu_*^0 \mathcal{O}_{\mathsf{F}}(\mu^* E) = \nu_*^0 \mu^*(\mathcal{O}_{\mathbf{P}}(E)),$$

i.e.,

$$\begin{aligned}
\widetilde{E}_x &:= H^0(\nu^{-1}(x), \mathcal{O}_{\nu^{-1}(x)}(\mu^* E)) \\
&= H^0(\hat{x}, \mathcal{O}_{\hat{x}}(E)) \\
&= \Gamma(\hat{x}, E),
\end{aligned}$$

where $\hat{x} = \mu \circ \gamma^{-1}(x)$ as usual, and where $\Gamma(\hat{x}, E)$ is simply the vector space of global sections of E over $\hat{x} \subset \mathbf{P}$. Recall that we have the relative de Rham sequence on F

$$0 \longrightarrow \mu^{-1}(E) \longrightarrow \Omega_\mu^0(E) \xrightarrow{d_\mu} \Omega_\mu^1(E) \xrightarrow{d_\mu} \Omega_\mu^2(E),$$

and the induced mappings

$$\nu_*^0 \Omega_\mu^0(E) \xrightarrow{\tilde{d}_\mu} \nu_*^0 \Omega_\mu^1(E) \xrightarrow{\tilde{d}_\mu} \nu_*^0 \Omega_\mu^2(E), \qquad (10.1.7)$$

(see (7.1.17)). It is not difficult to see that we have the isomorphisms

$$\begin{aligned}
\nu_*^0 \Omega_\mu^0(E) &= \mathcal{O}_{\mathbf{M}}(\widetilde{E}) \quad \text{(by definition)} \\
\nu_*^0 \Omega_\mu^1(E) &\cong \Omega_{\mathbf{M}}^1(\widetilde{E}) \\
\nu_*^0 \Omega_\mu^2(E) &\cong \Omega_{\mathbf{M},+}^2,
\end{aligned}$$

where the subscript '+' on the two-forms indicates the self-dual two-forms on $\mathbf{M}$. Thus we see that $\tilde{d}_\mu$ induces a connection $D = D_A$ on $\widetilde{E}$

$$D_A : \mathcal{O}_{\mathbf{M}}(\widetilde{E}) \to \Omega_{\mathbf{M}}^1(\widetilde{E}),$$

and moreover, the composition $\tilde{d}_\mu \circ \tilde{d}_\mu$ in (10.1.7) can be identified with the curvature of the connection

$$F_A = D_A \circ D_A : \mathcal{O}(\widetilde{E}) \to \Omega^2(\widetilde{E}). \qquad (10.1.8)$$

The image of the curvature mapping in (10.1.8) being in $\Omega^2_+(\widetilde{E})$ is equivalent to the fact that the connection is anti-self-dual. This is a reformulation of the basic correspondence between anti-self-dual connections on open subsets U of M and holomorphic Yang–Mills vector bundles on $\widehat{U}$ (see Theorem 8.1.2, Remark (5) after this theorem, and Exercise 8.1.4).

We can formulate the same thing at the line-space level. Consider (10.1.1), and let U be open in M, and let

$$\widehat{U} := \sigma \circ \rho^{-1}(U), \qquad \hat{x} = \sigma \circ \rho^{-1}(x), \quad \text{for } x \in U,$$

as before. Then we have the following generalization of Theorem 8.1.2.

Theorem 10.1.2. *There is a one-to-one correspondence between:*
   (a) *holomorphic Yang–Mills vector bundles E on $\widehat{U} \subset$ L,*
       and
   (b) *holomorphic vector bundles with connection $(\widetilde{E}, A)$ on $U \subset$ M.*

Remark: Note that there is *no* restriction on the vector bundle with connection $(\widetilde{E}, A)$ on U in this theorem in contrast to the self-dual and anti-self-dual cases (i.e., when $(\widetilde{E}, A)$ corresponded to bundles E on P or P*).

We shall not prove this theorem here; it is quite similar to the proof of Theorem 8.1.2 (see, e.g., Eastwood 1981, and Pool 1987). We shall only indicate the nature of the mapping $E \mapsto (\widetilde{E}, A)$ which we shall denote by $\mathcal{P}$ (the Penrose transform from line space L to Minkowski space M). We define $\widetilde{E}$ by the locally free sheaf $\mathcal{O}(\widetilde{E})$ corresponding to it, as usual, namely,

$$\mathcal{O}_M(\widetilde{E}) := \rho^0_* \sigma^* \mathcal{O}_L(E),$$

from the double fibration (10.1.1), and the connection D_A on $\widetilde{E}$ is induced from the mapping

$$\rho^0_* \Omega^0_\mu(E) \xrightarrow{\tilde{d}_\mu} \rho^0 \Omega^1_\mu(E),$$

just as in (10.1.7) above, where we note that there is an embedding

$$\rho^0_* \Omega^1_\mu(E) \to \Omega^1_M(\widetilde{E});$$

this gives $D_A : \mathcal{O}_{\mathbf{M}}(\widetilde{E}) \to \Omega^1_{\mathbf{M}}(\widetilde{E})$, the connection on $\widetilde{E}$. We note that (see Theorem 1.4.1 and its proof)

$$\hat{x} \cong \mathbf{P}_1 \times \mathbf{P}_1, \quad x \in U,$$
$$\tilde{p} \cong \mathbf{P}_1 \cap U, \quad p \in \widehat{U}.$$

The curvature $F_A = D_A \circ D_A = D_A^2$ restricted to the line $\tilde{p} \cong \sigma^{-1}(p) \subset \mathbf{G}$ is trivial, but it is *no* restriction to require that a curvature restricted to a one-dimensional object be trivial. Thus we obtain no natural restrictions, and, in fact, there are none at all, as the theorem asserts. We can now formulate a fundamental result due to Henkin and Manin (1980). Consider the inhomogeneous Yang–Mills equations on an open set U in $\mathbf{M}$, namely,

$$D_A F_A = 0,$$
$$D_A^* F_A = J.$$

The axial current

$$*J \in \Omega^3(U, \operatorname{End}(\widetilde{E})),$$

and is the analogue of the axial current in the inhomogeneous Maxwell equations.

Theorem 10.1.3. *Let U be open in $\mathbf{M}$, and let E be a holomorphic Yang–Mills bundle on $\widehat{U}$, then*

(a) *there is a unique extension $E^{(2)}$ of E to $\widehat{U}^{(2)}$,*
(b) *there is an isomorphism*

$$\mathcal{P} : H^2(\widehat{U}, \mathcal{O}(E \otimes \odot^3 N_{\mathbf{L}})) \xrightarrow{\cong} H^0(U, \Omega^3_{\mathbf{M}}(\operatorname{End}(\widetilde{E})),$$

such that if $\eta = \eta(E^{(2)})$ is the obstruction to extending $E^{(2)}$ to $\widehat{U}^{(3)}$, then $\tilde{J} = \mathcal{P}(\eta)$ satisfies

$$D_A^* F_A = \tilde{J}.$$

As a corollary, we have the following result due to Yasskin, Green and Isenberg (1978), and Witten (1978).

Corollary 10.1.4. *The vector bundle E extends to $\widehat{U}^{(3)}$ if and only if $(\widetilde{E}, A)$ satisfies*

$$D_A^* F_A = 0.$$

Thus we see that the geometric property of E extending to a third order neighborhood on $\widehat{U}$ is equivalent to $(\widetilde{E}, A)$ satisfying the full Yang–Mills equations. Theorem 10.1.3 was first formulated by Henkin and Manin (1980) with an outline of the proof. Detailed proofs of this specific result can be found in Pool (1987) and Buchdahl (1985).

In their paper Henkin and Manin also formulate a more general Penrose transform for coupled Yang–Mills–Higgs–Dirac fields in terms of algebraically coupled cohomology equations on $\mathbb{L}$. We shall not try to formulate the most general result, as its reformulation in terms of supergeometry is more appealing, and we shall give that in the next section.

Example 10.1.5 (Maxwell–Dirac coupled equations): In Eastwood, Pool, and Wells (1985) an example of the Henkin–Manin scheme is carried out, including a family of examples of solutions. There are, in fact, very few explicit examples of solutions of strongly coupled field equations, in contrast to the self-dual situation. Consider the equations on $\mathbf{M}^I$,

$$
\begin{aligned}
(\nabla^{AA'} - \lambda\Phi^{AA'})\varphi_A &= 0, \\
(\nabla^{AA'} + \lambda\Phi^{AA'})\psi_{A'} &= 0, \\
d^* F &= \lambda\varphi_A\psi_{A'}dz^{AA'},
\end{aligned}
\tag{10.1.9}
$$

where

$$\Phi = \Phi_{AA'}dz^{AA'}$$

is the Maxwell potential, $F = d\Phi$ is the Maxwell field strength, and λ is a coupling constant. We seek triples $(\varphi_A, \psi_{A'}, \Phi_{AA'})$ which satisfy these equations. The equations (10.1.9) arise from the Lagrangian on M^4

$$
\begin{aligned}
\mathcal{A} = \int_{M^4} \{ |F_\Phi|^2 + [\psi_{A'}(\nabla^{AA'} - \lambda\Phi^{AA'})\varphi_A \\
- \varphi_A(\nabla^{AA'} + \lambda\Phi^{AA'})\psi_{A'}] \} d^4 x.
\end{aligned}
\tag{10.1.10}
$$

We want to describe certain linear Penrose transforms in this line space setting. Let $\mathcal{O}(p, q)$ denote the line bundle on $\mathbf{P} \times \mathbf{P}^*$ which is parametrized by the integers p and q, where

$$\mathcal{O}(p, q) := \pi_1^* \mathcal{O}_{\mathbf{P}}(p) \otimes \pi_2^* \mathcal{O}_{\mathbf{P}^*}(q),$$

where π_1, and π_2 are the natural projections of $\mathbf{P} \times \mathbf{P}^*$ onto the first and second factors of this cartesian product. We let $\mathcal{O}_{\mathbf{L}}(p, q)$ be the restriction of $\mathcal{O}(p, q)$ to $\mathbf{L}$, i.e.,

$$\mathcal{O}_{\mathbf{L}}(p, q) := \mathcal{O}(p, q)/\mathcal{I}_{\mathbf{L}}.$$

Now let $E \to \mathbf{L}^I$ be a holomorphic Yang–Mills line bundle, then we have the following generalization of Theorem 7.2.3. There are isomorphisms

$$\mathcal{P} : H^1(\mathbf{L}^I, \quad \mathcal{O}(E)(-3, 0)) \to \ker \mathcal{D}_A : \Gamma(\mathbf{M}^I, S_+) \to \Gamma(\mathbf{M}^I, S^-),$$
$$\mathcal{P} : H^1(\mathbf{L}^I, \mathcal{O}(E^*)(0, -3)) \to \ker \mathcal{D}_A^* : \Gamma(\mathbf{M}^I, S_-) \to \Gamma(\mathbf{M}^I, S^+),$$
$$\mathcal{P} : \{E\} \to \{\widetilde{E}, A\},$$

$$(10.1.11)$$

where $\mathcal{D}_A$ and its adjoint $\mathcal{D}_A^*$ are Dirac operators as in (10.1.9). Let

$$\widehat{\psi} \in H^1(\mathbf{L}^I, \mathcal{O}(E)(-3, 0)), \quad \mathcal{P}(\widehat{\psi}) = \psi_{A'},$$
$$\widehat{\varphi} \in H^1(\mathbf{L}^I, \mathcal{O}(E^*)(0, -3)), \quad \mathcal{P}(\widehat{\varphi}) = \varphi_A.$$

As we indicated above, E has a unique extension $E^{(2)}$ to $(\mathbf{L}^I)^{(2)}$, and we let $\eta(E) = \eta(E^{(2)})$ be the obstruction to extending to third order. Then

$$\eta(E) \in H^2(\mathbf{L}^I, \mathcal{O}(\odot^3 N_{\mathbf{L}})),$$

and one can calculate that $\odot^3 N_{\mathbf{L}} = \mathcal{O}(-3, -3)$, so

$$\eta(E) \in H^2(\mathbf{L}^I, \mathcal{O}(-3, -3)).$$

We note that $\widehat{\varphi} \wedge \widehat{\psi}$ is also a well-defined element of this same cohomology group $H^2(\mathbf{L}^I, \mathcal{O}(-3, -3))$, using the fact that E and E^* are in duality. The basic result of Henkin and Manin (1980) in this case is the following theorem.

Theorem 10.1.6. *The triple $\{(\widetilde{E}, A), \varphi_A, \psi_{A'}\}$ satisfies (10.1.9) if and only if the triple $\{E, \widehat{\varphi}, \widehat{\psi}\}$ satisfies*

$$\eta(E) = \widehat{\varphi} \wedge \widehat{\psi}.$$

For details of the proof of this see Eastwood, Pool and Wells (1985).

We shall now give a family of simple solutions to (10.1.9). Let $v^{AA'}$ be a fixed null direction in $\mathbf{M}^I$, so $v^{AA'}$ is the product of constant spinors $v^{AA'} = \varphi^A \psi^{A'}$. Consider potentials $\Phi_{AA'}$ of the form

$$\Phi_{AA'}(z) = \varphi_A \psi_{A'} U(z),$$

where $U(z)$ is a scalar field. Then choose coordinates

$$z^{AA'} = \begin{pmatrix} u & \zeta \\ \tilde{\zeta} & v \end{pmatrix},$$

so that u is the coordinate in the $v^{AA'}$-direction. If $U(z) = U(u, v, \zeta, \tilde{\zeta})$ satisfies

$$\frac{\partial U}{\partial u} = 0$$

$$\frac{\partial^2 U}{\partial \zeta \partial \tilde{\zeta}} = 1, \qquad (10.1.12)$$

then the triple

$$\{\varphi_A, \psi_{A'}, \Phi_{AA'} = \varphi_A \psi_{A'} U(z)\}$$

satisfies (10.1.9) for $\lambda = 1$. This is easy to verify. If we let $n = n^a = n^{AA'}$ be a vector satisfying

$$n^a n_a = 1,$$
$$n^a v_a = 0,$$

then

$$\{\varphi_A, \psi_{A'}, \Phi_{AA'} = \varphi_A \psi_{A'} (n^a z_a)^2\}$$

satisfies (10.1.9). We note that (10.1.9) is conformally invariant, so any conformal transformation of $\mathbf{M}^I$ will yield solutions $\check{\varphi}_A, \check{\psi}_{A'}$, $\check{\Phi}_{AA'}$, where the first two parts of the solution are not necessarily constant. The inverse Penrose transform of the special solution (10.1.13), $\{\Phi, \Psi, E, \eta(E)\}$ as in Theorem 10.1.3 are computed explicitly in Eastwood, Penrose and Wells (1981) in algebraic terms, where $\eta(E)$ is the obstruction to third order extension of E corresponding to

$$*(\varphi_A \psi_{A'}(n^a v_a)^2 dz^{AA'}).$$

Here we have taken $\tilde{E}$ to be the trivial bundle on $\mathbf{M}^I$, as usual. The components of the inverse Penrose transform have the form

$$\hat{\psi} = \{\hat{\psi}_{\alpha\beta}\}, \qquad \hat{\varphi} = \{\hat{\varphi}_{\alpha\beta}\}, \qquad E = \{g_{\alpha\beta}\},$$

where $\{U_\alpha\}$ is a suitable covering of $\mathbf{L}^I$ with four open sets (we can *not* get by with two open sets as in the case of $\mathbf{P}^I$), and the functions $\hat{\varphi}_{\zeta\beta}$, $\hat{\psi}_{\alpha\beta}$, are *rational* functions of the natural coordinates on $\mathbf{L} \subset \mathbf{P} \times \mathbf{P}^*$, and $g_{\alpha\beta} = e^{f^{\alpha\beta}}$, where $f_{\alpha\beta}$ is also a rational function of these same coordinates, obtained by explicit integration of the potential $\Phi_{AA'}$. The most difficult calculation was the obstruction of third order η, but the result obtained has the simple form $\eta = \{\eta_{\alpha\beta\gamma}\}$, where

$$\eta_{012} = \eta_{023} = 0$$

$$\eta_{013} = -\eta_{123} = \frac{i}{6} t_0 t_1 t_2 \{\varphi_0 \psi_{0'} + \frac{\pi_{0'}}{\pi_{1'}} \varphi_0 \psi_{1'} \qquad (10.1.14)$$
$$+ \frac{\eta_0}{\eta_1} \varphi_1 \psi_{0'} + \frac{\eta_0 \pi_{0'}}{\eta_1 \pi_{1'}} \varphi_1 \psi_{1'}\},$$

where

$$([\omega^A, \pi_{A'}], [\eta_{A'}, \zeta^{A'}])$$

are the coordinates for $\mathbf{P} \times \mathbf{P}^*$,

$$\mathbf{P}^I = \{(\omega^A, \pi_{A'}): \pi_{A'} \neq 0\},$$
$$\mathbf{P}^{*I} = \{(\eta_{A'}, \zeta^{A'}): \eta_{A'} \neq 0\},$$

and

$$V_0 = \{\pi_{0'} \neq 0\}, \qquad V_1 = \{\pi_{1'} \neq 0\},$$
$$V_0^* = \{\eta_0 \neq 0\}, \qquad V_1^* = \{\eta_1 \neq 0\},$$

is a covering of $\mathbf{P}^I \times \mathbf{P}^{*I}$, and

$$U_0 = (V_0 \times V_0^*) \cap \mathbf{L},$$
$$U_1 = (V_0 \times V_1^*) \cap \mathbf{L},$$
$$U_2 = (V_1 \times V_0^*) \cap \mathbf{L},$$
$$U_3 = (V_1 \times V_1^*) \cap \mathbf{L},$$

is the covering of L^1 for which (10.1.14) was calculated. The functions t_α in (10.1.14) are defined by $t_\alpha = (Z^\alpha W_\alpha)|_{U_\alpha}$, i.e., t_α is the normal coordinate in that coordinate chart with respect to the embedding of L in $P \times P^*$. It is interesting to note that η is defined by only *one* rational function, instead of the four distinct ones which were possible. This is quite reminiscent of the single transition function which defines a Yang–Mills potential in the self-dual case. We do not know if this is spurious, or if there is something deeper here. It would be very interesting to understand deformations of this example. Note that this example is like a plane wave in Minkowski space (or indeed a nonlinear plane wave in an Einstein space-time, see Penrose and Rindler 1986). It has a preferred null direction, and it has superharmonic behavior transversal to the null direction ($\frac{\partial^2 U}{\partial \zeta \partial \tilde\zeta} > 0$ means superharmonic on the $(\zeta, \tilde\zeta)$-plane). The quadratic behavior (or more generally superharmonic behavior) would seem to imply that there is no finite action for these solutions when restricted to real Minkowski space, but one can calculate that, letting $U_a(z) = \frac{\partial U}{\partial z^a}$,

$$F = dA = v_a dU \wedge dz^a$$
$$= v_a U_b dz^a \wedge dz^b,$$

and

$$|F|^2 = F_{ab}F^{ab} = (-v_a U_b)(-v^a U_b)$$
$$= -(v_a v^a)(U_b U^b) = 0,$$

since v_a was a null vector. Thus we see from (10.1.10) that the action integral of these solutions is actually zero.

10.2   Supergeometry and Yang–Mills fields

Supersymmetric transformations of physical field theories were introduced as a 'super group' of the classical Lorentz transformations which would allow transformations from particles of half-integral spin to integral spin and vice-versa. In contemporary physics there are a number of supersymmetric field theories, as well as the rapidly evolving supersymmetric string theories, which are being used to model elementary particles as well as gravity. The supergravity theories are generalizations of Einstein's model of gravity which include

Fermionic fields as a part of the model, and are attempts to quantize gravity as well as to unify gravity with particle physics. There is an extensive literature on the subject, and for a recent survey from the physical point of view we recommend Wess and Bagger (1983). There are two major developments from the mathematical point of view: supermanifolds and super Lie algebras (or graded Lie algebras, Kac-Moody algebras). We refer to the book of Kac (1985) for a review of these generalizations of Lie algebras. The notion of supermanifold (a geometric space on which supersymmetry transformations will make sense) goes back to Berezin, Kostant, and Leites (see the reviews by Berezin and Leites 1975; Kostant 1977; Leites 1974, 1980, 1983; and Manin 1988, for further references to the historic and technical developments of this subject). These spaces play for the group of supersymmetry transformations the same role that Minkowski space and its curved version, Lorentzian manifolds, play for the interaction of physics with the classical Lorentz group. The basic point of view adopted by Berezin, Leites, and Kostant was that a supermanifold is a ringed space of a specific type; that is, it is a topological space with a structure sheaf which is a sheaf of Z_2-graded rings with specific local properties. From this point of view, the development of supermanifolds is very similar to that of complex or algebraic geometry in its modern sheaf-theoretic form. An alternative approach to supermanifolds was initiated by Rogers (1980), and further developed in the papers of Boyer and Gitler (1984), Volovich (1983), and Rothstein (1986), among others. This is similar to a third development due to B. De Witt (1984). We shall leave the discussion of all of these variants of supergeometry to the literature (cf. a recent symposium proceedings in which there were some excellent surveys: Seifert, Clarke and Rosenblum 1984). In this section we want to introduce briefly enough of the theory of supermanifolds in order to be able to describe Witten's fundamental ideas for representing solutions of coupled Yang–Mills–Higgs–Dirac equations in terms of a Penrose transform of super vector bundles on a super twistor manifold in the spirit of Theorem 8.1.14. Part of this is carried out in the book by Manin (1988), and the remainder is contained in a paper by Harnad et al. (1985), as we shall describe in more detail below (see also the survey article by Harnad, Hurtubise and Shnider 1989, and the lecture-note volume by Shnider and Wells 1989). We shall first give the rudiments of superalgebra and supergeometry, and return to the field theory questions somewhat later.

Let $A = A_0 \oplus A_1$ be a $\mathbf{Z}_2$-graded ring. This means that the products in the ring A satisfy $A_0 \cdot A_0 \subset A_0$, $A_1 \cdot A_0 \subset A_1$, $A_1 \cdot A_0 \subset A_1$, and $A_1 \cdot A_1 \subset A_0$. We call elements of A_0 *even* and elements of A_1 *odd*. We call an element a of A *homogeneous* if a is either even or odd, and we define the *degree* of a homogeneous element a to be 0 if a is even and 1 if a is odd. We denote the degree of a homogeneous element a by $|a|$ (not to be confused with the absolute value of a, but it will be clear from the context). A $\mathbf{Z}_2$-graded ring A is *supercommutative* if it satisfies

$$ab = (-1)^{|a||b|}ba,$$

for all homogeneous elements a and b of A. An *A-module T* is a $\mathbf{Z}_2$-graded bimodule which satisfies

$$at = (-1)^{|t||a|}ta,$$

for $a \in A, t \in T$, where $T = T_0 \oplus T_1$, and we are using the same conventions for degree, etc., as in the case of the ring A. There is a natural mapping from A-modules to A-modules called the *parity change* Π (or parity change functor mapping the category of A-modules to the category of A-modules, to use the language of category theory, which we shall not really need). It is defined by

$$\Pi : A\text{-modules} \longrightarrow A\text{-modules}$$

defined by setting

$$(\Pi T)_0 := T_1,$$
$$(\Pi T)_1 := T_0.$$

Note that the ring A is an A-module itself, and as such ΠA is an A-module, but ΠA is no longer a $\mathbf{Z}_2$-graded ring (since $(\Pi A)_1 \cdot (\Pi A)_1 \subset (\Pi A)_1$, for instance).

An example of such a $\mathbf{Z}_2$-graded ring (and its most important example in this context) is given by a Grassmann algebra over a vector space V. Let

$$A = A^*(V) = \sum_p \bigwedge^p(V),$$

for some vector space V defined over the real or complex numbers. In this form A is a *graded* algebra, graded by $p \in \mathbf{Z}$. Let the $\mathbf{Z}_2$-grading of A be defined by

$$A = \sum_{p \text{ even}} \bigwedge^p(V) \oplus \sum_{p \text{ odd}} \bigwedge^p(V)$$
$$= A_0 \oplus A_1.$$

A second, much more trivial example is given by

$$A = A_0 \oplus A_1,$$

where $A_0 = \mathbb{C}$, $A_1 = 0$, i.e., there is no odd part. Any commutative ring can in this same way be considered as a supercommutative $\mathbb{Z}_2$-graded ring.

A *free module of rank $p|q$ over A* is defined by

$$T = A^p \oplus (\Pi A)^q,$$

where $A^p = \bigoplus_{j=1}^p A$ is the direct sum of p copies of the $\mathbb{Z}_2$-module A, and so forth. As an example we could consider $\mathbb{C}^{m|n}$, where we consider, as above, the complex numbers as a $\mathbb{Z}_2$-graded ring. We call this *super complex Euclidean space* of dimension or rank $m|n$.

We want to give some of the formalism which allows us to define supermanifolds in a reasonable fashion. We recall that a ring is a *local ring* if it has a unique maximal ideal. For instance, the rings

$$R_1 = \mathbb{C}[x_1, \ldots, x_n], \quad \text{(polynomials)},$$
$$R_2 = \mathbb{C}[[x_1, \ldots, x_n]], \quad \text{(formal power series)},$$
$$R_3 = \{\text{convergent power series at a point}\}.$$

are all local rings, in each case the maximal ideal $\mathfrak{m}_j$ has the form $\{f \in R_j : f(0) = 0\}$. A *ringed space* is a topological space X equipped with a sheaf of rings $\mathcal{R}$ on X, where the stalks $\mathcal{R}_x$ of $\mathcal{R}$ at the point $x \in X$ are local rings. The sheaf $\mathcal{R}$ is referred to as the *structure sheaf* of the ringed space $(X, \mathcal{R})$ and will be denoted by $\mathcal{R} = \mathcal{R}_X$ if there is more than one topological space involved, as usual. We shall often denote the ringed space simply by X when the structure sheaf is understood. A *morphism* of ringed spaces $(X, \mathcal{R}_X) \longrightarrow (Y, \mathcal{R}_Y)$ is a pair of mappings $(f, f^\#)$ where $f : X \longrightarrow Y$ is a continuous mapping, and $f^\# : \mathcal{R}_Y \longrightarrow f_*^0 \mathcal{R}_X$ is a sheaf homomorphism, where f_*^0 is the 0th direct image under the mapping f. In addition, we require that the induced homomorphism of rings

$$f_x^\# = \mathcal{R}_{Y, f(x)} \longrightarrow \mathcal{R}_{X, x}$$

is local. Namely, if A and B are two local rings with maximal ideals m_A and m_B, then a homomorphism $h : A \longrightarrow B$ is *local* if $h^{-1}(m_B) \subset m_A$.

This is an abstract model for building spaces of various kinds. Let us now give some examples. If X is a topological space, a differentiable manifold, or a complex manifold, then X has the natural structure of a ringed space. We simply let $\mathcal{R}$ be either $\mathcal{C}$, $\mathcal{E}$, or $\mathcal{O}$, the sheaves of continuous, differentiable, or holomorphic functions on X, respectively. We see that in each case here the morphism has the form (f, f^*), where f is either a continuous, differentiable, or holomorphic mapping, and $f^\# := f^*$ is the usual induced pullback mapping on functions. If we only dealt with these three cases, we would not need to introduce the abstract notion of ringed space. However, we can use the ringed space notion to define more complicated spaces, which we can treat like ordinary manifolds in many ways, using the ringed space formalism. These include spaces with singularities, formal neighborhoods (discussed in §10.1), schemes in algebraic geometry, and supermanifolds. We shall discuss formal neighborhoods and singularities as prototypes of our definition of supermanifolds. We shall not discuss schemes, as that would take us too far afield at the present time.

If Y is a complex submanifold of X, the spaces

$$(Y, \mathcal{O}_Y^{(k)}) = (Y, \mathcal{O}_X / \mathcal{I}_Y^{k+1})$$

are a countable sequence of ringed spaces which have nilpotent elements in their stalks (for $k > 0$). Namely, if f is a function on X which vanishes on Y to first order, then f will define a nonzero element f_x in $\mathcal{O}_{Y,x}^{(k)}$, and this is an example of a nilpotent element in the local ring $\mathcal{O}_{Y,x}^{(k)}$. So the topological space Y stays the same, but the sheaves $\mathcal{O}_Y^{(k)}$ are a sequence of sheaves with increasingly more information about the behavior of Y in the normal direction with respect to its embedding in X.

In a similar vein, if we have a complex manifold X, then the sheaves $\mathcal{O}_X \subset \mathcal{E}_X \subset \mathcal{C}_X$ each determine the complex, differentiable, and topological structures on the same topological space X. We shall make this more precise in the following manner. We say that two ringed spaces $(X, \mathcal{R}_X)$ and $(Y, \mathcal{R}_Y)$ are *isomorphic* if there is a morphism $(f, f^\#)$ such that f is a homeomorphism of spaces and $f^\#$ is an isomorphism of sheaves. We state the following proposition in the context of complex manifolds, but it is also true in the category of differentiable, real-analytic, and other classes of manifolds.

Proposition 10.2.1. *Let* $(X, \mathcal{O}_X)$ *be a ringed space with the property that for each point* $x \in X$ *there is a neighborhood* U *of* x *such that there is a ringed space isomorphism,*

$$(U, \mathcal{O}_X|_U) \cong (U', \mathcal{O}_{\mathbf{C}^n}),$$

where U' *is an open set in* $\mathbf{C}^n$, *and* $\mathcal{O}_{\mathbf{C}^n}$ *is the sheaf of holomorphic functions on* $\mathbf{C}^n$, *then* X *can be given the structure of a complex manifold, and every complex manifold arises in the manner.*

This is not difficult to prove, and we leave it to the reader. On any complex manifold X the sheaf of holomorphic functions will give X the structure of a ringed space, as we have observed before. The local isomorphism hypothesized in Proposition 10.2.1 gives a system of local coordinates, and it is necessary to check that the induced changes of coordinates for two such local isomorphisms are holomorphic mappings. This is not difficult, when you unravel the definitions.

The proposition says, in effect, that we can define a complex manifold by simply giving a global sheaf and a local (topologically local!) isomorphism to a *specific model* of a ringed space.

Let us give a local model for a space with a singularity. Let U be an open set in $\mathbf{C}^n$, and let $f_1, \ldots, f_r$ be holomorphic functions defined in U. Let

$$V = \{x \in U : f_1 = \cdots = f_r = 0\},$$

then we say that V is an *analytic subvariety* of U. If $df_1 \wedge \cdots \wedge df_r \neq 0$ on V, then V would be nonsingular and a submanifold of U. But, at points where $df_1 \wedge \cdots \wedge df_r = 0$, then the subvariety has a singular point. For instance, if $U = \mathbf{C}^n, f = z_1^2 - z_2^3$, then $0 = (0,0)$ is a singular point (this is depicted in Figure 10.2.1). We let $\mathcal{I}_V := \{f \in \mathcal{O}_U : f|_V = 0\}$ be the ideal sheaf of all functions in $\mathcal{O}_U$ which vanish on V. We define

$$\mathcal{O}_V := \mathcal{O}_U / \mathcal{I}_V$$

to be the *structure sheaf* of the subvariety V. Now the stalks of $\mathcal{O}_V$ at points $x \in V$ are still local rings, but the local ring $\mathcal{O}_{V,x}$ at a singular point will be more complicated than the ring of power series in a certain number of variables (which is the case at the nonsingular points). The structure of the ring *reflects* in an algebraic manner, the character of the singularity. Modern algebraic and analytic geometry analyze the structure of singularities by analyzing the structure of the local ring of the structure sheaf at the singular point.

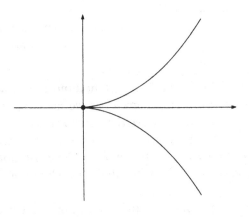

Fig. 10.2.1. The analytic space $z_1^2 - z_2^3 = 0$.

An *analytic space* is a ringed space $(X, \mathcal{O}_X)$ which is locally isomorphic to the local model $(V, \mathcal{O}_V)$, where V is a subvariety of an open set $U \subset \mathbf{C}^n$. This is exactly the same, using Proposition 10.2.1, as the definition of a complex manifold, but allowing for singularities in the local model. Since an analytic space *does not* have local coordinates at every point, it is more difficult to define it in terms of local coordinate systems and overlapping mappings which are holomorphic, the original definition of complex manifolds which goes back to Weyl in the 1920s. A *scheme* is similar to an analytic space, but the local model is different, and the underlying topological space need not be Hausdorff. On an analytic subvariety, one can identify the points of the space with the maximal ideals defined by functions vanishing at a point. In a scheme the points of the underlying space are the prime ideals in a ring, thus endowing it with a richer structure. The open sets are defined as complements of zero sets of polynomials.

The arithmetic nature of the solutions of the polynomial equations plays a role in the geometry of the space, and the ringed space point of view allows for a strong interaction of the arithmetic and geometric aspects of varieties defined by polynomial equations (the concept goes back to Grothendieck in the 1950s; see Hartshorne 1977 for a detailed discussion of schemes).

After a considerable digression we now want to give a local model for a supermanifold. Let V be a complex vector space, let $\bigwedge^* V$ be the graded exterior algebra of V, and let $\bigwedge^* V = \bigwedge_0 \oplus \bigwedge_1 V$ be the decomposition into forms of even and odd degree, making $\bigwedge^* V$ into a $\mathbf{Z}_2$-algebra. Let U be an open set in $\mathbf{C}^n$, and define

$$\mathcal{A} = \mathcal{A}_{\mathbf{C}^n, V} := \mathcal{O}_{\mathbf{C}^n} \otimes_{\mathbf{C}} \textstyle\bigwedge^* V = \mathcal{O}_{\mathbf{C}^n}(\textstyle\bigwedge^* V).$$

Thus $\mathcal{A} = \mathcal{A}_0 \oplus \mathcal{A}_1$, and is a $\mathbf{Z}_2$-graded sheaf of supercommutative rings (in fact, algebras), consisting of local $\bigwedge^* V$-valued holomorphic functions. If $(z^1, \ldots, z^n)$ are coordinates in $\mathbf{C}^n$, and $\xi_1, \ldots, \xi_m$ are generators for $V \cong \bigwedge^1 V$, then $(z^1, \ldots, z^n; \xi_1, \ldots, \xi_m)$ are an even-odd system of coordinates for the space $(U, \mathcal{O}_U(\bigwedge^* V))$, where U is an open set in $\mathbf{C}^n$. If we have any $f \in \Gamma(U, \mathcal{O}_U(\bigwedge^* V))$, then we can express

$$f(z, \xi) = \sum_\alpha f_\alpha(z)\xi^\alpha, \quad f_\alpha(z) \in \mathcal{O}(U),$$

where $\xi^\alpha = \xi_1^{\alpha_1} \ldots \xi_m^{\alpha_m}$, the product being in the (noncommutative) ring $\bigwedge^* V$. These are the fundamental functions of supergeometry. We could define the same thing for an open set in $\mathbf{R}^n$, replacing holomorphic by differentiable functions.

Let X be a topological space, and let $\mathcal{O}$ be a sheaf of $\mathbf{Z}_2$-graded supercommutative rings on X. Let $\mathcal{N}$ be the ideal sheaf in $\mathcal{O}$ of all nilpotent elements in $\mathcal{O}$, and let

$$\mathcal{O}_{\mathrm{red}} := \mathcal{O}/\mathcal{N}.$$

Then we say that $(X, \mathcal{O})$ is a *complex supermanifold of dimension* $n|m$ if:

(a) $(X, \mathcal{O}_{\mathrm{red}})$ is a complex manifold, and

(b) for each point $x \in X$, there is a neighborhood U of x such that

$$\mathcal{O}|_U \cong \mathcal{O}_{\mathrm{red}}|_U(\textstyle\bigwedge^* \mathbf{C}^m).$$

We can also give the same definition in the smooth or real-analytic case. Let us give a simple example. Let $E \to X$ be a holomorphic vector bundle over a complex manifold X with structure sheaf $\mathcal{O}_{\text{hol}}$, and define $\mathcal{O} := \mathcal{O}_{\text{hol}}(\bigwedge^* E)$ to be a $\mathbb{Z}_2$-graded sheaf over X, then $(X, \mathcal{O})$ is a supermanifold, since locally $E \cong U \times \mathbb{C}^m$, for some m, and hence locally $\mathcal{O} \cong \mathcal{O}_{\text{hol}}(\bigwedge^* \mathbb{C}^m)$ as desired. Clearly in this case we have $\mathcal{O}/\mathcal{N} = \mathcal{O}_{\text{hol}}$, the usual holomorphic functions on X. Note that there is always a well-defined morphism of supermanifolds (as ringed spaces) of the form $i : X_{\text{red}} \to X$, where

$$i : (X, \mathcal{O}_{\text{red}}) \longrightarrow (X, \mathcal{O}),$$

where i is induced by the natural sheaf projection $\mathcal{O} \longrightarrow \mathcal{O}/\mathcal{N}$ and the identity mapping on X. If $(X, \mathcal{O})$ is a supermanifold of the form $(X, \mathcal{O}_{\text{red}}(\bigwedge^* E))$, for some vector bundle E over X, then there is a mapping of supermanifolds (i.e., as ringed spaces) of the form $p : X \longrightarrow X_{\text{red}}$, given by the pair

$$(\text{id}_X, \mathcal{O}_X \overset{\iota}{\to} \mathcal{O}_X(\bigwedge^* E)),$$

where ι is the natural injection of 0-forms into the algebra of all forms. We call such a manifold a *split* manifold and ι is the splitting mapping. We see that a supermanifold is by definition a locally split manifold. If we consider differentiable supermanifolds, then we have the following theorem due to Batchelor (1979).

Theorem 10.2.2. *Any differentiable supermanifold $(X, \mathcal{E})$ is of the form $(X, \mathcal{E}_{\text{red}}(\bigwedge^* E))$ for some differentiable vector bundle E over X.*

The proof of this depends on the existence of a partition of unity. It is not true in the holomorphic category. There are examples of supermanifolds which are not globally split in a holomorphic fashion (Green 1982, Rothstein 1985, Manin 1988).

Starting with a super Euclidean space, say $\mathbb{C}^{n|N}$, then we can construct a large class of supermanifolds whose reductions will be the classical Grassmannian and flag manifolds.

Let

$$\mathbf{G}(p|q; \mathbb{C}^{n|N})$$

be the set of all submodules of the module $\mathbb{C}^{n|N}$ of rank $p|q$. This set can be given the structure of a supermanifold of dimension

$$p(n - p)|q(N - q),$$

and its reduction is the usual Grassmannian $G(p; \mathbf{C}^n)$ of p-planes in n-space. More generally, we can define the super flag manifolds

$$F(p_1|q_1, \ldots, p_r|q_r; \mathbf{C}^{n|m}),$$

the set of all r-tuples $(L_1, \ldots, L_r)$ of free submodules of $\mathbf{C}^{n|m}$ satisfying

$$L_1 \subset \cdots \subset L_r \subset \mathbf{C}^{n|m},$$

and

$$\operatorname{rank} L_j = p_j|q_j.$$

These can also be given suitable coordinate systems (and hence a suitable structure sheaf!), which makes this set into a supermanifold (these constructions are described more fully in Ferber 1978, Manin 1988, Harnad, Hurtubise and Shnider 1989, and Shnider and Wells 1989).

The basic double fibration for the representation of Yang–Mills fields in general (§10.1) was

$$
\begin{array}{ccc}
 & \mathsf{F} & \\
\swarrow & & \searrow \\
\mathsf{L} & & \mathsf{M},
\end{array}
\tag{10.2.1}
$$

and we can ask for all double fibrations of supermanifolds

$$
\begin{array}{ccc}
 & L & \\
\swarrow & & \searrow \\
L & & M,
\end{array}
\tag{10.2.2}
$$

whose reduction will agree with (10.2.1). Manin gives a complete classification of such double fibrations which arise naturally as flag manifolds from a given super twistor space, and we shall describe some of these now.

Let $\mathsf{T}^{4|N}$ be a free C-module of rank $4|N$, and we call this a *super twistor space of rank* $4|N$ (or N-extended twistor space). We may require $\mathsf{T}^{4|N}$ to be equipped with specific quadratic forms of various types, just as we did for the usual twistor spaces in Chapter 1. Manin has a description of the quadratic forms one needs. We shall use this super twistor space to create associated super twistor manifolds just as in Chapter 1. First we shall list some super twistor manifolds whose reduction is $\mathsf{M} = F(2; \mathsf{T}^4) = \mathsf{F}_2$, our usual complexified compactified Minkowski space.

Namely, one has

$$F(2|0, 2|N; \mathsf{T}^{4|N}), \qquad \dim 4|4N,$$
$$F(2|0, 4|0; \mathsf{T}^{4|N}), \qquad \dim 4|4N,$$
$$F(0|N, 2|N; \mathsf{T}^{4|N}), \qquad \dim 4|2N,$$
$$F(2|0; \mathsf{T}^{4|N}), \qquad \dim 4|2N,$$
$$F(2|N; \mathsf{T}^{4|N}), \qquad \dim 4|4N.$$

One can add to this list by using the parity change mapping Π, as well as by looking at isotropic submanifolds of larger twistor flag manifolds defined by certain quadratic forms. The basic pattern which emerges is quite clear: if one ignores the 'odd part' of the super flag manifold under consideration, and takes into account the obvious degeneracies, the spaces reduce mnemonically to the usual space $F(2; \mathsf{T}^4)$, e.g.,

$$F(2|0, 2|0; \mathsf{T}^{4|0}) = F(2; \mathsf{T}^4).$$

There is one particular triple, which appears in Witten's paper (1978) which was the impetus for a lot of work. Namely,

$$L = L^{5|2N} := F(1|0, 3|N; \mathsf{T}^{4|N}),$$
$$F = F^{6|4N} := F(1|0, 2|0, 3|N; \mathsf{T}^{4|N}),$$
$$M = M^{4|4N} := F(2|0, 2|N; \mathsf{T}^{4|N}).$$

This particular double fibration (10.2.2) has the reduction to (10.2.1), and corresponds to the most important supersymmetry and supergeometry models of physicists. In particular, $M^{4|4N}$ is compactified super Minkowski space. The group of matrices $GL(\mathsf{T}^{4|N})$ act naturally on super twistor space. By looking at a suitable subgroup G which preserves a specific quadratic form Φ on $\mathsf{T}^{4|N}$ (a generalization of the action of $SU(2,2)$ acting on the classical twistor space), we obtain the super conformal group acting on $M^{4|4N}$, a subgroup of which is the super Poincaré group. This is the group of supersymmetry transformations acting on affine super Minkowski space $M_{\text{Aff}}^{4|4N}$ with local coordinates $(x^{AA'}, \theta^{iA}, \theta_j^{B'})$, $i, j = 1, \ldots, N$, where $x^{AA'}$ are the normal Minkowski space coordinates, and θ^{iA} and $\theta_j^{B'}$ are odd spinorial coordinates. These spinor coordinates come naturally from the same type of universal bundle construction for the spin bundles as given in §7.1. Recall how $\mathbb{F}_{12}$ had local coordinates of the form $(z^{AA'}, \pi_{A'})$. This is a similar type of construction in this case (see Manin 1988, Harnad, Hurtubise and Shnider 1989, and Shnider and Wells 1989 for more details concerning super Minkowski space, its local structure, and the groups which act on them).

A holomorphic vector bundle on a supermanifold $(X, \mathcal{A})$ is defined simply as a locally free sheaf of $\mathcal{A}$-modules. In the double fibration discussed above,

$$
\begin{array}{ccc}
 & F^{6|4N} & \\
\mu \swarrow & & \searrow \nu \\
L^{5|2N} & & M^{4|4N},
\end{array}
$$

the fibers of μ are of dimension $1|2N$, with reduction isomorphic to $\mathbf{P}_1$, while the fibers of ν are of dimension $2|0$ and are isomorphic to $\mathbf{P}_1 \times \mathbf{P}_1$ (no reduction is necessary). Let U be an open set in $M^{4|4N}$, and let $\widehat{U}$ be $\mu \circ \nu^{-1}(U)$, as usual, then a Yang–Mills holomorphic vector bundle $E \to \widehat{U}$ will have a Penrose transform $(\widetilde{E}, D)$, of the form

$$
\widetilde{E} = \nu^0_* \mu^* E,
$$

and D is the induced superconnection.

$$
D : \Gamma(U, \widetilde{D}) \to \Omega^1(U, \widetilde{E}),
$$

where Ω^1 is the sheaf of one-forms in the setting (defined as the dual to the tangent sheaf of X, which in turn is defined as the sheaf of superderivations of $\mathcal{A}$).

Thus the superconnection D has the local form $\nabla + \Phi$ and the *components* of Φ and its curvature F_D can be identified with fields which satisfy certain differential equations on U. The components of the curvature satisfy the *constraint equations* of Witten (1978). They are derived by observing that the curvature $F_D = D \circ D$ must vanish on the super null lines (the fibers of μ). These fibers have one-dimensional reduction (the usual null lines), but have $2N$-dimensional odd part. Thus, requiring the curvature to vanish on a $2N$-extended one-dimensional manifold is a nontrivial condition, as opposed to the lack of such constraint in the classical null line setting (Theorem 10.1.2). Witten's major observation is that the components of the superconnection, suitably identified (and classically denoted by

$$
\{A_a(x, \theta), \psi(x, \theta), \varphi(x, \theta)\},
$$

where these are Lie algebra-valued fields) must satisfy the supersymmetric analogy of the Yang–Mills–Dirac–Higgs equations on super Minkowski space.

The leading components of these superfields have expansions of the form

$$A_a(x,\theta) = A_a^0(x) + A_a^\alpha(x)\theta_\alpha + \cdots,$$
$$\psi(x,\theta) = \psi^0(x) + \psi^\alpha(x)\theta_\alpha + \cdots,$$
$$\varphi(x,\theta) = \varphi^0(x) + \varphi^\alpha(x)\theta_\alpha + \cdots$$

and will satisfy the same field equations on the classical Minkowski space M^4. The paper by Harnad et al. (1985) shows that any solution of the classical equations on M^4 consists of the leading terms of a superfield expansion on $M^{4|4N}$ which satisfy the supersymmetric Yang–Mills equations in the larger space. Moreover, this solution corresponds precisely to the same fields which satisfy Witten's constraint equation (cf. Harnad, Hurtubise and Shnider 1989 and Shnider and Wells 1989). Such a solution corresponds, by the work of Witten and Manin, to a holomorphic super vector bundle of Yang–Mills type, that is, it is trivial on the fibers $\hat{x} = \mu \circ \nu^{-1}(x)$, for $x \in U$. Manin gives some examples of constructions of such bundles in his book (1988), utilizing a generalization of the Barth–Horrocks monad construction.

References

Adler, M. and van Moerbeke, P. (1980), Completely integrable systems, Euclidean Lie algebras, and curves, *Adv. Math.* **38**, 257–317.

Adler, M. and van Moerbeke, P. (1980), Linearization of Hamiltonian systems, Jacobi varieties and representation theory, *Adv. Math.* **38**, 318–379.

Ahlfors, L. V. (1953), *Complex Analysis*, McGraw-Hill, New York.

Athorne, C. (1982), Factorization of an SU(3) monopole patching function, *Phys. Lett.* **B117**, 55–58.

Athorne, C. (1983), Cylindrically and spherically symmetric monopoles in SU(3)-gauge theory, *Comm. Math. Phys.* **88**, 43–62.

Atiyah, M. F. (1979), *Geometry of Yang–Mills Fields*, Scuola Normale Superiore, Pisa.

Atiyah, M. F. (1979a), Real and complex geometry in four dimensions, *in: The Chern Symposium 1979*, Springer-Verlag, Berlin, New York.

Atiyah, M. F. (1981), Green's functions for self-dual 4-manifolds, *in: Mathematical Analysis and Applications*, Acad. Press, New York, pp. 129–158.

Atiyah, M. F. (1982), Geometry of monopoles, *in: Monopoles in Quantum Field Theory*, N. S. Craigie, P. Goddard, and W. Nahm, eds., World Scientific, Singapore, pp. 3–20.

Atiyah, M. F. (1984), Instantons in two and four dimensions, *Comm. Math. Phys.* **93**, 437–451.

Atiyah, M. F., Bott, R. and Shapiro, A. (1964), Clifford modules, *Topology* **3**, 3–38.

Atiyah, M. F., Hitchin, N. J., Drinfeld, V. G. and Manin, Yu. I. (1978), Construction of instantons, *Phys. Lett.* **A65**, 185–187.

Atiyah, M. F., Hitchin, N. J. and Singer, I. M. (1978), Self-duality in four-dimensional Riemannian geometry, *Proc. Roy. Soc.* **A362**, 425–461.

Atiyah, M. F. and Ward, R. S. (1977), Instantons and algebraic geometry, *Comm. Math. Phys.* **55**, 111–124.

Bailey, T. N. (1985), Twistors and fields with sources on worldlines, *Proc. Roy. Soc. London* **A397**, 143–155.

Bailey, T., Ehrenpreis, L. and Wells, Jr., R. O. (1982), Weak solutions of the massless field equations, *Proc. Roy. Soc. London* A **384**, 403–425.

Barth, W. (1977), Some properties of stable rank-2 vector bundles on P_n, *Math. Ann.* **226**, 125–150.

Barth, W. (1982), Lectures on mathematical instanton bundles, *in: Gauge Theories: Fundamental Interactions and Rigorous Results*, P. Dita, V. Georgescu, and R. Purice, eds., Birkhäuser, Boston, pp. 177–206.

Barth, W. and Hulek, K. (1978), Monads and moduli of vector bundles, *Manuscr. Math.* **25**, 323–347.

Batchelor, M. (1979), The structure of supermanifolds, *Trans. Amer. Math. Soc.* **253**, 329–338.

Bateman, H. (1904), The solution of partial differential equations by means of definite integrals, *Proc. Lond. Math. Soc. (2)* **1**, 451–458.

Beem, J. K. and Ehrlich, P. E. (1981), *Global Lorentzian Geometry*, Marcel Dekker, New York.

Belavin, A. A., Polyakov, A. M., Schwarz, A. S. and Tyupkin, Yu. S. (1975), Pseudoparticle solutions of the Yang–Mills equations, *Phys. Lett.* **59B**, 85–87.

Berezin, F. A. and Leites, D. A. (1975), Supermanifolds, *Soviet. Math. Dokl.* **16**, 1218–1222.

Bernard, C. W., Christ, N. H., Guth, A. H. and Weinberg, E. J. (1977), Pseudoparticle parameters for arbitrary gauge groups, *Phys. Rev.* **D16**, 2967–2977.

Bjorken, J. D. and Drell, S. D. (1965), *Relativistic Quantum Fields*, McGraw-Hill, New York.

Blanchard, A. (1956), Sur les variétés analytiques complexes, *Ann. Sci. Norm. Sup. Paris* **73**, 157–202.

Bochner, S. and Martin, W. T. (1948), *Several Complex Variables*, Princeton University Press, Princeton, N.J.

Bourguignon, J.-P. and Lawson, Jr., H. B. (1981), Stability and isolation phenomena for Yang–Mills fields, *Comm. Math. Phys.* **79**, 189–230.

Bowman, M. C. (1983), A description of E. Weinberg's continuous family of monopoles using the ADHMN formalism, *Phys. Lett.* **133B**, 344–346.

Bowman, M. C., Corrigan, E., Goddard, P., Puaca, A. and Soper, A. (1984), The construction of spherically symmetric monopoles using the ADHMN formalism, *Phys. Rev.* **D12**, 3100–3112.

Boyer, C. P., Finley, III, J. D. and Plebanski, J. F. (1980), Complex general relativity, H and HH spaces—a survey of one approach, *in: General Relativity and Gravitation, Vol. II*, A. Held, ed., Plenum, New York, pp. 241–281.

Boyer, Charles P. and Gitler, Samuel (1984), The theory of G-infinity supermanifolds, *Trans. Amer. Math. Soc.* **285**, 241–267.

Bredon, G. E. (1967), *Sheaf Theory*, McGraw-Hill, New York.

Brown, L., Carlitz, R. and Lee, C. (1977), Massless excitations in pseudoparticle fields, *Phys. Rev.* **D16**, 417–422.

Brown, S. A., Panagopoulos, H. and Prasad, M. K. (1982), Two separated SU(2) Yang–Mills–Higgs monopoles in the Atiyah–Drinfeld–Hitchin–Manin–Nahm construction, *Phys. Rev.* **D26**, 854–863.

Bryant, R. L. (1982), Conformal and minimal immersions of compact surfaces into the 4-sphere, *J. Diff. Geom.* **17**, 455–473.

Bryant, R. L. (1985), Lie groups and twistor spaces, *Duke Math. J.* **52**, 223–261.

Buchdahl, N. P. (1985), Analysis on analytic space and non-self-dual Yang–Mills fields, *Trans. Amer. Math. Soc.* **288**, 431–469.

Burzlaff, J. (1982), Transition matrices of self-dual solutions to SU(N)-gauge theory, *Lett. Math. Phys.* **6**, 141–146.

Campbell, P. (1979), Projective geometry, Lagrangian subspaces, and twistor theory, *Internat. J. Theor. Phys.* **18**, 9–15.

Cheng, T. P. and Li, L. F. (1984), *Gauge Theory of Elementary Particles*, Clarendon Press, Oxford.

Chern, S. S. (1967), *Complex Manifolds Without Potential Theory*, Van Nostrand Reinhold, New York.

Christ, N. H. (1979), Self-dual Yang–Mills solutions, *in: Complex Manifold Techniques in Theoretical Physics*, D. E. Lerner and P. D. Sommers, eds., Pitman, London, pp. 45–54.

Christ, N. H., Weinberg, E. J. and Stanton, N. K. (1978), General self-dual Yang–Mills solutions, *Phys. Rev.* **D18**, 2013–2025.

Cochina, A. (1972), On the null hypersurfaces of twistor metric, *Rev. Roumaine Math. Pures Appl.* **17**, 857–863.

Corrigan, E. F. (1982), Monopole Solitons, *Geometric Techniques in Gauge Theories (Scheveningen, 1981)*, Springer-Verlag, Berlin, New York, pp. 160–178.

Corrigan, E. and Fairlie, D. B. (1977), Scalar field theory and exact solutions to a classical SU(2)-gauge theory, *Phys. Lett.* **67B**, 69–71.

Corrigan, E. F., Fairlie, D. B., Templeton, S. and Goddard, P. (1978a), A Green function for the general self-dual gauge field, *Nucl. Phys.* **B140**, 31–44.

Corrigan, E. F., Fairlie, DM. F. B., Yates, R. G. and Goddard, P. (1978b), The construction of self-dual solutions to $SU(2)$-gauge theory, *Comm. Math. Phys.* **58**, 223–240.

Corrigan, E. and Goddard, P. (1981), An n monopole solution with $4n - 1$ degrees of freedom, *Comm. Math. Phys.* **80**, 575–587.

Corrigan, E. F. and Goddard, P. (1984), Construction of instanton and monopole solutions and reciprocity, *Ann. Phys.* **154**, 253–279.

Corrigan, E., Goddard, P. and Kent, A. (1985), Some comments on the ADHM construction in $4k$ dimensions, *Comm. Math. Phys.* **100**, 1–14.

Courant, R. and Hilbert, D. (1965), *Methods of Mathematical Physics, 2 vols.*, Interscience, New York.

Curtis, G. E. (1978), Twistors and multipole moments, *Proc. Roy. Soc. London* **A359**, 133–149.

Curtis, W. D., Lerner, D. E. and Miller, F. R. (1978), Complex pp waves and the nonlinear graviton construction, *J. Math. Phys.* **19**, 2024–2027.

Curtis, W. D., Lerner, D. E. and Miller, F. R. (1979), Some remarks on the nonlinear graviton, *Gen. Rel. Grav.* **10**, 557–565.

Daniel, M., Mitter, P. K. and Viallet, C. M. (1978), Local stability of deformations of self-dual Yang–Mills fields on S^4, *Phys. Lett.* **77B**, 77–79.

De Witt, B. S. (1984), *Supermanifolds*, Cambridge University Press.

De Rham, Georges (1960), *Variétés Différentiable*, Hermann, Paris.

Dodziuk, J. and Min-Oo, (1982), An L_2-isolation theorem for Yang–Mills over complete manifolds, *Compositio Math.* **47**, 165–169.

Donaldson, S. K. (1983), Self-dual connections and the topology of smooth 4-manifolds, *Bull. AMS* **8**, 81–83.

Donaldson, S. K. (1984a), Instantons and geometric invariant theory, *Comm. Math. Phys.* **93**, 453–460.

Donaldson, S. K. (1984b), Nahm's equations and the classification of monopoles, *Comm. Math. Phys.* **96**, 387–407.

Drinfeld, V. G. and Manin, Yu. I. (1978a), Self-dual Yang–Mills fields over a sphere, *Funct. Anal. Appl.* **12**, 140–142.

Drinfeld, V. G. and Manin, Yu. I. (1978b), A description of instantons, *Comm. Math. Phys.* **63**, 177–192.

Drinfeld, V. G. and Manin, Yu. I. (1979), Instantons and bundles on CP^3, *Funct. Anal. Appl.* **13**, 124–134.

Eastwood, M. G. (1979), On raising and lowering helicity, *Twistor Newsletter* **8**, 37–38.

Eastwood, M. G. (1981), On the twistor description of massive fields, *Proc. Roy. Soc. London* **A374**, 431–445.

Eastwood, M. G. (1985), The generalized Penrose–Ward transform, *Math. Proc. Camb. Phil. Soc.* **97**, 165–187.

Eastwood, M. G. and Ginsberg, M. L. (1981), Duality in twistor theory, *Duke Math. J.* **48**, 177–196.

Eastwood, M. G., Penrose, R. and Wells, Jr., R. O. (1981), Cohomology and massless fields, *Comm. Math. Phys.* **78**, 305–351.

Eastwood, M. G., Pool, R. and Wells, Jr., R. O. (1985), The inverse Penrose transform of a solution to the Maxwell–Dirac–Weyl field equations, *J. Funct. Anal.* **60**, 16–35.

Eastwood, M. G. and Tod, P. (1982), Edth–a differential operator on the sphere, *Math. Proc. Cambridge Philos. Soc.* **92**, 317–330.

Eguchi, T. and Freund, P.G.O. (1976), Quantum gravity and world topology, *Phys. Rev. Letts.* **37**, 1251–1254.

Eguchi, T., Gilkey, P. B. and Hanson, A. J. (1978), Topological invariants and absence of an axial anomaly for a Euclidean Taub-NUT metric, *Phys. Rev.* **D17**, 423–427.

Eguchi, T. and Hanson, A. J. (1978), Asymptotically flat self-dual solutions to Euclidean gravity, *Phys. Lett.* **74B**, 249–251.

Ferber, Alan (1978), Supertwistors and conformal supersymmetry, *Nuclear Phys.* **B132**, 55–64.

Flaherty, Jr., E. J. (1978), The nonlinear graviton in interaction with a photon, *Gen. Rel. Grav.* **9**, 961–978.

Flume, R. (1978), A local uniqueness theorem for (anti-) self-dual solutions of SU(2) Yang–Mills equations, *Phys. Lett.* **76B**, 593–596.

Forger, M. (1979), Gauge theories, instantons and algebraic geometry, *Rep. Math. Phys.* **16**, 359–384.

Fröhlicher, A. (1955), Relations between the cohomology groups of Dolbeault and topological invariants, *Proc. Nat. Acad. Sci., U.S.A.* **41**, 641–644.

Fulton, W. (1969), *Algebraic Curves*, W. A. Benjamin, New York.

Ganoulis, N., Goddard, P. and Olive, D. (1982), Self-dual monopoles and Toda molecules, *Nucl. Phys.* **B205 [FS5]**, 601–636.

Gelfand, I. M., Gindikhin, S. G. and Graev, M. I. (1980), Integral geometry in affine and projective spaces, *Akad. Nauk SSSR,*

Vsesoyuz. Inst. Nauchn. i Tekhn. Informatsii, Moscow.

Gibbons, G. W. and Hawking, S. W. (1977), Action integrals and partition functions in quantum gravity, *Phys. Rev.* **D15**, 2752–2756.

Gibbons, G. W. and Hawking, S. W. (1978), Gravitational multi-instantons, *Phys. Lett.* **78B**, 430–432.

Gibbons, G. W. and Perry, M. J. (1980), New gravitational instantons and their interactions, *Phys. Rev.* **D22**, 313–321.

Gibbons, G. W. and Pope, C. N. (1978), CP2 as a gravitational instanton, *Comm. Math. Phys.* **61**, 239–248.

Gindikhin, S. G. (1980), Twistor representation of the solutions of Maxwell's system (Russian), *Funkcional. Anal. i Prilozhen.* **14**, 64–66.

Gindikhin, S. G. (1981), Twistor representation of a non-self-dual metric, *Funkcional. Anal. i Prilozhen.* **15**, 69–71.

Gindikhin, S. G. (1982a), On a description of the self-dual solutions of Einstein's equation with a cosmological term, *Funkcional. Anal. i Prilozhen* **16**, 64–65.

Gindikhin, S. G. (1982b), *Integral Geometry and Twistors*, Springer, Berlin, New York.

Gindikhin, S. G. and Henkin, G. M. (1981), The Penrose transform and complex integral geometry, *Akad. Nauk SSSR, Vsesoyuz. Inst. Nauchn. i Tekhn. Informatsii*, Moscow.

Gindikin, S. G. and Henkin, G. M. (1980), Complex integral geometry and the Penrose transformation for solutions of Maxwell equations (Russian), *Teoret. Mat. Fiz.* **43**, 18–31.

Ginsberg, M. L. (1983), Scattering theory and the geometry of multitwistor spaces, *Trans. Amer. Math. Soc.* **276**, 789–815.

Glimm, J. and Jaffe, A. (1981), *Quantum Physics*, Springer-Verlag, New York.

Goddard, P. and Olive, D. I. (1978), Magnetic monopoles in gauge field theories, *Rep. Prog. Phys.* **41**, 1357–1437.

Godement, R. (1964), *Topologie Algébrique et Théorie des Faisceaux*, Hermann, Paris.

Goldberg, J. N. (1979), Self-dual gauge fields, *Phys. Rev.* **D20**, 1909–1914.

Goldstein, N. (1982), Conull hypersurfaces in Minkowski space, *Proc. Amer. Math. Soc.* **85**, 531–532.

Grauert, H. (1960), *Ein Theorem der analytischen Garbentheorie und die Modulräume komplexer Strukturen*, #5, Inst. Hautes Etudes.

Green, P. (1982), On holomorphic graded manifolds, *Proc. Amer. Math. Soc.* **85**, 587–590.

Greenberg, M. J. and Harper, John R. (1981), *Algebraic Topology*, Benjamin/Cummings, Reading, Mass.

Griffiths, P. A. (1965), The extension problem for compact submanifolds of complex manifolds *I*, *Proc. Conf. Complex Analysis, Minneapolis 1964*, Springer-Verlag, New York, pp. 113–142.

Griffiths, P. A. (1966), The extension problem in complex analysis II; embeddings with positive normal bundle, *Amer. J. Math.* **88**.

Griffiths, P. A. and Harris, J. (1978), *Principles of Algebraic Geometry*, Wiley, New York.

Guillemin, Victor W., Kashiwara, Masaki and Kawai, Takahiro (1979), *Seminar on Micro-Local Analysis*, Princeton University Press, Princeton, N.J.

Gunning, R. C. and Rossi, Hugo (1965), *Analytic Functions of Several Complex Variables*, Prentice-Hall, Englewood Cliffs, N.J.

Hansen, R. O., Newman, E. T., Penrose, R. and Tod, K. P. (1978), The metric and curvature properties of *H*-space, *Proc. Roy. Soc. London* **A363**, 445–468.

Harnad, J., Hurturbise, J., Legaré, M. and Shnider, S. (1985), Constraint equations and field equations in supersymmetric $N = 3$ Yang–Mills theory, *Nucl. Phys.* **B256**, 609.

Harnad, J., Hurturbise, J., Shnider, S. (1989), Supersymmetric Yang–Mills Equations and Supertwistors, *Ann. Phys.* **(in press, June 1989)**.

Hartshorne, R. (1977), *Algebraic Geometry*, Springer-Verlag, Berlin, New York.

Hartshorne, R. (1978a), Stable vector bundles and instantons, *Comm. Math. Phys.* **59**, 1–15.

Hartshorne, R. (1978b), Stable vector bundles of rank 2 on P3, *Math. Ann.* **238**, 229–280.

Harvey, R. (1989), *Spinors and Calibrations*, Academic Press, Boston.

Hawking, S. W. (1977), Gravitational instantons, *Phys. Lett.* **60A**, 81–83.

Hawking, S. W. (1979), The path-integral approach to quantum gravity, *in: General Relativity*, S. W. Hawking and W. Israel, eds., Cambridge University Press, Cambridge, pp. 746–789.

Hawking, S. W. and Ellis, G. F. R. (1973), *The Large-Scale Structure of Space-Time*, Cambridge University Press, Cambridge.

Helgason, S. (1978), *Differential Geometry, Lie Groups, and Symmetric Spaces*, Academic Press, New York.

Henkin, G. M. (1980), Representation of Yang–Mills equations in the form of Cauchy–Riemann equations on the twistor space, *Dokl. Akad. Nauk SSSR* **255**, 844–847.

Henkin, G. M. (1981), Representation of the solutions to the φ^4-equation in the form of holomorphic bundles over twistor spaces, *Dokl. Akad. Nauk SSSR* **260**, 1086–1089.

Henkin, G. M. (1982), Yang–Mills–Higgs fields as holomorphic vector bundles (Russian), *Doklad. Akad. Nauk. SSSR* **265**, 1081–1085.

Henkin, G. M. (1985), Tangential Cauchy–Riemann equations and Yang–Mills–Higgs–Dirac fields, *Proc. Int. Congr. Math., Warsaw 1982*, Warsaw.

Henkin, G. M. and Manin, Yu. I. (1980), Twistor description of classical Yang–Mills–Dirac fields, *Phys. Lett.* **B 95**, 405–408.

Henkin, G. M. and Manin, Yu. I. (1981), On the cohomology of twistor flag spaces, *Compositio Math.* **44**, 103–111.

Hill, C. D. and MacKichan, Barry (1977), Hyperfunction cohomology classes and their boundary values, *Ann. Scuola Norm. Sup. Pisa Ser IV* **4**, 577–597.

Hirzebruch, F. (1966), *Topological Methods in Algebraic Geometry*, Springer-Verlag, New York.

Hitchin, N. J. (1979), Polygons and gravitons, *Math. Proc. Camb. Phil. Soc.* **85**, 465–476.

Hitchin, N. J. (1980), Linear field equations on self-dual spaces, *Proc. Roy. Soc. London* **A370**, 173–191.

Hitchin, N. J. (1981), Kaehlerian twistor spaces, *Proc. London Math. Soc.* **43**, 133–150.

Hitchin, N. J. (1982a), Monopoles and geodesics, *Comm. Math. Phys.* **83**, 579–602.

Hitchin, N. J. (1982b), Complex manifolds and Einstein's equations, in: *Twistor Geometry and Non-Linear Systems*, H. D. Doebner and T. D. Palev, eds., Springer Lecture Notes in Mathematics **970**, Berlin, New York, pp. 73–99.

Hitchin, N. J. (1983), On the construction of monopoles, *Comm. Math. Phys.* **89**, 145–190.

Hodges, A. (1982), Twistor diagrams, *Phys.* **A114**, 157–175.

Hörmander, L. (1973), *An Introduction to Complex Analysis in Several Variables*, North Holland, Amsterdam.

Hörmander, L. (1985), *The Analysis of Linear Partial Differential Operators*, 4 volumes, Springer-Verlag.

Horrocks, G. (1964), Vector bundles on the punctured spectrum of a local ring, *Proc. Lond. Math. Soc.* **14**, 684–713.

Horvath, Z. and Rouhani, S. (1984), SU(3) monopoles of unit charge, *Z. für Phys.* **C22**, 261–264.

Huang, K. (1982), *Quarks, Leptons and Gauge Fields*, World Scientific, Singapore.

Hughston, L. P. (1972), *On an Einstein–Maxwell Field With a Null Source*, Springer-Verlag, Berlin, New York.

Hughston, L. P. (1979a), *Twistors and Particles*, Lecture Notes in Physics, No. 97, Springer-Verlag, New York, Berlin.

Hughston, L. P. (1979b), *Some New Contour Integral Formulae*, Pitman, San Francisco.

Hughston, L. P. and Hurd, T. R. (1981), A cohomological description of massive fields, *Proc. Roy. Soc. London* **A378**, 141–154.

Hughston, L. P., Penrose, R., Sommers, P. and Walker, M. (1972), On a quadratic first integral for the charged particle orbits in the charged Kerr solution, *Comm. Math. Phys.* **27**, 303–308.

Hughston, L. P. and Ward, R. S. eds. (1979), *Advances in Twistor Theory*, Pitman, London.

Hurtubise, J. (1983), SU(2) monopoles of charge two, *Comm. Math. Phys.* **92**, 195–202.

Hurtubise, J. (1985), Monopole and rational maps: a note on a theorem of Donaldson, *Comm. Math. Phys.* **100**, 191–196.

Hurtubise, J. (1986), Instantons and jumping lines, *Comm. Math. Phys.* **105**, 107–122.

Husemoller, Dale (1975), *Fibre Bundles*, Springer-Verlag, New York.

Isenberg, J. and Yasskin, P. B. (1979), *Twistor Description of Non-self-dual Yang–Mills Fields*, Pitman, San Francisco.

Isenberg, J., Yasskin, P. B. and Green, P. S. (1978), Non-self-dual gauge fields, *Phys. Lett.* **B78**, 462–464.

Jackiw, R., Nohl, C. and Rebbi, C. (1977), Conformal properties of pseudoparticle configurations, *Phys. Rev.* **D15**, 1642–1646.

Jackiw, R. and Rebbi, C. (1977), Degrees of freedom in pseudoparticle systems, *Phys. Lett.* **67B**, 189–192.

Jaffe, A. and Taubes, C. (1980), *Vortices and Monopoles*, Birkhäuser, Boston.

Jones, P. E. and Tod, K. P. (1985), Minitwistor spaces and Einstein–Weyl spaces, *Classical and Quantum Gravity* **2**, 565–577.

Jozsa, R. (1979), *Sheaves in Physics—Twistor Theory*, SpringerVerlag, Berlin.

Kac, V. G. (1985), *Infinite Dimensional Lie Algebras*, Cambridge University Press, Cambridge.

Kent, S. L. and Newman, E. T. (1983), Yang–Mills theory in null-path space, *J. Math. Phys.* **24**, 949–959.

Ko, M., Ludvigsen, M., Newman, E. T. and Tod, K. P. (1981), The theory of H-space, *Phys. Rep.* **71**, 51–139.

Ko, M., Newman, E. T. and Penrose, R. (1977), The Kähler structure of asymptotic twistor space, *J. Math. Phys.* **18**, 58–64.

Kobayashi, S. and Nomizu, K. (1963), *Foundations of Differential Geometry, Vol. I*, Wiley, New York.

Kobayashi, S. and Nomizu, K. (1969), *Foundations of Differential Geometry, Vol. II*, Wiley, New York.

Kodaira, K. (1962), A theorem of completeness of characteristic systems for analytic families of compact submanifolds of complex manifolds, *Ann. Math.* **75**, 146–162.

Kodaira, K. (1963), On stability of compact submanifolds of complex manifolds, *Am. J. Math.* **85**, 79–94.

Komatsu, H. (1973), *Hyperfunctions and Pseudo-Differential Equations*, Springer-Verlag, Berlin, Heidelberg.

Kostant, Bertram (1977), Graded manifolds, graded Lie algebras, and prequantization, *Lecture Notes in Mathematics* **570**, 177–300, *Differential-Geometric Methods in Physics*, Springer-Verlag, Berlin, New York, pp..

Krantz, S. G. (1982), *Function Theory of Several Complex Variables*, Wiley-Interscience, New York.

Kronheimer, P. (1986), *D. Phil. thesis*, Oxford.

Kuiper, N., (1949), On conformally flat spaces in the large, *Ann. of Math* **50**, 916–924.

Kurke, H. (1981), *Application of Algebraic Geometry to Twistor Spaces*, Teubner, Leipzig.

Lang, S. (1971), *Algebra*, Addison-Wesley, Reading, Mass.

Law, P. R. (1975), Twistor theory and the Einstein equations, *Proc. Roy. Soc. London* **A399**, 111–134.

Lawson, B. and Michelson, M. L. (1989), *Spin Geometry*, to be published, Princeton University Press, Princeton, N.J.

LeBrun, C. (1982), The first formal neighbourhood of ambitwistor space for curved space-time, *Lett. Math. Phys.* **6**, 345–354.

LeBrun, C. (1983), Spaces of complex null geodesics in complex-Riemannian geometry, *Trans. Amer. Math. Soc.* **278**, 209–231.

Leiterer, J. (1983), The Penrose transform for bundles non-trivial on the general line, *Math. Nachr.* **112**, 35–67.

Leites, D. A. (1974), The spectra of graded-commutative rings (Russian), *Uspekhi Mat. Nauk, Ser. 2* **29**, 209–210.

Leites, D. A. (1980), Introduction to the theory of supermanifolds, *Russian Math. Surveys* **35**, 1–64.

Leites, D. A. (1983), *Theory of Supermanifolds*, (Russian), Karelskii Affiliate of Acad. Sciences USSR, Petrozavodsk.

Leznov, A. N. and Saveliev, M. V. (1980), Representation theory and integration of nonlinear spherically symmetric equations to gauge theories, *Comm. Math. Phys.* **74**, 111–118.

Luehr, C. P. and Rosenbaum, P. M. (1982), Twistor bundles and gauge action principles of gravitation, *J. Math. Phys.* **23**, 1471–1488.

Lugo, G. (1982), Structure of asymptotic twistor space, *J. Math. Phys.* **23**, 276–282.

Lukacs, B., et al., (1982), Structure of three-twistor particles, *J. Math. Phys.* **23**, 2108–2115.

Lukierski, J. (1980), Superconformal group and curved fermionic twistor space, *J. Math. Phys.* **21**, 561–567.

Madore, J., Richard, J. L. and Stora, R. (1979), An introduction to the twistor programme, *Phys. Rep.* **49**, 113–130.

Manin, Yu. I. (1981), Gauge fields and holomorphic geometry (Russian), *Current Problems in Mathematics* **17**, 3–55, Akad. Nauk. USSR, Moscow.

Manin, Yu. I. (1982), Gauge fields and cohomology of analytic sheaves, Springer-Verlag, Berlin, New York.

Manin, Yu. I. (1983), Gauge fields and holomorphic geometry, *J. Soviet Math.* **21**, 465–507.

Manin, Yu. I. (1988), *Gauge Field Theory and Complex Geometry*, Springer-Verlag, Berlin, Heidelberg, New York.

Manton, N. S. (1978), Complex structure of monopoles, *Nucl. Phys.* **B135**, 319–332.

McCarthy, P. J. (1981a), Explicit non-'t Hooft rational solutions of the Atiyah–Drinfeld–Hitchin–Manin instanton matrix equations for SU(2r), *Phys. Lett.* **A84**, 155–158.

McCarthy, P. J. (1981b), Rational parametrisation of normalised Stiefel manifolds and explicit non-'t Hooft solutions of the ADHM instanton matrix equations for Sp(n), *Lett. Math. Phys.* **5**, 255–261.

McCarthy, P. J. (1983), General solution of certain quadratic matrix equations and new families of solutions of the ADHM instanton matrix equations, *Proc. Roy. Ir. Acad.* **83A**, 1–16.

Meyers, C. and De Roo, M. (1979), New explicit instantons and the geometry of the parameter space, *Nucl. Phys.* **B148**, 61–73.

Milnor, John W. (1963), Spin structures on manifolds, *Enseign. Math.* **9**, 198–203.

Milnor, John W. and Stasheff, James (1974), *Characteristic Classes*, Princeton University Press, Princeton, N.J.

Min-Oo, (1982), An L_2-isolation theorem for Yang–Mills fields, *Compositio Math.* **47**, 153–163.

Morrow, J. and Kodaira, K. (1971), *Complex Manifolds*, Holt, Rinehart and Winston, New York.

Murray, M. K. (1983), Monopoles and spectral curves for arbitrary Lie groups, *Comm. Math. Phys.* **90**, 263–272.

Murray, M. K. (1984), Non-abelian magnetic monopoles, *Comm. Math. Phys.* **96**, 539–565.

Nahm, W. (1981), All self-dual multimonopoles for arbitrary gauge groups, *in: Structural Elements in Particle Physics and Statistical Mechanics*, J. Honerkamp et al., eds., NATO ASIS B82:301.

Nahm, W. (1982a), The construction of all self-dual multimonopoles by the ADHM method, *in: Monopoles in quantum field theory*, N. S. Craigie, P. Goddard, and W. Nahm, eds., World Scientific, Singapore, pp. 87–94.

Nahm, W. (1982b), The algebraic geometry of multimonopoles, *in: Group-Theoretical Methods in Physics*, M. Serdaroglu et al., eds., Springer Lecture Notes in Physics **180**, pp. 45.

Nahm, W. (1983), Self-dual monopoles and calorons, G. Denardo et al., eds., Springer Lecture Notes in Physics **201**, Trieste.

Newlander, A., Nirenberg, L. (1957), Complex analytic coordinates in almost-complex manifolds, *Annals of Math* **(2)65**, 391–404.

Newman, E. T. (1979), Deformed twistor space and H-space, *in: Complex Manifold Techniques in Theoretical Physics*, D. E. Lerner and P. D. Sommers, eds., Pitman, San Francisco, pp. 154–165.

Newman, E. T., Porter, J. R. and Tod, K. P. (1978), Twistor surfaces and right–flat spaces, *Gen. Rel. Grav.* **9**, 1129–1142.

O'Raifeartaigh, L. and Rouhani, S. (1981), Twisted axial symmetry and finitely separated monopoles, *Phys. Lett.* **105B**, 177–181.

O'Raifeartaigh, L. and Rouhani, S. (1982), Rings of monopoles with discrete axial symmetry: explicit solution for $N = 3$, *Phys. Lett.* **B112**, 143–147.

O'Raifeartaigh, L. and Rouhani, S. (1983), On the stability of the $SU(2)$ separated two-monopole configuration, *Phys. Lett.* **B121**, 151–155.

O'Raifeartaigh, L., Rouhani, S. and Singh, L. P. (1982a), Explicit solution of the Corrigan–Goddard conditions for n monopoles for small values of the parameters, *Phys. Lett.* **B112**, 369–372.

O'Raifeartaigh, L., Rouhani, S. and Singh, L. P. (1982b), On the finitely separated two-monopole solution, *Nucl. Phys.* **B206**, 137–151.

Okonek, C., Schneider, M. and Spindler, H. (1980), *Vector Bundles on Complex Projective Spaces*, Birkhäuser, Boston.

Osborn, H. (1982), On the Atiyah–Drinfeld–Hitchin–Manin construction for self-dual gauge fields, *Comm. Math. Phys.* **86**, 195–219.

Panagopoulos, H. (1983), Multimonopoles in arbitrary gauge groups and the complete $SU(2)$ two-monopole system, *Phys. Rev.* **D28**, 380–384.

Patton, C. M. (1979), *Zero Rest Mass Fields and the Bargmann Complex Structure*, Pitman, San Francisco.

Penkov, I. B. (1982), Linear differential operators and the cohomology of analytic spaces, *Uspekhi Mat. Nauk* **37**, 171–172.

Penrose, R. (1967), Twistor algebra, *J. Math. Phys.* **8**, 345–366.

Penrose, R. (1968a), Twistor quantization and curved space-time, *Intern. J. of Theor. Phys.* **1**, 61–99.

Penrose, R. (1968b), The structure of space-time, *in: Battelle Rencontres*, C. M. De Witt and J. A. Wheeler, eds., Benjamin, New York, pp. 121–235.

Penrose, R. (1969), Solutions to the zero-rest-mass equations, *J. Math. Phys.* **10**, 38–39.

Penrose, R. (1975), Twistor theory, its aims and achievements, *in: Quantum Gravity*, C. J. Isham, R. Penrose, and D. W. Sciama, eds., Clarendon Press, Oxford, pp. 268–407.

Penrose, R. (1976), Nonlinear gravitons and curved twistor theory, *Gen. Rel. Grav.* **7**, 31–52.

Penrose, R. (1977a), The twistor progamme, *Rep. Math. Phys.* **12**, 65–76.

Penrose, R. (1977b), Massless fields and sheaf cohomology, *Twistor Newsletter* **5**, 9–13.

Penrose, R. (1979a), On the twistor description of massless fields, *in: Complex Manifold Techniques in Theoretical Physics*, D. Lerner and P. Sommers, eds., Pitman, London.

Penrose, R. (1979b), A Googly Graviton?, *in: Advances in Twistor Theory*, L. P. Hughston and R. S. Ward, eds., Pitman, London, pp. 168–176.

Penrose, R. (1980), *The Complex Geometry of the Natural World*, Acad. Sci. Fennica, Helsinki.

Penrose, R. (1982), Quasi-local mass and angular momentum in general relativity, *Proc. Roy. Soc. London* **A381**, 53–63.

Penrose, R. (1983), Physical space-time and nonrealizable CR-structures, *Bull. Amer. Math. Soc.* **N.S. 8**, 427–448.

Penrose, R. and MacCallum, M. A. H. (1973), Twistor theory: an approach to the quantisation of fields and space-time, *Phys. Rep.* **6C**, 241–316, *241–316*, pp..

Penrose, R. and Rindler, W. (1984), *Spinors and Space-Time. Vol. I: Two-Spinor Calculus and Relativistic Fields*, Cambridge University Press, Cambridge.

Penrose, R. and Rindler, W. (1986), *Spinors and Space-Time. Vol. II: Spinor and Twistor Methods in Space-Time Geometry*, Cambridge University Press, Cambridge.

Penrose, R. and Sparling, G. A. J. (1979), The twistor quadrille: a line bundle based on the Coulomb field, *in: Advances in Twistor Theory*, L. P. Hughston and R. S. Ward, eds., Pitman, London.

Penrose, R. and Ward, R. S. (1980), Twistors for flat and curved space-time, *in: General Relativity and Gravitation, vol. II*, A. Held, ed., Plenum, New York, pp. 283–328.

Perjes, Z. (1975), Twistor variables of relativistic mechanics, *Phys. Rev.* **D11**, 2031–2041.

Perjes, Z. (1982), Introduction to twistor particle theory, Springer-Verlag, Berlin, New York.

Polking, J. and Wells, Jr., R. O. (1978), Boundary values of Dolbeault cohomology classes and a generalized Bochner–Hartogs Theorem, *Abhand. Math. Sem. Univ. Hamburg* **48**, 1–24.

Pool, Robert (1987), Yang–Mills fields and extension theory, *Memoirs Amer. Math. Soc.* **65**, 358.

Porter, J. R. (1982), Left-flat plane waves as H spaces, *Gen. Rel. Grav.* **14**, 53–59.

Porter, J. R. (1983), Self-dual Yang–Mills fields on Minkowski space-time, *J. Math. Phys.* **24**, 1233–1239.

Prasad, M. K. (1981), Yang–Mills–Higgs monopole solutions of arbitrary topological charge, *Comm. Math. Phys.* **80**, 137–149.

Prasad, M. K. and Rossi, P. (1981), Construction of exact Yang–Mills–Higgs multimonopoles of arbitrary charge, *Phys. Rev. Lett.* **46**, 806–809.

Prasad, M. K. and Sommerfield, C. M. (1975), Exact classical solution for the 't Hooft monopole and the Julia–Zee dyon, *Phys. Rev. Lett.* **35**, 760–762.

Qadir, A. (1980), Field equations in twistors, *J. Math. Phys.* **21**, 514–520.

Range, R. Michael (1986), *Holomorphic Functions and Integral Representations in Several Complex Variables*, Springer-Verlag, New York.

Rajaraman, R. J. (1982), *Solitons and Instantons*, North-Holland, Amsterdam.

Rawnsley, J. H. (1979), On the Atiyah–Hitchin–Drinfeld–Manin vanishing theorem for cohomology groups of instanton bundles, *Math. Ann.* **241**, 43–56.

Rawnsley, J. H. (1982), On the rank of horizontal maps, *Math. Proc. Cambridge Philos. Soc.* **92**, 485–488.

Rogers, Alice (1980), A global theory of supermanifolds, *J. Math. Phys.* **21**, 1352–1365, London.

Rothstein, Mitchell J. (1985), Deformations of complex supermanifolds, *Proc. Amer. Math. Soc.* **95**, 255–260.

Rothstein, Mitchell J. (1986), The axioms of supermanifolds and a new structure arising from them, *Trans. Amer. Math. Soc.* **297**, 159–180.

Rouhani, S. (1984), Patching matrix for SU(2) monopoles of arbitrary charge, *Phys. Rev.* **D30**, 819–822.

Salamon, S. (1982), Quaternionic Kähler manifolds, *Invent. math.* **67**, 143–171.

Sato, M., Kawai, T. and Kashiwara, M. (1973), Microfunctions and pseudo-differential equations, hyperfunctions and pseudo-differential equations, H. Kometsu, ed., *Lecture Notes in Math* **287**, 265–529, Springer-Verlag, Berlin, Heidelberg, New York.

Schapira, Pierre (1970), *Théorie des Hyperfonctions*, Springer-Verlag, Berlin, New York.

Schwartz, L. (1959), *Théorie des Distributions*, 2 vols, Hermann, Paris.

Schwarz, A. S. (1977), On regular solutions of Euclidean Yang–Mills equations, *Phys. Lett.* **67B**, 172–174.

Schwarz, A. S. (1983), Supergravity, complex geometry and G-structures, *Comm. Math. Phys.* **87**, 37–63.

Seifert, H.-J., Clarke, C. J. S. and Rosenblum, A. (1984), *Mathematical Aspects of Superspace*, Reidel, Dordrecht-Boston.

Serre, J.-P. (1956), Géométrie algébrique et géométrie analytique, *Ann. Inst. Fourier* **6**, 1–42.

Shirokov, P. A. and Shirokov, A. P. (1962), *Affine Differentialgeometrie*, Teubner, Leipzig.

Shnider, S.. and Wells, R. O., Jr. (1989), *Supermanifolds, Super Twistor Spaces and Super Yang–Mills Fields*, Les Presses de l'Université de Montréal, Montréal, Canada.

Sibner, L. M., Sibner, R. J. and Uhlenbeck, K. K. (1989), Solutions to Yang–Mills equations which are not self-dual, *Proc. Nat. Acad. Sci., U.S.A.*, (to appear).

Simms, D. J. (1976), *A Survey of the Application of Geometric Quantization*, Springer-Verlag, Berlin, New York.

Spanier, E. H. (1966), *Algebraic Topology*, McGraw-Hill, New York.

Sparling, G. A. J. (1978), Dynamically broken symmetry and global Yang–Mills theory in Minkowsi space, (Preprint), Pittsburgh University.

Steenrod, N. (1951), *The Topology of Fibre Bundles*, Princeton University Press, Princeton, N.J.

Sternberg, Shlomo (1965), *Lectures on Differential Geometry*, Prentice-Hall, Englewood Cliffs, N.J.

Streater, R. F. and Wightman, A. S. (1978), *PCT, Spin and Statistics, and All That*, Benjamin-Cummins, Reading, Mass.

Taubes, C. H. (1980), On the equivalence of the first and second order equations for gauge theories, *Comm. Math. Phys.* **75**, 207–227.

Taubes, C. H. (1981), The existence of multi-monopole solutions to the non-Abelian, Yang–Mills–Higgs equations for arbitrary simple gauge groups, *Comm. Math. Phys.* **80**, 343–367.

Taubes, C. H. (1982), The existence of a non-minimal solution to the SU(2) Yang–Mills–Higgs equations on R^3: I and II, *Comm. Math. Phys.* **86**, 257–298, 299–320.

Tod, K. P. (1977), Some symplectic forms arising in twistor theory, *Rep. Math. Phys.* **11**, 339–346.

Tod, K. P. (1982), Self-dual Kerr–Schild metrics and null Maxwell fields, *J. Math. Phys.* **23**, 1147–1148.

Tod, K. P. and Perjes, Z. (1976), Two examples of massive scattering using twistor Hamiltonians, *Gen. Rel. and Grav.* **7**, 903–913.

Tod, K. P. and Ward, RYu. I. S. (1979), Self-dual metrics with self-dual Killing vectors, *Proc. Roy. Soc. London* **A368**, 411–427.

Topiwala, P. (1987), A new proof of the existence of Kähler–Einstein metrics on $K3$: I and II, *Invent. math.* **89**, 425–448, 449–454.

Trautman, A. (1977), Solutions of the Maxwell and Yang–Mills equations associated with Hopf fibrings, *Int. J. Theor. Phys.* **16**, 561–565.

Treves, Francois (1975), *Basic Linear Partial Differential Equations*, Academic Press, New York.

Uhlenbeck, K. K. (1982), Removable singularities in Yang–Mills fields, *Comm. Math. Phys.* **83**, 11–29.

Volovich, I. V. (1983), Lambda-supermanifolds and fibrations, *Dokl. Akad. Nauk SSSR* **269**, **(3)**, 524–527.

Wallach, N. R. (1973), *Harmonic Analysis on Homogeneous Spaces*, M. Dekker, New York.

Ward, R. S. (1977), On self-dual gauge fields, *Phys. Lett.* **61A**, 81–82.

Ward, R. S. (1979a), Sheaf cohomology and an inverse twistor function, in Hughston and Ward, eds., section 2.8.

Ward, R. S. (1979b), The self-dual Yang–Mills and Einstein equations, *in: Complex Manifold Techniques*, Lerner and Sommers eds., Pitman, San Francisco.

Ward, R. S. (1978), A class of self-dual solutions of Einstein's equations, *Proc. Roy. Soc. London* **A363**, 289–295.

Ward, R. S. (1980), Self-dual space-times with cosmological constant, *Comm. Math. Phys.* **78**, 1–17.

Ward, R. S. (1981a), A Yang–Mills–Higgs monopole of charge 2, *Comm. Math. Phys.* **79**, 317–325.

Ward, R. S. (1981b), Ansätze for self-dual Yang–Mills fields, *Comm. Math. Phys.* **80**, 563–574.

Ward, R. S. (1981c), Two Yang–Mills–Higgs monopoles close together, *Phys. Lett.* **B102**, 136–138.

Ward, R. S. (1981d), Magnetic monopoles with gauge group $SU(3)$ broken to $U(2)$, *Phys. Lett.* **B107**, 281–284.

Ward, R. S. (1982), Deformations of the embedding of the $SU(2)$ monopole solution in $SU(3)$, *Comm. Math. Phys.* **86**, 437–448.

Ward, R. S. (1983), Stationary axisymmetric space-time, *Gen. Rel. Grav.* **15**, 105–109.

Ward, R. S. (1984), Completely solvable gauge-field equations in dimension greater than four, *Nucl. Phys.* **B236**, 381–396.

Weil, A. (1958), *Introduction à l'Etude des Variétés Kaehleriennes*, Hermann, Paris.

Weinberg, E. J. (1979), Parameter counting for multimonopole solutions, *Phys. Rev.* **D20**, 936–944.

Weinberg, E. J. (1980), Fundamental monopoles and multimonopole solutions for arbitrary simple gauge groups, *Nucl. Phys.* **B167**, 500–524.

Weinberg, E. J. (1982), A continuous family of magnetic monopole solutions, *Phys. Lett.* **B119**, 151–154.

Wells, Jr., R. O. (1979a), Complex manifolds and mathematical physics, *Bull. Amer. Math. Soc. (NS)* **1**, 296–336.

Wells, Jr., R. O. (1979b), Cohomology and the Penrose transform, *in: Complex Manifold Techniques in Theoretical Physics*, D. Lerner, and P. Sommers, eds., Pitman, London, pp. 92–114.

Wells, Jr., R. O. (1980), *Differential Analysis on Complex Manifolds*, 2nd ed., Springer-Verlag, Berlin, New York.

Wells, Jr., R. O. (1981), Hyperfunction solutions of the zero-rest-mass field equations, *Comm. Math. Phys.* **78**, 567–600.

Wells, Jr., R. O. (1982a), The conformally invariant Laplacian and the instanton vanishing theorem, *in: Seminar on Differential Geometry*, S.-T. Yau, ed., Princeton University Press, Princeton, N.J., pp. 483–498.

Wells, Jr., R. O. (1982b), *Complex Geometry in Mathematical Physics*, Presses de l'Université de Montréal, Montréal.

Wess, J. and Bagger, J. (1983), *Supersymmetry and Supergravity*, Princeton University Press, Princeton, N.J.

Whittaker, E. T. (1903), On the partial differential equations of mathematical physics, *Math. Ann.* **57**, 333–355.

Witten, E. (1977), Some exact multi-instanton solutions of classical Yang–Mills theory, *Phys. Rev. Letts.* **38**, 121–124.

Witten, E. (1978), An interpretation of classical Yang–Mills theory, *Phys. Lett.* **B77**, 394–398.

Witten, E. (1979), Some comments on the recent twistor space constructions, *in: Complex Manifold Techniques in Theoretical Physics*, D. E. Lerner and P. D. Sommers, eds., Pitman, San Franciso-London, pp. 207–218.

Witten, L. (1979), Static axially symmetric solutions of self-dual SU(2)-gauge fields in Euclidean four-dimensional space, *Phys. Rev.* **D19**, 718–720.

Woodhouse, N. M. J. (1976), Twistor theory and geometric quantization, Springer-Verlag, Berlin, New York.

Woodhouse, N. M. J. (1983), On self-dual gauge fields arising from twistor theory, *Phys. Lett.* **A94**, 269–270.

Woodhouse, N. M. J. (1985), Real methods in twistor theory, *Class. Quantum Grav.* **2**, 257–291.

Wu, H. (1982), The Bochner technique, *in: Proceedings of the 1980 Beijing Symposium on Differential Geometry and Differential Equations, Vol. II*, S. S. Chern and Wu Wen-tsün, eds., Gordon and Breach, New York, pp. 929–1071.

Yang, C. N. (1977), Condition of self-duality for SU(2)-gauge fields on Euclidean four-dimensional space, *Phys. Rev. Letts.* **38**, 1377–1379.

Yasskin, P. B. and Isenberg, J. A. (1982), Non-self-dual nonlinear gravitons, *Gen. Rel. Grav.* **14**, 621–627.

Yau, S. T. (1978), On the Ricci curvature of a compact Kähler manifold and the complex Monge–Ampère equation, I, *Comm. Pure Appl. Math.* **31**, 339–411.

Subject and author index